AF781336

New Frontiers in Atmospheric Science

New Frontiers in Atmospheric Science

Edited by Smith Paul

SYRAWOOD
PUBLISHING HOUSE
New York

Published by Syrawood Publishing House,
750 Third Avenue, 9th Floor,
New York, NY 10017, USA
www.syrawoodpublishinghouse.com

New Frontiers in Atmospheric Science
Edited by Smith Paul

© 2018 Syrawood Publishing House

International Standard Book Number: 978-1-68286-529-3 (Hardback)

This book contains information obtained from authentic and highly regarded sources. Copyright for all individual chapters remain with the respective authors as indicated. All chapters are published with permission under the Creative Commons Attribution License or equivalent. A wide variety of references are listed. Permission and sources are indicated; for detailed attributions, please refer to the permissions page and list of contributors. Reasonable efforts have been made to publish reliable data and information, but the authors, editors and publisher cannot assume any responsibility for the validity of all materials or the consequences of their use.

The publisher's policy is to use permanent paper from mills that operate a sustainable forestry policy. Furthermore, the publisher ensures that the text paper and cover boards used have met acceptable environmental accreditation standards.

Trademark Notice: Registered trademark of products or corporate names are used only for explanation and identification without intent to infringe.

Cataloging-in-Publication Data

New frontiers in atmospheric science / edited by Smith Paul.
p. cm.
Includes bibliographical references and index.
ISBN 978-1-68286-529-3
1. Atmospheric physics. 2. Atmosphere. I. Paul, Smith.
QC861.3 .N49 2018
551.5--dc23

Printed in China.

TABLE OF CONTENTS

PREFACE

Atmospheric science studies the atmosphere and the effect that natural or human activities have on it. It uses a number of instruments like weather balloons, rocketsondes and satellites to study and forecast the weather and atmospheric patterns. The umbrella of atmospheric science encompasses meteorology, climatology, atmospheric chemistry, atmospheric physics, etc. This book unravels the recent studies in the field of atmospheric science. Coherent flow of topics, student-friendly language and extensive use of examples make this book an invaluable source of knowledge. It aims to serve as a resource guide for students and experts alike and contribute to the growth of the discipline.

Significant researches are present in this book. Intensive efforts have been employed by authors to make this book an outstanding discourse. This book contains the enlightening chapters which have been written on the basis of significant researches done by the experts.

Finally, I would also like to thank all the members involved in this book for being a team and meeting all the deadlines for the submission of their respective works. I would also like to thank my friends and family for being supportive in my efforts.

Editor

1

Size dependence of phase transitions in aerosol nanoparticles

Yafang Cheng[1], Hang Su[1], Thomas Koop[2], Eugene Mikhailov[3] & Ulrich Pöschl[1]

Phase transitions of nanoparticles are of fundamental importance in atmospheric sciences, but current understanding is insufficient to explain observations at the nano-scale. In particular, discrepancies exist between observations and model predictions of deliquescence and efflorescence transitions and the hygroscopic growth of salt nanoparticles. Here we show that these discrepancies can be resolved by consideration of particle size effects with consistent thermodynamic data. We present a new method for the determination of water and solute activities and interfacial energies in highly supersaturated aqueous solution droplets (Differential Köhler Analysis). Our analysis reveals that particle size can strongly alter the characteristic concentration of phase separation in mixed systems, resembling the influence of temperature. Owing to similar effects, atmospheric secondary organic aerosol particles at room temperature are expected to be always liquid at diameters below ~20 nm. We thus propose and demonstrate that particle size should be included as an additional dimension in the equilibrium phase diagram of aerosol nanoparticles.

[1] Multiphase Chemistry Department, Max Planck Institute for Chemistry, 55128 Mainz, Germany. [2] Faculty of Chemistry, Bielefeld University, 33615 Bielefeld, Germany. [3] Institute of Physics, St. Petersburg State University, 198904 St. Petersburg, Russia. Correspondence and requests for materials should be addressed to H.S. (email: h.su@mpic.de) or to Y.C. (email: yafang.cheng@mpic.de) or to U.P. (email: u.poschl@mpic.de).

The phase state of aerosol particles is a key determinant of their physico-chemical interactions, climate and health effects[1–4]. Chemical composition, relative humidity and temperature have been recognized as the main factors controlling aerosol phase state[1,2,5], but they are insufficient to explain observations for nanometer-sized particles, even for well-studied inorganic reference substances[3,6–10], such as ammonium sulphate (AS).

Progress in resolving these discrepancies has been hampered by a lack of reliable thermodynamic data, such as water and solute activities and interfacial energies. This lack of data exists because nanodroplets can become more highly supersaturated (with respect to a crystal) when compared with bulk solution samples, but traditional measurement techniques and data, especially the liquid-vapour interfacial energy (σ_{lv}, also called surface tension), are mostly limited to sub-saturated and saturated solutions. The existing measurement techniques for σ_{lv} often require touching the solution by needles or other intermediates, such as a plate or capillary tube[11]. Solute crystallization induced by heterogeneous nucleation can take place in supersaturated solution on the contact surface, which makes the precise measurements difficult. Measurement of σ_{lv} using the vibration modes of a magnetically levitated drop could avoid the contact but is limited to supermicrometre particles owning to the stability in the magnetic trap[12]. Measurements of water activities (a_w) have been extended to supersaturated range by Electrodynamic Balance (EDB) methods[13,14] (for example, up to $\sim 30\,\mathrm{mol\,kg^{-1}}$ for AS), but it is not applicable for the nanosize range either.

To overcome these limitations, we develop a new method for the determination of these specific thermodynamic parameters, closing the gap between experimental and modelled results. Further analysis reveals that particle size can strongly alter the phase separation process in mixed systems. We thus propose and demonstrate that particle size should be included as an additional dimension in the equilibrium phase diagram of aerosol nanoparticles.

Results

Thermodynamic properties of supersaturated nanodroplets. According to theoretical predictions, AS nanoparticles are expected to show a strong size dependence of deliquescence relative humidity (DRH) similar to sodium chloride (NaCl), with differences up to $\sim 10\%$ RH between particles with dry diameters of 6 and 60 nm (refs 6,8; Fig. 1a). Experimental data, however, reveal a much weaker size dependence for AS, that is, similar DRH values for 6-nm and 60-nm particles[9] (Fig. 1b). Moreover, model calculations based on previously available thermodynamic data agree reasonably well with the hygroscopic growth observed for both 60-nm and 6-nm NaCl particles and also for 60-nm AS particles, but deviate strongly for 6-nm AS particles as shown in Fig. 1a,b. Further investigations of these discrepancies, however, were impeded by a lack of reliable thermodynamic data.

In this study, a new method, termed Differential Köhler Analysis (DKA), is developed for the determination of thermodynamic properties from hygroscopic growth measurement data of aerosolized nanodroplets. Differential application of the classical Köhler equation to aerosol particles with the same hygroscopic growth factor but different dry diameters enables the simultaneous determination of a_w and σ_{lv} as a function of solute concentration (see Methods and Supplementary Note 1). By applying the DKA approach to high precision measurement data from hygroscopic tandem differential mobility analyser (HTDMA) experiments[9] (see Methods), we obtained consistent and robust estimates for a_w and σ_{lv} of aqueous AS nanoparticles with molalities up to $b \sim 160\,\mathrm{mol\,kg^{-1}}$ (solute mass fractions up to $x_s \sim 0.96$) (Fig. 2). The DKA results agree well with literature data in the previously accessible range of concentrations and enable accurate model calculations of the hygroscopic growth of highly concentrated AS nanoparticles (purple line in Fig. 1b); for DKA results of NaCl nanoparticles see Supplementary Fig. 1.

Measured and predicted phase transitions and equilibria. For model predictions of the DRH and ERH (efflorescence relative humidity), we calculated solute activity (a_s) and obtained the interfacial energy of the salt in its saturated aqueous solution upon deliquescence (σ_{sl}) and the interfacial energy of salt embryos (γ_{sl}) in highly supersaturated aqueous solution upon efflorescence based on the DKA-derived a_w and σ_{lv} (Supplementary Notes 2,3). With this approach, we are able to determine σ_{sl} and γ_{sl} at one particular diameter and then apply them for predictions over the entire size range of 6–60 nm (see Methods). As shown in Fig. 1c,d, the DKA-based thermodynamic data allow quantifying and reconciling the contrasting size dependencies in the relative humidity thresholds of the deliquescence and efflorescence transitions for both AS and NaCl nanoparticles. The experimental DRH values observed for NaCl and AS nanoparticles are in a good agreement with our Köhler-Ostwald-Freundlich model[15] using DKA data (solid lines in Fig. 1c) but not with the modelling approach neglecting the effect of particle size on solubility (dashed lines in Fig. 1c). The solute molality on deliquescence (b^*) represents the size-dependent water solubility of nanoparticles. It is much higher than the solubility of bulk material ($b_{bulk}{}^*$) and agrees well with the Ostwald-Freundlich model prediction (Fig. 1e). Owing to a lack of reliable a_s and σ_{sl} data, previous studies failed to reproduce the size-dependent solubility of AS, and therefore, were not able to explain the different size dependence of DRH for AS and NaCl[6,7].

The observed ERH values also agree very well with our predictions using DKA data (Fig. 1d) in a model combining the Köhler equation and classical nucleation theory[16,17] (see Methods). The solute molality upon efflorescence (b_e) represents the threshold concentration triggering solute nucleation and liquid–solid phase separation. It is not sensitive to particle size for NaCl, where a 30% increase in b_e was sufficient to maintain the same nucleation rate despite a 1,000-fold volume reduction between 60-nm and 6-nm particles (Fig. 3a and red diamonds in Fig. 1f). In contrast, an about 12-fold increase in b_e was required to trigger the nucleation and phase separation when going from 60-nm to 6-nm AS particles (Fig. 3b and blue circles in Fig. 1f). The different behaviour of NaCl and AS can be attributed largely to different relations between solute activity and concentration, that is, for the same change of activity, a stronger increase in molality is required for AS than for NaCl (Supplementary Fig. 2).

It has been shown that water activity rather than molality is the determinant for the homogenous nucleation of ice in solutions[18]. Similarly, our results demonstrate that solute nucleation and phase separation are determined by solute activity, implying that the non-ideality of solution properties become vital for the calculation of nucleation rates and understanding of phase transitions, especially for nanodroplets (Fig. 3). Indeed, the weaker size dependence of the threshold relative humidity (DRH, ERH) of AS is a consequence of its stronger size dependence of the threshold solute concentration (b^*, b_e) compared with NaCl (Supplementary Fig. 3).

The existence of metastable AS droplets with molalities up to $\sim 380\,\mathrm{mol\,kg^{-1}}$ (AS:H_2O molar ratio $\sim 7{:}1$) upon efflorescence (Fig. 1f) provides new insight on the phase state of salt nanoparticles. Such a high solute-to-solvent ratio implies that the droplet is no longer a typical aqueous solution, but may be regarded as a 'molten' salt particle with a few water molecules as a

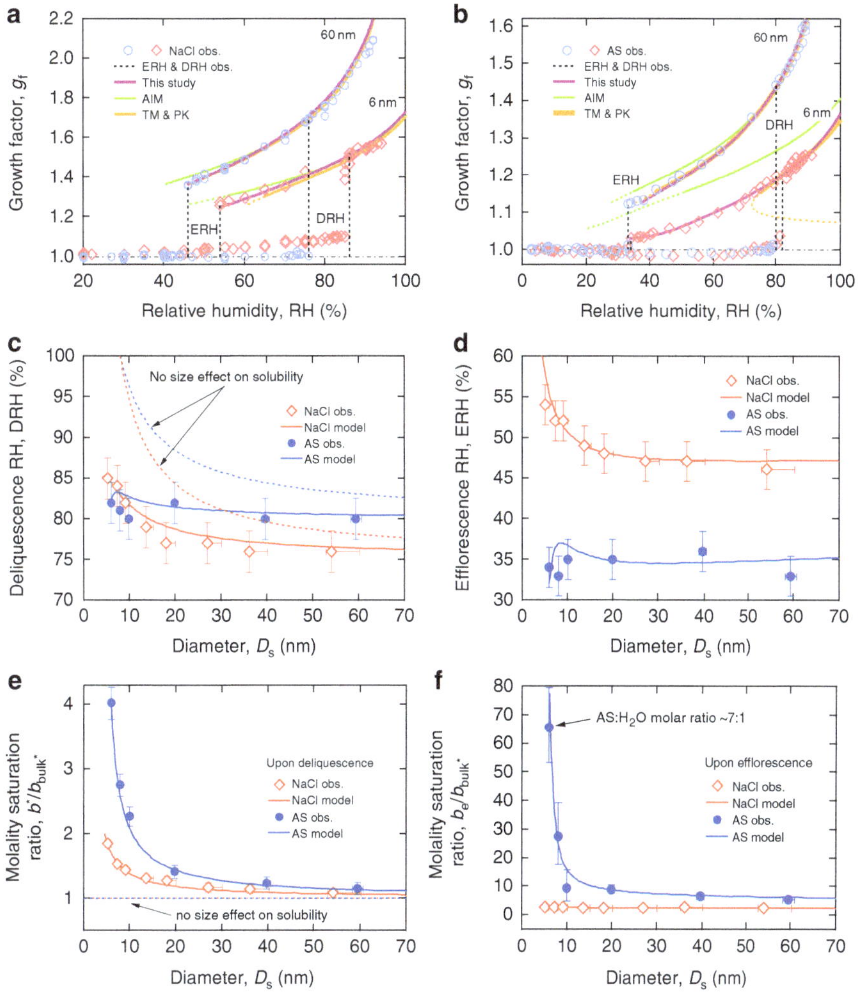

Figure 1 | Hygroscopic growth, deliquescence and efflorescence of nanoparticles. (**a**,**b**) Hygroscopic growth curves of sodium chloride (NaCl) and ammonium sulphate (AS). Growth factor data are taken from HTDMA experiments[8,9] for particles with dry diameters of 60 nm (blue circle) and 6 nm (red diamond). Köhler model curves are based on different thermodynamic data sets: in this study, purple line with DKA (Differential Köhler Analysis) derived data for AS and modified TM & SP (Tang-Munkelwitz[35,36] and Seinfeld-Pandis[37]) for NaCl; AIM (Aerosol Inorganics Model[38]), green line; TM & PK (Tang-Munkelwitz[35,36] and Pruppacher-Klett[39]) as in Biskos *et al.*[8,9], orange line (Supplementary Note 3). Dashed lines indicate extrapolation beyond validated concentration range. Size-dependent threshold values of relative humidity and solute saturation ratio on deliquescence (**c**,**e**) and efflorescence (**d**,**f**). Saturation ratios refer to bulk water solubility expressed in solute molality (b_{bulk}*). Model curves were obtained by combination of the Köhler equation with the Ostwald-Freundlich equation for deliquescence and with classical nucleation theory for efflorescence. Dashed lines indicate model predictions without consideration of size-dependent solubility. Error bars are estimated by considering measurement uncertainties.

'solute' or 'impurity' (here 'molten' does not imply a high melting point). By further reducing particle size, we expect an increase in the AS mole fraction, such that the particle ultimately becomes a pure molten salt at ambient temperature. This is also supported by the fact that the interfacial energy σ_{lv} retrieved here at a temperature $T = 298$ K approaches a value of 0.182 N m^{-1} (Fig. 2b), which is consistent with the value of 0.185 N m^{-1} predicted for a molten AS at the same temperature[19].

Size effects on phase transitions. Although the Kelvin effect is commonly considered for liquid-vapour equilibrium and phase partitioning, size effects on the liquid-solid equilibrium upon phase transitions have been rarely taken into account in atmospheric thermodynamic and kinetic models. On the basis of the above results, we propose to include the particle dry diameter (D_s) as the third dimension in the liquid–solid equilibrium phase diagrams of aerosol particles. Figure 4a shows such a 3-D phase diagram for the AS-water system. In Fig. 4a, the surfaces represent the equilibrium between liquid and crystalline phases. It is estimated from polynomial fitting to literature data (solid circles). A slice on the surface at $D_s^{-1} = 0$ ($D_s \rightarrow \infty$) equals the traditional T-x_s diagram for bulk material as illustrated in Fig. 4b.

A slice on the surface at constant T provides a D_s^{-1}-x_s phase diagram showing the influence of particle size on equilibrium composition. As an example, Fig. 4c shows a D_s^{-1}-x_s slice at 215 K, which exhibits a similar pattern as the traditional T-x_s diagram in Fig. 4b. Figure 4d displays the D_s^{-1}-x_s diagram at 298 K, which contains the observed size-dependent solubility of AS particles reflecting the Ostwald–Freundlich effect[15]. Note that, at this temperature, the ice phase does not exist. Extrapolation of the experimental data for aqueous AS to $x_s = 1$ leads to a critical melting diameter of ~4 nm, that is, pure AS particles below this size are expected to be in a liquid state at 298 K.

Finally, a slice at constant x_s gives a T-D_s^{-1} phase diagram showing the effect of particle size on phase transition temperature (Fig. 4e). For example, in a system with an AS mass fraction of $x_s = 0.63$, the characteristic temperature for complete dissolution

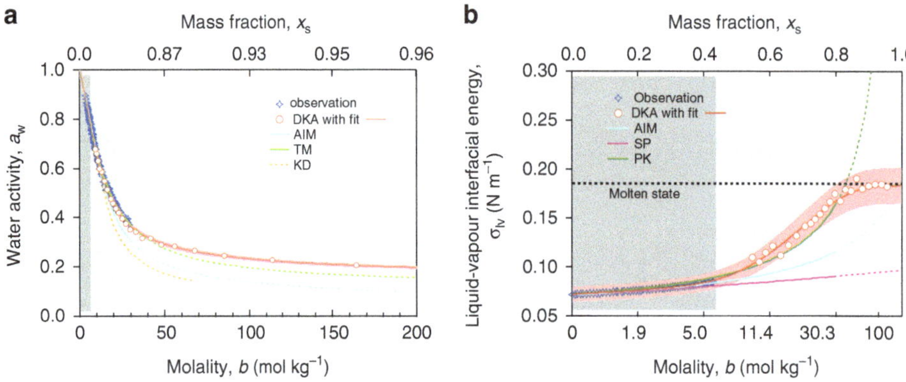

Figure 2 | Concentration-dependent thermodynamic properties of ammonium sulphate solution. (**a**,**b**) Water activity (a_w, on a mole fraction basis) and liquid-vapour interfacial energy (σ_{lv}) are plotted against solute molality (b) and mass fraction (x_s). The DKA-derived a_w and σ_{lv} (red open circles) are compared with observations (blue stars)[39,40], AIM (Aerosol Inorganic Model[38], blue line), and other parameterizations for a_w (TM = Tang-Munkelwitz[35], green line; KD = Kreidenweis[41], orange line) and σ_{lv} (PK = Pruppacher-Klett[39], dark green line; SP = Seinfeld-Pandis[37], purple line) (Supplementary Note 3). Dashed lines indicate extrapolation beyond validated concentration range. The equations for the best fit of DKA results (red lines) can be found in Supplementary Table 1. Red shaded areas indicate the uncertainties in the DKA retrieval, estimated by Monte Carlo analyses (Supplementary Note 1). The dark and light grey shaded areas mark the sub-saturated and saturated concentration with respect to bulk solution and supersaturated concentration before efflorescence of supermicrometre droplets, respectively. The white area marks the highly supersaturated concentration where the a_w data are not available in the literature.

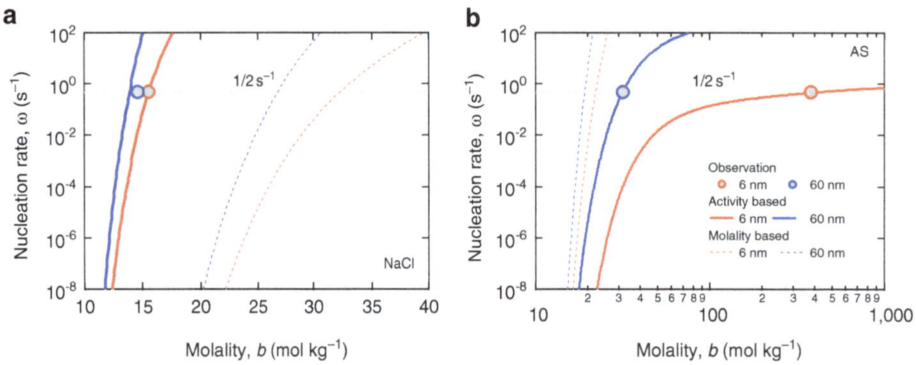

Figure 3 | Homogeneous nucleation rate as a function of solute molality. (**a**) Sodium chloride (NaCl) and (**b**) ammonium sulphate (AS). The nucleation rates (ω) are calculated from solute activity (thick lines) and concentration (thin dashed lines) at selected particle sizes (6 and 60 nm). Grey dashed line corresponds to ω of 0.5 s^{-1}, at which at least one nucleation event is triggered during the expected induction time interval (t_i and $\omega t_i = 1$). Here, t_i is estimated as 2 s according to the residence time in the HTDMA experiments[9]. The circles represent the observed nucleation (efflorescence) concentrations.

decreases from ~470 K for bulk material to ~298 K for 10-nm particles. At $x_s = 0$ or 1, the diagram becomes a T-D_s^{-1} diagram for a pure substance, reflecting the Gibbs–Thomson effect[20]. Related effects of particle size on melting, freezing and glass transitions have been reported and quantified for a wide range of materials such as water, organic compounds and metals[21–23] (Supplementary Fig. 4). Our results demonstrate that similar effects and formalisms also apply to other phase transition processes and mixed systems, such as liquid–solid phase separation in salt-water droplets. The size-dependent phase transition temperature and increased solubility of nanoparticles reflect a tendency of smaller particles for staying in liquid and mixed phase compared with bulk materials (Supplementary Fig. 5). This may partly explain the unresolved size-dependent morphology of organic-AS particles, that is, smaller particles are internally mixed while larger particles adopt a de-mixed partially engulfed structure[10].

Discussion

According to the size effects outlined above, aerosol particles of identical chemical composition may coexist in different states in the atmosphere depending on particle diameter. Thus, we are interested in the critical diameter ($D_{s,c}$) below which aerosol particles are expected to be liquid and well-mixed at ambient temperature. Using T-D_s^{-1} diagrams like Fig. 4e for particles with different chemical composition (Supplementary Fig. 6), we obtained critical dry diameters as a function of bulk phase transition temperature (T_{bulk}). As shown in Fig. 5, all data for aqueous AS and NaCl particles available at 298 K converge onto a compact near-linear correlation of $1/D_{s,c}$ versus T_{bulk}. The glass transition of low chain length polystyrene[22] as the only organic reference material (with low molecular weight) for which similar substrate-free data are currently available falls onto the same correlation line. These findings suggest that the influence of particle size on the liquefaction of substances is related to bulk phase transition temperature in a similar way for different salts as well as organic compounds, implying a close relationship between interfacial energy and the enthalpy of phase transition similar to the Turnbull relation[24] (see Supplementary Fig. 7 and Supplementary Discussion for more details).

Chamber experiments with pine emissions-derived secondary organic aerosol (SOA) as well as field measurements of

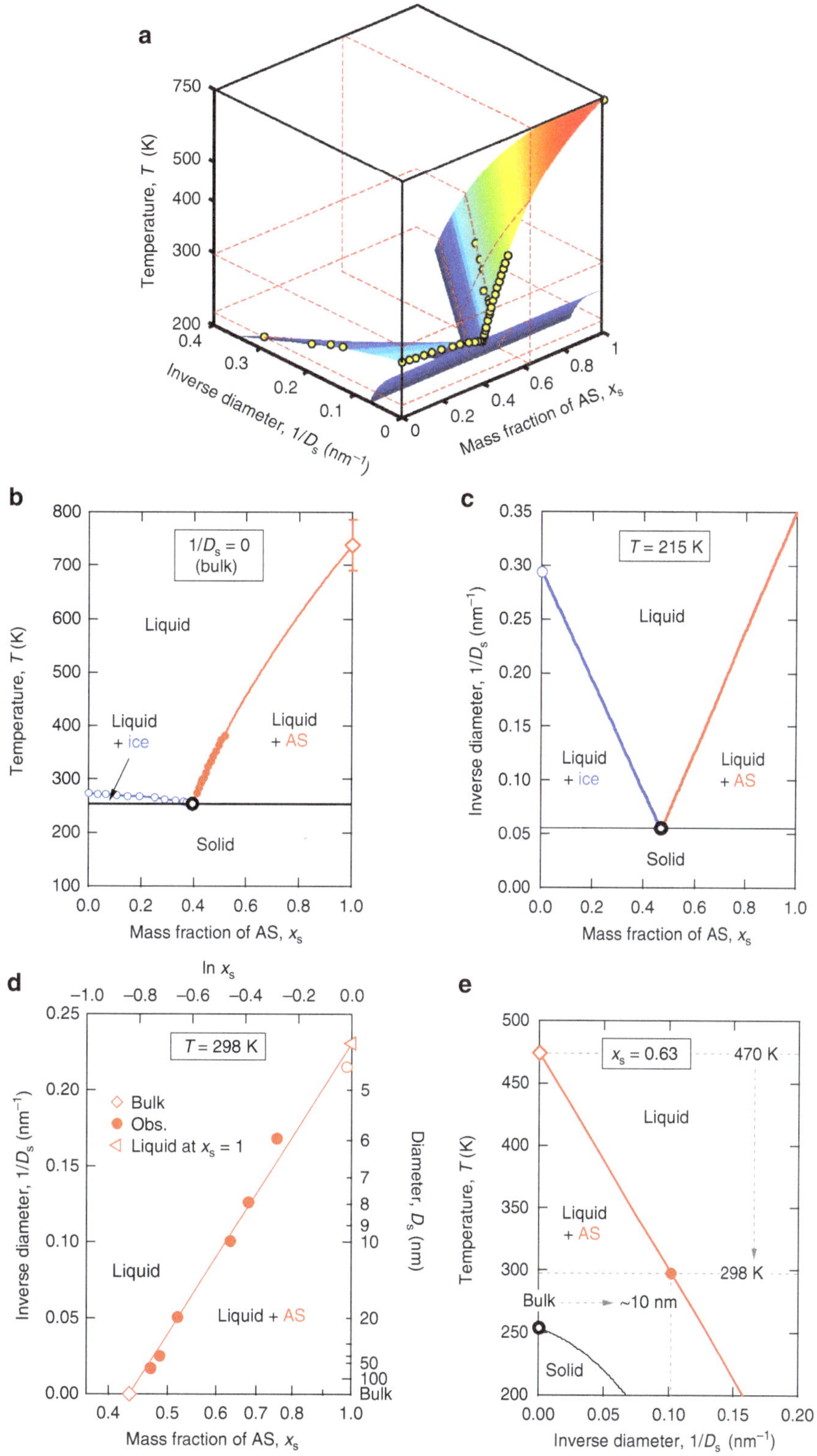

Figure 4 | Liquid-solid equilibrium phase diagrams for the ammonium sulphate—water system. (**a**) 3D phase diagrams in the coordinates of inverse diameter $1/D_s$, temperature T and ammonium sulphate (AS) mass fraction x_s. The solid circles represent the available data of bulk phase diagram of aqueous AS solution[42], size-dependent melting temperature of ice[23] and solubility of AS. The surfaces, coloured by the corresponding temperature, are estimated from polynomial fitting, showing the equilibrium between liquid and crystalline phases. The dashed planes illustrate the following slices. (**b**) The T-x_s phase diagram for bulk solution. The temperature-dependent phase separation of aqueous AS bulk solution (red line) is constrained by the melting temperature of AS at $x_s = 1$ (open diamond). The error bar shows the uncertainty of AS melting temperature according to close-cell measurements[43]. The D_s^{-1}-x_s phase diagram at 215 K (**c**) and at 298 K (**d**). In (**d**), the solid circles represent the observations for nano-sized droplets. The open diamond is the bulk solubility and the open circle is predicted by the DKA-constrained solute activity. The critical melting diameter (open triangle) is obtained by linear extrapolation to $x_s = 1$ (molten state). (**e**) The T-D_s^{-1} phase diagram at $x_s = 0.63$. The black open circle represents the eutectic point for bulk solution (also in **b** and **c**). The black line shows the size dependence of eutectic points.

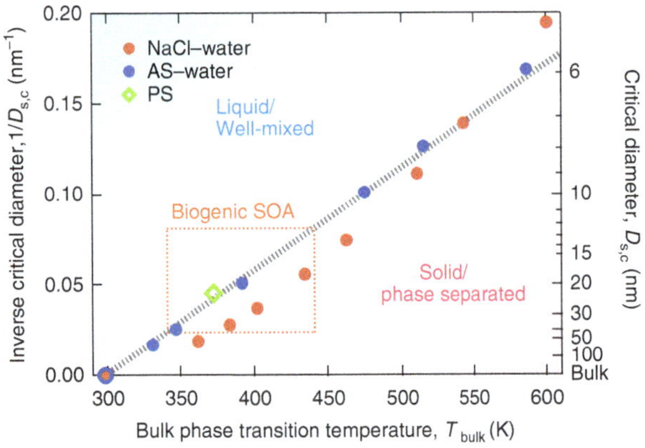

Figure 5 | Dependence of critical diameter on bulk phase transition temperature. Inverse critical diameters of liquefaction at 298 K ($D_{s,c}^{-1}$) are plotted against bulk phase transition temperatures (T_{bulk}) for aqueous ammonium sulphate (AS, blue solid circle), aqueous sodium chloride (NaCl, red solid circle) and low chain length polystyrene (PS, green open diamond)[22]. The data points are observations and the dotted line is a linear fit to all data through the point of [298 K, $D_s^{-1} = 0$]. The orange dashed line bounded area indicates the parameter range estimated for atmospheric biogenic secondary organic aerosol (SOA).

atmospheric biogenic SOA indicate a change of particle phase state at ~20–30 nm (refs 3,25), suggesting that particles below this limit are in the liquid-like state. This observation appears consistent with a critical diameter around ~20 nm (12–40 nm), which can be derived from the estimated bulk melting temperatures of individual functionalized acids found in biogenic SOA (~340–440 K, orange dashed line bounded area in Fig. 5, see Supplementary Note 4)[2]. This consistency increases confidence in the similarity of T_{bulk}-$D_{s,c}$ relations. SOA particles usually consist of mixtures of multiple compounds, which reduces the bulk melting temperature[26], thus increasing the critical diameter and implying that SOA particles in the nucleation mode are very likely always liquid (of potentially high viscosity). The liquid-to-solid transition of SOA particles observed at ~20–30 nm was suggested to be a glass transition[3,25], for which the transition temperature is usually by a factor of ~0.7 lower than for melting[2]. Explaining the observed transition would thus require a fairly high bulk glass transition temperature (~370 K), suggesting that the SOA particles may contain oligomers or polymers, in agreement with independent observations[27].

The methods and results presented in this study advance the understanding and challenge the traditional modelling of atmospheric aerosol phase state and processing. The size effects on the phase states may have important implications for a number of atmospheric interfacial and condensed phase processes such as nucleation, water/gas uptake, polymerization and heterogeneous reactions[2,27–29]. The combination of the DKA method and tandem differential mobility analyser technique can be used to study the thermodynamic properties of nanoparticles from a wide range of materials and enables substrate-free experimental investigation of nanoparticles not only on humidification and drying but also with regard to other types of conditioning, for example, exposure to organic solvents, heat treatment and volatilization[30,31]. Thus, the scientific approach presented here may be useful for nanoparticle studies in various fields, including the environmental, biomedical and materials sciences[20,32].

Methods

HTDMA experiments. Our analysis is based on the Harvard hygroscopic tandem differential mobility analyser (HTDMA) measurements[8,9], where a mono-disperse dry particle fraction of dry diameter D_s is selected by the first DMA, equilibrated to a defined relative humidity (equivalent to water vapour saturation ratio $s_w = \text{RH}/100\%$) and finally goes through a second DMA for the determination of the humidified size distribution. The HTDMA results were usually reported in the form of the growth factor g_f as a function of s_w at each D_s. Details of shape factor correction of HTDMA data can be found in Supplementary Note 1. Data from the efflorescence mode experiments (containing the most concentrated droplet solutions) are adopted for the DKA analysis.

Differential Köhler analysis (DKA). The theoretical basis of the DKA method is the Köhler equation[33], which describes the equilibrium s_w over a spherical droplet as a function of a_w, σ_{lv} and the droplet size D_{sol}

$$s_w = a_w \exp\left(\frac{4\sigma_{lv} v_w}{RTD_{sol}}\right) = a_w \exp\left(\frac{4\sigma_{lv} v_w}{RTD_s g_f}\right) \tag{1}$$

where v_w is the partial molar volume of water (Supplementary Note 1), R is the universal gas constant and T is temperature. D_{sol} in equation (1) is often replaced by the product of D_s and g_f, in which D_s is the dry diameter of spherical solute particles and g_f is the particle geometrical diameter growth factor, that is, $g_f = D_{sol}/D_s$. As parameters s_w, D_s and g_f can be determined from the HTDMA measurements, σ_{lv} and a_w (considered as size-independent) become the only unknowns in equation (1). Note that σ_{lv} and a_w are both functions of g_f (g_f represents the solute concentration x_s, see Supplementary Note 1). For each x_s (or g_f), we need to construct at least two independent equations by performing HTDMA measurements at different D_s to realize the differential analysis as follows. Converting equation (1) into logarithmic form, we have

$$\ln s_w = \ln a_w(g_f) + \frac{A}{D_s}\sigma_{lv}(g_f) \tag{2}$$

where $A = 4v_w/(RTg_f)$. With measurements at two dry diameters D_{s1} and D_{s2}, we have

$$a_w = \frac{S_{w1}^{\left(\frac{D_{s1}}{D_{s1}-D_{s2}}\right)}}{S_{w2}^{\left(\frac{D_{s2}}{D_{s1}-D_{s2}}\right)}} \quad \text{and} \quad \sigma_{lv} = \left(\frac{A}{D_{s1}} - \frac{A}{D_{s2}}\right)^{-1}(\ln s_{w1} - \ln s_{w2}) \tag{3}$$

where s_{w1} and s_{w2} are measured at the same g_f but at D_{s1} and D_{s2}, respectively. When measurements at the same g_f are available for more than two initial particle dry diameters, a multiple regression was applied to all sizes resulting in a best estimate of a_w and σ_{lv}. More details about the DKA method can be found in Supplementary Note 1.

Modelling of deliquescence and efflorescence. Deliquescence of salt particle of dry diameter D_s happens when the ambient s_w equals the equilibrium s_w over a droplet of the wet diameter equal to D_s and saturated with respect to crystalline salt particles of size D_s. The deliquescence concentration b^* was determined by the Ostwald–Freudlich equation[15] through corresponding deliquescence solute activity a_s^*

$$a_s^* = a_{s,bulk}^* \exp\frac{4\sigma_{sl} M_s}{m_i \rho_s RTD_s} \tag{4}$$

in which a_s^* is the solute activity in a solution saturated with respect to salt particles of diameter D_s, $a_{s,bulk}^*$ is the solute activity in saturated bulk solution, M_s is molar weight of solute, ρ_s is the density of solute and m_i is the stoichiometric number of dissociated ions. The σ_{sl} is determined to be 0.156 and 0.265 N m^{-1} for AS and NaCl, respectively (Supplementary Note 2). Substituting b^* into the Köhler equation (1) gives the DRH.

The efflorescence is predicted based on classical nucleation theory[16,17,34] along with the Köhler equation[33]. Given a Köhler curve, the volume of the droplet V_{sol} and corresponding b at each point can be easily determined. Then we can calculate the nucleation rate ω at each point by[16]

$$\omega = JV_{sol} = V_{sol} K \exp\left(-\frac{4\left(\sum_j k_j\right)\alpha^2\gamma_{sl}^3}{27\Delta G_v^2 k_b T}\right) \tag{5}$$

where V_{sol} is the volume of the droplet, K is the pre-exponential factor, k_j and α are geometrical constants dependent on the morphology of the particular crystal, ΔG_v is the excess free energy of solute per unit volume in the crystalline phase over that in solution, k_b is the Boltzmann constant. We assumed that the formation of a single critical-sized nucleus is sufficient to initiate the crystallization of the entire droplet and trigger the efflorescence. Then ERH and b_e can be determined from the point where $\omega t_i = 1$ ($t_i = 2$ s is the induction time interval for the investigated system[9]) (Fig. 3). We performed model predictions of DRH and ERH over the entire size range of 6–60 nm for AS and NaCl by this approach with σ_{sl} and γ_{sl} determined at one particular diameter (6 nm, Supplementary Note 2).

References

1. Martin, S. T. Phase transitions of aqueous atmospheric particles. *Chem. Rev.* **100,** 3403–3454 (2000).
2. Koop, T., Bookhold, J., Shiraiwa, M. & Poschl, U. Glass transition and phase state of organic compounds: dependency on molecular properties and implications for secondary organic aerosols in the atmosphere. *Phys. Chem. Chem. Phys.* **13,** 19238–19255 (2011).
3. Virtanen, A. *et al.* An amorphous solid state of biogenic secondary organic aerosol particles. *Nature* **467,** 824–827 (2010).
4. Pöschl, U. Atmospheric aerosols: composition, transformation, climate and health effects. *Angew. Chem. Int. Ed.* **44,** 7520–7540 (2005).
5. Krieger, U. K., Marcolli, C. & Reid, J. P. Exploring the complexity of aerosol particle properties and processes using single particle techniques. *Chem. Soc. Rev.* **41,** 6631–6662 (2012).
6. Russell, L. M. & Ming, Y. Deliquescence of small particles. *J. Chem. Phys.* **116,** 311–321 (2002).
7. Orr, Jr C., Hurd, F. K. & Corbett, W. J. Aerosol size and relative humidity. *J. Colloid Sci.* **13,** 472–482 (1958).
8. Biskos, G., Russell, L. M., Buseck, P. R. & Martin, S. T. Nanosize effect on the hygroscopic growth factor of aerosol particles. *Geophys. Res. Lett.* **33,** L07801 (2006).
9. Biskos, G., Paulsen, D., Russell, L. M., Buseck, P. R. & Martin, S. T. Prompt deliquescence and efflorescence of aerosol nanoparticles. *Atmos. Chem. Phys.* **6,** 4633–4642 (2006).
10. Veghte, D. P., Altaf, M. B. & Freedman, M. A. Size dependence of the structure of organic aerosol. *J. Am. Chem. Soc.* **135,** 16046–16049 (2013).
11. Damson, A. W. & Gast, A. P. *Physical Chemistry of Surfaces*, 6th edn (John Wiley & Sons, Inc., 1997).
12. Vicente, C., Yao, W., Maris, H. J. & Seidel, G. M. Surface tension of liquid ^{4}He as measured using the vibration modes of a levitated drop. *Phys. Rev. B* **66,** 214504 (2002).
13. Tang, I. N., Munkelwitz, H. R. & Wang, N. Water activity measurements with single suspended droplets: The NaCl-H_2O and KCl-H_2O systems. *J. Colloid Interf. Sci.* **114,** 409–415 (1986).
14. Chan, C. K., Liang, Z., Zheng, J., Clegg, S. L. & Brimblecombe, P. Thermodynamic properties of aqueous aerosols to high supersaturation: I-measurements of water activity of the system $Na^+ - Cl^- - NO_3^- - SO_4^{2-} - H_2O$ at ~298.15K. *Aerosol Sci. Tech.* **27,** 324–344 (1997).
15. Ostwald, W. Über die vermeintliche isomerie des roten und gelben Quecksilbersoxyds und die Oberflächenspannung fester Körper. *Z. Phys. Chem.* **34,** 495–503 (1900).
16. Cohen, M. D., Flagan, R. C. & Seinfeld, J. H. Studies of concentrated electrolyte solutions using the electrodynamic balance. 3. Solute nucleation. *J. Phys. Chem.* **91,** 4583–4590 (1987).
17. Gao, Y., Chen, S. B. & Yu, L. E. Efflorescence relative humidity of airborne sodium chloride particles: A theoretical investigation. *Atmos. Environ.* **41,** 2019–2023 (2007).
18. Koop, T., Luo, B., Tsias, A. & Peter, T. Water activity as the determinant for homogeneous ice nucleation in aqueous solutions. *Nature* **406,** 611–614 (2000).
19. Dutcher, C. S., Wexler, A. S. & Clegg, S. L. Surface tensions of inorganic multicomponent aqueous electrolyte solutions and melts. *J. Phys. Chem. A* **114,** 12216–12230 (2010).
20. Alcoutlabi, M. & McKenna, G. B. Effects of confinement on material behaviour at the nanometre size scale.. *J. Phys.: Condens. Matter* **17,** R461 (2005).
21. Couchman, P. R. & Jesser, W. A. Thermodynamic theory of size dependence of melting temperature in metals. *Nature* **269,** 481–483 (1977).
22. Forrest, J. A. & Mattsson, J. Reductions of the glass transition temperature in thin polymer films: Probing the length scale of cooperative dynamics. *Phys. Rev. E* **61,** R53–R56 (2000).
23. Pan, D., Liu, L.-M., Slater, B., Michaelides, A. & Wang, E. Melting the ice: on the relation between melting temperature and size for nanoscale ice crystals. *ACS Nano* **5,** 4562–4569 (2011).
24. Turnbull, D. Formation of crystal nuclei in liquid metals. *J. Appl. Phys.* **21,** 1022–1028 (1950).
25. Virtanen, A. *et al.* Bounce behavior of freshly nucleated biogenic secondary organic aerosol particles. *Atmos. Chem. Phys.* **11,** 8759–8766 (2011).
26. Marcolli, C., Luo, B. & Peter, T. Mixing of the organic aerosol fractions: liquids as the thermodynamically stable phases. *J. Phys. Chem. A* **108,** 2216–2224 (2004).
27. Kalberer, M. *et al.* Identification of polymers as major components of atmospheric organic aerosols. *Science* **303,** 1659–1662 (2004).
28. Murray, B. J. *et al.* Heterogeneous nucleation of ice particles on glassy aerosols under cirrus conditions. *Nat. Geosci.* **3,** 233–237 (2010).
29. Shiraiwa, M., Ammann, M., Koop, T. & Poschl, U. Gas uptake and chemical aging of semisolid organic aerosol particles. *Proc. Natl Acad. Sci. USA* **108,** 11003–11008 (2011).
30. Joutsensaari, J., Vaattovaara, P., Vesterinen, M., Hämeri, K. & Laaksonen, A. A novel tandem differential mobility analyzer with organic vapor treatment of aerosol particles. *Atmos. Chem. Phys.* **1,** 51–60 (2001).
31. Cheng, Y. F. *et al.* Size-resolved measurement of the mixing state of soot in the megacity Beijing, China: diurnal cycle, aging and parameterization. *Atmos. Chem. Phys.* **12,** 4477–4491 (2012).
32. Rabinow, B. E. Nanosuspensions in drug delivery. *Nat. Rev. Drug Discov.* **3,** 785–796 (2004).
33. Köhler, H. The nucleus in and the growth of hygroscopic droplets. *Trans. Faraday Soc.* **32,** 1152–1161 (1936).
34. Gao, Y., Chen, S. B. & Yu, L. E. Efflorescence relative humidity for ammonium sulfate particles. *J. Phys. Chem. A* **110,** 7602–7608 (2006).
35. Tang, I. N. & Munkelwitz, H. R. Water activities, densities, and refractive indices of aqueous sulfates and sodium nitrate droplets of atmospheric importance. *J. Geophys. Res-Atmos.* **99,** 18801–18808 (1994).
36. Tang, I. N. Chemical and size effects of hygroscopic aerosols on light scattering coefficients. *J. Geophys. Res-Atmos.* **101,** 19245–19250 (1996).
37. Seinfeld, J. H. & Pandis, S. N. *Atmospheric Chemistry and Physics, from Air Pollution to Climate Change* (John Wiley, 2006).
38. Clegg, S. L., Brimblecombe, P. & Wexler, A. S. Thermodynamic Model of the System $H^+ - NH_4^+ - Na^+ - SO_4^{2-} - NO_3^- - Cl^- - H_2O$ at 298.15K. *J. Phys. Chem. A* **102,** 2155–2171 (1998).
39. Pruppacher, H. R. & Klett, J. D. *Microphysics of clouds and precipitation* (Kluwer Academic Publishers, 1997).
40. Clegg, S. L., Ho, S. S., Chan, C. K. & Brimblecombe, P. Thermodynamic properties of aqueous $(NH_4)_2SO_4$ to high supersaturation as a function of temperature. *J. Chem. Eng. Data* **40,** 1079–1090 (1995).
41. Kreidenweis, S. M. *et al.* Water activity and activation diameters from hygroscopicity data-Part I: Theory and application to inorganic salts. *Atmos. Chem. Phys.* **5,** 1357–1370 (2005).
42. Xu, J., Imre, D., McGraw, R. & Tang, I. Ammonium sulfate: equilibrium and metastability phase diagrams from 40 to − 50°C. *J. Phys. Chem. B* **102,** 7462–7469 (1998).
43. Kendall, J. & Davidson, A. W. The melting point of ammonium sulfate. *J. Ind. Eng. Chem.* **13,** 303–304 (1921).

Acknowledgements

This study was supported by the Max Planck Society (MPG), National Natural Science Foundation of China (41330635), the EU project PEGASOS (265148), Russian Foundation for Basic research (12-05-00620-à), and RC 'Geomodel' of St Petersburg State University.

Author contributions

Y.C. and H.S. designed and performed the study. T.K., E.M. and U.P. discussed the results, interpretation and implications and commented on the manuscript at all stages. Y.C. and H.S. wrote the supplement. Y.C., H.S. and U.P. wrote the manuscript with inputs from all the co-authors.

Additional information

Competing financial interests: The authors declare no competing financial interests.

2

Thermodynamic controls of the Atlantic Niño

Hyacinth C. Nnamchi[1], Jianping Li[2,3], Fred Kucharski[4], In-Sik Kang[5], Noel S. Keenlyside[6], Ping Chang[7,8] & Riccardo Farneti[4]

Prevailing theories on the equatorial Atlantic Niño are based on the dynamical interaction between atmosphere and ocean. However, dynamical coupled ocean-atmosphere models poorly simulate and predict equatorial Atlantic climate variability. Here we use multi-model numerical experiments to show that thermodynamic feedbacks excited by stochastic atmospheric perturbations can generate Atlantic Niño s.d. of $\sim 0.28 \pm 0.07$ K, explaining $\sim 68 \pm 23\%$ of the observed interannual variability. Thus, in state-of-the-art coupled models, Atlantic Niño variability strongly depends on the thermodynamic component ($R^2 = 0.92$). Coupled dynamics acts to improve the characteristic Niño-like spatial structure but not necessarily the variance. Perturbations of the equatorial Atlantic trade winds ($\sim \pm 1.53\,\mathrm{m\,s^{-1}}$) can drive changes in surface latent heat flux ($\sim \pm 14.35\,\mathrm{W\,m^{-2}}$) and thus in surface temperature consistent with a first-order autoregressive process. By challenging the dynamical paradigm of equatorial Atlantic variability, our findings suggest that the current theories on its modelling and predictability must be revised.

[1] Department of Geography, University of Nigeria, Nsukka 410001, Nigeria. [2] College of Global Change and Earth System Science, Beijing Normal University, Beijing 100875, China. [3] Joint Center for Global Change Studies, Beijing 100875, China. [4] Earth System Physics Section, Abdus Salam International Centre for Theoretical Physics, Strada Costiera, 11, Trieste 34151, Italy. [5] School of Earth and Environmental Sciences, Seoul National University, Seoul 151-742, South Korea. [6] Geophysical Institute, University of Bergen and Bjerknes Centre for Climate Research, Allégaten 70, 5007 Bergen, Norway. [7] Department of Oceanography, Texas A&M University, College Station, Texas 77843-3160, USA. [8] Collaborative Innovation Center of Marine Science and Technology, Ocean University of China, Qingdao 266100, China. Correspondence and requests for materials should be addressed to H.C.N. (email: hyacinth.nnamchi@unn.edu.ng) or to J.P.L. (email: ljp@bnu.edu.cn).

The equatorial Atlantic Ocean exhibits sporadic interannual fluctuations in zonal sea surface temperature (SST) gradients associated with westerly wind perturbations[1–4]. This dominant coupled ocean-atmosphere mode of behaviour is often considered the Atlantic equivalent of the Pacific El Niño and is hence referred to as the Atlantic Niño[1,2]. The Atlantic Niño peaks during the boreal summer (June-July-August), when intensified equatorial upwelling typically leads to the development of the cold tongue[5]. Atmospheric convection and circulation changes due to the Atlantic Niño can cause increased precipitation across the equatorial Atlantic and decreases over the Sahel[6]. Beyond the Atlantic sector, the Atlantic Niño can excite near-global scale tropical atmospheric teleconnections, causing shifts in weather regimes in parts of the Americas, Africa, the Mediterranean, India and beyond[6–9]. Indeed, a growing body of evidence shows that the equatorial Atlantic modulates, and could therefore enhance the predictability of, the equatorial Pacific events[2,10–13].

These far-reaching effects call for an improved understanding of the physical processes that underlie equatorial Atlantic variability. As in the case of the Pacific El Niño, the leading theory for the equatorial Atlantic Niño is based on Bjerknes dynamical feedbacks[1–3,14,15]. The theory postulates that zonal wind perturbations $\{u\}'$ over the western equatorial Atlantic can drive ocean currents and equatorial baroclinic waves that redistribute upper ocean heat content, thereby modulating SST change in the eastern basin. Studies motivated by the Bjerknes theory show that the peak phase of the Atlantic Niño in June-July-August is preceded by equatorial westerly wind perturbations ($\{u\}'>0$) (refs 1,2). However, compared with that of the Pacific, the Bjerknes coupling strength is of the order of 50% weaker[2,16] and $\{u\}'$ accounts for only ~15–35% of observed equatorial Atlantic variability[17]. In addition to the Bjerknes theory, recent studies propose that intrinsic dynamics of equatorial deep zonal jets[18] and meridional advection of warm water from the north tropical Atlantic[17] contribute to Atlantic Niño variability. To date, all existing theories on the Atlantic Niño are largely inspired by dynamical thinking.

But how essential is dynamical coupling between the ocean and atmosphere to equatorial Atlantic variability? We explore this question by comparing simulations of two sets of experiments using 12 different climate models[19] (see Methods). The first set of experiments is based on state-of-the-art, fully coupled general circulation models (CGCMs) (full-CGCMs, hereafter), and in the second set, dynamical feedbacks are disabled by thermodynamically coupling the atmosphere to a 50-m deep slab of motionless ocean (slab-CGCMs). We show that thermodynamic mechanisms play dominant roles in both sets of experiments, accounting for 68 ± 23% of observed equatorial Atlantic interannual SST variability in the slab-CGCMs, through the wind-evaporation-SST feedbacks[20]. The impact of coupling to a dynamical ocean model is to amplify the variability and the Niño-like spatial structure. We conclude that, driven largely by stochastic atmospheric forced heat and moisture fluxes, the Atlantic Niño is not different from a first-order autoregressive (AR(1)) process.

Results

Equatorial winds preconditioning of the Atlantic Niño. We first examine the relationship between the Atlantic Niño SST anomaly $\{SST\}'$ during its peak season of boreal summer and the preceding season near surface $\{u\}'$ over the western equatorial Atlantic Ocean as a measure of Bjerknes feedback in the full-CGCMs. We then introduce the slab-CGCMs and describe the contributions from thermodynamic feedbacks only.

The relationship between $\{u\}'$ and $\{SST\}'$ exhibits wide variations across the full-CGCMs (Fig. 1). The $\{u\}'|\{SST\}'$ correlation is robust ($P \leq 0.001$) and of the same sign as observations for only six of the 12 models. As expected, models of overly strong $\{u\}'|\{SST\}'$ coupling (for example, GFDL-CM2.0, GFDL-CM2.1 and UKMO-HadGEM1) tend to overestimate equatorial Atlantic SST variability represented by the s.d. ($\sigma\{SST\}'$) of the Atlantic Niño index. However, it is rather surprising that some models (CSIRO-Mk3.0 and MRI-CGCM2.3.2) with weak $\{u\}'|\{SST\}'$ coupling exhibit realistic $\sigma\{SST\}'$. Indeed, even models with weak (ECHAM5/MPI-OM) and statistically significant (GISS-ER) but incorrect $\{u\}'|\{SST\}'$ correlation signs, can generate realistic $\sigma\{SST\}'$ of the Atlantic Niño.

Furthermore, some numerical models (for example, CGCM3.1_T47, GISS-ER and MRI-CGCM2.3.2) that simulate the correct structure of seasonality (Supplementary Fig. 1)—usually considered a distinctive feature of the Atlantic Niño[1,2]—fail to predict the expected $\{u\}'|\{SST\}'$ relationship. Indeed, the multi-model variance of equatorial Atlantic $\sigma\{SST\}'$ explained by the $\{u\}'|\{SST\}'$ coupling is only ~37%, suggesting that processes

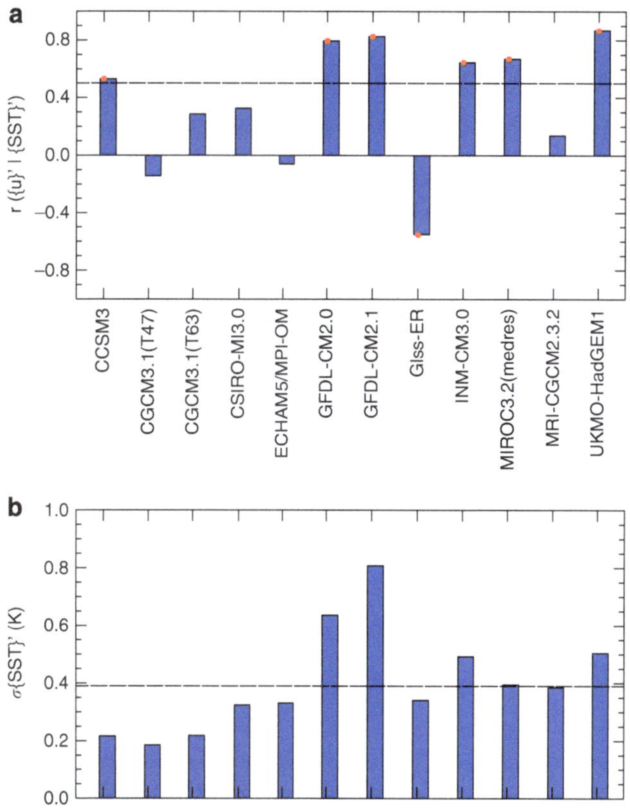

Figure 1 | Effect of the equatorial Atlantic westerly wind perturbations on Atlantic Niño variability in fully coupled models. (**a**) Correlation between the $\{u\}'$ averaged over the western equatorial Atlantic Ocean (2°N-2°S, 10°W-40°W) during March-April-May (MAM) and $\{SST\}'$ averaged over 3°N-3°S, 5°W-15°W during June-July-August (JJA), following the approach of ref. 17. The $\{u\}'$ at 925-hPa was analysed for models (CCSM3, CSIRO-Mk3.0 and ECHAM5/MPI-OM) for which the surface winds were not available from the Coupled Models Intercomparison Project, phase 3 archive. Red dots denote statistical significance at $P \leq 0.001$. Horizontal black line denote mean observational results[17]. (**b**) Standard deviations of the Atlantic Niño SST index ($\sigma\{SST\}'$) in JJA. Horizontal black line denote observational data sets estimates. In each panel, the abscissa shows the model names.

other than the Bjerknes feedback may be key to equatorial Atlantic variability.

Thermodynamic forcing of the equatorial Atlantic Niño. In reality, SST variability is caused by a combination of dynamical (three-dimensional temperature advection) and thermodynamic (surface net heat flux, in part driven by stochastic atmospheric perturbations) feedbacks on the ocean mixed layer. Indeed, recent studies have questioned the essence of ocean dynamics in generating the equatorial Pacific El Niño, concluding that thermodynamic feedbacks may be more important than previously thought[21–24]. However, the effect of stochastic atmospheric induced heat and moisture fluxes in the origin of the equatorial Atlantic Niño remains largely unknown and is thus examined in the present study.

The monthly Atlantic Niño index of the 12 slab-CGCMs exhibits strong variability ($\sigma = \sim 0.28 \pm 0.07$ K; for the full-CGCMs, $\sigma = \sim 0.47 \pm 0.09$ K). Figure 2 shows that the $\sigma\{SST\}'$ of the Atlantic Niño of the full-CGCMs (representing the total variability of the equatorial Atlantic ocean-atmosphere coupled system) is linearly dependent on the $\sigma\{SST\}'$ of the slab-CGCMs with no interactive ocean dynamics. The multi-model correlation coefficient is 0.96, corresponding to 92% of the explained variance. In comparison, the equatorial Pacific variability represented by the Niño-3 index exhibits more scatter with an explained variance of only 3%. This implies that while dynamical coupling in the equatorial Atlantic acts to amplify the $\sigma\{SST\}'$, coupled dynamics in the Pacific tends to modify the SST variability.

We determined the proportions of the observed SST variability over the eastern equatorial Atlantic Ocean resulting from thermodynamic processes by computing the boreal summer ratios of the modelled $\sigma\{SST\}'$ to recent historical observations[25–27] (1984–2013). The results show that 10 models have ratios of >0.50, with the 12 slab-CGCMs ensemble-mean of $\sim 68 \pm 23\%$, suggesting that thermodynamic processes dominate the equatorial Atlantic interannual SST variability (Fig. 2c). To put this result in context, similar ratios were calculated for the boreal winter peak season of the equatorial Pacific El Niño using the Niño-3 index. The corresponding ensemble-mean contribution of thermodynamic feedbacks for the Niño-3 region is considerably lower at $\sim 32 \pm 11\%$ (Fig. 2d).

The full-CGCM ratios were also plotted on the same axes of Fig. 2c,d. The observed ocean mixed layer in the equatorial Atlantic exhibits marked annual cycles and is essentially shallow (<50 m) in the Atlantic Niño region (Supplementary Fig. 2). Thus, some models with deep mixed layer of generally >50 m

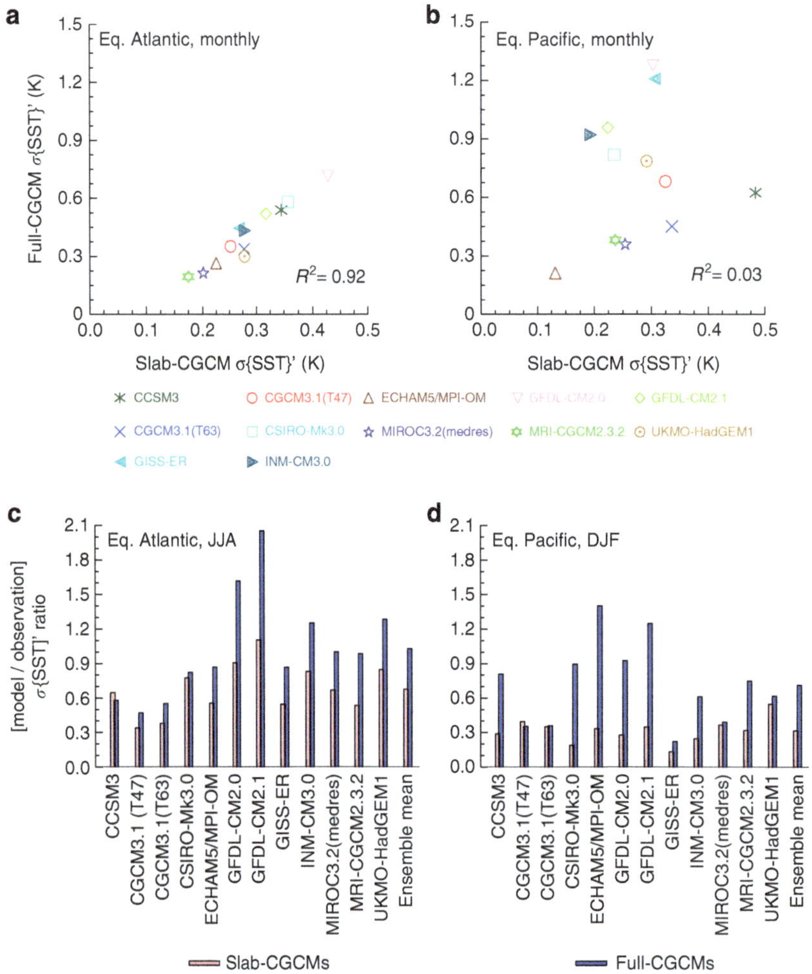

Figure 2 | Multi-model linearity of thermodynamic feedbacks in the eastern equatorial Atlantic and Pacific oceans and modelled proportions of the observed variability. (**a**,**b**) Relationship between the monthly Atlantic Niño (3°S-3°N, 0-20°W) and Pacific Niño-3 (5°S-5°N, 90-150°W) $\sigma\{SST\}'$ from slab-CGCMs and full-CGCMs. (**c**,**d**) Atlantic Niño and Pacific Niño-3 model/observation ratio of the $\sigma\{SST\}'$. Blue bars denote the ratios for full-CGCMs; light-pink, slab-CGCMs. A ratio of >1.0 denotes an overestimation of observed variability of ~ 0.40 K and ~ 0.95 K for the Atlantic Niño and Pacific Niño-3, respectively. Panels c and d are based on the JJA and December-January-February (DJF) seasonal means, respectively.

(for example, CCSM3, CGCM3.1_T47 and CGCM3.1_T63) tend to underestimate the observed $\sigma\{SST\}'$. In contrast, for those models in which the thermodynamic component generate $\sigma\{SST\}'$ close to observations (for example, GFDL-CM2.0, GFDL-CM2.1, INM-CM3.0 and UKMO-HadGEM1), coupling to ocean dynamics leads to unrealistically strong amplitudes of the Atlantic Niño $\sigma\{SST\}'$—given that the mixed layer depths are not much different from the observed. The ensemble-mean of the full-CGCMs reproduces the equatorial Atlantic SST temporal variability as may be expected from the consistency of the modelled mean mixed layer depths with observations.

To determine the comparative fidelity of the full-CGCM and slab-CGCM models in reproducing the observed spatial structure of the Atlantic Niño, we extracted the associated SST variability over the equatorial Atlantic Ocean (10°N–10°S, 20°E–60°W) as the leading empirical orthogonal function (Supplementary Fig. 3). This analysis essentially precludes that a model may still have an Atlantic Niño that is not the leading mode; we assume that if this occurs, it means that the model is unrealistic. The accuracy of the models is evaluated in relation to observations and the results quantified via second-order statistics[28].

Among the six models exhibiting realistic representation of equatorial Atlantic SST variability, defined as those falling within the third arc from observations, four (CCSM3, CSIRO-Mk3.0, GFDL-CM2.0 and GFDL-CM2.1) are slab-CGCMs and only two

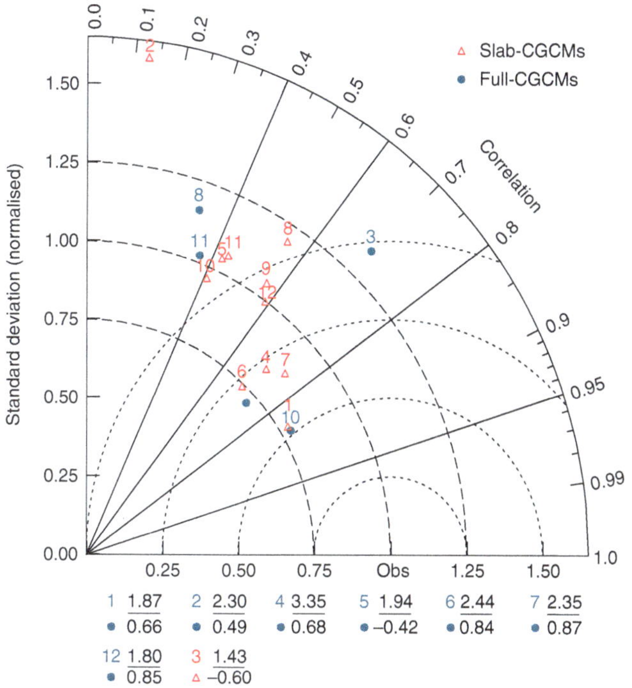

Figure 3 | Comparative evaluation of the spatial patterns of equatorial Atlantic $\{SST\}'$ simulated in thermodynamic and fully coupled control experiments. For each model, the leading empirical orthogonal function of the $\{SST\}'$ is computed for the equatorial Atlantic Ocean (10°N–10°S, 20°E–60°W), and the spatial pattern determined by regressing the $\{SST\}'$ onto the associated time series. The radial coordinate denotes the magnitude of total s.d., normalized by the observed value; the angular coordinate represents the correlation with observations[25–27]. Distances between 'obs' denoting observed point and any model point is proportional to the r.m.s. model error. Outliers (defined as models with negative correlations and/or $\sigma > 1.65$) are listed below the plot. 1-CCSM3; 2-CGCM3.1(T47); 3-CGCM3.1(T63); 4-CSIRO-Mk3.0; 5-ECHAM5/MPI-OM; 6-GFDL-CM2.0; 7-GFDL-CM2.1; 8-GISS-ER; 9-INM-CM3.0; 10-MIROC3.2(medres); 11-MRI-CGCM2.3.2; 12-UKMO-HadGEM1.

(INM-CM3.0 and MIROC3.2_medres) are fully coupled (Fig. 3). These two full-CGCMs with a more realistic leading mode have both realistic $\sigma\{SST\}'$ and $\{u\}'|\{SST\}'$ coupling (Fig. 1).

On the other hand, those full-CGCMs with realistic $\sigma\{SST\}'$ despite weak or incorrect $\{u\}'|\{SST\}'$ correlation (for example, CSIRO-Mk3.0, ECHAM5/MPI-OM, GISS-ER and MRI-CGCM2.3.2), fail to reproduce the observed spatial structure of the equatorial Atlantic SST variability. These results suggest that while thermodynamic feedbacks can generate strong $\sigma\{SST\}'$, the characteristic spatial structure of the Atlantic Niño could arise from coupled dynamics. Indeed, in most models (for example, CGCM3.1_T47, CGCM3.1_T63, GFDL-CM2.0, GFDL-CM2.1, INM-CM3.0, MIROC3.2_medres and UKMO-HadGEM1), coupling to ocean dynamics actually improves the spatial correlation with observations, but not the amplitudes of the SST variability. The implication is that, overall, the fully coupled models are no better than their thermodynamic counterparts in simulating the spatial structure of the Atlantic Niño.

Previous studies suggest that the Atlantic Niño exhibits interannual periodicity (see ref. 18 and references therein). However, as shown in Fig. 4, the supposed interannual peaks of the observed and modelled monthly Atlantic Niño spectrum are hardly statistically different from an AR(1) process consistent with the spectrum in an earlier analysis[29]. A comparison of the modelled spectra reveals that rather than the widely discussed coupled dynamics, the AR(1) spectral properties of the equatorial Atlantic SST variability are well reproduced with thermodynamic feedbacks alone (Fig. 4b–m), as predicted by simple stochastic climate models in which the ocean mixed layer integrates random atmospheric perturbations[30,31]. Coupling to ocean dynamics tends to increase the interannual, but not the low-frequency, variance of the Atlantic Niño, but the spectra retain their AR(1) character (Fig. 4n–y).

Evolution and heat budget of the Atlantic Niño. The above analysis clearly shows that thermodynamic processes are alone able to reproduce aspects of Atlantic Niño variability. Nevertheless, there remains a wide spread of model performance that is further exacerbated by coupling to a dynamical ocean model, as may be expected from larger mean biases of the full-CGCMs. Our aim is not to fully account for inter-model differences, but instead to investigate the physical processes involved. In this respect, we further analyse GFDL-CM2.0 (ref. 32) as its thermodynamic configuration captures the space-time variability of the Atlantic Niño fairly well (Figs 2–4). The fully coupled configuration of GFDL-CM2.0 simulates strong $\{u\}'|\{SST\}'$—suggesting that Bjerknes feedbacks operate in the model—in the equatorial Atlantic Ocean.

As shown in Fig. 5, initial $\{SST\}'$ forms off southeastern tropical Atlantic Ocean; that is, the Benguela Niño sector at *t*-6 leading to the peak phase of the Atlantic Niño. In subsequent months, the maximum $\{SST\}'$ propagates north-westwards, amplifies to cover the equatorial Atlantic Ocean and then begins to decay. In the absence of ocean dynamics, the ocean mixed layer warming and cooling rates (the time derivative of temperature, $\partial T/\partial t$) are governed entirely by net surface heat flux anomaly ($\{Q_{NET}\}'$) driven by stochastic atmospheric perturbations. Thus, at *t*-6, the $\{Q_{NET}\}'$ in the Benguela Niño area is associated with weakening of the trade winds and westerly wind perturbations ($\{u\}' > 0$) in the equatorial region. The following months are marked by a progressive northward intensification of the westerlies and $\{Q_{NET}\}'$ that drives the $\{SST\}'$.

Observational studies show that net heat flux plays a significant role in the annual cycle of equatorial Atlantic Ocean SST[33–35]. Similarly, a modelling analysis shows that the interaction of SST

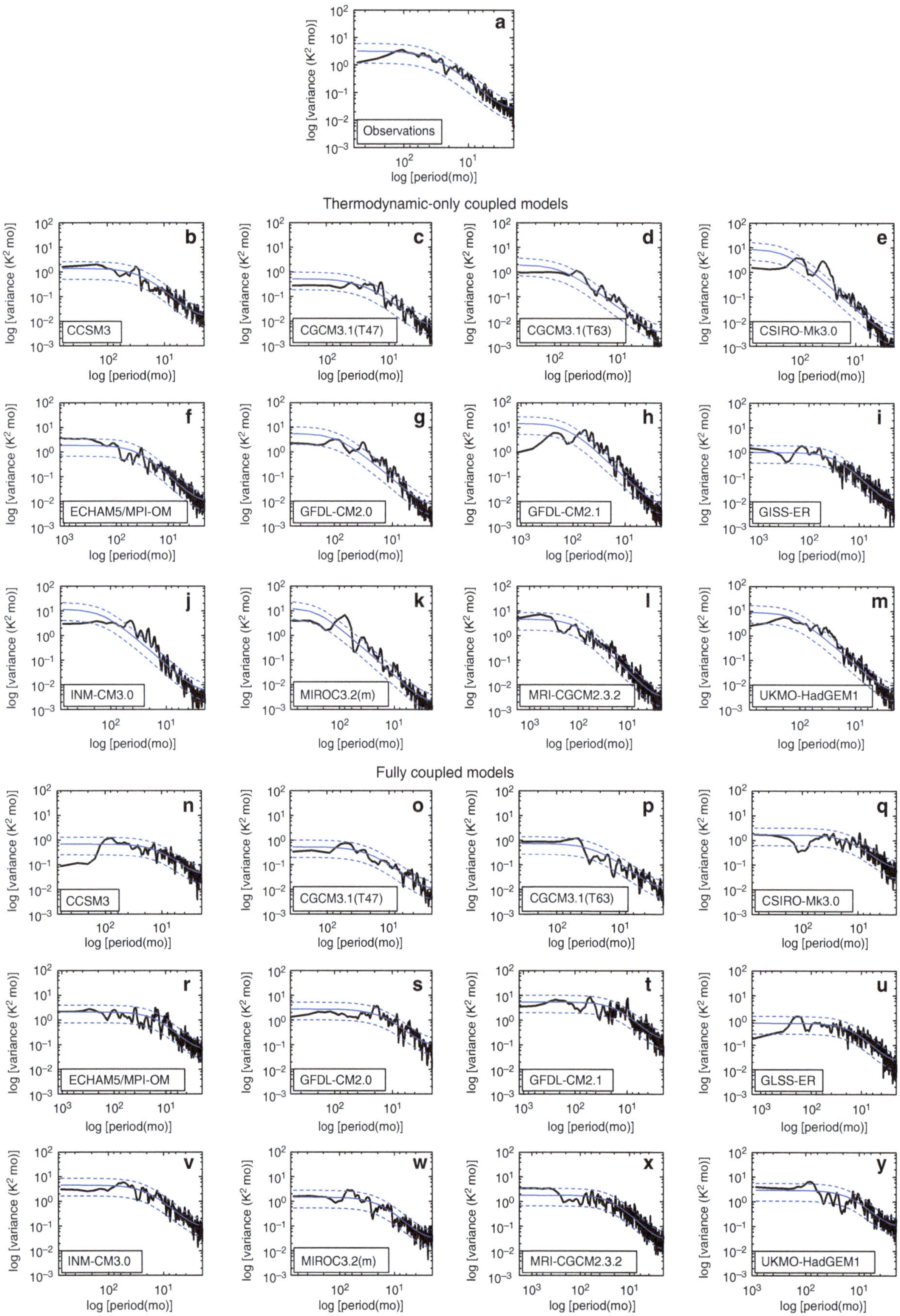
a
Observations
Thermodynamic-only coupled models
b
CCSM3
c
CGCM3.1(T47)
d
CGCM3.1(T63)
e
CSIRO-Mk3.0
f
ECHAM5/MPI-OM
g
GFDL-CM2.0
h
GFDL-CM2.1
i
GISS-ER
j
INM-CM3.0
k
MIROC3.2(m)
l
MRI-CGCM2.3.2
m
UKMO-HadGEM1
Fully coupled models
n
CCSM3
o
CGCM3.1(T47)
p
CGCM3.1(T63)
q
CSIRO-Mk3.0
r
ECHAM5/MPI-OM
s
GFDL-CM2.0
t
GFDL-CM2.1
u
GLSS-ER
v
INM-CM3.0
w
MIROC3.2(m)
x
MRI-CGCM2.3.2
y
UKMO-HadGEM1
log [variance (K^2 mo)]
log [period(mo)]

Figure 4 | Observed and modelled spectra of Atlantic Niño. (**a**) Observed[25] monthly spectrum of the Atlantic Niño index for 1984–2013. Panels (**b–m**) Atlantic Niño spectra simulated by the slab-CGCMs. (**n–y**), Atlantic Niño spectra simulated by the full-CGCMs. The model names are indicated on the bottom left corner of each panel. For the observations and models, the thick back curves represent the Atlantic Niño spectra; solid blue curves denote the AR(1) spectra; dashed blue curves, the 95% confidence bounds of the AR(1) spectra.

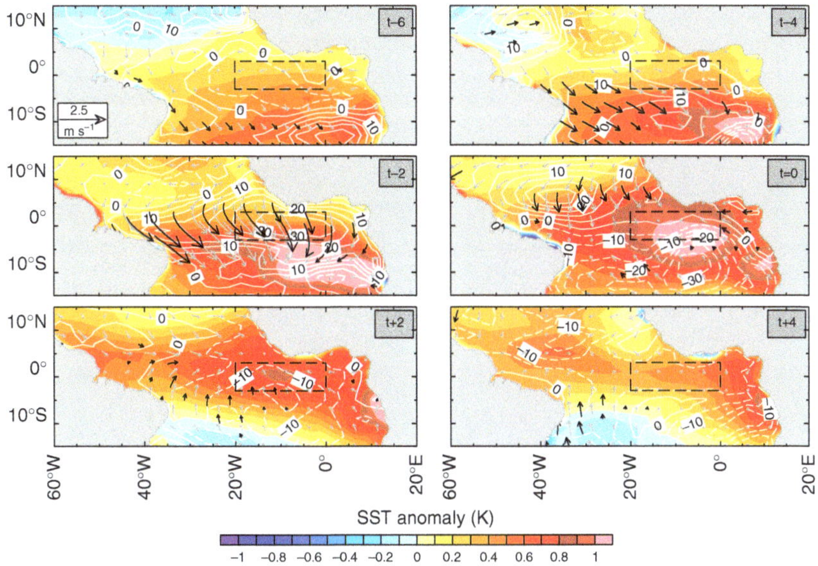

Figure 5 | Surface net heat flux and winds associated with Atlantic Niño in the thermodynamic configuration of GFDL-CM2.0. Evolution maps of monthly anomalies of SST (colour scale: unit, K), Q_{NET} (contours: unit, W m^{-2}) and winds (arrows: unit, m s^{-1}) regressed on the Atlantic Niño index fixed at $t = 0$, corresponding to July. Dashed black box indicate the Atlantic Niño domain defined by ref. 1. Thick black arrows represent statistically significant ($P \leq 0.05$) wind anomalies. The net heat flux is contoured at an interval of 5 W m^{-2}; positive (negative) values indicate that the ocean mixed layer is gaining (loosing) heat.

with radiation and evaporation feedbacks can produce quasi-biennial equatorial Atlantic variability[36]. Some previous studies, however, suggest that interannual $\{Q_{NET}\}'$ dampens rather than drive the $\{SST\}'$ in the region, although the explained variance is relatively low[37–39]. The upper ocean heat budget in the tropical Atlantic involves exceedingly complex feedback processes[4,17,18,36,40], and outcomes may depend on specific domains analysed. Furthermore, the evolution maps indicate that the relationship between $\{Q_{NET}\}'$ and $\{SST\}'$ actually undergoes a phase change during the life cycle of the Atlantic Niño (Fig. 5). The $\{Q_{NET}\}'$ drives $\{SST\}'$ during the development stages. Once the peak phase is reached and the associated westerly and northerly wind anomalies decline, the relation is reversed and $\{Q_{NET}\}'$ begins to reduce $\{SST\}'$, leading to the decay of an Atlantic Niño event. Thus, the $\{Q_{NET}\}'$ averaged in the Atlantic Niño region is $+28.58$ W m^{-2} at *t*-2, -4.08 W m^{-2} at $t = 0$ and -16.15 W m^{-2} at $t+3$.

The $\{Q_{NET}\}'$ is determined by ocean-atmosphere feedbacks resulting from four different flux components, expressed here as anomalies: shortwave radiation $\{Q_{SW}\}'$, longwave radiation $\{Q_{LW}\}'$, latent heat $\{Q_{LH}\}'$ and sensible heat $\{Q_{SH}\}'$ (see Methods). The evolution of the $\{\partial T/\partial t\}'$ over the equatorial Atlantic Ocean during a typical Atlantic Niño year is closely reproduced by the $\{Q_{NET}\}'$ such that the residual term is quite small (Fig. 6a). Here $\{\partial T/\partial t\}'$ is largely controlled by shortwave radiation ($\{Q_{SW}\}' = +15.83$ W m^{-2}) and by forcing from negative latent heat flux ($\{Q_{LH}\}' = -14.35$ W m^{-2}) peaking at *t*-2 during the development phase of the Atlantic Niño. The $\{Q_{LW}\}'$ and $\{Q_{SH}\}'$ appear to play comparatively minor roles. Superimposed on the climatological-mean state, negative $\{Q_{LH}\}'$ represents a suppression of evaporation due to a weakening of the trade winds, giving rise to net surface heating. Thus, the $\{Q_{LH}\}'$ and $\{Q_{NET}\}'$ tend to evolve together (Supplementary Fig. 4). Once the mature phase of Atlantic Niño is reached, $\{Q_{LH}\}'$ also appears vital to the decay of the $\{SST\}'$. The evolution of $\{Q_{LH}\}'$ from negative to positive suggests that as the equatorial westerly perturbations weakens, the trade winds strengthen, thereby intensifying evaporative cooling and the decay of Atlantic Niño $\{SST\}'$.

While these feedback mechanisms dominate in slab-CGCMs, in which ocean-atmosphere coupling is solely determined by thermodynamics, it is imperative to understand their relative contributions in a fully coupled model that also includes mixed layer temperature advection. Unlike the slab-CGCMs and earlier analysis[17] based on some assumed constant depths, we diagnose the heat budget of 500-year integration of the fully coupled configuration of GFDL-CM2.0 by explicitly accounting for the time evolution of the ocean mixed layer depth (see Methods). The GFDL-CM2.0 partly captures the annual cycle of the equatorial Atlantic thermocline consistent with observations[41] and reanalysis[42] data sets, although its shoaling is somewhat delayed (Supplementary Fig. 5).

Figure 7a shows that the peak of the sum of temperature advection and $\{Q_{NET}\}'$ terms at *t*-2 results in the peak of $\{\partial T/\partial t\}'$ after a month, this phase-lag may be attributed to the inability of the monthly time series analysed to fully resolve the heat budget in time and neglecting other terms such as vertical advection and mixing. Nevertheless, the $\{Q_{NET}\}'$ terms drive the SST change; the time average of $\{\partial T/\partial t\}'$ for the 2 months leading to the peak of the Atlantic Niño ($2.1\,e^{-7}$ K s^{-1}) is comparable with the sum of the advection and $\{Q_{NET}\}'$ terms ($2.0\,e^{-7}$ K s^{-1}). On the other hand, the peak contribution of the zonal $\{u\partial T/\partial x\}'$ and meridional $\{v\partial T/\partial y\}'$ advection components in driving the ocean mixed layer temperature is delayed until $t = 0$, suggesting that the Atlantic Niño event may be essentially triggered by the $\{Q_{NET}\}'$

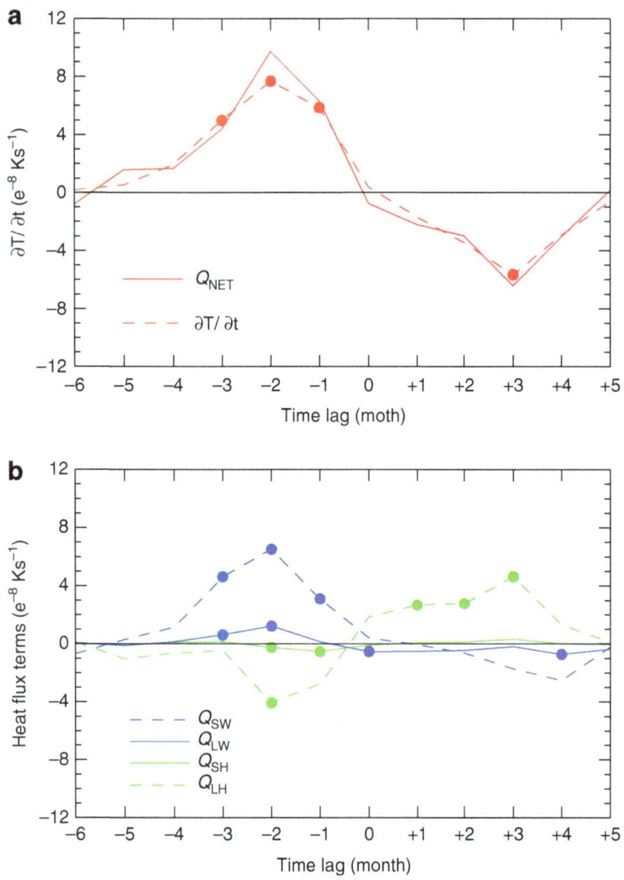

Figure 6 | Atlantic Niño heat budget in the thermodynamic configuration of GFDL-CM2.0. (**a**) Composite evolution of anomalies of the $\partial T/\partial t$ (dashed curve) and Q_{NET} (solid curve). (**b**) Individual terms of Q_{NET}: composite anomalies of Q_{SW} (blue, dashed), Q_{LW} (blue, solid), Q_{SH} (green, solid) and Q_{LH} (green, dashed). Composites are based on a $+1.0\sigma$ of the Atlantic Niño SST index at lag $t=0$ for July. Note that Q_{SW} is defined positive downward, while the other fluxes are defined positive upwards. Dots denote statistical significance at $P \leq 0.05$.

before coupled dynamics sets in (Fig. 7c). Consistent with the results of the thermodynamic configuration of GFDL-CM2.0, the $\{Q_{NET}\}'$ is largely controlled by $\{Q_{SW}\}'$ and $\{Q_{LH}\}'$ (Fig. 7e).

We also conducted heat budget analysis for 500-year integration of the fully coupled configuration of MIROC3.2_medres which exhibits realistic Atlantic Niño seasonality, $\sigma\{SST\}'$ and $\{u\}'|\{SST\}'$ coupling and, among the 12 models studied, the 'best' spatial representation of equatorial Atlantic SST variability. As shown in the right panels (Fig. 7b,d,f), the results of MIROC3.2_medres are basically consistent with those of the GFDL-CM2.0 model. Here again, the missing terms (seen as the residuals in Methods) seem to delay the development of the peak $\{SST\}'$—nonetheless, the $\{Q_{NET}\}'$ terms appear as the driver, and in an integrated sense can explain most of the modelled $\{\partial T/\partial t\}'$.

Discussion

An analysis of numerical models of two different levels of ocean-atmosphere coupling suggests a dominant role of stochastic atmospheric forced heat and moisture flux in equatorial Atlantic variability. Recent studies based on *in situ* and satellite observations show that net surface heat flux is important for the development of the climatological-mean cold tongue[33–35]. Thus, it is not entirely surprising that interannual fluctuations of net heat flux could drive cold tongue SST variability.

The Atlantic Niño is typically preceded by changes in the southeast trade winds[1–4] associated with perturbations of the St Helena subtropical anticyclone and Benguela Niño evolution[43–46]. Here we hypothesize that these atmospheric fluctuations can generate the space-time structure and AR(1) spectrum of the equatorial Atlantic Niño via the wind-evaporation-SST feedback mechanisms[20,47], even under motionless ocean conditions. Coupled dynamics can further enhance the characteristic Niño-like structure and energize the interannual variance of the Atlantic Niño, via partly resolving the Bjerknes feedback and changes in mixed layer depth, and in some models this results in an overestimation of the observed variability.

The tropical Atlantic Ocean has experienced profound climate shifts[48–50] characterized by Niño-like warming trends along the climatological-mean axis of the cold tongue[50]. However, projections of tropical Atlantic climate change, which has implications for the marine ecosystems and socio-economy of the adjacent countries, are uncertain[50] partly due to major systematic biases of state-of-the-art models in reproducing the basic state[51–54]. Like the conventional Atlantic Niño, both the warming trends[50] and model systematic biases are most pronounced during the boreal summer[52] and exhibit Niño-like spatial structure[51–54]. The severe biases in mean state can cause unrealistic seasonal cycle in the full-CGCMs, whereas the seasonal cycle in the slab-CGCMs is realistic since it is heavily constrained by the Q_{flux} (see Methods), which is constructed using the observed SST. The Atlantic Niño may therefore, in reality, result largely from perturbations of the mean seasonal cycle[55,56]. Although we have not considered seasonality explicitly, we do find that a slab-CGCM can simulate the correct structure (not shown). This suggests that further work is required to identify the underlying mechanisms for Atlantic Niño seasonality.

The mean state biases can impact equatorial Atlantic variability, as has been discussed in some recent studies[54,57]. However, this relation is non-trivial and there is little evidence for a relation between the mean state biases and variability in our ensemble of coupled models ($R^2 = 0.04$). The mean state biases have often been attributed to the atmospheric component of the coupled models; leading hypothesis implicates weakening of the trade winds[51,52]. By highlighting the importance of thermodynamic processes, in addition to the more often discussed dynamical feedbacks, our findings may allow for an improved understanding of these biases, and detection and attribution of tropical Atlantic climate change.

Aspects of the thermodynamically driven equatorial Atlantic variability discussed in this study—including its evolution and spatial characteristics—appear similar to the results reported for the Pacific El Niño[21–24]. Yet, we have shown that thermodynamic processes contribute far less proportion of the overall variance and have less linear control on the state-of-the-art model simulation of the equatorial Pacific event, compared to the Atlantic counterpart. Perhaps, the leading role of atmospheric induced heat and moisture flux explain why the predictability of equatorial Atlantic SST variability, considered a key variable for seasonal climate predictions[4,6], is comparatively low. While the predictability limit of tropical Atlantic SST variability estimated from observations is of the order of 5–8 months, that for the Pacific generally spans >8 months[58].

If thermodynamic feedbacks constitute the main source of Atlantic Niño variability as described in this study, then potential dynamical predictability may actually be lower than currently thought. Thus, our results call for a re-thinking of the procedures for improving the modelling and predictions of equatorial

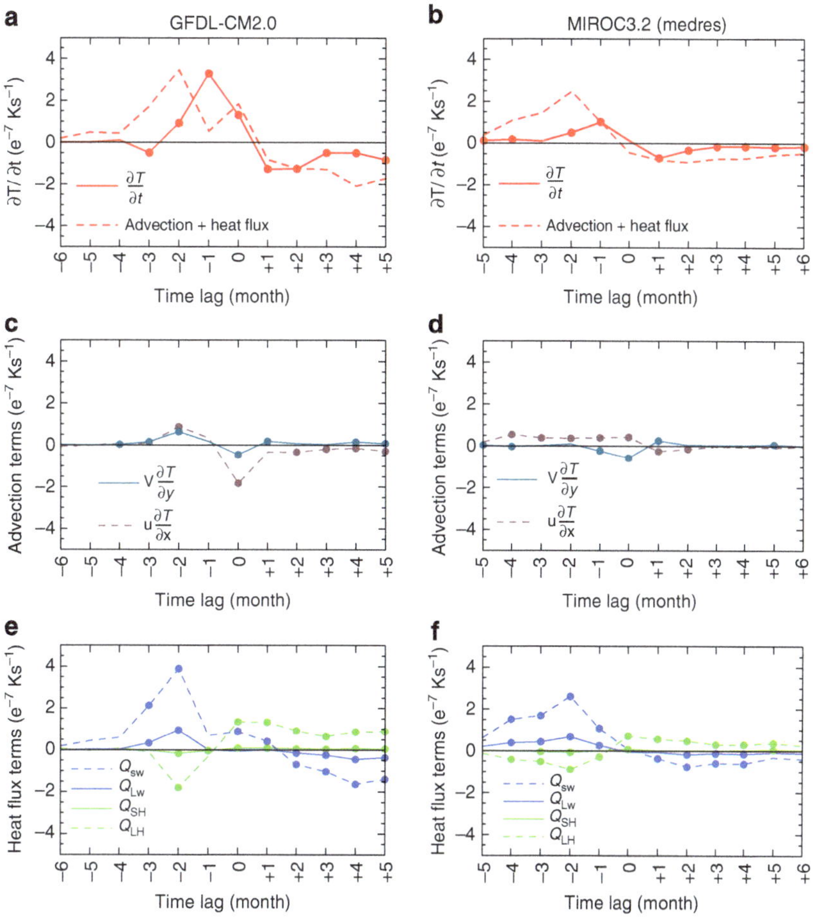

Figure 7 | Atlantic Niño heat budget in fully coupled configurations of GFDL-CM2.0 and MIROC3.2_medres models. (**a**) Composite evolution of anomalies of $\partial T/\partial t$ (solid curve) and the sum of advection and Q_{NET} terms (dashed curve). (**c**) individual temperature advection terms: composite anomalies of the $u\partial T/\partial x$ (pink, dashed) and $v\partial T/\partial y$ (blue, solid). (**e**) Q_{NET} terms: composite anomalies of Q_{SW} (blue, dashed), Q_{LW} (blue, solid), Q_{SH} (green, solid) and Q_{LH} (green, dashed). Note that Q_{SW} is defined positive downward, while the other fluxes are defined positive upwards. The composites are based on a $+1.0\sigma$ of the Atlantic Niño SST index at lag $t=0$ for July; dots denote statistical significance at $P \leq 0.05$. Similar plots are shown on the right panels (**b**,**d**,**f**) based on the MIROC3.2_medres for $+1.0\sigma$ composites of the Atlantic Niño SST index at lag $t=0$ for the month of peak seasonality, June.

Atlantic climate that may necessarily be thermodynamically motivated[59–61], and involving improved representation of atmospheric variability in the coupled models. In this regard, further modelling studies and enhanced observations are needed to confirm the relative importance of dynamical and thermodynamic processes in equatorial Atlantic variability.

Methods

Observational SST data sets. The National Oceanic Atmospheric Administration Extended Reconstructed SST version 3b available for 2- × 2- longitude–latitude grids[25], the Hadley Centre Global Sea Ice and SST[26] version 1 (1- × 1-) and the Kaplan Extended SST[27] version 2 (5- × 5-) were analysed. Each data set was linearly de-trended via the least squares method and then the three bi-linearly interpolated onto 1- × 1- longitude–latitude grids to create an ensemble-mean.

Model selection. Twelve numerical models that contributed skin temperature—a proxy for SST—to the 'slab ocean control experiment' of the CMIP3 and have integration lengths of ≥30 years were selected for analysis. The slab-CGCM integration lengths (in years) studied are as follows: CCSM3 (51), CGCM3.1_T47 (30), CGCM3.1_T63 (30), CSIRO-Mk3.0 (60), ECHAM5/MPI-OM (100), GFDL-CM2.0 (50), GFDL-CM2.1(100), GISS-ER (120), INM-CM3.0 (60), MIROC3.2_medres(60), MRI-CGCM2.3.2 (150) and UKMO-HadGEM1 (70). Description of these models and experimental designs are available from the CMIP website (http://www-pcmdi.llnl.gov/ipcc/about_ipcc.php).

For each model, the final years of ('pre-industrial control experiment') integration of the fully coupled configuration (full-CGCM) corresponding to the available length of the slab-CGCM were studied. As in the slab control experiment, greenhouse forcing was held constant in the pre-industrial control experiment and this allows us to explore the intrinsic variability, as the tropical Atlantic Ocean has exhibited marked warming trends[48–50].

Thermodynamic coupled model heat budget. In this modelling set-up, the time evolution of the SST is determined by the net heat flux (Q_{NET}) based on the following thermodynamic equation:

$$\frac{\partial T}{\partial t} = \left[\frac{Q_{SW} - Q_{LW} - Q_{LH} - Q_{SH}}{\rho C_w h}\right] + Q_{flux} + R \qquad (1)$$

where ρ, h and C_w are constants representing the sea water density, ocean mixed layer depth and specific heat capacity of ocean water ($\rho = 10^3\,\mathrm{kg\,m^{-3}}$; $h = 50\,\mathrm{m}$; $C_w = 4 \times 10^3\,\mathrm{J\,kg^{-1}K^{-1}}$), respectively, and where T is the SST. Q_{SW}, Q_{LW}, Q_{LH} and Q_{SH} denote net downward shortwave radiation at the ocean surface, net upward longwave radiation, surface latent and sensible heat flux, respectively. Note that Q_{SW} is positive downward, while the other fluxes are positive upwards. Q_{flux} is prescribed monthly climatological-mean ocean heat flux, which contributes to the annual cycle but does not drive SST variability. R is a residual term that denotes the sum of boundary flux errors from averages over the Atlantic Niño region and high frequency variability not resolved by the monthly time series analysed. Thus, the variability of $\partial T/\partial t$ is governed by the terms on the right-hand side of the equation enclosed in square bracket [], which denotes the Q_{NET}.

Fully coupled model heat budget. Here in addition to the Q_{NET} terms, coupling to interactive ocean dynamics denotes the introduction of three-dimensional ocean water advection and thus the exclusion of the restoration term (Q_{flux}). The temperature tendency is determined as follows:

$$\frac{\partial[T]}{\partial t} = -\left[u\frac{\partial T}{\partial x}\right] - \left[v\frac{\partial T}{\partial y}\right] + \frac{Q_{SW} - Q_{LW} - Q_{LH} - Q_{SH}}{\rho C_w h} + R \qquad (2)$$

where u and v are the horizontal ocean current velocities; and where the first two terms on the right-hand side represent zonal and meridional advection terms. R is a residual term that represents the sum of boundary flux errors from averages over the Atlantic Niño region, vertical temperature advection $[w\partial T/\partial z]$ and unresolved physical processes (for example, diffusion, turbulent mixing and high frequency variability not resolved by the monthly time series analysed). In the fully coupled configuration, the mixed layer depth (h) is defined as the depth at which ocean temperatures are 0.5 K lower than those at the surface; thus, h changes in space and time. Following the approach of ref. 62, [•] is computed for each variable as the vertical average for the ocean mixed layer:

$$[\bullet] = \frac{1}{h}\int_0^h \bullet dz \tag{3}$$

Once the heat budget terms were calculated, each was linearly de-trended, and the climatological-mean annual cycle removed to create an anomaly field {•}' via the least squares method and domain-average was then computed for the Atlantic Niño region.

Mixed layer depth and thermocline. Observed data sets on the climatological-mean ocean mixed layer depth estimated using the temperature criterion described above and thermocline (represented by the depth of the 20 °C isotherm) were taken from the World Ocean Atlas 1994 (WOA94) (ref. 41), available online from: http://www.esrl.noaa.gov/psd/data/gridded/data.nodc.woa94.html. These data sets were analysed alongside those obtained from the European Centre for Medium-range Weather Forecasts Operational Ocean Reanalysis System 3 (ORAS3) (ref. 42) for the period 1959–2009 (http://apdrc.soest.hawaii.edu/datadoc/ecmwf_oras3.php). Modelled mixed layer depth and thermocline were then computed from the historical ('climate of the twentieth century') experiment of CMIP3 and analysed for the 45-year period from 1955 to 1999.

Atlantic and pacific Niño indices. The Atlantic Niño index was computed as the domain averaged SST anomaly over the equatorial Atlantic Ocean (3°N–3°S, 0–20W). The Niño-3 index, determined as the domain-average at 5-N–5-S, 90-W–150-W, was used to represent eastern equatorial Pacific El Niño.

Significance tests. We determined the statistical significance levels based on the two-tailed P values using Student's t-test, except for the Atlantic Niño spectrum in which significance was estimated as the (lower and upper) 95% confidence bounds of the theoretical AR(1) spectrum using the lag-1 autocorrelation function.

References

1. Zebiak, S. E. Air–sea interaction in the equatorial Atlantic region. *J. Clim.* **6,** 1567–1586 (1993).
2. Keenlyside, N. S. & Latif, M. Understanding equatorial Atlantic interannual variability. *J. Clim.* **20,** 131–142 (2007).
3. Carton, J. A. *et al.* Decadal and interannual SST variability in the tropical Atlantic Ocean. *J. Phys. Oceanogr.* **26,** 1165–1175 (1996).
4. Chang, P. *et al.* The cause of the fragile relationship between the Pacific El Niño and the Atlantic El Niño. *Nature* **443,** 324–328 (2006).
5. Xie, S.-P. & Carton, J. A. (eds Wang, C., Xie, S.-P & Carton, J. A.) *AGU Geophysical Monograph* 147, 121–142 (2004).
6. Giannini, A., Saravanan, R. & Chang, P. Oceanic forcing of Sahel rainfall on interannual to interdecadal time scales. *Science* **302,** 1027–1030 (2003).
7. Losada, T., Rodríguez-Fonseca, B. & Kucharski, F. Tropical influence on the summer mediterranean climate. *Atmos. Sci. Lett.* **13,** 36–42 (2012).
8. Kucharski, F. *et al.* A Gill–Matsuno-type mechanism explains the tropical Atlantic influence on African and Indian monsoon rainfall. *Q. J. R. Meteorol. Soc.* **135,** 569–579 (2009).
9. Losada, T. *et al.* Tropical response to the equatorial mode: AGCM multimodel approach. *Clim. Dyn.* **35,** 45–52 (2009).
10. Rodríguez-Fonseca, B. *et al.* Are Atlantic Niños enhancing Pacific ENSO events in recent decades? *Geophys. Res. Lett.* **36,** L20705 (2009).
11. Keenlyside, N. S. *et al.* Potential of equatorial Atlantic variability to enhance El Niño prediction. *Geophys. Res. Lett.* **40,** 2278–2283 (2013).
12. Polo, I. *et al.* Processes in the Pacific La Nina onset triggered by the Atlantic Niño. *Clim. Dyn.* **44,** 115–131 (2015).
13. Kucharski, F. *et al.* Tropical Atlantic influence on Pacific variability and mean state in the twentieth century in observations and CMIP5. *Clim. Dyn.* **44,** 881–896 (2015).
14. Bjerknes, J. Atmospheric teleconnections from the equatorial Pacific. *Mon. Weather Rev.* **97,** 163–172 (1969).
15. Wyrtki, K El Niño—The dynamic response of the equatorial Pacific Ocean to atmospheric forcing. *J. Phys. Oceanogr.* **5,** 572–584 (1975).
16. Lübbecke, J. F. & McPhaden, M. J. A comparative stability analysis of Atlantic and Pacific Niño modes. *J. Clim.* **26,** 5965–5980 (2013).
17. Richter, I. *et al.* Multiple causes of interannual sea surface temperature variability in the equatorial Atlantic ocean. *Nat. Geosci.* **6,** 43–47 (2013).
18. Brandt, P. *et al.* Interannual atmospheric variability forced by the deep equatorial Atlantic ocean. *Nature* **473,** 497–500 (2011).
19. Meehl, G. A. *et al.* The WCRP CMIP3 multimodel dataset: a new era in climate change research. *Bull. Amer. Meteorol. Soc.* **88,** 1383–1394 (2007).
20. Xie, S.-P. & Philander, S. G. H. A coupled ocean-atmosphere model of relevance to the ITCZ in the eastern Pacific. *Tellus A* **46,** 340–350 (1994).
21. Dommenget, D. The slab ocean El Niño. *Geophys. Res. Lett.* **37,** L20701 (2010).
22. Clement, A. *et al.* Rethinking the ocean's role in the southern oscillation. *J. Clim.* **24,** 4056–4072 (2011).
23. Zhang, H., Clement, A. & Di Nezio, P. The South Pacific meridional mode: a mechanism for ENSO-like variability. *J. Clim.* **27,** 769–783 (2013).
24. Dommenget, D. *et al.* Analysis of the slab ocean El Niño atmospheric feedbacks in observed and simulated ENSO dynamics. *Clim. Dyn.* **42,** 3187–3205 (2014).
25. Smith, T. M. *et al.* Improvements to NOAA's historical merged land–ocean surface temperature analysis (1880–2006). *J. Clim.* **21,** 2283–2296 (2008).
26. Rayner, N. A. *et al.* Global analyses of sea surface temperature, sea ice, and night marine air temperature since the late nineteenth century. *J. Geophys. Res.* **108,** 4407 (2003).
27. Kaplan, A. *et al.* Analyses of global sea surface temperature 1856-1991. *J. Geophys. Res.* **103,** 567–589 (1998).
28. Taylor, K. E. Summarizing multiple aspects of model performance in a single diagram. *J. Geophys. Res.* **106,** 7183–7192 (2001).
29. Latif, M. & Grötzner, A. The equatorial Atlantic oscillation and its response to ENSO. *Clim. Dyn.* **16,** 213–218 (2000).
30. Hasselmann, K. Stochastic climate models Part I. Theory. *Tellus* **28,** 473–485 (1976).
31. Frankignoul, C. & Hasselmann, K. Stochastic climate models. Part 2. Application to sea-surface temperature variability and thermocline variability. *Tellus* **29,** 284–305 (1977).
32. Delworth, T. L. *et al.* GFDL's CM2 global coupled climate models. Part I: Formulation and simulation characteristics. *J. Clim.* **19,** 643–674 (2006).
33. Wade, M., Caniaux, G. & du Penhoat, Y. Variability of the mixed layer heat budget in the eastern equatorial Atlantic during 2005–2007 as inferred using Argo floats. *J. Geophys. Res.* **116,** C08006 (2011).
34. Foltz, G. R. *et al.* Seasonal cycle of the mixed layer heat budget in the northeastern tropical Atlantic Ocean. *J. Clim.* **26,** 8169–8188 (2013).
35. Hummels, R. *et al.* Diapycnal heat flux and mixed layer heat budget within the Atlantic cold tongue. *Clim. Dyn.* **43,** 3179–3199 (2014).
36. Trzaska, S. *et al.* South Atlantic variability arising from air–sea coupling: local mechanisms and tropical–subtropical interactions. *J. Clim.* **20,** 3345–3365 (2007).
37. Hu, Z. & Huang, B. Physical processes associated with the tropical Atlantic SST meridional gradient. *J. Clim.* **19,** 5500–5518 (2006).
38. Hu, Z. & Huang, B. Physical processes associated with the tropical Atlantic SST gradient during the anomalous evolution in the southeastern ocean. *J. Clim.* **20,** 3366–3378 (2007).
39. Ding, H., Keenlyside, N. S. & Latif, M. Equatorial Atlantic interannual variability: role of heat content. *J. Geophys. Res.* **115,** C09020 (2010).
40. Chang, P. *et al.* Climate fluctuations of tropical coupled systems—The role of ocean dyynamics. *J. Clim.* **19,** 5122–5174 (2006).
41. Monterey, G. I. & Levitus, S. *Climatological cycle of mixed layer depth in the world ocean* pp 5 (U.S. Gov. Printing Office, Washington, D.C., 1997).
42. Balmaseda, M. A. *et al.* The ECMWF ocean analysis system: ORA-S3. *Mon. Weather Rev.* **136,** 3018–3034 (2008).
43. Lübbecke, J. F. *et al.* On the connection between Benguela and equatorial Atlantic Niños and the role of the South Atlantic Anticyclone. *J. Geophys. Res.* **115,** C09015 (2010).
44. Richter, I. *et al.* On the triggering of Benguela Niños: Remote equatorial versus local influences. *Geophys. Res. Lett.* **37,** L20604 (2010).
45. Nnamchi, H. C., Li, J. P. & Anyadike, R. N. C. Does a dipole mode really exist in the South Atlantic Ocean? *J. Geophys. Res.* **116,** D15104 (2011).
46. Lübbecke, J. F. *et al.* Variability in the South Atlantic anticyclone and the Atlantic Niño mode. *J. Clim.* **27,** 8135–8150 (2014).
47. Chang, P., Ji, L. & Li, H. A decadal climate variation in the tropical Atlantic Ocean from thermodynamic air-sea interactions. *Nature* **385,** 516–518 (1997).
48. Deser, C. *et al.* Twentieth century tropical sea surface temperature trends revisited. *Geophys. Res. Lett.* **37,** L10701 (2010).
49. Xie, S.-P. *et al.* Global warming pattern formation: Sea surface temperature and rainfall. *J. Clim* **23,** 966–986 (2010).
50. Tokinaga, H. & Xie, S.-P. Weakening of the equatorial Atlantic cold tongue over the past six decades. *Nat. Geosci.* **4,** 222–226 (2011).
51. Richter, I. & Xie, S.-P. On the origin of equatorial Atlantic biases in coupled general circulation models. *Clim. Dyn.* **31,** 587–598 (2008).
52. Grodsky, S. A. *et al.* Tropical Atlantic Biases in CCSM4. *J. Clim.* **25,** 3684–3701 (2012).

53. Richter, I. *et al.* Equatorial Atlantic variability and its relation to mean state biases in CMIP5. *Clim. Dyn* **42,** 171–188 (2014).
54. Ding, H. *et al.* The impact of mean state errors on equatorial Atlantic interannualvariability in a climate model. *J. Geophys. Res.* **120,** 1133–1151 (2015).
55. Burls, N. J., Reason, C. J. C., Penven, P. & Philander, S. G. Similarities between the tropical Atlantic seasonal cycle and ENSO: an energetics perspective. *J. Geophys. Res.* **116,** C11010, doi:10.1029/2011JC007164 (2011).
56. Burls, N. J. *et al.* Energetics of the tropical Atlantic zonal mode. *J. Clim.* **25,** 7442–7466 (2012).
57. Munoz, E. *et al.* Mean and variability of the tropical Atlantic Ocean in the CCSM4. *J. Clim.* **25,** 4860–4882 (2012).
58. Li, J. P. & Ding, R. Q. Temporal-spatial distribution of the predictability limit of monthly sea surface temperature in the global oceans. *Int. J. Climatol.* **33,** 1936–1947 (2013).
59. Chang, P. *et al.* Tropical Atlantic seasonal predictability: the roles of El Niño remote influence and thermodynamic air-sea feedback. *Geophys. Res. Lett.* **30,** 1501 (2003).
60. Barreiro, M. & Chang, P. A linear tendency correction technique for improving seasonal prediction of SST. *Geophys. Res. Lett.* **31,** L23209 (2004).
61. Batté, L. & Déqué, M. A stochastic method for improving seasonal predictions. *Geophys. Res. Lett.* **39,** L09707 (2012).
62. Li, T. *et al.* Relative role of dynamic and thermodynamic processes in the development of the Indian Ocean dipole: an OGCM diagnosis. *Geophys. Res. Lett.* **29,** 2110 (2002).

Acknowledgements

H.C.N. conducted parts of the study during research stays at the International Centre for Theoretical Physics, Trieste, Italy through the Junior Associateship Scheme and at the Geophysical Institute, University of Bergen, Norway through funding from the Strategic Partnership for International Research and Education (SPIRE) and the Research Council of Norway (233680/E10). H.C.N. and N.S.K. were supported by the EU FP7/2007-2013 PREFACE Project (603521); H.C.N. and J.P.L. were supported by China 973 Program (2010CB950400) and NSFC Key Project (41030961). P.C. was supported by U.S. National Science Foundation Grants OCE-1334707 and AGS-1462127, and National Oceanic and Atmospheric Administration Grant NA11OAR4310154. P.C. also acknowledges the support from China's National Basic Research Priorities Programme (2013CB956204) and the Natural Science Foundation of China (41222037 and 41221063).

Author contributions

H.C.N. conceived the study, conducted the analysis and wrote the initial draft of the manuscript under the supervision of J.P.L. H.C.N., J.P.L., F.K., I.S.K., N.S.K, P.C. and R.F discussed further analysis and interpreted the results. All authors contributed to improve the manuscript.

Additional information

Competing financial interests: The authors declare no competing financial interests.

3

Connecting the oxidation of soot to its redox cycling abilities

María Antiñolo[1,2], Megan D. Willis[1], Shouming Zhou[1] & Jonathan P.D. Abbatt[1]

Although it is known that soot particles are emitted in large quantities to the atmosphere, our understanding of their environmental effects is limited by our knowledge of how their composition is subsequently altered through atmospheric processing. Here we present an on-line mass spectrometric study of the changing chemical composition of hydrocarbon soot particles as they are oxidized by gas-phase ozone, and we show that the surface-mediated loss rates of adsorbed polycyclic aromatic hydrocarbons in soot are directly connected to a significant increase in the particle redox cycling abilities. With redox cycling implicated as an oxidative stress mechanism that arises after inhalation of atmospheric particles, this work draws a quantitative connection between the detailed heterogeneous chemistry occurring on atmospheric particles and a potential toxic mechanism attributable to that aerosol.

[1] Department of Chemistry, University of Toronto, 80 St George Street, Toronto, Ontario, M5S 3H6 Canada. [2] Facultad de Ciencias y Tecnologías Químicas, Universidad de Castilla-La Mancha, Avda. Camilo José Cela s/n, Ciudad Real 13071, Spain. Correspondence and requests for materials should be addressed to J.A. (email: jabbatt@chem.utoronto.ca).

One of the main toxicological mechanisms resulting from atmospheric particulate matter (PM) inhalation is oxidative stress, arising from an imbalance of oxidants and antioxidants[1,2]. This imbalance can be produced by catalytic redox-cycling reactions, through species with labile oxidation states, which result in the depletion of reducing agents within the cell and the generation of reactive oxygen species (ROS). Likely catalysts that have been identified are transition metals and quinones, although other redox classes of compounds are also possible[3–5]. It is also shown that PM can lead to increased negative health effects after it has been exposed to atmospheric oxidants. For example, greater lung injury and inflammation were observed when rats were exposed to diesel exhaust particles (DEP) oxidized by O_3 compared with non-oxidized DEP[6]. Other studies reported an increase of the response in redox cycling rates when different combustion particles were exposed to O_3 (refs 7–9). Photoxidized DEP also caused greater inflammation and showed an increase in the redox activity[10,11]

Whereas these studies have identified important correlations, to move forward in this field it is necessary to develop a molecular-level understanding of the composition and atmospheric processing mechanisms of particles. As well, we must connect this detailed surface chemistry to the routes for PM toxicity, such as the ability of PM to promote redox cycling and ROS generation. To that end, the central hypothesis in this work is that reactive oxidation chemistry with soot particles involves interactions with either polycyclic aromatic hydrocarbons (PAHs), which are known to coat combustion particles, or with the chemically unsaturated graphitic soot backbone that has PAH-like character. In particular, it is well known that quinones are oxidation products of PAHs[12–15], which are potential redox-cycling agents. Indeed, gas-surface oxidation of adsorbed anthracene by ozone has identified anthraquinone as a major oxidation product[16].

The motivation to study soot particles is multifold. In addition to established correlations between black carbon and negative health outcomes[17], from a chemical perspective quinones and hydroquinones have been detected on carbon black surfaces and in engine exhaust emissions[18–20], and their high efficiency as electron transfer agents has been demonstrated[3]. As well, Shinyashiki *et al.*[21] reported that most of the redox and electrophilic activity of the organic extracts of DEP is associated with quinone-like substances. Past studies have reported an increase in PM redox cycling activity when carbonaceous particles have been exposed to ozone[7–9].

There are many kinetic studies addressing the reaction between soot and O_3 where the primary focus has been on characterizing the overall loss rate of gaseous O_3. In particular, the uptake coefficient, γ, describes the kinetics of this heterogeneous process, being the net probability that a molecule (O_3 in this case) undergoing a gas-kinetic collision with a surface (soot in this case) is lost to the surface. γ values of $\sim 10^{-3}$ have been reported for freshly emitted soot and $\sim 10^{-6}$–10^{-4} for aged soot[22]. There are also kinetic studies for the exposure of various PAHs adsorbed on different surfaces to gaseous O_3 (refs 13,23–25). For example, the reported uptake coefficient for O_3 with benzo[*a*]pyrene (BaP) adsorbed on soot was in the range 10^{-6}–10^{-5} for O_3 concentrations between 5.0×10^{11} and 2.5×10^{13} molecule cm^{-3} (ref. 23). Notably, the oxidation kinetics of PAHs that are formed as an intrinsic component of soot aerosol have not been studied.

In this work, we determine a relationship between the oxidation-driven chemical change of hydrocarbon soot particles and their redox cycling activity, demonstrating a quantitative connection of the atmospheric oxidative processing kinetics and mechanism to potential toxic effects of PM. Using a new aerosol mass spectrometric technique, we monitor the real-time loss of soot-bound PAHs as they are oxidized within an environmental chamber and determine the O_3 uptake coefficient from the observed chemistry. We also make measurements of the redox activity of the increasingly oxidized soot particles[26]. In addition, we evaluate the redox cycling activity that arises when soot particles that are partially coated with benzo[*e*]pyrene (BeP), a PAH detected in ambient particles[27], are exposed to O_3. The aim of these latter experiments is to compare the surface chemistry observed for a single PAH species with the behaviour of a collective ensemble of PAHs present within combustion soot.

Results

Size characterization of soot particles. The size distribution of soot particles inside the chamber was measured with a scanning mobility particle sizer (SMPS) between 14.6 and 710.5 nm during the experiments for both bare soot and soot coated with BeP. No significant change in the size distribution was observed after BeP coating, consistent with the coating being sub-monolayer in average thickness. In Supplementary Table 1, average values for the mean mobility diameter, particle number concentration and total mass loading inside the chamber before starting the oxidation are presented with the mixing ratios of gas-phase species. The observed high mean diameter (>300 nm) arises from coagulation of smaller soot particles occurring during the time required to fill the chamber with the required mass loading to run the DTT assay (around 1 h 30 min). The soot particle effective density was also calculated from mobility and aerodynamic measurements and is shown in Supplementary Table 1[28]. When comparing mobility and aerodynamic diameters of the soot particles, it was found that the mobility diameter was always much higher than the aerodynamic diameter, indicating that the soot particles have a low effective density.

During an experimental run, the particles became on average 11% larger in mobility diameter, and their number concentration decreased on average 45%, with both effects likely arising from preferential loss or coagulation of small particles. Oxidation of the particles is unlikely to alter their size significantly, given that the oxidation occurs only at the surface, unless there is some morphological change occurring as well.

Oxidation-driven compositional changes of soot particles. To monitor the composition of the particles, measurements with a recently developed instrument, the Soot-Particle Aerosol Mass Spectrometer (SP-AMS)[29,30], were performed. The SP-AMS detects refractory black carbon (rBC) and associated non-refractory species via a two-step process involving, first, laser vaporization and, second, electron impact ionization. In this manner, most species detected by the SP-AMS were elemental carbon, detected as C_x ions where $x \geq 1$, and organic molecules, appearing as both molecular ions and organic fragment ions. Some experiments were done having the thermal vaporizer removed from the SP-AMS, so only particles containing rBC could be detected. This confirmed that the organic species seen in the spectra were internally mixed with rBC. The mass spectrum of bare soot is shown in Fig. 1a, where only organic and elemental carbon mass-to-charge ratios are included. Although other species such as nitrate, sulfate and ammonium were also detected, they have not been included in this spectrum as they are negligible. Using standard relative ionization efficiencies for these two species (see Methods), the ratio of the mass of organic species to the mass of elemental carbon species was 0.59 ± 0.12.

High resolution mass spectral analysis was performed to identify molecular ions of PAHs in the spectrum, as shown in Fig. 1a, that is, these peaks can only arise from unsaturated

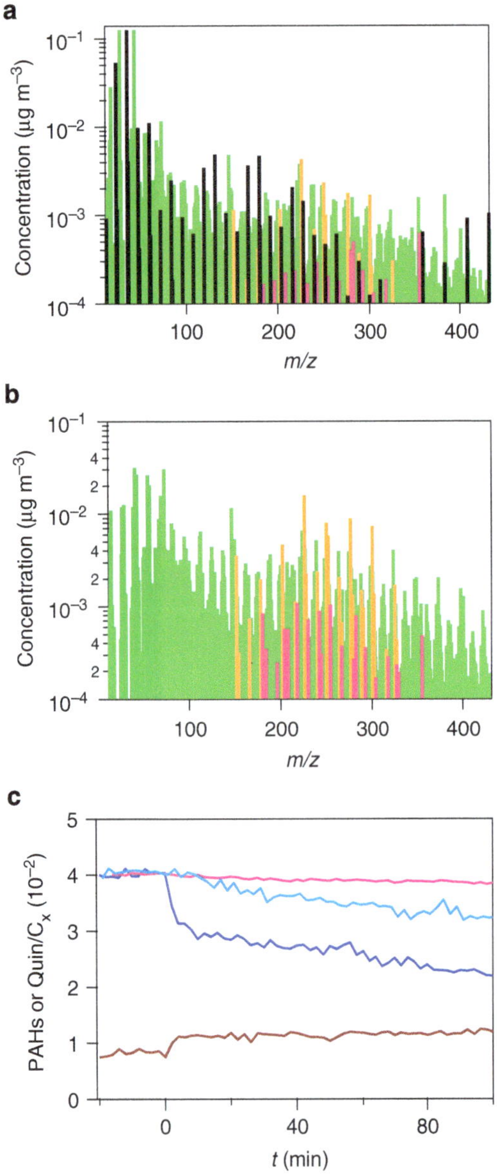

Figure 1 | Polycyclic aromatic hydrocarbons (PAHs) on uncoated soot particles. (**a**) Typical SP-AMS (soot-particle aerosol mass spectrometer) spectrum (obtained with the laser vaporizer) of organic species (green) and elemental carbon species (C_x, black) in soot particles used in this work. Highlighted in orange are the species considered to be PAHs and in pink those considered to be quinones. (**b**) AMS (aerosol mass spectrometer) spectrum (obtained with the thermal vaporizer) for the soot, (**c**) Typical time evolution of the SP-AMS signals normalized to elemental carbon of PAHs at 1.5×10^{12} molecule cm^{-3} O_3 (pink), at 3.4×10^{12} molecule cm^{-3} O_3 (light blue) and at 1.1×10^{13} molecule cm^{-3} (dark blue), and quinones at 1.1×10^{13} molecule cm^{-3} O_3 (red); $t = 0$ min is the time at which O_3 was added to the chamber.

C_xH_y species. PAHs are a commonly identified constituent class within AMS spectra and very likely to be present in soot particles. Confidence that these peaks arise from PAHs is enhanced through the measurements of kinetics to be described below. PAH molecular ions (general formula C_xH_y, with $x>y$) were chosen because PAHs are not highly fragmented by electron impact ionization. The species highlighted in Fig. 1a were selected based on the PAHs that were measured on ambient particles in Mexico City using aerosol mass spectrometry[27] and are listed in Supplementary Table 2. This is probably a subset of all PAHs in the soot, given that other species are likely present including alkylated species that will fragment to a larger degree under electron impact conditions. The chosen PAHs are generally large PAHs that partition principally to particles. On average, the ratio of the mass of these PAHs to the mass of elemental carbon in our soot particles was 0.04 ± 0.02.

Also shown in Fig. 1a are the ions that are likely arising from quinones that are structurally similar to the PAHs under consideration. Quinones fragment more than PAHs, so besides the molecular ions, $M^{\bullet+}$ ($C_xH_{y-2}O_2^{\bullet+}$), M-CO$^{\bullet+}$ ($C_{x-1}H_{y-2}O^{\bullet+}$) ions were considered as markers for these species. This fragmentation pattern was confirmed by SP-AMS observation of the oxidation of adsorbed benzo[*e*]pyrene, in the experiments described below. We cannot rule out that some of the peaks that have been attributed to quinones have contributions due to other species or mass spectral fragments with the same molecular formula.

The time evolution of the signals attributed to total PAHs and quinones is plotted in Fig. 1c. The signals were normalized to the total elemental carbon signal to take into account particle loss during the course of the experiment. It can be seen that the signal attributed to PAHs decreased when soot particles were exposed to O_3, whereas the signal corresponding to quinones simultaneously increased. Decays of PAHs are shown in Fig. 1c at three O_3 concentrations: 1.5×10^{12}, 3.4×10^{12} and 1.1×10^{13} molecule cm^{-3}, with slower decays observed at lower O_3 levels.

A standard AMS spectrum (that is, obtained by using only the thermal vaporizer, with the rBC volatilization laser off) for non-refractory species coated on soot particles is shown in Fig. 1b. Molecular ions for potential PAHs were also detected in the spectra, in many cases rising above the organic background in a more pronounced manner than in the SP-AMS spectrum where the whole particle is volatilized. This might indicate that, along with other organics, PAHs are preferentially on the surface of soot, whereas other organics (and presumably some PAHs too) are buried within soot and are only detected when the particle is fully volatilized by the laser of the SP-AMS. Ions for potential quinones were also detected in the AMS spectrum, indicating that these species can also be located on the surface of soot particles.

For those experiments in which soot particles were coated with BeP, the peak at *m/z* 252 that corresponds to the BeP molecular ion ($C_{20}H_{12}^{\bullet+}$) was one of the most intense peaks in the SP-AMS spectrum (see Supplementary Fig. 1a). Other peaks of high intensity include fragments of ionized BeP. The time evolution of the peak at *m/z* 252 depicted in Supplementary Fig. 1b illustrates a clear decrease of the BeP signal when O_3 is added to the chamber. At the same time, the peak at *m/z* 282 that corresponds to the molecular formula of the BeP quinone ion, $C_{20}H_{10}O_2^+$, rises in intensity. In this manner, the experiments with BeP-coated soot confirm that particle-bound PAHs are oxidized under exposure to ozone in the reaction chamber, and that among likely products are the corresponding particle-bound quinones.

To support these observations, diffuse reflectance infrared fourier transform (DRIFT) spectra were recorded to observe functional group changes occurring on the surface of soot particles when they were exposed to ozone. In Supplementary Fig. 2, spectra are included for bare soot before oxidation (where the reference spectrum is that of KBr powder) and soot after 15 and 360 min of exposure to ozone at 4.9×10^{14} molecule cm^{-3} (where the reference spectrum is that of the un-oxidized soot mixed into KBr powder). These spectra were recorded in experiments done in the reaction chamber of the DRIFT set-up. Although not shown here because of their lower quality, the spectra recorded for soot taken from the chamber exhibited the same results.

Before oxidation, soot exhibits three regions of interest in its IR spectrum: a band centred at 1,600 cm^{-1} that can be attributed to the stretching of C = O bond in carbonyl groups of quinone-like species, as Sutherland *et al.*[31] suggested for carbon blacks and Gulyás *et al.*[32] for carbon fibres, another band centred at 1,250 cm^{-1} that is due to the stretching of C-O in various chemical environments, also observed by Sutherland *et al.*[31], and a group of small peaks, very much influenced by the noise from the detector, between 940 and 550 cm^{-1} that could correspond to the C-H bending mode of aromatic groups.

When the soot was exposed to O_3, the spectra show clear indication of oxidation. In particular, there is a small increase of the 1,600 and 1,250 cm^{-1} bands observed, and the appearance of bands centred at 1,700, 1,370 and 1,000 cm^{-1}. The 1,700-cm^{-1} band is typical of C = O bond stretching in ketones or carboxylic acids. In particular, Sutherland *et al.*[31] also observed by transmission FT-IR the appearance of a band at around 1,700 cm^{-1} when they oxidized carbon blacks with O_3. They found that the absorbance of that band was related to the surface oxygen concentration measured by X-ray photoelectron spectroscopy. Holder *et al.*[8] oxidized methane soot with a mixture of O_3 and NO_x, and they also detected the increase of the band at 1,700 cm^{-1}, assigned to carbonyl groups. We believe the increase of the 1,600 cm^{-1} band is likely due to the increase of quinone-like functional groups on the surface of the particles, and the 1,370 and 1,000 cm^{-1} bands are probably due to C-O bonds. DRIFT spectra for soot coated with BeP showed no observable difference when compared with the spectra obtained for bare soot, due to the thin coatings used.

Uptake coefficient of O_3 on PAHs linked to soot particles. Decays of the PAH or BeP signal from O_3 exposure to soot particles (Fig. 1c and Supplementary Fig. 1b) were fitted to a double-exponential equation. This double time constant behaviour was previously observed during the photodegradation of PAHs[33]. The loss may be attributed to a mechanism in which near-surface-bound PAHs react with O_3 and are depleted in the first stage of reaction, and, on a slower timescale, losses of PAHs also arise from diffusion within the particle. Two pseudo-first order rate constants were obtained for each decay: k_1', which corresponds to the fast loss that is observed at the beginning of each decay, and k_2', which describes the slow loss during the latter part of each decay. In the subsequent analysis only k_1' is considered, as it corresponds to the dominant process. Results for k_2' are presented in Supplementary Fig. 3.

An enhancement of k_1' was observed when the O_3 concentration increased, as plotted in Fig. 2a. This enhancement became less important at high O_3 concentration until k_1' reached a maximum value, which is consistent with the Langmuir-Hinshelwood mechanism as previously reported for the reaction of BaP and O_3 (refs 23,24). A reaction via this mechanism involves a species, such as the total PAHs or BeP, that is strongly adsorbed to the particle surface and a gas-phase species, O_3, which is at equilibrium with the gas phase and the particle surface. The reaction is a two-step process, with the adsorption of O_3 followed by a surface reaction. According to this model, the number of sites that are available for the reaction with O_3 is limited so that when a certain O_3 concentration is reached, the surface is saturated. The data shown in Fig. 2a were fitted to the Langmuir–Hinshelwood equation[24]:

$$k_1' = \frac{k'_{max,1} K_{O_{3,1}} [O_3]}{1 + K_{O_{3,1}} [O_3]} \quad (1)$$

where $K_{O_{3,1}}$ is the ozone gas-to-surface equilibrium constant and $k'_{max,1}$ is the maximum rate coefficient that would be observed at high ozone concentration. The subscript 1 indicates that these constants refer to the fast loss observed at the beginning of the double exponential decay. $K_{O_{3,1}}$ and $k'_{max,1}$ were determined from this fitting. For the slower loss observed in the decay, as it is due to a diffusive process that may not be described by Langmuir–Hinshelwood kinetics, we do not fit the data in Supplementary Fig. 3a to this mechanism.

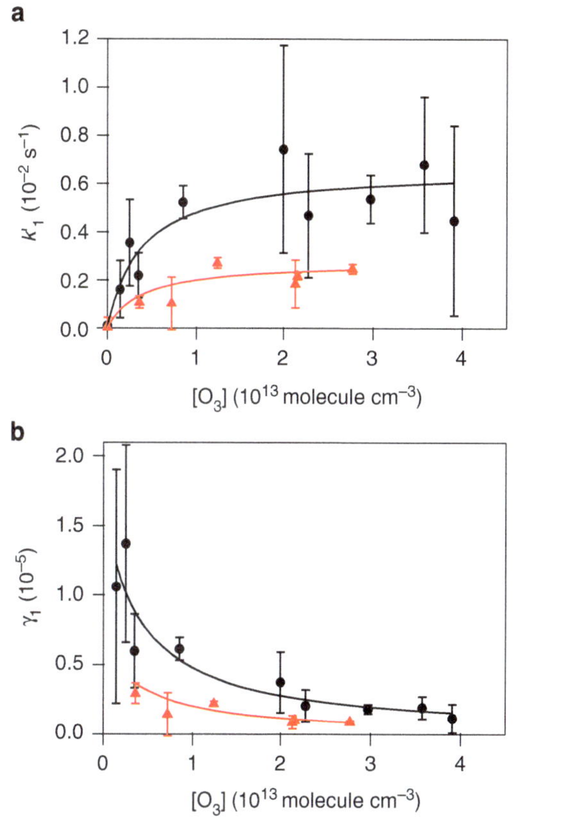

Figure 2 | Kinetic results for uncoated and benzo[e]pyrene (BeP)-coated soot. (**a**) Pseudo-first order rate constants for the fast loss process (1) observed in PAH (polycyclic aromatic hydrocarbon) decay as a function of O_3 concentration for the reactions with total PAHs on soot (black circles) and BeP on soot (red triangles). Data were fitted to equation 1. (**b**) Uptake coefficients calculated from the data in the upper graph using equation 2 as a function of O_3 concentration. Data were fitted to equation 3. Every point corresponds to one experiment. Error bars represent one standard deviation.

The experimental $K_{O_{3,1}}$ values determined in this work are very similar for total PAHs and BeP on soot: $(2.6 \pm 1.6) \times 10^{-13}$ and $(2.7 \pm 1.3) \times 10^{-13}$ cm^3, respectively, similar to that previously reported for BaP on soot: $(2.8 \pm 0.2) \times 10^{-13}$ cm^3 (ref. 23). This indicates that for all these processes, O_3 is partitioned between the soot surface and the gas phase to a similar extent. The values of $k'_{max,1}$ obtained in this study differ somewhat for total PAHs and BeP ($(6.5 \pm 3.1) \times 10^{-3}$ s^{-1} and $(2.7 \pm 1.1) \times 10^{-3}$ s^{-1}, respectively) and the previously reported value ($(1.5 \pm 0.1) \times 10^{-2}$ s^{-1}) for BaP[23]. These results indicate that BaP is more reactive than BeP, as already reported by Perraudin *et al.*[25] $k'_{max,1}$ for total PAHs corresponds to a mixture of the rate constants of the different PAHs that are on soot particles.

For a bimolecular reaction on the surface, the uptake coefficient is calculated with the following equation:

$$\gamma_1 = \frac{4k_1'}{\sigma \omega_{O_3} [O_3]} \quad (2)$$

where σ is the molecular cross-section of a PAH, that was considered to be the same as for BaP, 10^{-14} cm^2 (refs 23,24), and ω_{O_3} is the mean velocity of O_3. The resulting γ_1 values are on the order of 10^{-5}–10^{-6} for total PAHs and 10^{-6} for BeP on soot. The uptake coefficients determined in this work for the first phase of the decay are plotted as a function of O_3 concentration in Fig. 2b, and with the solid line determined by equation 3, which is the result of combining equations 1 and 2.

$$\gamma_1 = \frac{4k'_{\max,1}K_{O_{3,1}}}{\sigma\omega_{O_3}\left(1+K_{O_{3,1}}[O_3]\right)} \tag{3}$$

Changes in the soot redox cycling activity due to oxidation. To measure the redox cycling rates of the soot particles, we employ the DTT assay[26] where dithiothreitol (DTT) acts as the reducing agent in place of biological molecules, such as glutathione and ascorbate. Despite it being an indirect measure of one aspect of toxicity, several studies have used the DTT assay to evaluate the toxicity of different chemical species[7–9,26,34–38].

The mass-normalized DTT decay rates of oxidized particles are shown as a function of O_3 exposure ($[O_3]t$) in Fig. 3 for both bare soot and BeP-coated soot. In all cases the data plotted are the total DTT activity, without filtering the buffer extract solutions prior to analysis. We estimate a $\pm 15\%$ systematic uncertainty as typical for each DTT decay point in the figure that incorporates the uncertainties in the mass of soot measured on the filter and from the statistical analysis of the decay rates. An additional variability (on the order of $\pm 40\%$) sometimes arises due to the assay response itself when conducted with soot, as compared with the assay with a fully soluble species. In particular, the soot mixtures were quite heterogeneous in nature with different degrees of coagulation, thus affecting their response.

Before oxidation, the redox activity of bare soot was 49 ± 7 pmol min^{-1} µg^{-1}. Figure 3 demonstrates that at O_3 exposures lower than 10^{12} molecule cm^{-3} h, the increase in DTT activity is very low compared with what is observed at higher exposures. This indicates that the number of reactive encounters of ozone at lower exposures is not sufficient to significantly oxidatively process the surface of soot. However, at higher exposures, it is possible to link the kinetics of PAHs on soot to the activation of the DDT response of soot. In particular, the concentration of O_3 used for the experiment at which the O_3 exposure was $\sim 10^{12}$ molecule cm^{-3} h was 3×10^{12} molecule cm^{-3} (see Supplementary Table 3). At this concentration, the rate coefficient k_1' is $\sim 2 \times 10^{-3}$ s^{-1} (see Fig. 2a), corresponding to a time constant for the loss of PAHs of ~ 10 min. This is in excellent agreement with the 15 min of time required to generate the exposure of $\sim 10^{12}$ molecule cm^{-3} h that leads to the increase in the DTT activity in Fig. 3. Moreover, the decays present in Fig. 1c corroborate that, at low O_3 levels, the oxidation of PAHs is minor. The fact that the DTT activity increases at an O_3 exposure that matches the kinetics for the oxidation of a PAH with O_3 indicates that the oxidation of PAHs (or PAH-like structures) present on soot particles, and the formation of their oxidation products, is responsible for the increased redox activity of these particles. Among the oxidation products that contribute to the increase of the redox activity are quinones, although it is likely that these are not the only oxidation products that contribute to the DTT activity. In additional experiments, we removed quinones present in soot by the derivatization process that has been widely used in quinone analysis[39]. When the DTT assay was performed on the samples that were derivatized, the decrease of the redox activity was 76% for the non-oxidized soot, and 45% for the oxidized soot, indicating the possible contribution of other species than quinones to the increase in the redox activity.

DTT activity keeps increasing at high O_3 exposures, whereas k_1' levels off at $[O_3] \sim 10^{13}$ molecule cm^{-3}. The additional DTT activity either arises from the oxidation of the buried PAHs or else it arises from more redox active species formed by subsequent oxidation of the first generation products. Indeed, in previous work from our laboratory involving the redox cycling activity of naphthalene secondary organic aerosol, we observed that a multiple generation oxidation product, a hydroxylated napthaquinone, contributed significantly to the overall redox cycling activity of the particles formed[40].

Our redox cycling rates are in qualitative agreement with previous studies. In particular, Holder *et al.*[8] observed an increase in the DTT activity when they oxidized methane soot at very high O_3 concentrations in the presence of NO_x. Li *et al.*[9] exposed DEP to lower levels of O_3 and NO_x, and also observed an increase of the DTT activity. McWhinney *et al.*[7] found an increase in redox cycling rates when combustion particles were exposed to different O_3 concentrations. However, they attributed this behaviour to the presence of SOA forming on the surface of the particles due to the reaction of engine emission gases and O_3. Rattanavaraha *et al.*[10] also detected an increase of the redox activity when DEP were exposed to sunlight in an outdoor chamber, in both the presence and absence of a volatile hydrocarbon mixture. In our work, we have a less complicated mixture of gases due to the nature of the fuel, and denuders were used to remove possible interferences from gas-phase species. Therefore, the formation of SOA in our system is less important, with the ratio of the mass of organic to elemental carbon changing by $< 15\%$ after oxidation.

A negligible difference between the DTT decay rate for bare soot and soot coated with BeP was observed, which is consistent with PAHs not being redox active. For the oxidized particles, it is not possible to say within the uncertainties whether an increase in activity was observed when BeP is present. Although an increase of redox activity could be expected, it is possible that the redox activity of BeP oxidation products was similar to the activity of bare oxidized soot, yielding no change between uncoated soot and soot coated with BeP. We note that DTT decay rates obtained in this work for oxidized soot are roughly an order of magnitude higher than those determined for ambient PM[26,34–37], indicating the potential significance for aged soot particles to contribute to the overall redox cycling abilities of ambient PM.

Discussion

This study provides the first mechanistic connection between the increased redox cycling abilities of oxidized soot particles and

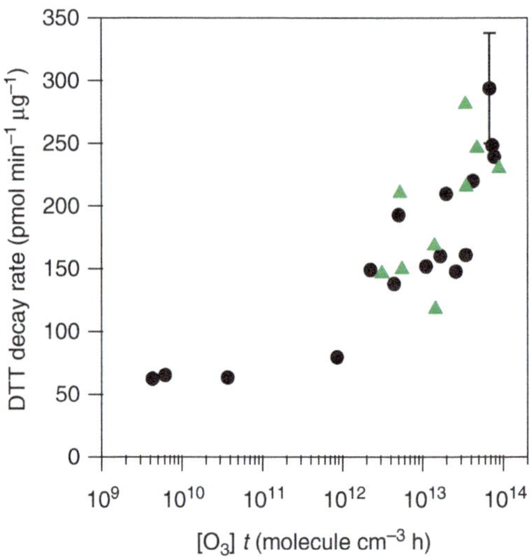

Figure 3 | Redox cycling results. Mass-normalized DTT (dithiothreitol) decay rate measured for bare soot (black circles) and soot coated with BeP (green triangles) as a function of O_3 exposure. A systematic error bar of $\pm 15\%$ for the DTT decay rate was included for one point.

their chemical composition. In particular, novel aerosol mass spectrometry measurements indicate that PAH-like species within the soot particles decay on ozone exposure, with a corresponding increase in the mass spectrometric signals attributed to quinones. The formation of surface-bound quinone-like species is confirmed by infrared analysis. These findings were corroborated by oxidation experiments performed with soot particles coated with BeP, in which BeP clearly decayed while the mass spectrometric peak of its corresponding quinone increased. The capability of the oxidized particles to be catalysts of electron-transfer processes increased, consistent with quinones being good redox cycling agents.

The increases of the redox cycling activity of combustion particles when exposed to O_3 were conducted in a low-NO_x and low-volatile organic compound (VOC) environment. Although in the ambient atmosphere levels of NO_x and VOCs are much higher, the results obtained in this work isolate the effects of heterogeneous ozone oxidation on the soot surface and the corresponding changes in redox cycling ability. To measure a significant change in the redox cycling rate, an ozone exposure equivalent to 3 h at 7.4×10^{11} molecule cm^{-3} (30 p.p.b. by volume), a typical background level[41], was required. In more polluted environments, where soot particles are more likely to be emitted, and where O_3 mixing ratios can reach over 2.5×10^{12} molecule cm^{-3} (100 p.p.b. by volume)[41], the redox cycling activity would reach its maximum value in less time.

Methods

Reaction chamber set-up. Experiments were performed in batch mode in a 1-m^3 environmental Teflon bag full of clean air. Soot particles were generated by a soot generator (miniCAST 6203A, Jing) that uses a propane-air diffusion flame. A flow of particles diluted in air (6 slpm, standard liter per minute) was emitted by the soot generator, but only 1 slpm was passed through a denuder filled with Carulite 200 (MnO_2/CuO catalyst, Carus Corporation) before entering the chamber. This denuder is usually used to remove O_3 in the gas phase but it also decreased by 98% the levels of NO and NO_2 generated in the combustion process. In some experiments, an activated charcoal (Norit PK 3–5, Sigma-Aldrich) denuder was used to remove VOCs, although no changes in the results were observed when it was employed.

For the experiments in which the soot particles were coated by BeP, the flow of particles was also passed through a glass tube ($l = 35$ cm, $d = 1$ cm) with the inner walls covered by BeP and heated using external heating tapes to 375 K. To cover the walls of the tube, BeP was dissolved in CH_2Cl_2 and the solution was spread on the inner walls of the tube by rotating and inclining it while the solvent was evaporating. These conditions made possible an estimated average BeP coating thickness of 3×10^{-2} nm on the soot particles, that is, well less than that of a monolayer. This estimation was done following the same method that Zhou *et al.*[42] followed in which assumes a monodisperse size distribution of particles, a uniform coating thickness of the BeP over the particles, an assumption that the particle surface area is that of a sphere with diameter set by its mobility diameter and the amounts of BeP as measured by aerosol mass spectrometry (see below). Given that the surface area of soot will be far higher than that of a smooth sphere, we are quite confident that the coatings of BeP are very thin, and that the soot is likely not fully coated by the PAH.

Once soot particles were in the chamber, O_3 was introduced at concentrations ranging 1.75×10^{12} and 1×10^{14} molecule cm^{-3} (70–4,000 p.p.b.). It was generated by passing 1 lpm of air (Linde, Grade 0.1) or O_2 (Linde, Grade 2.6) over a 185 nm Hg pen-ray lamp (UVP). Five gas analysers were used to measure the mixing ratios inside the chamber of O_3 (Model 49*i*, Thermo Scientific), NO_x ($= NO + NO_2$) (Model 42*i*, Thermo Scientific), SO_2 (Model 43*i*, Thermo Scientific), CO (Model 48*i*, Thermo Scientific) and CO_2 (Model 410*i*, Thermo Scientific). The number densities of soot particles and their size distributions were determined by a SMPS consisting of a differential mobility analyser (DMA, TSI) to size-select the particles and a condensation particle counter (CPC, TSI) to measure their number concentration. Flow rates going to the gas analysers and SMPS were 3.5 slpm and 0.3 slpm, respectively.

Soot-particle aerosol mass spectrometer. A SP–AMS (Aerodyne Research; more details in Onasch *et al.*[29]) was used to observe chemical changes in soot particles when they were exposed to O_3. The SP–AMS is a high resolution time-of-flight aerosol mass spectrometer (HR-ToF-AMS) with an intracavity Nd:YAG laser vaporizer ($\lambda = 1{,}064$ nm). In a standard HR-ToF-AMS, a beam of particles is first generated with an aerodynamic lens and then flash vaporized on a heated tungsten surface. The vapours are ionized by electron impact, and the resulting ions are detected with high mass resolution in a time-of-flight mass spectrometer. The tungsten furnace (~ 600 °C) only vaporizes non-refractory material, whereas the laser vaporizer allows the detection of refractory black carbon particles, and non-refractory materials that are internally mixed with soot. Data were analysed using SQUIRREL (v1.51H) and its corresponding module for high-resolution data, PIKA (v1.10H), running within Igor Pro (v6.22A, WaveMetrics). The SP–AMS was calibrated with 300 nm Regal Black particles (Regal 400R, Cabot Corp.) and with NH_4NO_3 as described by Onasch *et al.*[28] Standard relative ionization efficiencies of 0.2 and 1.4 were used for elemental carbon species and organic species, respectively[29,43].

Collection of filters and DRIFTS. Once the exposure to ozone was complete, the particles in the chamber were pulled onto a polytetrafluoroethylene (PTFE) filter (Pall Life Sciences, 2 µm pore size, 47 mm diameter). Ozone exposure was changed by modifying either time or $[O_3]$. Some of the samples collected on the filters were used to further characterize the chemical composition of the particles by DRIFT spectroscopy (DRIFTS). A FT–IR spectrometer (Magna-IR 550 Series II, Nicolet) accumulated 1,000 scans per spectrum at a resolution of 2 cm^{-1}, and a diffusion reflection accessory (Praying Mantis, Harrick) was located in the sample compartment to measure DRIFT spectra. Reference spectra using dried KBr powder (<10 µm particle size, FT-IR grade, Sigma-Aldrich) were always recorded before spectra of the samples were measured, and samples were always diluted by adding KBr until the content of soot was around 0.2% of the total mass. Complementary experiments of soot oxidation were done in a reaction chamber inside the sample compartment of the diffuse reflection accessory (Praying Mantis diffuse reflection environmental chamber) to better characterize the DRIFT spectrum. A 100 sccm (standard cubic centimetre per minute) flow of 20 p.p.m. O_3 was passed through the reaction chamber, and spectra were recorded

DTT assay. Particles collected on the filters were also used to measure the redox cycling activity of the particles immediately after collection. Filters were placed in 10 ml phosphate buffer solution (0.1 M, pH 7.4) and sonicated for 15 min to obtain suspensions of soot that underwent the DTT assay[26] after an appropriate dilution (soot concentration in the diluted samples ranged between 1.2 and 14 µg ml^{-1}). The extraction efficiency from the filter to the buffer solution was observed to be around 80%. Control experiments were done sonicating clean filters to corroborate that the influence on the DTT response of peroxides generated in this process is negligible. For some of these suspensions, aliquots were filtered (PTFE VWR syringe filters, 0.2 µm pore size, 13 mm diameter) before measuring their DTT activity to determine the activity due to water-soluble species only. The assay of the soot suspension measured the redox activity due to both soluble species and species present at the surface of the soot particles. For the assay, 1 ml aliquots of sample were combined with 0.25 ml of 0.5 mM DTT in pH 7.4 phosphate buffer, and kept at 310 K to simulate human-body temperature. The reaction was quenched over the course of 1 h by adding 0.25 ml of the reaction mixture to 1 ml of 5,5′-dithiobis-(2-nitrobenzoic acid) (DTNB). A yellow-coloured compound (2-nitro-5-thiobenzoate) was formed, and its absorption was measured at 412 nm. The DTT decay rate of a blank solution was subtracted every time the assay was performed, and the results were mass-normalized to make them comparable from sample to sample. In Supplementary Fig. 4, examples of the decay of DTT for non-filtered solutions are shown at three different exposures to O_3 together with the decay observed for a quinone that is known to be active in the DTT assay, 1,4-naphthoquinone (DTT decay rate $= 3{,}100 \pm 280$ pmol min^{-1} µg^{-1}). The measurements of the DTT activity were repeated for one solution in different days, and reproducible results were observed, showing a negligible effect of partitioning between particle and aqueous phase.

References

1. Gurgueira, S. A., Lawrence, J., Coull, B., Murthy, G. G. K. & Gonzalez-Flecha, B. Rapid increases in the steady-state concentration of reactive oxygen species in the lungs and heart after particulate air pollution inhalation. *Environ. Health Perspect.* **110,** 749–755 (2002).
2. González-Flecha, B. Oxidant mechanisms in response to ambient air particles. *Mol. Aspects Med.* **25,** 169–182 (2004).
3. Squadrito, G. L., Cueto, R., Dellinger, B. & Pryor, W. A. Quinoid redox cycling as a mechanism for sustained free radical generation by inhaled airborne particulate matter. *Free Radical Biol. Med.* **31,** 1132–1138 (2001).
4. Xia, T. *et al.* Quinones and aromatic chemical compounds in particulate matter induce mitochondrial dysfunction: Implications for ultrafine particle toxicity. *Environ. Health Perspect.* **112,** 1347–1358 (2004).
5. Netto, L. E. S. & Stadtman, E. R. The iron-catalyzed oxidation of dithiothreitol is a biphasic process: hydrogen peroxide is involved in the initiation of a free radical chain of reactions. *Arch. Biochem. Biophys.* **333,** 233–242 (1996).
6. Madden, M. C., Richards, J. H., Dailey, L. A., Hatch, G. E. & Ghio, A. J. Effect of ozone on diesel exhaust particle toxicity in rat lung. *Toxicol. Appl. Pharmacol.* **168,** 140–148 (2000).

7. McWhinney, R. D., Gao, S. S., Zhou, S. M. & Abbatt, J. P. D. Evaluation of the effects of ozone oxidation on redox-cycling activity of two-stroke engine exhaust particles. *Environ. Sci. Technol.* **45,** 2131–2136 (2011).
8. Holder, A. L., Carter, B. J., Goth-Goldstein, R., Lucas, D. & Koshland, C. P. Increased cytotoxicity of oxidized flame soot. *Atmos. Pollut. Res.* **3,** 25–31 (2012).
9. Li, Q. F., Wyatt, A. & Kamens, R. M. Oxidant generation and toxicity enhancement of aged-diesel exhaust. *Atmos. Environ.* **43,** 1037–1042 (2009).
10. Rattanavaraha, W. *et al.* The reactive oxidant potential of different types of aged atmospheric particles: An outdoor chamber study. *Atmos. Environ.* **45,** 3848–3855 (2011).
11. Lichtveld, K. M. *et al. In vitro* exposures in diesel exhaust atmospheres: resuspension of PM from filters versus direct deposition of PM from air. *Environ. Sci. Technol.* **46,** 9062–9070 (2012).
12. Sasaki, J., Aschmann, S. M., Kwok, E. S. C., Atkinson, R. & Arey, J. Products of the gas-phase OH and NO3 radical-initiated reactions of naphthalene. *Environ. Sci. Technol.* **31,** 3173–3179 (1997).
13. Mmereki, B. T., Donaldson, D. J., Gilman, J. B., Eliason, T. L. & Vaida, V. Kinetics and products of the reaction of gas-phase ozone with anthracene adsorbed at the air-aqueous interface. *Atmos. Environ.* **38,** 6091–6103 (2004).
14. Kautzman, K. E. *et al.* Chemical composition of gas- and aerosol-phase products from the photooxidation of naphthalene. *J. Phys. Chem. A* **114,** 913–934 (2010).
15. Van Cauwenberghe, K. A. in *Handbook of Polycyclic Aromatic Hydrocarbons.* (eds Bjørseth, Ž. A. & Ramdahl, T.) 351–384 (Dekker, 1985).
16. Kwamena, N.-O. A., Earp, M. E., Young, C. J. & Abbatt, J. P. D. Kinetic and product yield study of the heterogeneous gas – surface reaction of anthracene and ozone. *J. Phys. Chem. A* **110,** 3638–3646 (2006).
17. Shrestha, G., Traina, S. & Swanston, C. Black carbon's properties and role in the environment: a comprehensive review. *Sustainability* **2,** 294–320 (2010).
18. Cho, A. K. *et al.* Determination of four quinones in diesel exhaust particles, SRM 1649a, an atmospheric PM2.5. *Aerosol Sci. Technol.* **38,** 68–81 (2004).
19. Jakober, C. A. *et al.* Quinone emissions from gasoline and diesel motor vehicles. *Environ. Sci. Technol.* **41,** 4548–4554 (2007).
20. Hallum, J. V. & Drushel, H. V. The organic nature of carbon black surfaces. *J. Phys. Chem.* **62,** 110–117 (1958).
21. Shinyashiki, M. *et al.* Electrophilic and redox properties of diesel exhaust particles. *Environ. Res.* **109,** 239–244 (2009).
22. Sander, S. P. *et al.* Chemical Kinetics and Photochemical Data for Use in Atmospheric Studies, Evaluation No. 17. JPL Publication 10-6, Jet Propulsion Laboratory, Pasadena (2011).
23. Pöschl, U., Letzel, T., Schauer, C. & Niessner, R. Interaction of ozone and water vapor with spark discharge soot aerosol particles coated with benzo[a]pyrene: O_3 and H_2O adsorption, benzo[*a*]pyrene degradation, and atmospheric implications. *J. Phys. Chem. A* **105,** 4029–4041 (2001).
24. Kwamena, N.-O. A., Thornton, J. A. & Abbatt, J. P. D. Kinetics of surface-bound benzo[*a*]pyrene and ozone on solid organic and salt aerosols. *J. Phys. Chem. A* **108,** 11626–11634 (2004).
25. Perraudin, E., Budzinski, H. & Villenave, E. Kinetic study of the reactions of ozone with polycyclic aromatic hydrocarbons adsorbed on atmospheric model particles. *J. Atmos. Chem.* **56,** 57–82 (2007).
26. Cho, A. K. *et al.* Redox activity of airborne particulate matter at different sites in the Los Angeles Basin. *Environ. Res.* **99,** 40–47 (2005).
27. Dzepina, K. *et al.* Detection of particle-phase polycyclic aromatic hydrocarbons in Mexico City using an aerosol mass spectrometer. *Int. J. Mass Spectrom.* **263,** 152–170 (2007).
28. DeCarlo, P. F., Slowik, J. G., Worsnop, D. R., Davidovits, P. & Jiménez, J. L. Particle morphology and density characterization by combined mobility and aerodynamic diameter measurements. part 1: theory. *Aerosol Sci. Technol.* **38,** 1185–1205 (2004).
29. Onasch, T. B. *et al.* Soot particle aerosol mass spectrometer: development, validation, and initial application. *Aerosol Sci. Tech.* **46,** 804–817 (2012).
30. Willis, M. D. *et al.* Collection efficiency of the soot-particle aerosol mass spectrometer (SP-AMS) for internally mixed particulate black carbon. *Atmos. Meas. Tech.* **7,** 4507–4516 (2015).
31. Sutherland, I., Sheng, E., Bradley, R. H. & Freakley, P. K. Effects of ozone oxidation on carbon black surfaces. *J. Mater. Sci.* **31,** 5651–5655 (1996).
32. Gulyas, J., Foldes, E., Lazar, A. & Pukanszky, B. Electrochemical oxidation of carbon fibres: Surface chemistry and adhesion. *Compos. Part A-Appl. S* **32,** 353–360 (2001).
33. Kim, D., Kumfer, B. M., Anastasio, C., Kennedy, I. M. & Young, T. M. Environmental aging of polycyclic aromatic hydrocarbons on soot and its effect on source identification. *Chemosphere* **76,** 1075–1081 (2009).
34. Hu, S. *et al.* Redox activity and chemical speciation of size fractioned PM in the communities of the Los Angeles-Long Beach harbor. *Atmos. Chem. Phys.* **8,** 6439–6451 (2008).
35. Múgica, V. *et al.* PM composition and source reconciliation in Mexico City. *Atmos. Environ.* **43,** 5068–5074 (2009).
36. Ntziachristos, L., Froines, J. R., Cho, A. K. & Sioutas, C. Relationship between redox activity and chemical speciation of size-fractionated particulate matter. *Part. Fibre Toxicol.* **4,** 5 (2007).
37. Verma, V. *et al.* Redox activity of urban quasi-ultrafine particles from primary and secondary sources. *Atmos. Environ.* **43,** 6360–6368 (2009).
38. McWhinney, R. D., Badali, K., Liggio, J., Li, S. M. & Abbatt, J. P. D. Filterable redox cycling activity: a comparison between diesel exhaust particles and secondary organic aerosol constituents. *Environ. Sci. Technol.* **47,** 3362–3369 (2013).
39. Delgado-Saborit, J. M., Alam, M. S., Godri Pollitt, K. J., Stark, C. & Harrison, R. M. Analysis of atmospheric concentrations of quinones and polycyclic aromatic hydrocarbons in vapour and particulate phases. *Atmos. Environ.* **77,** 974–982 (2013).
40. McWhinney, R. D., Zhou, S. & Abbatt, J. P. D. Naphthalene SOA: redox activity and naphthoquinone gas-particle partitioning. *Atmos. Chem. Phys.* **13,** 9731–9744 (2013).
41. Finlayson-Pitts, B. & Pitts, J. J. *Chemistry of the Upper and Lower Atmosphere* (Academic Press, 2000).
42. Zhou, S., Lee, A. K. Y., McWhinney, R. D. & Abbatt, J. P. D. Burial effects of organic coatings on the heterogeneous reactivity of particle-borne benzo a pyrene (BaP) toward ozone. *J. Phys. Chem. A* **116,** 7050–7056 (2012).
43. Jiménez, J. L. *et al.* Ambient aerosol sampling using the Aerodyne Aerosol Mass Spectrometer. *J. Geophys. Res.: Atmos.* **108,** 8425 (2003).

Acknowledgements

We thank NSERC, the Sloan Foundation, and Environment Canada for providing operational funding for this project. M.A. would like to thank Fundación Ramón Areces for providing her a grant.

Author contributions

M.A., S.Z. and J.A. conceived and designed the experiments; M.A., M.W. and S.Z. performed the experiments; M.A. and M.W. analysed the data; M.A., M.W., S.Z. and J.A. co-wrote the paper.

Additional information

Competing financial interests: The authors declare no competing financial interests.

4

Recent decrease in typhoon destructive potential and global warming implications

I-I Lin[1] & Johnny C.L. Chan[2]

Typhoons (tropical cyclones) severely impact the half-billion population of the Asian Pacific. Intriguingly, during the recent decade, typhoon destructive potential (Power Dissipation Index, PDI) has decreased considerably (by ~35%). This decrease, paradoxically, has occurred despite the increase in typhoon intensity and ocean warming. Using the method proposed by Emanuel (in 2007), we show that the stronger negative contributions from typhoon frequency and duration, decrease to cancel the positive contribution from the increasing intensity, controlling the PDI. Examining the typhoons' environmental conditions, we find that although the ocean condition became more favourable (warming) in the recent decade, the atmospheric condition 'worsened' at the same time. The 'worsened' atmospheric condition appears to effectively overpower the 'better' ocean conditions to suppress PDI. This stronger negative contribution from reduced typhoon frequency over the increased intensity is also present under the global warming scenario, based on analysis of the simulated typhoon data from high-resolution modelling.

[1]Department of Atmospheric Sciences, National Taiwan University, No.1, Sec. 4, Roosevelt Road, Taipei 106, Taiwan. [2]Guy Carpenter Asia-Pacific Climate Impact Centre, School of Energy and Environment, City University of Hong Kong, Hong Kong, Hong Kong. Correspondence and requests for materials should be addressed to I.-I.L. (email: iilin@as.ntu.edu.tw).

Despite the severe impact of individual tropical cyclones such as Sandy (in 2012)[1] and Haiyan (in 2013)[2,3], recent global tropical cyclone activity as a whole has actually decreased considerably since the early 1990s[4]. An evident decrease in tropical cyclone destructive potential (that is, PDI) has been observed in the most active and hazardous tropical cyclone basin on the Earth, the Western North Pacific Main Development Region (MDR, Supplementary Fig. 1)[5–7]. Because these tropical cyclones (referred to as typhoons in the region) are severe threats to the half-billion population and the huge volume of economic activities in the Asia Pacific regions (for example, China, Taiwan, Japan, Philippines, Hong Kong and Korea) each year[5], it is of great interest to study this observed decrease in typhoon destructive potential[5–7].

The PDI (Power Dissipation Index), Proposed by Emanuel in 2005 (ref. 8), is a widely used parameter to characterize the destructive potential of tropical cyclones. The PDI of a tropical cyclone is defined as the sum of the 6-hourly maximum surface wind speed (cyclone intensity) cube over the lifetime of a cyclone. The annual PDI of a specific tropical cyclone basin is simply the sum of the PDIs of all tropical cyclones passing that basin in a year (or in the cyclone-active season of that year). As this research does not involve with individual cyclone cases, for convenience, the annual PDI will subsequently be simplified to 'PDI'.

In this study, we found that in the recent decade, the PDI-contributing factors (typhoon frequency, duration and intensity) had made opposite contributions to PDI in the recent decade. Although there was some increase in typhoon intensity, the typhoon frequency and duration decreased at the same time. The negative contributions from the reduction in typhoon frequency and duration overpowered the positive contribution from the increased intensity; therefore, the PDI still decreased. Similar opposite contributions to PDI were suggested in the global warming scenario[9]. Using the projected typhoon data from high-resolution modelling from Zhao and Held[10,11], we found that although typhoon intensity may increase (relative to the present intensity) under the global warming scenario, typhoon frequency could decrease even more notably. The PDI is the residual after the offset from these opposing contributors. Therefore, based on the above projection, the western North Pacific typhoon PDI under global warming was estimated to be ~85% of the current value, equivalent to a 15% reduction in destructive potential.

Results

The observed recent PDI decrease. The value of the typhoon-season (July–October) PDI in the recent pentad (2008–2012) was only ~63% that of the 1993–1997 period (see Fig. 1a). To understand the cause of this decrease, we first examined the ocean environment because the ocean thermal conditions, including both the sea surface temperature (SST) and the upper ocean heat content (UOHC), are important factors for tropical cyclone intensity change and could impact PDI[8,12–18]. The UOHC (also called the tropical cyclone heat potential) is defined as the integrated heat content from the SST down to the 26 °C isotherm (D26, the measure of the subsurface warm ocean layer thickness) in the subsurface ocean[14,16,17].

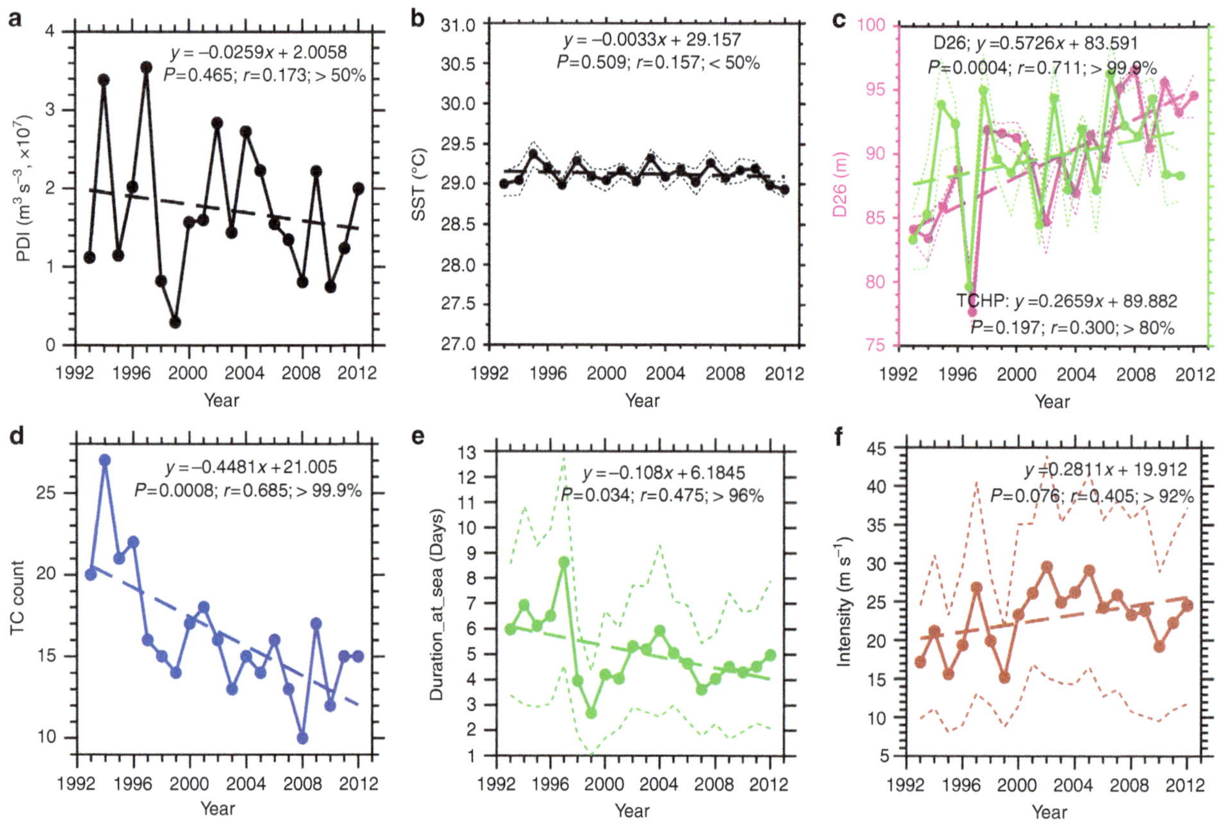

Figure 1 | Time series of PDI and related parameters. Time evolution of the observed PDI and other parameters over the western North Pacific MDR in the past two decades. The trend line for each time series, based on linear regression is also depicted. Standard deviations are depicted by dotted curves. (**a**) PDI, (**b**) SST, (**c**) the depth of the 26 °C isotherm (D26) and TCHP (Tropical Cyclone Heat Potential or upper ocean heat content (UOHC)). (**d**) Typhoon case number in the typhoon season (July-October) of a year, (**e**) as in **d**, but for the averaged typhoon duration, (**f**) as in **d**, but for the averaged typhoon intensity.

Figure 1b,c depicts the typhoon season (July–October) SST and UOHC over the MDR. In the past two decades, although the SST exhibited little change, the UOHC and D26 increased considerably[3,19] (Fig. 1c). In comparison to the 1990s, recent UOHC and D26 both exhibited clear increases of ~10%, which opposed the decrease in PDI. The origin of these increases in UOHC and D26 were thought to be associated with the recent strengthening of the easterly trade winds, which pile up warm surface ocean water towards the western Pacific[19,20]. As a result, a thicker layer of warm water accumulates in the western North Pacific typhoon MDR[3,19,20]. Therefore, the objective of this study was to understand why typhoon PDI has actually decreased despite evident warming in the upper ocean.

The three PDI contributing factors. PDI is determined not only by typhoon intensity (that is, maximum surface wind speed) but also by duration (lifetime) and typhoon occurrence frequency (case number). The evolution of all of these parameters (Fig. 1d–f) shows that although the intensity increased somewhat, both the duration and number decreased considerably. Thus, the reduction in PDI was not due to the intensity but the duration and number.

This result is further quantified in Table 1, following the method proposed by Emanuel[21], to separate the different contributions of the annual typhoon case number (N), annual-averaged weighted typhoon duration over ocean (D_{wt}, see Methods section for details) and annual-averaged intensity (I, in wind speed cube) to PDI. Supplementary Fig. 2 shows these terms with their respective long-term mean (1993–2012) removed. The data were based on the annual typhoon season (July–October) over the MDR domain during 1993–2012.

Table 1 summarizes the different contributions in each pentad. In the most recent pentad (2008–2012, 2nd-last row), PDI decreased by approximately −17% with respect to climatology (1993–2012 mean). Although the intensity contributed to an increase in PDI by ~6–7%, the negative contributions of number and duration are approximately −16% and −7%, respectively (jointly approximately −23 %). Therefore, PDI still decreased because the joint contributions of number and duration were much larger than those of intensity. Relative to the pentad in 1993–1997, the −45% PDI decrease in the latest pentad (2008–2012) was also attributed to the −74% drop in number and duration, with a 29% offset from the positive contribution from the intensity (last row in Table 1; for more details, see Supplementary Table 1).

Atmospheric and ocean environments. It is important to understand the evident decreases in typhoon number and duration[6,7] (Fig. 1d,e). Figure 2a,b shows that the decrease was accompanied by a strong increase in vertical wind shear (VWS, primarily contributed by the zonal VWS) and decrease in low-level relative vorticity in the typhoon genesis region (150–180°E, 10–17.5°N) (ref. 6,7). Especially after 2008, the VWS reached 10–18 $m\,s^{-1}$, which represented an environment unfavourable for formation[6,7,22–25]. Consistent with these developments, the typhoon genesis position also shifted north-westwards towards land[26], reflecting the difficulty of formation at the usual genesis region (east of 150°E; Fig. 2b-right axis, Fig. 3 and Supplementary Figs 3–6). The weak VWS and high vorticity are the main atmospheric dynamic conditions necessary for formation[22–25], and it becomes more difficult for typhoons to form further east. Because they form further to the northwest towards land[26], their durations over the ocean are also shortened (Figs 1e, 2b and 3). This trend was confirmed by the strong correlations (r = 0.56–0.88) found between the increase in the zonal VWS, decrease in vorticity, westward shift of the genesis longitude (1st lon), the reduction in typhoon duration (D_{wt}) and ND_{wt} (case number times duration; boldface entries in Table 2 and Supplementary Table 2, see also Supplementary Note 1).

In fact, the above change in the atmospheric conditions is part of the large-scale environmental change over the Pacific in the recent decade[27–31]. Many recent studies[27–29,32] have emphasized the dominance of this large-scale change by discussing the evident strengthening of the atmospheric circulation (including the enhanced easterly trade wind and Walker cell) and a La Nina-like decadal phenomenon[27–29,32]. This phenomenon is clearly visible in Fig. 3, which shows a strengthening of the subtropical high pressure system[30,31,33,34] and the easterly wind anomaly at 850 hPa. In this research, we used three parameters to characterize the large-scale circulation condition: the 850 hPa easterly wind; the subtropical high area index (SHAI)[6] and a newly proposed related parameter called subtropical high-intensity index (SHII; see Methods).

The strengthening of the large-scale circulation can cause an increase in the VWS. This effect occurs because the VWS is the difference between the winds in the upper (200 hPa) and lower (850 hPa) troposphere. Enhanced circulation[29] can increase both the lower-tropospheric easterlies (trade wind) and upper-tropospheric westerlies, and thus the shear increases (see also Supplementary Figures 3 and 4). Climatologically, the low-level relative vorticity over the MDR is generally high[24]. With strengthening of the easterlies associated with large-scale circulation strengthening (Fig. 3), the magnitude of the relative vorticity decreases, as manifested in a weakened monsoon trough condition.

This control of the large-scale circulation on typhoon-related atmospheric parameters (that is, VWS and vorticity) can be observed in the high correlations (r = 0.83–0.91) between the easterly wind, SHAI at 850 hPa and zonal VWS (see the boldface entries in Table 2 and Fig. 4). Furthermore, strong negative correlations (−0.77 to −0.83, see the boldface entries in Table 2 and Fig. 4) existed between these three parameters with PDI because an increase in these atmospheric parameters contributed to a reduction in PDI (Fig. 4). From an even broader perspective, the relative SST parameter (typhoon basin SST with respect to

Table 1 | Contributing factors to annual PDI.

Pentad	PDI (%)	I (%)	$N+D_{wt}$ (%)	N (%)	D_{wt} (%)
1993-1997 w.r.t. mean	28.6	−22.2	50.8	23.4	27.4
1998-2002 w.r.t. mean	−30	−15.8	−14.2	0.2	−14.4
2003-2007 w.r.t. mean	17.8	30.8	−13	−11.6	−1.4
2008-2012 w.r.t. mean	−16.6	6.8	−23.4	−15.8	−7.6
[2008-2012] w.r.t. [1993-1997]	−45.2	29	−74.2	−39.2	−35

PDI, Power Dissipation Index; w.r.t., with respect to. Contributions of the observed typhoon case number (N), duration (D_{wt}) and intensity (I) to the western North Pacific annual PDI, based on the method from Emanuel 2007 ref. 18. The top 4 rows are changes in each pentad w.r.t. the long-term mean (1993-2012). The last row is the change in the most-recent pentad (2008-2012) w.r.t. the 1st pentad in the early 1990s (1993-1997).

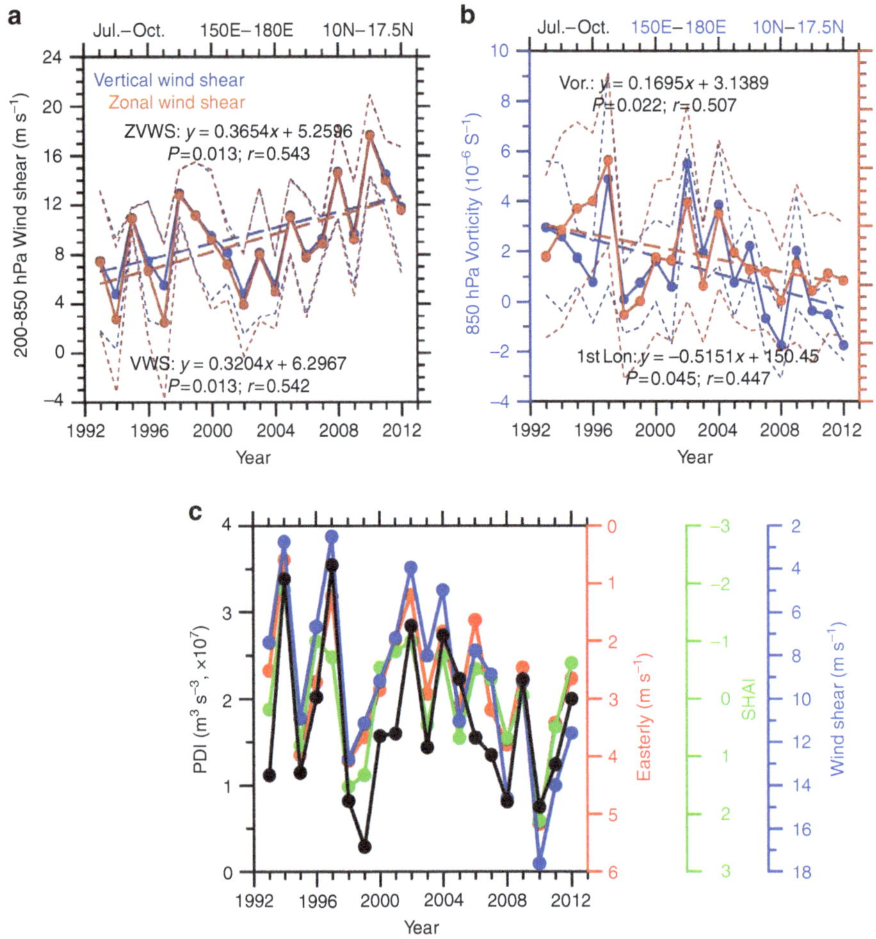

Figure 2 | Vertical wind shear (VWS) and other parameters at the typhoon genesis region. (**a**) Time evolution of the typhoon-season averaged VWS and zonal VWS in the typhoon genesis region in the past two decades. (**b**) As in **a**, but for the 850 hPa relative vorticity (left axis) and the genesis longitude (right axis). (**c**) Coherent variability between PDI, the easterly wind at 850 hPa, SHAI (subtropical height area index) at 850 hPa and the zonal VWS. Note that the three *y*-axes at right are reversed, so as to show the reduction in PDI (left axis) with the increase in these three suppressive parameters.

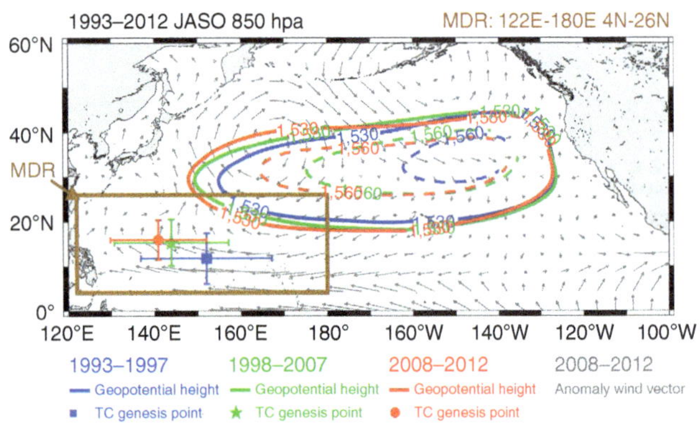

Figure 3 | Strengthening of the subtropical high. Strengthening of the subtropical high as depicted by the 1,560 and 1,530 geopotential lines at 850 hPa with anomalous wind vectors from the latest pentad (2008–2012) with respect to 20-year mean (1993–2012) overlaid. The averaged typhoon genesis positions (with 1 standard deviation) over three different periods in the past 20 years are also depicted. MDR = main development region.

global, tropical mean SST)[35,36] also showed a decreasing trend, supporting the decline in topical cyclone (TC) activity in the recent decade (other details see Supplementary Figs 7 and 8).

In addition to the correlations between the different parameter pairs, statistical analyses of each individual time series were also performed. As shown in Fig. 1d,e, the correlation was $\sim$0.7 for the typhoon count time series, with $P = 0.0008$. For the typhoon duration time series, $r \sim 0.5$ and $P = 0.034$. However, as the contributions from the typhoon intensity were opposite to the contributions from the duration and typhoon frequency (count), the P value for the PDI time series was larger ($P = 0.465$). It should be noted that due to the competing and offsetting impacts between typhoon duration/frequency and intensity, and between ocean and atmosphere, it was not possible for PDI to exhibit a clear trend because it was the residual result from opposing and competing contributors.

Therefore, from the typhoon perspective, the recent strengthening of the large-scale circulation induced a 'worsened' atmosphere condition for typhoons and PDI, even though such strengthening also provided a 'better' ocean because the increase in the easterly wind piled up warm surface water to the western North Pacific and increased D26 and UOHC, as discussed earlier[19,20,27,29]. However, the impact of the 'worsened' atmosphere appeared to dominate over the 'better' ocean to reduce the typhoon destructive potential through the strong suppression of typhoon number and duration. This interesting interplay can be observed between the ocean and the atmosphere. Apparently, for the present epoch in the western

Table 2 | Correlations among 3 groups of parameters.

Correlation	PDI group				Circulation group				TC-atm. group (genesis region)			
	PDI	D_{wt}	ND_{wt}	*N*	*I*	Easterly	SHAI	SHII	ZVWS	Vor.	1st lon	1st lat
PDI group												
PDI	1.00	0.86	0.72	0.43	0.58	−0.83	−0.77	−0.72	**−0.79**	0.64	0.77	−0.45
D_{wt}	**0.86**	1.00	0.83	0.50	0.29	−0.68	−0.60	−0.63	**−0.71**	0.70	0.88	−0.76
ND_{wt}	**0.72**	0.83	1.00	0.89	−0.06	−0.67	−0.67	−0.75	**−0.71**	0.57	0.76	−0.69
N	0.43	0.50	0.89	1.00	−0.31	−0.51	−0.57	−0.66	**−0.56**	0.37	0.51	−0.49
I	0.58	0.29	−0.06	−0.31	1.00	−0.49	−0.45	−0.22	−0.36	0.25	0.25	0.12
Circulation group												
Easterly	**−0.83**	−0.68	−0.67	−0.51	−0.49	1.00	0.90	0.78	0.91	−0.68	−0.61	0.42
SHAI	**−0.77**	−0.60	−0.67	−0.57	−0.45	0.90	1.00	0.91	0.83	−0.44	−0.61	0.32
SHII	**−0.72**	**−0.63**	**−0.75**	**−0.66**	−0.22	0.78	0.91	1.00	0.79	−0.46	−0.64	0.38
TC − atm. group (genesis region)												
ZVWS	**−0.79**	**−0.71**	**−0.71**	**−0.56**	−0.36	**0.91**	**0.83**	0.79	1.00	−0.79	−0.74	0.60
Vor.	**0.64**	**0.70**	**0.57**	0.37	0.25	−0.68	−0.44	−0.46	−0.79	1.00	0.72	−0.69
1st lon	**0.77**	**0.88**	**0.76**	0.51	0.25	−0.61	−0.61	−0.64	−0.74	0.72	1.00	−0.73
1st lat	−0.45	**−0.76**	**−0.69**	−0.49	0.12	0.42	0.32	0.38	0.60	−0.69	−0.73	1.00

lat, latitude; lon, longitude; PDI, Power Dissipation Index; SHAI, subtropical high area index; SHII, Subtropical High-Intensity Index; TC-atm., TC-atmosphere; Vor., Vorticity.
Correlations among the PDI, circulation and TC-atmosphere groups of parameters, based on observations in 1993-2012. The PDI group consists of PDI-related parameters including PDI, D_{wt} (duration), ND_{wt} (case number × duration), *N* (case number) and *I* (intensity). The circulation group consists of three parameters related to the large-scale circulation at 850 hPa, including the easterly wind, SHAI and SHII. The typhoon-atmosphere group at the genesis region consists of typhoon-related atmospheric parameters, including ZVWS (zonal vertical wind shear), 850 hPa vorticity, genesis longitude and genesis latitude. This table is part of the large 23 parameter by 23 parameter table in Supplementary Table 2.

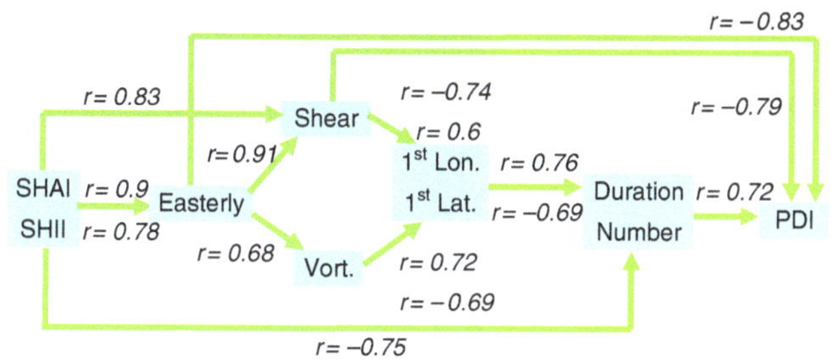

Figure 4 | Flow diagram of key correlations. Flow diagram showing a summary of the key correlations in Table 2.

North Pacific, the atmosphere dominates over the ocean in controlling typhoon PDI.

Although the possibility of an increase in PDI (and hence the destructiveness potential) due to ocean warming was suggested before[7], our results provided new evidence to show that such a situation was not always applicable because the ocean was not always dominant. We found that the situation was more complex, and it was also possible for PDI to decrease despite ocean warming over the western North Pacific Ocean. A 'worsened' atmosphere can effectively dominate over a 'warmer' ocean to decrease the typhoon destructiveness potential.

Because PDI depends not only on the intensity but also on typhoon duration and number, even if the intensity increases due to ocean warming, it does not mean that typhoon duration and number will increase in tandem. This is especially true over the western North Pacific Ocean, in which these parameters are very much controlled by the dynamical factors in the atmosphere (for example, shear) than the thermal factors (for example, SST, UOHC and humidity). As in Table 1, contributions from duration and number to PDI were often opposite to the contribution from intensity (three out of four pentads). Therefore, the final PDI was mostly the residual after the offset.

Discussion

These opposite contributions from the typhoon intensity versus the frequency/duration to PDI triggers an interesting question, that is, global warming implications[8–13,35–53]. Climate projections have reported that although the tropical cyclone intensity may increase under global warming, the occurrence frequency is projected to reduce[9–11,37,44,46–48]. As noted by Zhao and Held[10], the typhoon occurrence frequency over the western North Pacific could reduce under global warming, due to the possible impact from reduction in the mid-troposphere vertical ascending motion and mass flux.

It is intriguing to quantify and explore these possible cancellation effects on PDI. We conducted the PDI analyses using simulated typhoon data (for the late-twenty-first century projection) from a state-of-the-art high-resolution model, as in Zhao and Held[10] and Zhao *et al.*[11]. Table 3 presents the results. Based on this projection, the typhoon intensity (annual averaged in the western North Pacific domain) under global warming was ~4.8% (30.12 $m\,s^{-1}$ versus 28.74 $m\,s^{-1}$) higher than the current intensity. In addition, there was a slight increase of ~3.2% in the duration. Although both the intensity and duration increased under global warming, there was an even larger typhoon frequency reduction of ~25.7% (Table 3). As a result, the typhoon PDI decreased by ~15.2%. The annual typhoon PDI under global warming was ~85% ($1.62 \times 10^7\,m^3\,s^{-3}$/$1.91 \times 10^7\,m^3\,s^{-3}$) of the current value, showing a reduction in the typhoon destructive potential for the Asia Pacific region under global warming.

These results suggest that we are seeing an important cancellation effect from co-existing and opposing PDI contributors (that is, frequency reduction versus intensity increase) under both the global warming scenario and observations in the current climate epoch over the western North Pacific Ocean. In addition, results from the quantitative analyses suggested that the positive contribution from the increased intensity could be much smaller than the negative contribution from the frequency reduction and resultant PDI decrease.

Table 3 | PDI in current and global warming scenarios.

Western North Pacific Domain	PDI ($\times 10^7$ m^3s^{-3}; std)	I(m s^{-1}; std)	N(cases; std)	D(days; std)
Current	1.91 (0.38)	28.74 (1.27)	22.28 (4.18)	6.52 (0.71)
Global warming	1.62 (0.44)	30.12 (1.32)	16.55 (3.50)	6.73 (0.99)
Global warming—current	−0.29	+1.38	−5.73	+0.21
% Change with respect to current	−15.2%	+4.8%	−25.7%	+3.2%

PDI, Power Dissipation Index.
Comparison of PDI and the three contributing factors under current and global warming scenarios (late 21st century projection), based on the simulated typhoon data from Zhao and Held[10] and Zhao *et al.*[11] high resolution modelling[58-60].

However, it should also be noted that although the reduction in the western North Pacific typhoon PDI and the cancellation effect were observed under both the global warming projection and the current climate epoch observation, it did not necessarily mean that the associated climate forcing was of the same origin. The observed recent climate epoch over the Pacific (including the western North Pacific) was likely related to natural variability, whereas the simulation from Zhao and Held[10] and Zhao *et al.*[11] corresponded to the global warming scenario.

For the western North Pacific region, both types of climate conditions could result in the co-existence of an increase in typhoon intensity but a decrease in frequency. Another similarity is that although the ocean environment under both climate conditions could be more favourable, a more unfavourable atmospheric environment could co-exist to effectively suppress the occurrence frequency, resulting in the eventual reduction in destructive potential (that is, PDI).

Finally, although the reduction in PDI appeared to suggest a less destructive scenario for the Asian Pacific region under both the current climate epoch and global warming projection, it should be taken with caution. Despite its wide usage, PDI[8,9,21] is not an 'all-inclusive' parameter for tropical cyclone destructiveness. For example, PDI does not necessary reflect a typhoon's impacts associated with rainfall, storm surge[54,55] and landfall[9] wind speed. Another caveat concerns the global warming projection. Although our PDI analyses were based on typhoon data produced from one of the most reputable models currently available[10,11], model to model discrepancies could exist. Although the results from Zhao and Held[10] were more consistent with the projections from Sugi *et al.*[52] and Murakami *et al.*[46], the projection from Emanuel[43] suggested that both the typhoon intensity and frequency could increase under global warming. Because different models have different assumptions and associated uncertainties, future analyses across more models will be important for further assessing the uncertainties (Supplementary Figs 9–30 and Supplementary Note 3).

Methods

Study domains. The domain of the western North Pacific Ocean is defined as 110–180°E, 0–45°N. Typhoon MDR is defined as 122–180°E, 4–26°N and the typhoon genesis region is 150–180°E, 10–17.5°N (that is, the eastern MDR)[6].

PDI-related parameters. These parameters were obtained or calculated based on the US Joint Typhoon Warning Centre's best track data of the International Best Track Archive for Climate Stewardship (IBTrACS) database. For details, see below.

Typhoon-related atmospheric parameters. The European Centre for Medium-Range Weather Forecasts's (ECMWF) monthly Interim Reanalysis database at each 1° grid was used. The data at 850 hPa were used for the relative vorticity and humidity. The VWS was calculated based on the difference in the vector wind between 2 heights (200 and 850 hpa). Similarly, the zonal VWS was calculated, but for the zonal wind only. Validation of this data using the *in situ* radiosound data is shown in Supplementary Figs 31–33 and Supplementary Note 2.

Large-scale atmospheric circulation parameters. The recent strengthening of the atmospheric circulation was characterized by an evident increase in the easterly trade wind. For the study, winds at 850 hPa from the ECMWF Interim data were used. The SHAI proposed by Liu and Chan[6] was used to characterize the associated subtropical high strengthening. The SHAI quantified the regional coverage of the subtropical high in the WNP, originally defined as the normalized grid counts enclosed by the 5,880-gpm (geopotential metre) line at 500 hPa. In this research, we used both the 5,880-gpm line at 500 hPa and the 1,530-gpm line at 850 hPa to calculate the SHAI. The incorporation of the 850 hPa SHAI was due to a possible stronger association between the lower atmosphere (for example, vorticity, VWS) and typhoon activities. It could also better represent the atmospheric forcing on the ocean. The MDR was the region of interest; therefore, the SHAI was calculated within the MDR.

In addition, we defined an additional new index called the SHII. The SHII was associated with SHAI, but was based on the averaged geopotential height for the grids enclosed by the 5,880-gpm (for the 500 hPa level) or the 1,530-gpm (for the 850 hPa level) lines. Therefore, the SHII carried the intensity information of the subtropical high, whereas the SHAI carried the area coverage information. The results from these two indices were similar, although there was a higher correlation between the SHII (at 850 hPa) and the typhoon occurrence frequency ($r = -0.66$; Table 2).

Ocean condition time series. Monthly, one-degree SST data from the UK Hadley Centre are used. The UOHC (also known as the Tropical Cyclone Heat Potential[16]) and D26 were used to characterize the upper ocean thermal condition. They were based on the satellite altimetry at each of the 0.25° by 0.25° grid in the WNP MDR (from 1993 onwards, that is, since satellite altimetry observation became available). This method was widely applied and was validated by *in-situ* Argo float observations[14,56,57]). The altimetry Sea Surface Height Anomaly (SSHA) data source was the monthly, delayed-mode, gridded SSHA data from the Archiving, Validation and Interpretation of Satellite Oceanographic Data (AVISO) data base (http://www.aviso.oceanobs.com/). The SSHA contained both mass and thermal contributions. Therefore, Gravity Recovery and Climate Experiment (GRACE) satellite data were used to remove the mass contribution[19].

PDI analysis based on the methodology from the study by Emanuel. Emanuel[21] provided a very useful way to approximate the contributions of each of the three factors to PDI. For each individual typhoon, PDI is calculated as follows:

$$PDI = \int_0^{\tau} V_{max}^3 \, dt \tag{1}$$

where V_{max} is the maximum surface wind speed at each 6-hourly time interval (t) over the typhoon duration (τ). The annual typhoon season PDI is the summation of all of the cases (N) in a typhoon season in a year. To avoid the possible over-weighted contribution from the long-duration in the genesis period, Emanuel[21] proposed to use a velocity-weighted duration (D_{wti}) of a typhoon, i, as follows:

$$D_{\_wt_i} = \frac{\int_0^{\tau} V_{max} dt}{V_{smax}} \tag{2}$$

where V_{smax} is the lifetime peak intensity of the typhoon, i. As explained in Emanuel[21], the reason to use the velocity-weighted duration instead of the regular duration ($D_{\_i}$) was because the duration of a typhoon's genesis (spin up) period can be long. However, during the genesis period, the weak wind has little contribution to PDI. As such, a velocity-weighted duration was proposed.

Averaging D_{wti} over all of the typhoon cases in a typhoon season of a year yields the following:

$$D_{wt} = \frac{1}{N} \sum_1^N D_{wt_i} \tag{3}$$

Finally, an annual averaged intensity (in wind speed cube) is defined as follows:

$$I = \frac{\sum_1^N \int_0^{\tau_i} V_{max}^3 dt}{\sum_1^N D_{wt_i}} \quad (4)$$

From above, the annually accumulated PDI in a typhoon season in a year is $N \times D_{wt} \times I$. If we use the natural logarithm, then $\ln(PDI) = \ln(N) + \ln(D_{wt}) + \ln(I)$. Emanuel[21] provided a quantitative way to separate the different contributions of N, D_{wt} and I to PDI. The results are shown in Table 1 and Supplementary Fig. 2. In this research, the data were based on the annual typhoon season (July–October) over the WNP MDR domain during 1993–2012.

References

1. Fogarty, C. T. & Blake, E. S. The double life of hurricane Sandy—and a climatological perspective of these post-tropical giants, in 'State of the Climate in 2012'. *Bull. Am. Meteorol. Soc.* **94,** S109–S110 (2013).
2. Lander, M., Guard, C. & Camargo, S. J. Tropical cyclones, super-typhoon Haiyan, in 'State of the Climate in 2013'. *Bull. Am. Meteorol. Soc.* **95,** S112–S114 (2014).
3. Lin, I. I., Pun, I. F. & Lien, C. C. 'Category-6' Supertyphoon Haiyan in global warming hiatus: contribution from subsurface ocean warming. *Geophy. Res. Lett.* **41,** 8547–8553 (2014).
4. Maue, R. N. Recent historically low global tropical cyclone activity. *Geophy. Res. Lett.* **38,** L14803 (2011; Please kindly note that though the data of this reference ends in 2010, consistent result from the extended analysis to 2014 could be found in the author's web (http://policlimate.com/tropical/).
5. Peduzzi, P. *et al.* Global trends in tropical cyclone risk. *Nat. Clim. Chang.* **2,** 289–294 (2012).
6. Liu, K. S. & Chan, J. C. L. Inactive period of Western North Pacific tropical cyclone activity in 1998-2011. *J. Clim.* **26,** 2614–2630 (2013).
7. Emanuel, K. Increasing destructiveness of tropical cyclones over the past 30 years. *Nature* **436,** 686–688 (2005).
8. Hsu, P.-C., Chu, P.-S., Murakami, H. & Zhao, X. An abrupt decrease in the late-season typhoon activity over the western North Pacific. *J. Clim* **27,** 4296–4312 (2014).
9. Kim, H.-S. *et al.* Tropical cyclone simulation and response to CO2 doubling in the GFDL CM2.5 high-resolution coupled climate model. *J. Clim* **27,** 8034–8054 (2014).
10. Zhao, M. & Held, I. TC-permitting GCM simulations of hurricane frequency response to sea surface temperature anomalies projected for the late 21st century. *J. Clim* **25,** 2995–3009 (2012).
11. Zhao, M., Held, I., Lin, S.-J. & Vecchi, G. Simulations of global hurricane climatology, interannual variability, and response to global warming using a 50km resolution GCM. *J. Clim* **22,** 6653–6678 (2009).
12. Webster, P. J., Holland, G. J., Curry, J. A. & Chang, H.-R. Changes in tropical cyclone number, duration, and intensity in a warming environment. *Science* **309,** 1844–1846 (2005).
13. IPCC. *5th Assessment Report of the Intergovernmental Panel on Climate Change.* (eds Stocker, T. F. *et al.*) (Cambridge Univ., 2013).
14. Shay, L. K., Goni, G. J. & Black, P. G. Role of a warm ocean feature on Hurricane Opal. *Mon. Weath. Rev* **128,** 1366–1383 (2000).
15. Emanuel, K. A., DesAutels, C., Holloway, C. & Korty, R. Environmental control of tropical cyclone intensity. *J. Atmos. Sci.* **61,** 843–858 (2004).
16. Goni, G. *et al.* Applications of satellite-derived ocean measurements to tropical cyclone intensity forecasting. *Oceanography* **22** 3 190–197 (2009).
17. Lin, I. I., Wu, C. C. & Pun, I. F. Upper ocean thermal structure and the western North Pacific category-5 typhoons—part I: ocean features and the category 5 typhoons' intensification. *Mon. Wea. Rev.* **136,** 3288–3306 (2008).
18. Lin, I. I. *et al.* An ocean coupling potential intensity index for tropical cyclones. *Geophys. Res. Lett.* **40,** 1878–1882 (2013).
19. Pun, I. F., Lin, I. I. & Lo, M. H. Recent increase in high tropical cyclone heat potential area in the Western North Pacific Ocean. *Geophy. Res. Lett* **40,** 4680–4684 (2013).
20. Merrifield, M. A. & Maltrud, M. E. Regional sea level trends due to a Pacific trade wind intensification. *Geophys. Res. Lett.* **38,** L21605 (2011).
21. Emanuel, K. A. Environmental factors affecting tropical cyclone power dissipation. *J. Clim* **20,** 5497–5509 (2007).
22. Matsuura, T., Yumoto, M. & Iizuka, S. A mechanism of interdecadal variability of tropical cyclone activity over the western North Pacific. *Clim. Dynam.* **21,** 105–117 (2003).
23. Chan, J. C. L. Thermodynamic control on the climate of intense tropical cyclones. *Proc. R. Soc. A* **465,** 3011–3021 (2009).
24. Gray, W. M. *Hurricanes: Their Formation, Structure, and Likely role in the Tropical Circulation. Meteorology Over the Tropical Oceans.* (ed. Shaw, D. B.) 155–218 (Royal Meteorological Society, 1979).
25. Chan, J. C. L. Decadal variations of intense typhoon occurrence in the western North Pacific. *Proc. R. Soc. A* **464,** 249–272 (2008).
26. Kossin, J. P., Emanuel, K. A. & Vecchi, G. A. The poleward migration of the location of tropical cyclone maximum intensity. *Nature* **509,** 349–352 (2014).
27. Wang, B. *et al.* Northern Hemisphere summer monsoon intensified by mega-El Nino/southern oscillation and Atlantic multi-decadal oscillation. *Proc. Natl. Acad. Sci. USA* **110,** 5347–5352 (2013).
28. Kosaka, Y. & Xie, S. P. Recent global warming hiatus tied to equatorial Pacific surface cooling. *Nature* **501,** 403–407 (2013).
29. England *et al.* Recent intensification of wind-driven circulation in the Pacific and the ongoing warming hiatus. *Nat. Clim. Chang.* **4,** 222–227 (2014).
30. Zhou, T. *et al.* Why the western Pacific subtropical high has extended westward since the late 1970s. *J. Clim* **22,** 2199–2215 (2009).
31. Luo, J. J., Sasaki, W. & Masumoto, Y. Indian Ocean warming modulates Pacific Climate change. *Proc. Natl Acad. Sci. USA* **109,** 18701–18706 (2012).
32. Hong, C.-C., Wu, Y.-K., Li, T. & Chang, C.-C. The climate regime shift over the Pacific during 1996/1997. *Clim. Dyn* **43,** 435–446 (2014).
33. Nigam, S. & Chan, S. C. On the summertime strengthening of the Northern Hemisphere Pacific Sea Level Pressure Anticyclone. *J. Clim* **22,** 1174–1192 (2009).
34. Wang, B., Xiang, B. & Lee, J.-Y. Subtropical high predictability establishes a promising way for monsoon and tropical storm predictions. *Proc. Natl Acad. Sci. USA* **110,** 2718–2722 (2013).
35. Elsner, J. B., Strazzo, S. E., Jagger, T. H. & LaRow, T. Sensitivity of limiting hurricane intensity to SST in the Atlantic, from observations and GCMS. *J. Clim* **26,** 5949–5957 (2013).
36. Bender, M. A. *et al.* Modeled impact of anthropogenic warming on the frequency of intense Atlantic hurricanes. *Science* **327,** 454–458 (2010).
37. Murnane, R. J. & Elsner, J. B. Maximum wind speeds and US hurricane losses. *Geophys. Res. Lett.* **39,** L16707 (2012).
38. Knutson, T. R. *et al.* Tropical cyclones and climate change. *Nat. Geosci* **3,** 157–163 (2010).
39. Elsner, J. B., Kossin, J. P. & Jagger, T. H. The increasing intensity of the strongest tropical cyclones. *Nature* **455,** 92–95 (2008).
40. Vecchi, G. A. & Soden, B. J. Effect of remote sea surface temperature change on tropical cyclone potential intensity. *Nature* **450,** 1066–1070 (2007a).
41. Villarini, G. & Vecchi, G. A. Twenty-first-century projections of North Atlantic tropical storms from CMIP5 models. *Nat. Clim. Chang.* **2,** 604–607 (2012).
42. Camargo, S. J. Global and Regional Aspects of tropical cyclone activity in the CMIP5 models. *J. Clim* **26,** 9880–9902 (2013).
43. Emanuel, K. A. Downscaling CMIP5 climate models shows increased tropical cyclone activity over the 21st century. *Proc. Natl Acad. Sci. USA* **110,** 12219–12224 (2013).
44. Knutson, T. R. *et al.* Dynamical downscaling projections of twenty-first-century Atlantic hurricane activity: CMIP3 and CMIP5 model-based scenarios. *J. Clim* **26,** 6591–6617 (2013).
45. Li, T. *et al.* Global warming shifts Pacific tropical cyclone location. *Geophys. Res. Lett.* **37,** L21804 (2010).
46. Murakami, H. *et al.* Future changes in tropical cyclone activity projected by the new high-resolution MRI-AGCM. *J. Clim* **25,** 3237–3260 (2012).
47. Yokoi, S. & Takayabu, Y. N. Multi-model projection of global warming impact on tropical cyclone genesis frequency over the western North Pacific. *J. Meteor. Soc. Japan* **87,** 525–538 (2009).
48. Murakami, H., Wang, B. & Kitoh, A. Future change of western North Pacific typhoons: Projections by a 20-km-mesh global atmospheric model. *J. Clim* **24,** 3237–3260 (2011).
49. Vecchi, G. A. & Soden, B. J. Increased tropical Atlantic wind shear in model projections of global warming. *Geophys. Res. Lett* **34,** L08702 (2007b).
50. Vecchi, G. A. & Soden, B. J. Global warming and the weakening of the tropical circulation. *J. Clim* **20,** 4316–4340 (2007c).
51. Li, W., Li, L., Ting, M. & Liu, Y. Intensification of northern hemisphere subtropical highs in a warming climate. *Nat. Geosci* **5,** 830–834 (2012).
52. Sugi, M., Murakami, H. & Yoshimura, J. A reduction in global tropical cyclone frequency due to global warming. *SOLA* **5,** 164–167 (2009).
53. Xiang, B., Wang, B., Li, J., Zhao, M. & Lee, J. Y. Understanding the anthropogenically forced change of equatorial Pacific trade winds in coupled climate models. *J. Clim* **27,** 8510–8526 (2014).
54. Mori, N. *et al.* Local amplification of storm surge by Super Typhoon Haiyan in Leyte Gulf. *Geophys. Res. Lett.* **41,** 5106–5116 (2014).
55. Lin, I. I., Goni, G. J., Knaff, J., Forbes, C. & Ali, M. M. Ocean heat content for tropical cyclone intensity forecasting and its impact on storm surge. *Natural Hazards* **66,** 1481–1500 (2013).
56. Pun, I. F. *et al.* Validation and application of altimetry-derived upper ocean thermal structure in the western North Pacific ocean for typhoon intensity forecast. *IEEE Trans. Geosci. Remote Sens* **45,** 1616–1630 (2007).
57. Pun, I. F. *et al.* Typhoon-ocean interaction in the western North Pacific: Part 2. *Oceanography* **24,** 32–41 (2011).
58. Dee, D. *et al.* The ERA-interim reanalysis: configuration and performance of the data assimilation system. *Quart. J. Roy. Meteor. Soc* **137,** 553–597 (2011).
59. Thorne, P. & Vose, R. Reanalyses suitable for characterizing long-term trends: are they really achievable? *Bull. Am. Meteor. Soc* **91,** 353–361 (2010).

60. Tippett, M. K., Carmago, S. J. & Sobel, A. H. A poisson regression index for tropical cyclone genesis and the role of large-scale vorticity in genesis. *J. Clim.* **24,** 2335–2357 (2011).

Acknowledgements

Work by I.-I.L. is supported by Taiwan's Ministry of Science and Technology under grants: NSC 101-2628-M-002 -001 -MY4 and MOST 103-2111-M-002 -002 -MY3. Work by J.C.L.C. is supported by the Research Grants Council of the HKSAR Government Grant CityU100210. We sincerely thank Ming Zhao from the NOAA GFDL for providing the simulated tropical cyclone data and for very helpful discussions. Thanks to the US NOAA/IBTrACS data base, UK Hadley Centre, European Centre for Medium-Range Weather Forecast, NASA, the AVISO altimetry team and Remote Sensing Systems for essential data sets, to Jay Hsia, Iam-Fei Pun, Siao-Ching Huang, Chun-Chi Lien and Meiling Lam for help in data processing; to Ping Huang, for help in CMIP5 data provision; to S. Nigam and T. Zhou for helpful discussions.

Author contributions

I.-I.L. and J.C.L.C. conceived the project, designed and conducted the study and co-wrote the manuscript.

Additional information

Competing financial interests: The authors declare no competing financial interests.

5

Regional climate impacts of a possible future grand solar minimum

Sarah Ineson[1], Amanda C. Maycock[2,3], Lesley J. Gray[3,4], Adam A. Scaife[1], Nick J. Dunstone[1], Jerald W. Harder[5], Jeff R. Knight[1], Mike Lockwood[6], James C. Manners[1] & Richard A. Wood[1]

Any reduction in global mean near-surface temperature due to a future decline in solar activity is likely to be a small fraction of projected anthropogenic warming. However, variability in ultraviolet solar irradiance is linked to modulation of the Arctic and North Atlantic Oscillations, suggesting the potential for larger regional surface climate effects. Here, we explore possible impacts through two experiments designed to bracket uncertainty in ultraviolet irradiance in a scenario in which future solar activity decreases to Maunder Minimum-like conditions by 2050. Both experiments show regional structure in the wintertime response, resembling the North Atlantic Oscillation, with enhanced relative cooling over northern Eurasia and the eastern United States. For a high-end decline in solar ultraviolet irradiance, the impact on winter northern European surface temperatures over the late twenty-first century could be a significant fraction of the difference in climate change between plausible AR5 scenarios of greenhouse gas concentrations.

[1] Met Office Hadley Centre, Exeter EX1 3PB, UK. [2] Centre for Atmospheric Science, Department of Chemistry, University of Cambridge, Cambridge CB2 1EW, UK. [3] National Centre for Atmospheric Science, University of Oxford, Oxford OX1 3PU, UK. [4] Department of Atmospheric, Oceanic and Planetary Physics, University of Oxford, Oxford OX1 3PU, UK. [5] Laboratory for Atmospheric and Space Physics, University of Colorado, Boulder, Colorado 80303 USA. [6] Department of Meteorology, University of Reading, Reading RG6 6BB, UK. Correspondence and requests for materials should be addressed to S.I. (email: sarah.ineson@metoffice.gov.uk).

The past few decades have been characterized by a period of relatively high solar activity. However, the recent prolonged solar minimum and subsequent weak solar cycle 24 have led to suggestions that the grand solar maximum may be at an end[1]. Using past variations of solar activity measured by cosmogenic isotope abundance changes, analogue forecasts for possible future solar output have been calculated. An 8% chance of a return to Maunder Minimum-like conditions within the next 40 years was estimated in 2010 (ref. 2). The decline in solar activity has continued, to the time of writing, and is faster than any other such decline in the 9,300 years covered by the cosmogenic isotope data[1]. If this recent rate of decline is added to the analysis, the 8% probability estimate is now raised to between 15 and 20%.

A number of studies have indicated that the decreases in global mean temperature associated with a future decline in solar activity are likely to be relatively small[3–7]. However, variability in ultraviolet solar irradiance has been linked to changes in surface pressure that resemble the Arctic and North Atlantic Oscillations (AO/NAO)[8–10] and studies of both the 11-year solar cycle[11,12] and centennial timescales[13] suggest the potential for larger regional effects. The mechanism for these changes is via a stratospheric pathway, a so-called 'top-down' mechanism, and involves altered heating of the stratosphere by solar ultraviolet irradiance. Anomalous temperatures in the region of the tropical stratopause give rise to changes in the subtropical stratospheric winds, in geostrophic balance with the modified equator-to-pole temperature gradient. This signal then propagates poleward and downward and is amplified by altered planetary wave activity[8] before being communicated throughout the depth of the troposphere in the Pacific and Atlantic basins[14]. This mechanism can also drive changes in tropical lower stratosphere temperatures, which can additionally affect the troposphere[15].

There is currently uncertainty regarding the ultraviolet variability that accompanies changes in total solar irradiance (TSI)[16]. Recent satellite measurements from the Spectral Irradiance Monitor (SIM) on the Solar Radiation and Climate Experiment satellite[17] show that the variability in ultraviolet over part of the declining phase of solar cycle 23 might be considerably larger than indicated by previous estimates[17,18].

Here, our experiments aim to explore the impact of this uncertainty in ultraviolet in the future Maunder Minimum scenario[2], using both Naval Research Laboratory Spectral Solar Irradiance model output[19] and the SIM data as a basis to estimate lower (EXPT-A) and upper (EXPT-B) bounds for the change in ultraviolet.

Regional structure is found in the wintertime response in both experiments, with enhanced relative cooling over northern Eurasia and the eastern United States. For EXPT-B, the change in winter northern European surface temperatures over the late twenty-first century is found to be of similar magnitude to the difference between RCP6.0 and RCP4.5 projections for this region. We conclude that solar forcing may be an important source of uncertainty in regional climate projections.

Results

Solar forcing scenarios. The future solar forcing scenarios are imposed on an ocean–atmosphere climate model following the representative concentration pathway (RCP) 8.5 for the period 2005 to 2100. Information on the model, experiment ensemble sizes and details of the construction of the solar forcing scenarios can be found in Methods. The scenarios are illustrated in Fig. 1, which shows the time variation in TSI and ultraviolet irradiance (defined here to be in the range 200–320 nm) for the control (hereafter CTRL-8.5) and experiments with respect to the mean

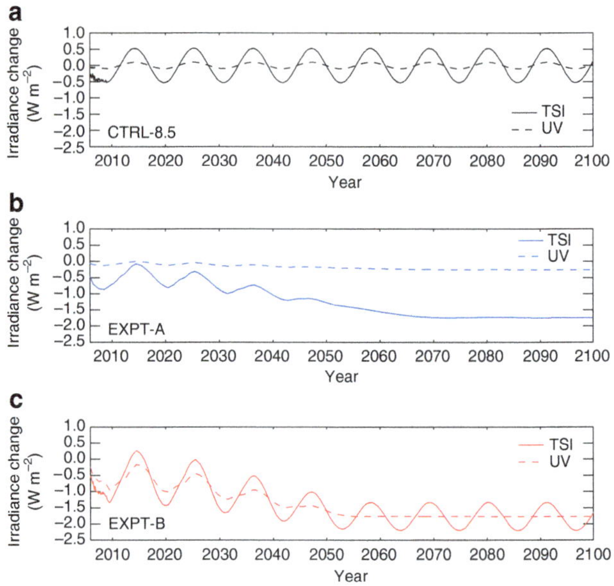

Figure 1 | Future solar forcing scenarios. Variations in solar forcing for Total Solar Irradiance ($W m^{-2}$) and ultraviolet irradiance in the 200–320 nm spectral band ($W m^{-2}$) relative to the mean of the repeated cycle in CTRL-8.5 for (**a**) CTRL-8.5 (black), (**b**) EXPT-A (blue) and (**c**) EXPT-B (red). The value of this mean is 1,366.2 $W m^{-2}$ for TSI and 27.4 $W m^{-2}$ for the ultraviolet band.

of CTRL-8.5. The mean difference relative to the repeated solar cycle in CTRL-8.5 is a reduction in TSI of 1.74 $W m^{-2}$ (0.127%) for EXPT-A and 1.76 $W m^{-2}$ (0.129%) for EXPT-B. For the ultraviolet component, the reduction is 0.25 $W m^{-2}$ (0.91%) for EXPT-A and 1.76 $W m^{-2}$ (6.43%) for EXPT-B.

Global mean temperature response. As a result of the decrease in solar irradiance, both experiments show widespread cooling with respect to CTRL-8.5 (Fig. 2). The relative annual global mean near-surface temperature change for the period 2050–2099 is a cooling of 0.13 and 0.12 °C for EXPT-A and EXPT-B, respectively. This offsets or delays the global warming trend by ~2 years and is small compared with the modelled global warming. This is consistent with other recently published results[3–7], which indicate that any change in global mean temperature due to a future prolonged solar minimum would do little to substantially offset or delay the warming due to projected increases in long-lived greenhouse gases. A comparison of forcings and responses can be found in ref. 20. In some of these studies, potential for regional variation is indicated[4,6,7], and we now explore this in more detail in our experiments.

Northern hemisphere winter response. In our experiments, while cooling is evident year round, the largest and most coherent anomalies are in the northern hemisphere in boreal winter and spring. Much of this structure can be explained by a relative change in atmospheric circulation. During December to February, a negative AO/NAO-like mean sea-level pressure pattern is seen in both experiments (Fig. 3a,e). Over the North Pacific, the deepened Aleutian low is consistent with the observed pressure response to changes in solar activity[10,21,22]. In the Atlantic sector, EXPT-B has a stronger and more significant response, showing a weakening in the NAO (defined for the model as the difference in sea-level pressure between gridpoints nearest to the Azores and southwest Iceland) of 1.8 hPa, which is a notable proportion of the interannual standard deviation of 7–8 hPa. Again, the NAO

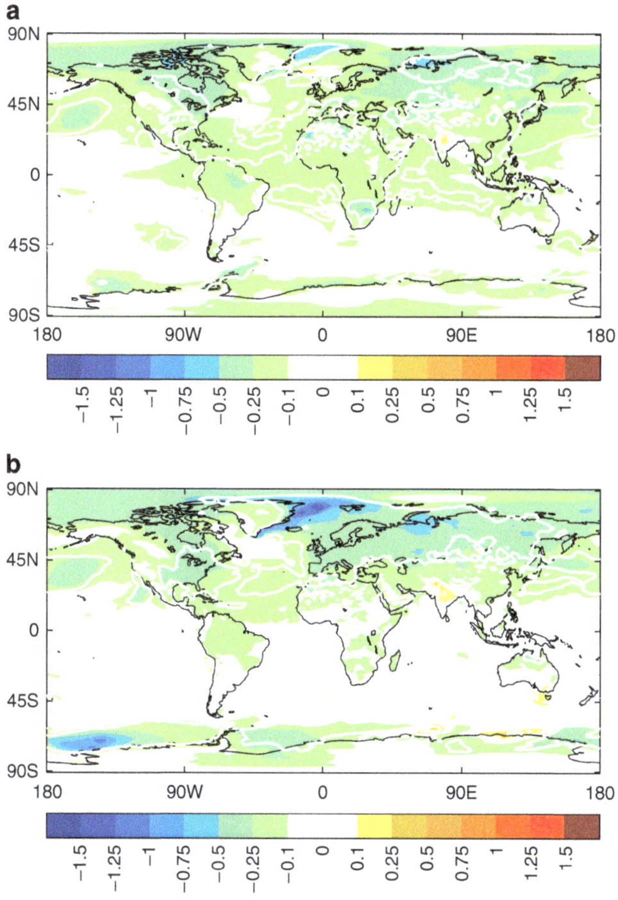

Figure 2 | Annual mean surface temperature response for future solar experiments. Difference in near-surface temperature (°C) between (**a**) EXPT-A and (**b**) EXPT-B and CTRL-8.5 for the period 2050-2099. Solid white contours indicate significance with a 95% confidence interval.

pattern is consistent with the observed response associated with the 11-year cycle over the recent re-analysis period[11], which has been found to be significant and maximizes at a lag of ~3–4 years for the period 1870–2010 (ref. 10). Relative to the historical control period (1971–2000), CTRL-8.5 has a negative shift in the NAO with a significant decrease of 3.2 hPa for the period 2050–2099 (not shown). This is broadly consistent with other modelling studies where the models have well resolved stratospheres[23,24].

There are marked regional differences in the wintertime pattern of cooling in the northern hemisphere with enhanced cooling over northern Eurasia and in the eastern U.S. in both experiments (Fig. 3b,f). Average temperature changes for northern hemisphere geographic regions[25] (Table 1) show overall cooling, but with significant and larger or comparable cooling for EXPT-B in the above regions consistent with the stronger negative NAO/AO in this experiment. Cooling to the east of Greenland in EXPT-B, and to a lesser extent in EXPT-A, is associated with a relative increase in the southward transport of sea ice by the East Greenland current (not shown); however, the observed relationship between the NAO and East Greenland sea ice is thought to be epoch dependent[26], and different models show a range of such relationships[27], so this element of our model's response may not be robust.

The quadrupole surface temperature response associated with a decrease in the AO/NAO[28] is evident when the global warming spatial pattern, defined for the purposes of this demonstration as the mean temperature in CTRL-8.5 at a lag of 2 years, has been removed (Fig. 3c,g). The general pattern of warming over northeastern North America/Greenland and southern Europe/northern Africa and cooling over southeastern North America and northern Europe can be seen more clearly in EXPT-B (Fig. 3g). Analysis of the stratospheric circulation[20] indicates that these changes are associated with a 'top-down' stratospheric pathway, as identified in a number of solar studies[8,9,12].

In EXPT-B, precipitation (Fig. 3h) increases in a wide band across the Atlantic and southern parts of Europe, extending eastwards across the Eurasian continent. There are decreases in precipitation over northern Europe. In general, EXPT-A (Fig. 3d) shows a similar but noisier response, with less significant signals. These changes are associated with a southward shift of the jet stream and decreased (increased) blocking over southern (northern) Europe in the experiments[20].

Changes in the frequency of extreme events are also affected. CTRL-8.5 shows a large decrease in frost days (20–35 days per winter) over much of southern and central Europe, northern Africa and southeast US for 2050–2099 relative to the historical simulation (1971–2000). There are also large changes over sea areas on the Canadian east coast and to the east of Greenland associated with the retreat of sea ice owing to anthropogenic climate change (Fig. 4a). In EXPT-B (Fig. 4c), and to a lesser extent in EXPT-A (Fig. 4b), we can see the clear signature of the negative AO/NAO with increases of a few frost days per winter in northern Europe and southeast US relative to CTRL-8.5, again consistent with NAO variability on other timescales[29].

European winter response. Examining the European response in more detail, average temperature changes for a northern European region are shown in Fig. 5. Relative to the historical period (1971–2000), RCP4.5 and CTRL-8.5 show substantial warming with a mean difference for 2050–2099 of 4.1 °C and 6.6 °C, respectively. EXPT-A and EXPT-B follow the same trajectory as CTRL-8.5, but with a reduced warming. Relative to CTRL-8.5, we find decreases in regional temperature for 2050–2099 of 0.4 °C (EXPT-A) and nearly 0.8 °C (EXPT-B). This regional cooling is therefore a notable fraction, ~30% for EXPT-B, of the difference between the temperature changes for CTRL-8.5 and RCP4.5. Indeed, compared with temperature changes for a more comprehensive range of scenarios run using a closely related model[30], the difference between CTRL-8.5 and EXPT-B is similar to the difference between RCP6.0 and RCP4.5 for this region (Fig. 5, inset).

Both RCP4.5 and CTRL-8.5 show significant increases in winter precipitation for northern Europe for 2050–2099 relative to the historical period, with a 12% increase for CTRL-8.5 (Fig. 6a). As expected with the more negative NAO-like pattern in the solar scenarios, the associated southward shift in the storm track acts to oppose this trend, with both EXPT-A and EXPT-B showing a relative decrease in precipitation compared with CTRL-8.5. In addition, the more southerly storm track enhances precipitation for southern Europe which tends to lessen the drying trend in this region in RCP4.5 and CTRL-8.5 (Fig. 6b). The CMIP5 multi-model ensemble projects drying for the Mediterranean region[31], so in this sensitive region, wintertime alleviation is likely to be important. However, we note that the changes in this region are small and the confidence is relatively low. As with temperature, the changes in precipitation associated with the solar scenarios, are again large fractions of the absolute difference between CTRL-8.5 and RCP4.5; ~65% for northern Europe and 90% for southern Europe in EXPT-B, although the uncertainty for southern Europe is large.

We note that low solar activity does not guarantee cold conditions in any specific European winter as additional

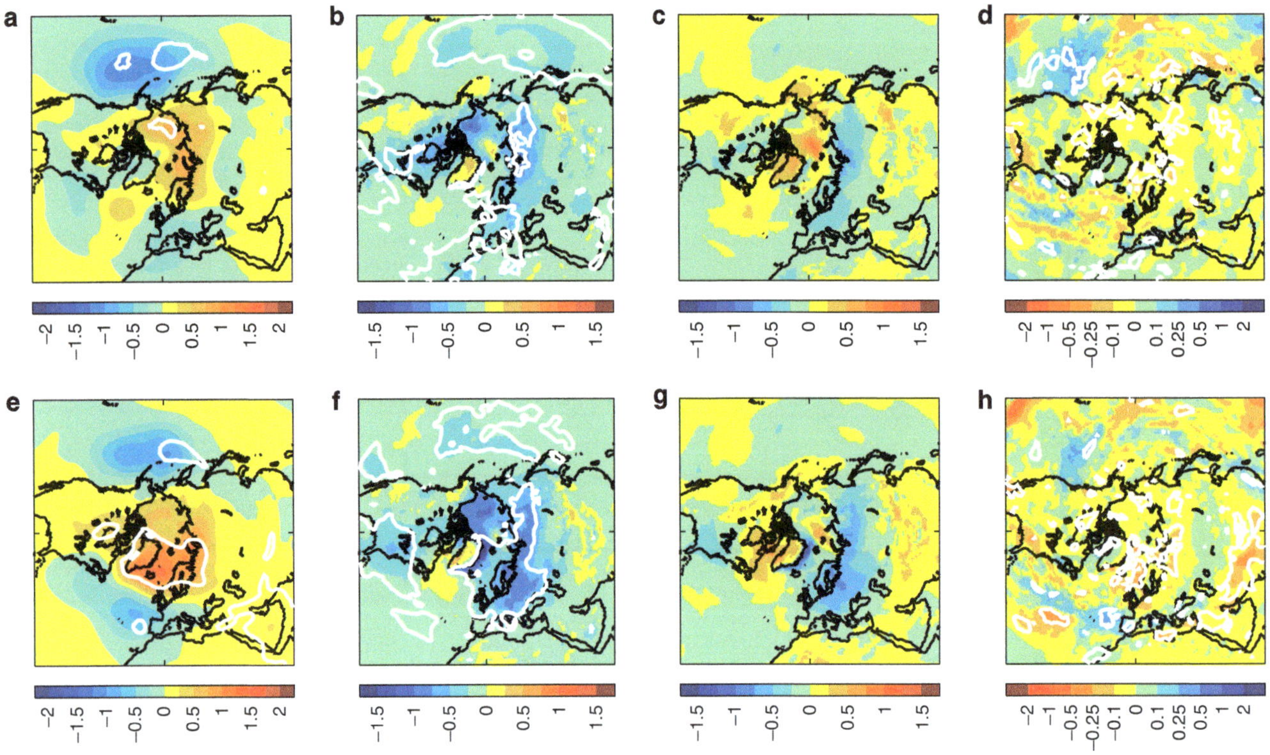

Figure 3 | Winter climate response for future solar minimum experiments. Differences between experiments and CTRL-8.5 for the period 2050–2099 for (**a**,**e**) sea-level pressure (hPa), (**b**,**f**) near-surface temperature (°C), (**d**,**h**) precipitation (mm day^{-1}). (**c**,**g**) They are as in **b** and **f** but with control lagged by 2 years to give similar global mean temperature. Upper panel plots are for EXPT-A and lower panel plots are for EXPT-B. Anomalies are for December to February. Solid white contours indicate significance with a 95% confidence interval.

Table 1 | Regional winter temperature response.

	EXPT-A	EXPT-B		EXPT-A	EXPT-B
North Europe	− **0.42**	− **0.75**	Greenland	− 0.26	− 0.13
Mediterranean	− **0.25**	− **0.23**	Western North America	− 0.05	− 0.11
North Asia	− 0.38	− **0.53**	Central North America	− 0.24	− **0.34**
			Eastern North America	− **0.45**	− **0.43**
			Alaska	− 0.31	− 0.17

Mean winter (December to February) regional near-surface temperature change (°C, land points only) for 2050-2099 for EXPT-A and EXPT-B relative to CTRL-8.5. Significance with a 90% confidence interval is shown in bold.

variability is introduced by other factors. The 360-year Central England Temperature record for December–February shows that the coldest winters in the UK occurred at low solar activity, but, for example, 1685/6, near the centre of the Maunder minimum, was the 5th warmest winter in the entire record[32]. This highlights the fact that solar variability acts only to bias the intrinsic year-to-year variability, which remains substantial for this region[33]. An example of this from all simulations is that for 12 of the 50 individual years in the period 2050–2099, the warmest ensemble member for northern European winter comes from EXPT-B (Fig. 7).

Possible feedback mechanisms. It has been suggested that solar variability may have affected past variations in North Atlantic temperature and salinity[34]. A positive ocean–atmosphere feedback on the cold northern European temperatures could occur through altered Atlantic heat content[35] and we see a clear signal of a reduced meridional surface temperature gradient over the west North Atlantic for EXPT-B (Fig. 3f). Feedbacks could also occur through decreased Atlantic meridional overturning circulation (AMOC) associated either with the persistent negative NAO forcing[36] or with more complex sea ice–ocean–atmosphere mechanisms[37–39]. However, there is uncertainty in the sign of the AMOC response to declining solar activity, with two recent studies suggesting a relative strengthening, associated with weaker radiative forcing and a possible change in stratosphere–troposphere coupling[6] or increased evaporation and densification in the North Atlantic[40]. In our experiments, we find little appreciable change in the AMOC relative to the anthropogenically forced decreasing trend in CTRL-8.5 (not shown). Both experiments show significant changes in tropical precipitation associated with shifts in the Intertropical Convergence Zone, with a southward shift in the Atlantic. These changes are of a pattern and magnitude that has been shown elsewhere[41] to be capable of inducing a compensating strengthening of the AMOC of the magnitude required to balance the effects of negative NAO forcing. However, detailed analysis of the mechanisms of AMOC response is beyond the scope of this paper.

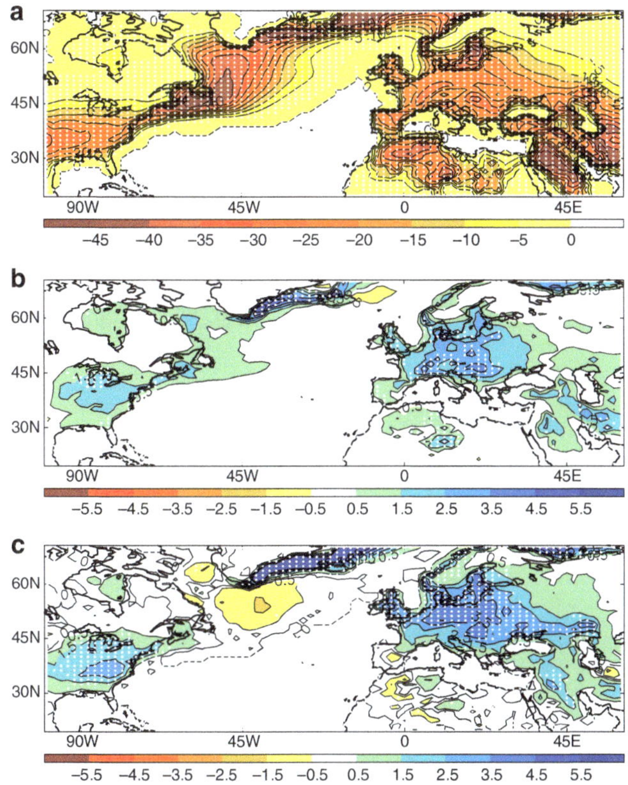

Figure 4 | Change in average number of frost days. Difference in winter (December to February) frost days (days) between (**a**) CTRL-8.5 (2050-2099) and the historical period (1971-2000), (**b**) EXPT-A and CTRL-8.5 (2050-2099) and (**c**) EXPT-B and CTRL-8.5 (2050-2099). Here, frost days correspond to a daily minimum surface air temperature <0 °C. White stippling indicates significance with a 95% confidence interval.

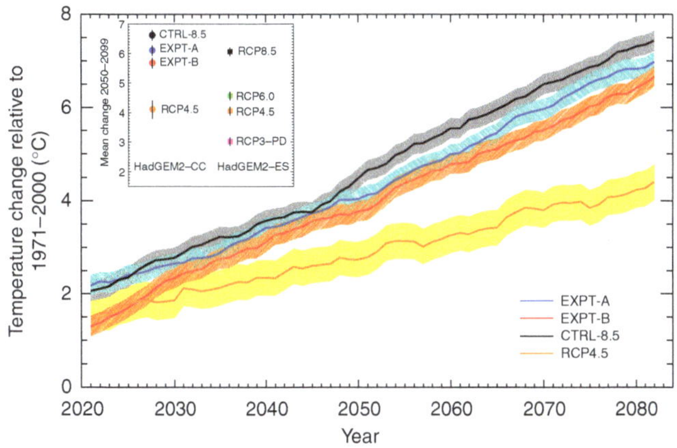

Figure 5 | Winter European surface temperature response. Mean winter (December to February) northern European (10° W-40° E, 48-75° N) near-surface temperature change (°C, land points only) for 2050-2099 for future scenarios relative to the historical simulation (1971-2000). Shading shows the standard error. The data has been smoothed by applying a 33-year running mean. The inset shows mean winter northern European near-surface temperature differences for 2050-2099 relative to 1971-2000 for the current experiments and for a range of RCP scenarios run with the closely related HadGEM2-ES model[30]. Whiskers show s.e.

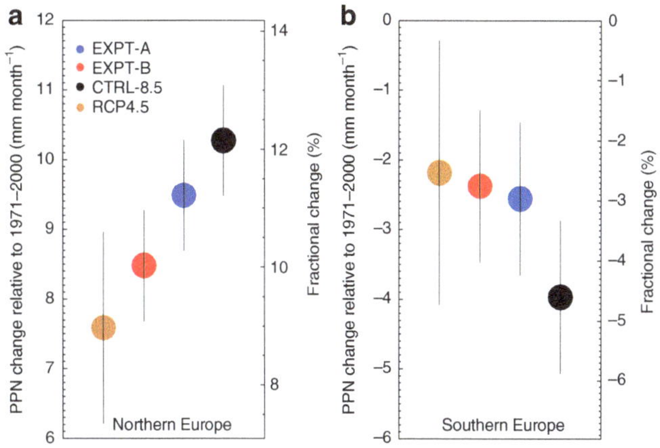

Figure 6 | Winter European precipitation response. Mean winter (December to February) European precipitation change (mm month^{-1}) for 2050-2099 for future scenarios relative to the historical simulation (1971-2000) for (**a**) northern Europe (10° W-30° E, 45-65° N) and (**b**) southern Europe (10° W-30° E, 35-45° N). Whiskers show the s.e. Differences between EXPT-B and CTRL-8.5 are significant with a 90% (75%) confidence interval for northern Europe (southern Europe).

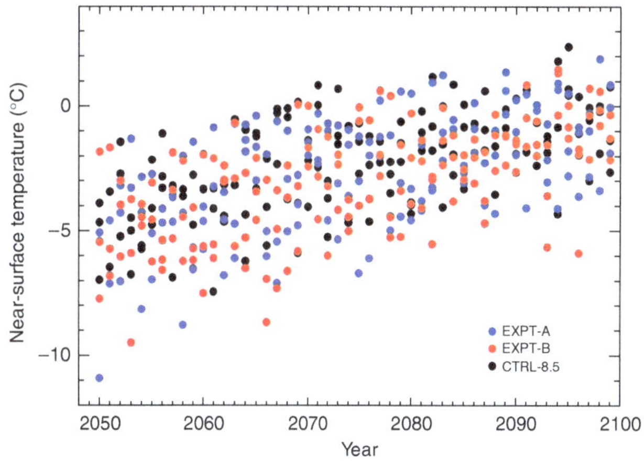

Figure 7 | Variability in modelled European winter temperatures. Winter (December to January) northern European (10° W-40° E, 48-75° N) near-surface temperatures (°C, land points only) for 2050 to 2099 for individual ensemble members of EXPT-A (blue), EXPT-B (red) and CTRL-8.5 (black).

Discussion

Numerous studies have identified links between past climate and solar variability[42,43]. During the Maunder Minimum (1645-1715), very few sunspots were seen despite regular observations[44]. If the past relationships between TSI and ultraviolet irradiance and sunspots are the same as are observed for modern solar variability, then a decline in both TSI and ultraviolet for this period can be assumed. The Maunder Minimum coincided with more severe winters in the UK and continental Europe[32] and many reconstructions[45,46] suggest atmospheric conditions were broadly comparable with the regional effects on European atmospheric circulation found here. Some modelling studies[13,47] also support the idea that similar regional cooling and circulation changes occurred during this period. On longer timescales, cosmogenic isotopes provide a proxy for TSI variations and only indirect meteorological information is available, nevertheless linkages can be made. For example, lake sediments have been analysed to demonstrate that a grand minimum of solar activity, the Homeric Minimum (~2,750–2,550 years before present), affected climate conditions through western Europe through altered regional circulation consistent with the negative phase of the NAO[48]. While these paleoclimatic studies appear largely consistent with our model result that weak solar ultraviolet irradiance can have a significant impact on European winter circulation through 'top-down' dynamical forcing, detailed modelling and attribution for these events is beyond the scope of this paper. We note that, to date, a variety of strengths of model response for past events have been obtained[13,47,49], which could, in part, be related to uncertainties in solar forcing and to the efficiency of the 'top-down' mechanism in different models[14].

We note that there are some uncertainties in our model response to solar forcing. The peak-to-trough variability in global mean temperature for the 11-year solar cycle is $\sim$0.06 °C but observational estimates suggest this may be too small. For example, a signal of $\sim$0.1 °C was obtained using a multivariate analysis with up-to-date estimates of natural (including volcanic aerosols) and anthropogenic forcing[50]. In general, climate models appear to underestimate this variability[5]. The modelled Atlantic atmospheric circulation response may also be larger if, as has recently been suggested, the ocean feedback is underestimated in the model[35].

The changes in TSI that are applied in our experiments are representative of current scientific understanding[31] and are smaller than those reported in earlier IPCC reports. The ultraviolet changes are designed to bracket the current uncertainty in spectral irradiance at these wavelengths, but we note that EXPT-B is likely to be at the high end of any estimate of ultraviolet change and the appreciable changes in climate that we find should therefore be viewed as a high estimate in this respect. This research highlights the need for accurate and long-term measurements of spectral solar irradiance to reduce current uncertainties.

In the solar scenario presented here, there is a gradual descent to Maunder Minimum-like conditions, which are then maintained for a period of 50 years until the end of the experiment. A modelling study which includes representation of the end of a solar minimum[7] indicates that global mean temperatures approach the reference model climatology within a few years of a return to the reference solar forcing. Although timescales may depend on the exact details of the forcing scenario and model, we would expect a reversal of the climate impacts described in our experiment following the end of the event. Estimates of the duration of grand solar minima, based on model reconstruction from ^{14}C data[51] and from ^{10}Be data[52], vary from a few decades to more than 100 years[53].

The latest data[1] (updated October 2013) on sunspot number, near-Earth interplanetary conditions, cosmic ray fluxes and geomagnetic activity are all consistent with the most rapid fall scenario used here[2]. All these data are in the upper 5 percentile of the distributions (lower 5 percentile in the case of cosmic rays which anti-correlate with solar activity) of predictions extended to these parameters[53]. Note also that these statistical predictions[51] are very similar to the predictions made using spectral techniques[54]. Thus all the current indicators are consistent with the most extreme decline discussed by Lockwood[2]. However, it should be remembered these predictions are statistical and not physical in nature, being based on past behaviour.

The northern hemisphere winter regional temperature and precipitation changes associated with the reduced solar forcing can amount to a substantial fraction of the difference between plausible RCP scenarios. Our results therefore suggest that uncertainties in natural forcing, in this case both the uncertainty in future solar output and uncertainty in spectral solar irradiance, should be taken into account alongside uncertainties in emissions in future climate forcing scenarios.

Methods

The control model. We use the Met Office Hadley Centre general circulation model, HadGEM2-CC (carbon cycle)[55], with historical and future forcings specified according to the Fifth Coupled Model Intercomparison Project (CMIP5) as used in HadGEM2-ES (Earth System)[30]. This version of HadGEM2 is a 'high top' model with 60 vertical levels and an upper boundary at 84 km, so is capable of resolving relevant stratospheric processes[56–58]. The horizontal resolution is 1.875° longitude by 1.25° latitude. The ocean resolution is 1° × 1°, increasing in the tropics to 0.3°, with 40 vertical levels. The TSI is partitioned across six shortwave spectral bands (0.2–10 μm) with spectral changes associated with changing TSI accounted for. For the historical period, the TSI and spectrally resolved irradiance (SSI) used is that recommended by CMIP5 (Lean, Calculations of solar irradiance, http://sparcsolaris.gfz-potsdam.de/cmip5.php, accessed 2 June 2009 from the earlier fuberlin site), with a repeating 11-year cycle, based on the period 1998–2008, imposed for future scenarios. Time-varying ozone distributions include a component related to solar variability[30].

Grand solar minimum scenario. The solar forcing is based on the most rapidly declining Lockwood[2] solar activity scenario and uncertainty in future levels of ultraviolet is explored with two experiments designed to bracket uncertainty in ultraviolet spectral irradiance. Experiments are initialized from 1 December 2005 with the same initial conditions as the parallel CTRL-8.5 ensemble. The derivation of the forcing is as follows:

EXPT-A. The future TSI for EXPT-A is constructed by relating changes in the historical TSI to the past solar modulation parameter, then applying this relationship to the future scenario for the parameter[1,2]. The SSIs are calculated by extrapolation of second-order polynomial fits of SSI versus TSI from the CMIP5 historical forcing data set. This experiment describes a relatively modest change in ultraviolet. The solar component in the prescribed ozone field also scales with the change in TSI.

EXPT-B. For EXPT-B, 200–320 nm data from the SIM instrument (2004–2009) are fitted to a typical solar cycle shape, giving an estimate for a peak-to-trough change in ultraviolet of 1.17 W m^{-2}. The SIM data used in this study is based on more reliable SORCE SIM Version 17 processing, but these results are consistent with the published version, Version 22, available at: http://lasp.colorado.edu/home/sorce/data/. Next, a scaling factor is derived between the Open Solar Flux (OSF) measure of solar activity[1,2] and ultraviolet. It is important to note that the use of the OSF proxy introduces a secular change in the past (and predicted future) ultraviolet irradiance that would not be obtained if other solar activity indices (such as the F10.7 radio flux) were used: however, ultraviolet-driven changes in the stratosphere are more highly correlated with OSF than indices such as F10.7 that do not show the same secular trends[59]. The scaling used here is based on calculating the average of peak-to-trough values in de-trended OSF data for the period 1982–2008 and taking this average to be equal to the SIM data estimate giving a relationship of 2.19 OSF units equalling 1 W m^{-2} of ultraviolet irradiance. The future projection in ultraviolet forcing is calculated by scaling the OSF for the grand solar minimum scenario, deriving a change in ultraviolet irradiance that would be consistent with the SIM data scaling linearly with the assumed future trend in OSF. The baseline period for normalizing the OSF with respect to the HadGEM2-CC historical forcing is 1964–2008. In EXPT-B, only the ultraviolet (200–320 nm) component of the shortwave irradiance is adjusted. Radiation bands 2 to 6 are unaltered, and so retain variability on the 11-year timescale out to 2100. This means that the reduction in TSI is by definition equal to the reduction in ultraviolet. In this experiment, the prescribed ozone is the same as that in CTRL-8.5. This enables direct comparison of the impact of UV changes with CTRL-8.5 and we believe it is justified as the large change in ultraviolet is likely to dominate the change in stratospheric heating in this case. We might expect inclusion of interactive ozone to enhance the response, but think this is likely to be of secondary importance for EXPT-B. However, we note that GCM studies have shown sensitivity to the magnitude and distribution of specified ozone changes[22,60].

Ensemble sizes and experiments used for comparison. HadGEM2-CC: CTRL-8.5, EXPT-B and EXPT-A have three ensemble members, while RCP4.5 has just a single member. The historical runs have three ensemble members.

HadGEM2-ES: All RCPs and historical runs have four ensemble members. All the results are for ensemble means, unless otherwise stated.

Statistics. Model statistical significance is tested using the Mann–Whitney test (a nonparametric test) with the null hypothesis that the difference in means are not significantly different from zero. One-tailed tests are used, given the prior evidence for a negative AO/NAO response[8,10].

Code availability. Due to intellectual property right restrictions, we cannot provide either the source code or the documentation papers for HadGEM2. The Met Office Unified Model (MetUM) is available for use under licence. A number of research organizations and national meteorological services use the MetUM in collaboration with the Met Office to undertake basic atmospheric process research, produce forecasts, develop the MetUM code and build and evaluate Earth system models. For further information on how to apply for a licence, see http://www.metoffice.gov.uk/research/collaboration/um-collaboration.

References

1. Lockwood, M. Reconstruction and prediction of variations in the open solar magnetic flux and interplanetary conditions. *Living Rev. Solar Phys.* **10,** 4 (2013).
2. Lockwood, M. Solar change and climate: an update in the light of the current exceptional solar minimum. *Proc. R. Soc. A* **466,** 303–329 (2010).
3. Feulner, G. & Rahmstorf, S. On the effect of a new grand minimum of solar activity on the future climate on Earth. *Geophys. Res. Lett.* **37,** L05707 (2010).
4. Rozanov, E. V., Egorova, T. A., Shapiro, A. I. & Schmutz, W. K. Modeling of the atmospheric response to a strong decrease of the solar activity. *Proc. Int. Astron. Union* **7,** 215–224 (2011).
5. Jones, G. S., Lockwood, M. & Stott, P. A. What influence will future solar activity changes over the 21st century have on projected global near-surface temperature changes? *J. Geophys. Res.* **117,** D05103 (2012).
6. Anet, J. G. *et al.* Impact of a potential 21st century "grand solar minimum" on surface temperatures and stratospheric ozone. *Geophys. Res. Lett.* **40,** 4420–4425 (2013).
7. Meehl, G. A., Arblaster, J. M. & Marsh, D. R. Could a future "Grand Solar Minimum" like the Maunder Minimum stop global warming? *Geophys. Res. Lett.* **40,** 1789–1793 (2013).
8. Kuroda, Y. & Kodera, K. Effect of solar activity on the polar-night jet oscillation in the Northern and Southern Hemisphere winter. *J. Meteorol. Soc. Jpn* **80,** 973–984 (2002).
9. Matthes, K., Kuroda, Y., Kodera, K. & Langematz, U. Transfer of the solar signal from the stratosphere to the troposphere: Northern winter. *J. Geophys. Res.* **111,** D06108 (2006).
10. Gray, L. J. *et al.* A lagged response to the 11 year solar cycle in observed winter Atlantic/European weather patterns. *J. Geophys. Res. Atmos.* **118,** 13,405–13,420 (2013).
11. Woollings, T., Lockwood, M., Masato, G., Bell, C. & Gray, L. Enhanced signature of solar variability in Eurasian winter climate. *Geophys. Res. Lett.* **37,** L20805 (2010).
12. Ineson, S. *et al.* Solar forcing of winter climate variability in the northern hemisphere. *Nat. Geosci.* **4,** 753–757 (2011).
13. Shindell, D. T., Schmidt, G. A., Mann, M. E., Rind, D. & Waple, A. Solar forcing of regional climate change during the Maunder Minimum. *Science* **294,** 2149–2152 (2001).
14. Gerber, E. P. *et al.* Assessing and understanding the impact of stratospheric dynamics and variability on the earth system. *Bull. Am. Meteorol. Soc.* **93,** 845–859 (2012).
15. Simpson, I. R., Blackburn, M. & Haigh, J. D. The role of eddies in driving the tropospheric response to stratospheric heating perturbations. *J. Atmos. Sci.* **66,** 1347–1365 (2009).
16. Ermolli, I. *et al.* Recent variability of the solar spectral irradiance and its impact on climate modelling. *Atmos. Chem. Phys.* **13,** 3945–3977 (2013).
17. Harder, J. W., Fontenla, J. M., Pilewskie, P., Richard, E. C. & Woods, T. N. Trends in solar spectral irradiance variability in the visible and infrared. *Geophys. Res. Lett.* **36,** L07801 (2009).
18. Haigh, J. D., Winning, A., Toumi, R. & Harder, J. W. An influence of solar spectral variations on radiative forcing of climate. *Nature* **467,** 696–699 (2010).
19. Lean, J. Evolution of the Sun's Spectral Irradiance since the Maunder Minimum. *Geophys. Res. Lett.* **27,** 2425–2428 (2000).
20. Maycock, A. C. *et al.* Possible impacts of a future Grand Solar Minimum on climate: stratospheric and global circulation changes. *J. Geophys. Res. Atmos.* (http://onlinelibrary.wiley.com/doi/10.1002/2014JD022022/full).
21. Roy, I. & Haigh, J. D. Solar cycle signals in sea level pressure and sea surface temperature. *Atmos. Chem. Phys.* **10,** 3147–3153 (2010).
22. Hood, L., Schimanke, S., Spangehl, T., Bal, S. & Cubasch, U. The surface climate response to 11-yr solar forcing during northern winter: observational analyses and comparisons with GCM simulations. *J. Clim.* **26,** 7489–7506 (2013).
23. Morgenstern, O. H. *et al.* Anthropogenic forcing of the northern annular mode in CCMVal2 models. *J. Geophys. Res.* **115,** D00M03 (2010).
24. Scaife, A. A. *et al.* Climate change projections and stratosphere-troposphere interaction. *Clim. Dyn.* **38,** 2089–2097 (2012).
25. Giorgi, F. & Francisco, R. Uncertainties in regional climate change predictions. A regional analysis of ensemble simulations with the HADCM2 GCM. *Clim. Dyn.* **16,** 169–182 (2000).
26. Jung, T. & Hilmer, M. The link between the North Atlantic Oscillation and Arctic Sea ice export through Fram Strait. *J. Clim.* **14,** 3932–3943 (2001).
27. Strong, C. & Magnusdottir, G. Modeled winter sea ice variability and the North Atlantic Oscillation: a multi-century perspective. *Clim. Dyn.* **34,** 515–525 (2010).
28. Thompson, D. W. J. & Wallace, J. M. *The Arctic Oscillation signature in the wintertime geopotential height and temperature fieldsGeophys. Res. Lett.* **25,** *1297–1300 (1998).*
29. Scaife, A. A., Folland, C. K., Alexander, L. V., Moberg, A. & Knight, J. R. European climate extremes and the North Atlantic Oscillation. *J. Clim.* **21,** 72–83 (2008).
30. Jones, C. D. *et al.* The HadGEM2-ES implementation of CMIP5 centennial simulations. *Geosci. Model Dev.* **4,** 543–570 (2011).
31. IPCC. *Climate Change 2013: The Physical Science Basis.* (eds Stocker, T. *et al.*) (Cambridge Univ. Press, 2013).
32. Lockwood, M., Harrison, R. G., Woollings, T. & Solanki, S. K. Are cold winters in Europe associated with low solar activity? *Environ. Res. Lett.* **5,** 024001 (2010).
33. Luterbacher, J., Dietrich, D., Xoplaki, E., Grosjean, M. & Wanner, H. European seasonal and annual temperature variability, trends, and extremes since 1500. *Science* **303,** 1499–1503 (2004).
34. Moffa-Sánchez, P., Born, A., Hall, I. R., Thornalley, D. J. R. & Barker, S. Solar forcing of North Atlantic surface temperature and salinity over the past millennium. *Nat. Geosci.* **7,** 275–278 (2014).
35. Scaife, A. A. *et al.* A mechanism for lagged North Atlantic climate response to solar variability. *Geophys. Res. Lett.* **40,** 434–439 (2013).
36. Eden, C. & Willebrand, J. Mechanism of Interannual to Decadal Variability of the North Atlantic Circulation. *J. Clim.* **14,** 2266–2280 (2001).
37. Miller, G. H. *et al.* Abrupt onset of the Little Ice Age triggered by volcanism and sustained by sea-ice/ocean feedbacks. *Geophys. Res. Lett.* **39,** L02708 (2012).
38. Lehner, F., Born, A., Raible, C. C. & Stocker, T. F. Amplified inception of European Little Ice Age by sea ice-ocean-atmosphere feedbacks. *J. Clim.* **26,** 7586–7602 (2013).
39. Schleussner, C. F. & Feulner, G. A volcanically triggered regime shift in the subpolar North Atlantic Ocean as a possible origin of the Little Ice Age. *Clim. Past* **9,** 1321–1330 (2013).
40. Menary, M. B. & Scaife, A. A. Naturally forced multidecadal variability of the Atlantic meridional overturning circulation. *Clim. Dyn.* **42,** 1347–1362 (2014).
41. Vellinga, M., Wood, R. A. & Gregory, J. M. Processes governing the recovery of a perturbed thermohaline circulation in HadCM3. *J. Clim.* **15,** 764–780 (2002).
42. Cubasch, U. *et al.* Simulation of the role of solar and orbital forcing on climate. *Adv. Space Res.* **37,** 1629–1634 (2006).
43. Gray, L. J. *et al.* Solar influences on climate. *Rev. Geophys.* **48,** RG4001 (2010).
44. Eddy, J. A. The Maunder Minimum. *Science* **192,** 1189–1202 (1976).
45. Luterbacher, J. *et al.* The late Maunder Minimum (1675-1715) - A key period for studying decadal scale climatic change in Europe. *Clim. Change* **49,** 441–462 (2001).
46. Pinto, J. G. & Raible, C. C. Past and recent changes in the North Atlantic oscillation. *WIREs Clim. Change* **3,** 79–90 (2012).
47. Spangehl, T. *et al.* Transient climate simulations from the Maunder Minimum to present day: role of the stratosphere. *J. Geophys. Res.* **115,** D00I10 (2010).
48. Martin-Puertas, C. *et al.* Regional atmospheric circulation shifts induced by a grand solar minimum. *Nat. Geosci.* **5,** 397–401 (2012).
49. Anet, J. G. *et al.* Impact of solar versus volcanic activity variations on tropospheric temperatures and precipitation during the Dalton Minimum. *Clim. Past* **10,** 921–938 (2014).
50. Lean, J. L. & Rind, D. H. How natural and anthropogenic influences alter global and regional surface temperatures: 1889 to 2006. *Geophys. Res. Lett.* **35** (2008).
51. Usoskin, I. G., Solanki, S. K. & Kovaltsov, G. A. Grand minima and maxima of solar activity: new observational constraints. *Astron. Astrophys.* **471,** 301–309 (2007).
52. Steinhilber, F., Abreu, J. A. & Beer, J. Solar modulation during the Holocene. *Astrophys. Space Sci. Trans.* **4,** 1–6 (2008).
53. Barnard, L. *et al.* Predicting space climate change. *Geophys. Res. Lett.* **38,** L16103 (2011).
54. Steinhilber, F. & Beer, J. Prediction of solar activity for the next 500 years. *J. Geophys. Res. Space Phys.* **118,** 1861–1867 (2013).
55. Martin, G. M. *et al.* The HadGEM2 family of Met Office Unified Model climate configurations. *Geosci. Model Dev.* **4,** 723–757 (2011).
56. Hardiman, S. C., Butchart, N., Hinton, T. J., Osprey, S. M. & Gray, L. J. The effect of a well-resolved stratosphere on surface climate: differences between CMIP5 simulations with high and low top versions of the Met Office Climate Model. *J. Clim.* **25,** 7083–7099 (2012).
57. Mitchell, D. M. *et al.* The effect of climate change on the variability of the Northern Hemisphere stratospheric polar vortex. *J. Atmos. Sci.* **69,** 2608–2618 (2012).
58. Osprey, S. M., Gray, L. J., Hardiman, S. C., Butchart, N. & Hinton, T. J. Stratospheric variability in twentieth-century CMIP5 simulations of the Met Office climate model: high top versus low top. *J. Clim.* **26,** 1595–1606 (2013).
59. Lockwood, M. *et al.* Top-down solar modulation of climate: evidence for centennial-scale change. *Environ. Res. Lett.* **5,** 034008 (2010).
60. Haigh, J. D. A. GCM study of climate change in response to the 11-year solar cycle. *Quart. J. Roy. Meteorol. Soc.* **125,** 871–892 (1999).

Acknowledgements
This work was supported by the Joint DECC/Defra Met Office Hadley Centre Climate Programme (GA01101) and also by the EU project SPECS funded by the European Commission's Seventh Framework Research Programme under the grant agreement 308378 (Met Office Hadley Centre authors), by the NERC National Centre for Atmospheric Science (NCAS) Climate directorate (L.J.G. and A.C.M.), an ERC ACCI grant (A.C.M) and an AXA Postdoctoral Fellowship (A.C.M.). We thank Neal Butchart, Steven Hardiman, Tim Hinton, Dan Mitchell and Scott Osprey for their work in developing the 'high top' version of HadGEM2 and support in running the simulations, and Neal Butchart and Gareth Jones for helpful discussions on the manuscript.

Author contributions
A.A.S., L.J.G. and S.I. planned the experiments. J.W.H. provided the SIM data, M.L. supplied the TSI/OSF scenario, and N.J.D. and J.C.M. adapted the data for use in the model. S.I. and A.C.M. ran the experiments and analysed the results. J.R.K. advised on statistical methods. All the authors contributed to discussing the results and writing the paper.

Additional information

Reprints and permission information is available online at http://npg.nature.com/reprintsandpermissions/

6

Aerosol generation by raindrop impact on soil

Young Soo Joung[1] & Cullen R. Buie[1]

Aerosols are investigated because of their significant impact on the environment and human health. To date, windblown dust and sea salt from sea spray through bursting bubbles have been considered the chief mechanisms of environmental aerosol dispersion. Here we investigate aerosol generation from droplets hitting wettable porous surfaces including various classifications of soil. We demonstrate that droplets can release aerosols when they influence porous surfaces, and these aerosols can deliver elements of the porous medium to the environment. Experiments on various porous media including soil and engineering materials reveal that knowledge of the surface properties and impact conditions can be used to predict when frenzied aerosol generation will occur. This study highlights new phenomena associated with droplets on porous media that could have implications for the investigation of aerosol generation in the environment.

[1]Department of Mechanical Engineering, Massachusetts Institute of Technology, 77 Massachusetts Avenue, Cambridge, Massachusetts 02139, USA. Correspondence and requests for materials should be addressed to Y.S.J. (email: ysjoung@mit.edu) or to C.R.B. (crb@mit.edu).

Aerosols, which are tiny liquid droplets or solid particles suspended in a gas[1–3], have been investigated because of their significant impact on the environment, human health and industrial applications[4–14]. Aerosol particles, larger than 1 μm, are known to originate from windblown dust and sea salt from sea spray resulting from bursting bubbles[6,15–20]. In addition, environmental aerosols are generated from both natural sources and anthropogenic activities such as volcanoes and combustion of fuels, respectively[3]. However, to the best of our knowledge, aerosol generation from raindrops hitting soil has not been reported and the origins of atmospheric bioaerosols containing elements of soil and environmental microorganisms remain illusive[1,8,21].

Aerosols generated from droplets hitting porous media can have significant impact on industrial applications as well as the environment[22–24]. There are several methods to generate aerosols such as electrospray, ultrasound and sprays; however, most methods use external energy sources to generate aerosol droplets. Because drop impingement-based aerosol generation can be employed using gravitational potential, it is promising for low-cost biological and environmental applications. Furthermore, this work could have widespread implications ranging from remediating the spread of disease-causing microbes to the earthy smell known as 'petrichor' present after a rain shower on a hot day[25].

In this work we provide evidence that rainfall on soil can also generate aerosols. High-speed imaging provides visual proof of aerosol generation when liquid water droplets hit soil at velocities consistent with rainfall. Within a specified range of impact velocities, we observe frenetic bubble generation and ejection of tiny droplets, producing aerosol above the surface (Supplementary Movie 1). We can predict when the frenetic aerosol generation occurs from the surface properties and impact conditions. This work demonstrates that aerosols can be generated on porous surfaces including soil when impinged by a liquid droplet.

Results

Aerosol generation from droplets hitting soils. Using different impact velocities, we observed drop impingement on soils using a high-speed camera (Fig. 1a–c; Supplementary Movie 2). When small water droplets, similar in size to raindrops, impinge on clay, sandy clay and clay loam[26], aerosol generation is observed. As reported in ref. 20, the diameter of a typical raindrop is 1–3 mm. The shape of a raindrop is not always spherical[27–29] because of the deformation in flight, but an assumed spherical shape is not unreasonable[30]. When a droplet hits the clay loam soil surface at a velocity consistent with light rainfall[31] (Fig. 1a), trapped gas bubbles appear inside the droplet (Fig. 1b). When the bubbles burst, multiple tiny jets of order tens of micrometres in diameter are ejected from the soil (Fig. 1c). It is believed that the bubbles pinned on the soil surface break when the bubbles reach the upper surface of the impinging droplet, resulting in bubble-bursting and liquid jets, a well-known occurrence during rainfall on the oceans, carbonated beverages and other problems in fluid dynamics[32–37].

Observation of aerosol generation on porous media. We first observed this phenomenon during drop impingement of liquid water on thin layer chromatography (TLC) plates. TLC plates can be considered ideal soil-like surfaces; they have similar water-absorption properties but they are topologically and chemically homogenous (Table 1). For example, when a water droplet at a speed of 0.4 $\mathrm{ms^{-1}}$ hits a TLC plate with similar water-absorption properties as clay, a large bubble was trapped inside the droplet (Fig. 1d). When the bubble bursts (Fig. 1e), a thin water jet emerges (Fig. 1f). Then, the jet breaks up into small droplets, which remain suspended in air (Fig. 1g) (Supplementary Movie 3). Frenetic water jets (aerosol dispersion) are observed in a specific range of impact velocities and surface properties, as shown in Fig. 1h (Supplementary Movie 4). In this case, hundreds of water jets were ejected from the droplet at speeds of $\sim 10\,\mathrm{ms^{-1}}$ in roughly 20 μs.

Delivery of materials in porous media with aerosols. From the environmental perspective, an aerosol refers to tiny suspensions that possess chemicals or solid particles[3]. Therefore, it is important to verify that materials within the porous media are entrained into the aerosols generated from droplets hitting porous media. To confirm that materials are transported with the aerosol, we conducted drop impingement on porous media with Rhodamine B on the pores. If the dye is suspended in the aerosol droplets, this can be observed under ultraviolet illumination. In a duct with a low-speed air flow, aerosols were generated from water (deionized water) droplets hitting porous surfaces permeated with Rhodamine B. Transparent glass is located above the porous media to collect the aerosol droplets generated from the porous media (Supplementary Fig. 1). The number of fluorescent spots on the transparent glass was measured with respect to distance from the location of aerosol generation. With increasing air-flow velocity, the mean of the particle distribution shifted away from the position of aerosol generation (Supplementary Fig. 2). This result shows that particulate matter within the porous surface can be convectively dispersed as aerosols generated from drop impingement.

Aerosol-generation process. Aerosol generation after drop impingement on porous media is a three-step process consisting of bubble formation, bubble growth and bubble bursting, as shown in Fig. 2. We performed experiments with different impact velocities and surfaces in order to elucidate the aerosol-generation mechanism (Supplementary Movies 5–7). High-speed images shown in Fig. 2a–c reveal the aerosol-generation process shown schematically in Fig. 2d–g. When a droplet hits the surface, the droplet flattens because of its kinetic energy. Tiny bubbles are trapped at the droplet/surface interface because the droplet expansion speed is faster than the water-absorption speed of the surface (Fig. 2d). It is known that a thin air film, with thickness 1 μm, forms on a flat surface when surface roughness is less than 2 nm (root-mean-square, RMS), for Weber number (We) from 700 to 900 and Re from 1,600 to 5,800 (refs 38,39). Subsequently, the air film evolves into a single bubble. On rough surfaces, the air film can be fragmented into small bubbles because of the irregularity of the surface. The period of the air film retraction is fast, given it's speed of roughly 5 $\mathrm{ms^{-1}}$ (ref. 39). The time required to form small air bubbles inside the droplet is shorter than hundreds of microseconds[39]. Compared with the bubble formation, the speed of droplet absorption into porous media is much slower because the porous media and soils, used in this work, have hydraulic diffusivities lower than $1.5 \times 10^{-4}\,\mathrm{m^2\,s^{-1}}$ (Table 1). A thin air film forms on rough surfaces such as soils during impact, and then fragments into small bubbles because of the irregularity of the porous surfaces during the air film retraction. The bubbles grow with air supplied from the porous media as it absorbs the liquid and displaces trapped air (Supplementary Movies 5 and 6). After the droplet reaches its maximum radius, the droplet oscillates, relieving its compressive pressure. At this time, bubbles grow by receiving air escaping from the pores of the surface as water is absorbed (Fig. 2e). The droplet height above the surface is reduced as the liquid enters the

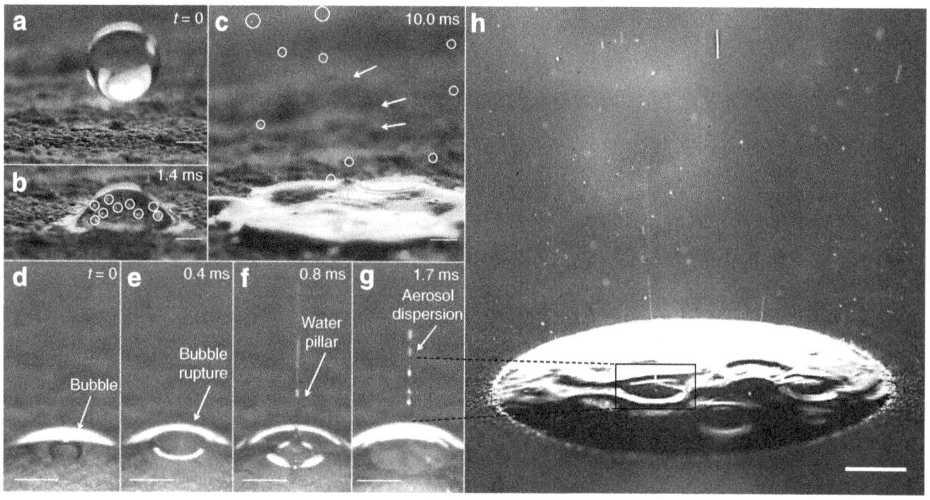

Figure 1 | Aerosol generation from droplets hitting soils and porous surfaces. (**a**) Clay loam with a rough surface before impingement with a droplet travelling at a speed of 2 ms^{-1}. (**b**) Tiny bubbles form under the droplet after impact (delineated by white circles in the image). (**c**) Tiny water jets are ejected from the droplet after impact. Scale bars in **a-c**, 1 mm. The white circles and arrows in the image highlight aerosols and jets ejected from the droplet. (**d-g**) High-speed images of the aerosol-generation process when a single bubble breaks inside the droplet. Scale bars, 1 mm. (**d**) A bubble is trapped inside the droplet. (**e**) The bubble breaks at the droplet/air interface. (**f**) A water jet emanates from the droplet. (**g**) The jet breaks up into tiny droplets. (**h**) Frenetic water jets and associated aerosols are generated on a thin layer chromatography plate after impact with a droplet travelling at 1.4 ms^{-1}. Scale bar, 1 mm. In the image, the white dots and lines are micron-scale water droplets ejected from the droplet on the surface. See Supplementary Movies 2-4.

porous media (Fig. 2f). Finally, the bubbles burst when they meet the air-water interface of the droplet (Fig. 2g), ejecting small jets that breakup into tiny droplets.

To generate aerosols from drop impingement on porous media, bubbles must form inside the initial large droplet after impact. In this work we consider droplets hitting dry soil, which is both wettable and firm enough to maintain its porous structure during the impingement. When droplets have an impact on powder surfaces or granular layers[40–44], we did not observe bubble formation during the drop impingement. We speculate that if the porous media deform considerably during impact there are low permeability paths for air to escape from beneath the droplet and so it is not trapped. However, more investigation is needed to justify this claim and illuminate the relationship between aerosol generation and mechanical strength of porous media.

Effects of impact conditions on aerosol generation. From the relationship between the kinetic energy and the surface energy of the droplet, we can estimate the size and number of bubbles after impact. It is known that jet diameter and velocity are strongly correlated with bubble size[15,34,45]. Previously, we have shown using energy conservation that the maximum diameter of a droplet after impact is linearly proportional to the impact velocity[46]. Therefore, the minimum droplet height (the droplet film thickness when the droplet reaches its maximum diameter after impact, Fig. 2d) is inversely proportional to the square of impact velocity as follows: $U_o^2 \sim d_{max}^2 \sim h_{min}^{-1}$, where U_o is the impact velocity, d_{max} is the maximum droplet diameter and h_{min} is the minimum droplet film thickness (Fig. 3a). Therefore, impact velocity governs the maximum bubble size inside the droplet because bubble size cannot exceed the droplet film thickness.

The number of bubbles and their diameter can be measured from the high-speed images (Fig. 3a,b). We find that the number of bubbles is proportional to the wetted area of the surface, which can be calculated from the maximum diameter of the spreading droplet. The relationship between the maximum number of bubbles and the impact velocity is as follows: $U_o^2 \sim A_{max} \sim N_{max}$, where A_{max} is the maximum wetted area of the surface, and N_{max} is the maximum number of bubbles formed inside the droplet (Fig. 3b). We employ the Weber number, $We = \rho d_o U_o^2 \gamma^{-1}$, to evaluate the effect of impact velocity on the number of bubbles and the bubble size, where ρ is the liquid density, d_o is the initial droplet diameter and γ is the liquid surface tension.

Effects of wetting properties on aerosol generation. The droplet permeation and the initial aerosol dispersion are affected by surface properties, mainly the wettability. The water-absorption speed of porous media can be characterized by the hydraulic diffusivity, $D_{cap} = r_c \gamma \cos\theta \mu^{-1}$, and r_c is the effective capillary radius, θ is the native contact angle of surface and μ is dynamic viscosity of the liquid[47]. If we assume that dry soil becomes fully wet after capillary suction (for example, there are no impenetrable pores), the hydraulic diffusivity can be found using the equation for capillary rise in the absence of gravitational effects: $H^2 \sim D_{cap} t$, where H is the rise height and t is time[48–50]. The hydraulic diffusivity can be experimentally estimated from capillary rise measurements[50]. We now define a modified Péclet number ($Pe = (U_o d_o) D_{cap}^{-1}$). The modified Péclet number is the ratio of the advective transport rate ($U_o d_o$) to the hydraulic diffusivity (D_{cap}) and used to represent the effects of surface wettability.

Figure 3c,d shows the effects of surface wettability and impact velocity on aerosol generation. In Fig. 3c, jetting initiation time is the time it takes to observe the first aerosol dispersion after the droplet reaches the maximum diameter after impact. The jetting initiation time is mainly governed by the wetting properties of the surface at low impact velocity (Fig. 3c). However, the droplet film thickness significantly affects the jetting initiation time at high impact velocities. TLC-A and TLC-C have the lowest and the highest hydraulic diffusivities, respectively. Therefore, the initial jetting occurs on TLC-C first and on TLC-A last at low impact velocities (less than 1.4 ms^{-1}). At higher impact velocities (higher than 2.0 ms^{-1}), however, the droplet height is the highest on TLC-B, as shown in Fig. 3d; therefore, the jetting initiation time is highest on TLC-B.

Table 1 | Characteristics and wetting properties of the media used to examine aerosol generation.

Sample	Layer thickness (μm)	Particle size (μm)	Surface roughness (RMS, μm)	Liquid	Hydraulic diffusivity ($mm^2 s^{-1}$)
Clay loam soil	>104	<500	16.9	Water	4.80
Loam soil	>104	<300	27.2	Water	3.11
Sandy soil	>104	300-600	35.4	Water	127.64
Clay soil	>104	<100	9.4	Water	2.61
Muck	>104	<500	62.1	Water	0.05
Sandy loam soil	>104	200-500	54.7	Water	24.58
Silt loam soil	>104	<400	26.8	Water	4.53
Peat	>104	<500	20.7	Water	0.13
Soil A (clay loam)	>104	<600	22.9	Water	4.58
Soil B (peat)	>104	<1,000	39.2	Water	0.04
Soil C (loam)	>104	<1,500	31.0	Water	4.89
Soil D (sandy loam)	>104	<1,000	46.2	Water	15.68
Soil E (sandy clay)	>104	<300	11.1	Water	12.01
Soil F (sand gravel)	>104	100-5,000	47.0	Water	32.83
Soil G (silt loam)	>104	<500	23.6	Water	4.20
Soil H (beach sand)	>104	100-400	30.3	Water	68.98
TLC-A (silica)	250	10	1.7	Water	9.5
TLC-B (cellulose)	100	2-20	5.0	Water	20.4
TLC-C (aluminium)	250	11-14	2.3	Water	22.4
TLC-D (silica)	200	25	1.6	Water	24.0
TLC-E (cellulose)	250	50	10	Water	37.0
TLC-A (silica)	250	10	1.7	Ethanol	2.3
TLC-B (cellulose)	100	2-20	5.0	Ethanol	14.7
TLC-C (aluminium)	250	11-14	2.3	Ethanol	4.7
TLC-D (silica)	200	25	1.6	Ethanol	4.7
TLC-E (cellulose)	250	50	10	Ethanol	26.4
TLC-A (silica)	250	10	1.7	Ethylene glycol	0.5
TLC-B (cellulose)	100	2-20	5.0	Ethylene glycol	2.2
TLC-C (aluminium)	250	11-14	2.3	Ethylene glycol	1.1
TLC-D (silica)	200	25	1.6	Ethylene glycol	1.2
TLC-E (cellulose)	250	50	10	Ethylene glycol	2.9
PTL-A	190	50-100	18.1	Water	85.6
PTL-B	330	20-200	11.00	Water	50.1
PTL-C	80	20-100	13.60	Water	52.4
PTL-D	150	<400	25.20	Water	144.6
PTL-E	140	10-200	16.4	Water	117.8
PTL-F	240	<600	16.5	Water	88.3
PTL-G	80	20-50	14.8	Ethanol	38.6
PTL-G	80	20-50	14.8	Ethylene glycol	21.0
PTL-G	80	20-50	14.8	Glycerol	0.4

PTL, porous titania layer; RMS, root-mean-square; TLC, thin layer chromatography.

Effect of porous structures on aerosol generation. Repeated drop impingements, which can fill up the pores with water, were conducted to examine the effect of porous structures. First, we periodically released water droplets within a period of 10 s. In this case, aerosol generation was not observed after the second drop impingement. However, when the period was 5 min, aerosol generation was observed in every drop impingement. These experiments show that bubbles are formed at the droplet/surface interface due to the inflation with air stored in pores of the surfaces. If the pores are filled with liquids, bubbles cannot form, resulting in no aerosol generation after impact. However, after the pores are dry, bubbles can be trapped again at the droplet/surface interface and generate aerosol when the bubbles break. It has been reported that bubbles can be generated and entrained in a droplet within the drop impact on solid and liquid[51,52]. However, the result of repeated drop impingements reveals that thin air films trapped at the droplet/solid or droplet/liquid interface is not the main reason of aerosol generation from droplets hitting porous surfaces. Inflation of air from porous structures has the important role of aerosol generation.

From Table 1, surface roughness of the porous media and soils used in this work ranges from 1 to 60 μm. Interestingly, aerosol generation was observed for surface roughness (RMS) ranging from 1 to 10 μm. This range is slightly higher than the air film thickness (1 μm) formed on solid surfaces during impact[38,53]. From this information we can speculate as to why aerosol is not generated on highly roughened surfaces. First, thin air films cannot form on surfaces with high surface roughness exceeding roughly 10 μm; therefore, small bubbles are not trapped on the surfaces during the air film retraction. Second, higher surface roughness has a tendency to increase the effective pore radius of the porous media, resulting in higher hydraulic diffusivity[50]. Indeed, air bubbles were not observed on porous media with hydraulic diffusivity and surface roughness higher than 50 mm^2s^{-1} and 20 μm, respectively.

Characterization of aerosol generation. On the basis of the drop impingement results, we found a domain that resulted in aerosol generation. We employ the We and a modified Péclet number (Pe) parameters in a two-dimensional (2D) map of our experimental conditions. We observed drop impingements on various types of porous media, including soils, with different liquids and impact velocities. In total we conducted roughly 600 experiments.

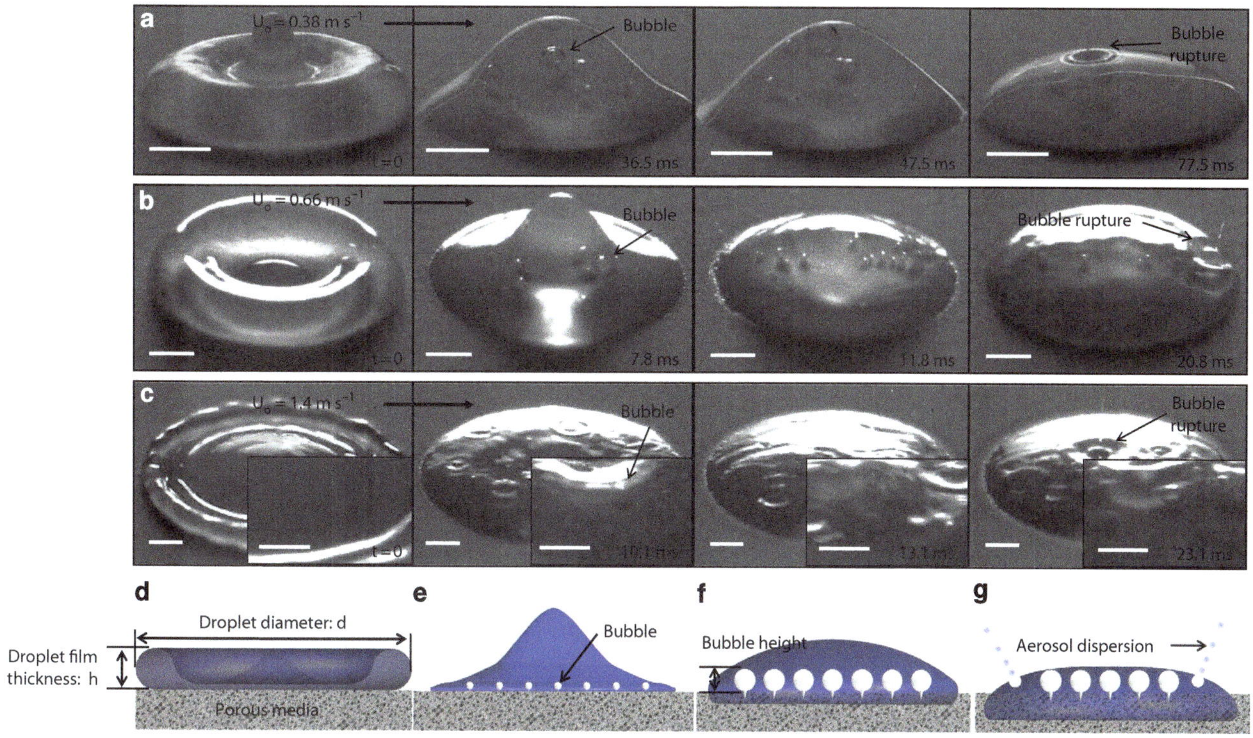

Figure 2 | Three-step process of aerosol generation after drop impingement on porous media. (**a-c**) High-speed images of bubble formation inside water droplets on a thin layer chromatography plate, composed of aluminium oxide particles (11-14 μm), with respect to impact velocity (U_o). Scale bars, 1 mm. (**d-g**) Schematic illustration of the aerosol-generation process. (**d**) A droplet influences a porous surface and expands. (**e**) After the droplet reaches maximum diameter, tiny bubbles are entrained at the interface of the droplet and the porous surface. (**f**) The droplet height decreases as water absorbs into the porous surface, while the bubbles grow due to air trapped in the porous surface. (**g**) Bubbles meet the air/water interface and rupture, generating tiny water jets. The number of bubbles created inside the droplet increases with impact velocity. See Supplementary Movies 5-7.

Experiments were performed on 16 different soils (eight commercial soils purchased from World's Science, and other soils collected from the Boston metropolitan area), five different TLC plates (Sigma-Aldrich, micro- and nanoporous surfaces: silica, cellulose and aluminium oxide with different pore sizes) and seven porous titania films fabricated in-house[50] (microporous surfaces fabricated with different conditions) with four different liquids (water, ethanol, ethylene glycol and glycerol; Table 1).

Figure 4 summarizes the observations with respect to Pe and We. We and Pe were varied from 1 to 10^4 and from 1 to 10^6, respectively. Aerosol generation is observed in a specific region of We and Pe, as shown in Fig. 4. There was no aerosol generation in the low and high We regions (We < 10 or We > 10^3). When We > 10^3, we do not observe aerosol generation because splashing in radial direction (Fig. 4a) leads to air entrainment. By contrast, when We < 10, few bubbles are entrained and they are much smaller than the droplet height, resulting in delayed bubble bursting and reabsorbed bubbles into the surface. In addition, the surfaces did not show aerosol generation at high or low hydraulic diffusivity (Pe < 10 or Pe > 10^4). When Pe > 10^4, bubbles cannot become sufficiently large to generate aerosols during droplet spreading because the droplet does not permeate into the surface. Conversely, if Pe < 10, the droplet quickly absorbs into the porous surface, faster than the droplet expands. Therefore, air is not trapped at the interface of the droplet and the surface, and there is no aerosol generation.

Discussion

Frenetic aerosol generation is observed at the core of the aerosol-generation region on the characterization map (c.f. yellow symbols in Fig. 4). To achieve frenetic aerosol generation, bubbles must not only reach sufficient size to disperse aerosol but also become similar in size to the droplet film thickness in order to break simultaneously when the bubbles meet the top surface of the droplet. Therefore, moderate Pe ($10^2 < \text{Pe} < 10^3$) is necessary for frenzied aerosol generation since this balances bubble size (favouring high Pe) while minimizing the time it takes for the initial droplet to spread and the bubbles to burst (favouring low Pe). According to the study by Lee *et al.*[34], bubbles must have a minimum size to generate liquid jets when they burst. The average bubble height and the average droplet height become similar within a specific range of We (Fig. 3a). Frenetic aerosol generation occurs when the bubble diameter is very close to the droplet height, leading multiple bubbles to burst at roughly the same time after impact (Fig. 4b). As shown in Fig. 4, frenetic aerosol generation occurs when We/Pe ~ 1. In previous work, we defined $\text{We/Pe} = \text{Re}_W = U_o \rho r_c \cos\theta \mu^{-1}$, where Re_W is the Washburn–Reynolds number. Re_W is the ratio of the droplet inertial to the capillary flow properties of the porous media as $\text{We/Pe} = \rho D_{cap} U_o \gamma^{-1}$ like the capillary number with ρD_{cap} replacing μ. This frenetic aerosol dispersion is very attractive because quite a number of aerosols can be produced within a few microseconds, which could be useful for applications requiring fast aerosol generation at ambient temperature.

It is well known that air films are trapped between droplets and solid surfaces during impact[38,39,51,54,55]. In general, the main features of drop impact dynamics are described by We and Re. It has been reported that the splashing condition on a solid surface can be defined as $K_d = \text{We}^{1/2}\text{Re}^{1/4} > 57.7$ (refs 55,56). In this work, drop impingement was conducted for Re from 60 to 10,244 and K_d from 7 to 325. The frenetic aerosol-generation region,

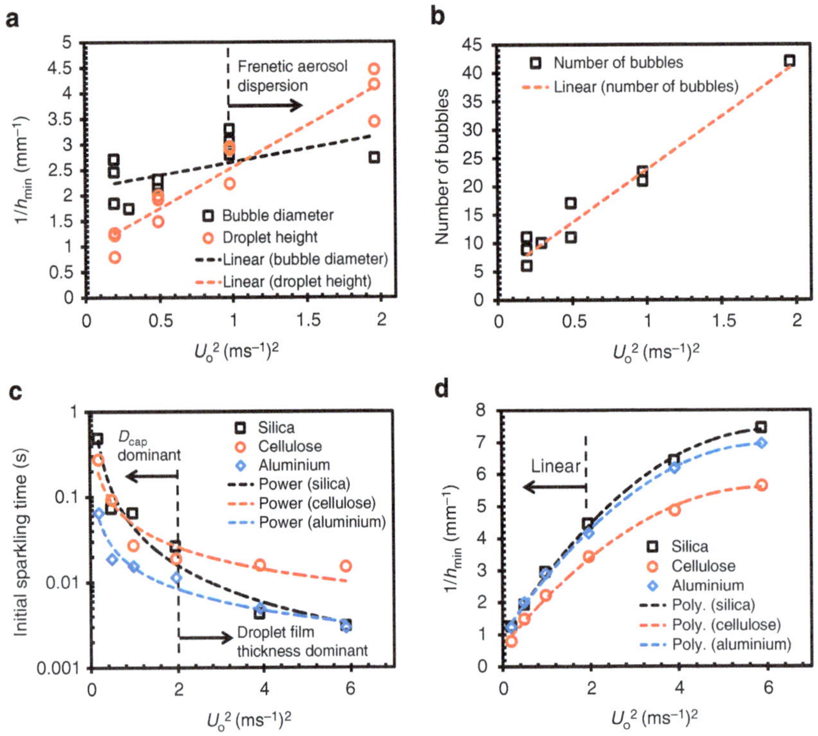

Figure 3 | Characteristics of droplets and bubbles generated after impingement on TLC plates. The hydraulic diffusivities of TLC-A (silica), TLC-B (cellulose) and TLC-C (aluminium) are 9.5, 20.4 and 22.4 mm^2 s^{-1}, respectively. The dot lines indicate the fitting lines of the experimental data marked with symbols. (**a**) The average bubble diameter does not change significantly with impact velocity U_o^2 but the average droplet film thickness is inversely proportional to U_o^2. (**b**) The maximum number of bubbles inside the droplet linearly increases with respect to U_o^2. (**c**) The time it takes to observe the first jet is affected by the surface wettability at low impact velocity but by the droplet film thickness at high impact velocity, since the droplet film thickness governs the maximum bubble size at high impact velocity. (**d**) The droplet film thickness h_{min} after hitting the surfaces is the highest on TLC-B.

occurring at $Re_W \sim 1$, corresponds to $K_d \sim 70$ and $Re \sim 5{,}000$. In our previous work[46], the splashing region (radial jetting) can be characterized as $Re_W > 5$, which corresponds to $K_d > 100$. Frenetic aerosol generation occurs at slightly lower Re_W than what is required for splashing. This result shows reasonable agreement with the other drop impingement studies because the splashing starts around $K_d \sim 70$. The advantage of using Re_W is that we can consider the impact conditions and the wetting properties simultaneously. Therefore, we can predict impingement modes considering various kinds of surfaces with a wide range of wetting properties.

Re_W can be used to predict the patterns of aerosol dispersion from droplets. Interestingly, when $Re_W = We/Pe < 0.2$, we observe aerosol dispersion at the edge of droplets. Supplementary Movie 8 shows edge aerosol dispersion when a silica TLC plate was used as the substrate with $We = 113$ and $Pe = 501$. This aerosol dispersion has a different mechanism from that of jets originating from the cusps of droplets hitting flat surfaces at relatively high impact velocity[55,57]. For the case of drop impingement on flat surfaces, Yarin *et al.*[57] suggested the dimensionless impact velocity $u = Ca\lambda^{3/4} = U_o(\rho\gamma^{-1})^{1/4}\nu^{-1/8}f^{-3/8}$, where Ca is the capillary number, ν is the kinematic viscosity and f is the drop frequency, to predict the splashing threshold. The splashing threshold is defined when $u = 17$–18 under the assumption that surface roughness is much less than droplet diameter. We cannot estimate the dimensionless impact velocity in our cases because we used a single droplet for the drop impingement rather than using successive droplets with drop frequency f. Compared with the minimum impact velocity of 5 ms^{-1} for splashing in successive drop impingement, we observe the side aerosol dispersion for impact velocities lower than 2 ms^{-1}. Edge aerosol dispersion occurs when the maximum bubble size is smaller than the droplet height. Under this condition the bubbles begin bursting at the edge of the droplet. We observe aerosol dispersion at the edge of droplets on both smooth and rough porous surfaces, as shown in Supplementary Movies 2 and 8, respectively.

From the characterization map we conclude that aerosol generation can occur during light and moderate rainfall on soils with hydraulic diffusivity less than sand. To locate the impingement domains for rainfall on soil (red, yellow and green regions in Fig. 4), the hydraulic diffusivity of the soils were obtained from the literature[58–60] and the impact velocity of raindrops was set from light rain ($U_o < 4.0$ ms^{-1}) to heavy rain ($U_o > 7.0$ ms^{-1})[31]. When the impingement domains are added to the characterization map, the overlapping regions of aerosol generation and rainfall are evident (Fig. 4). We verified aerosol generation experimentally for conditions consistent with light and moderate rain on soil with similar water-absorption properties as clay and sandy clay soil. Aerosol generation was not observed on sand or during conditions consistent with heavy rainfall.

In this work we demonstrate that aerosols can be generated from raindrops on soil. However, there are a few limitations of this work that have yet to be fully explored. First, raindrop–aerosol collisions are a major mechanism of aerosol removal[3]. Even though aerosols are generated from droplets, successive raindrops can eliminate the aerosols from the air; therefore, large-scale environmental effects might not be expected. Second, the gravitational settling velocity can be relatively fast when the droplet size is around a few tens of microns. Even when considering evaporation, the settling speed is not negligible. Therefore, further investigation of aerosol migration is necessary

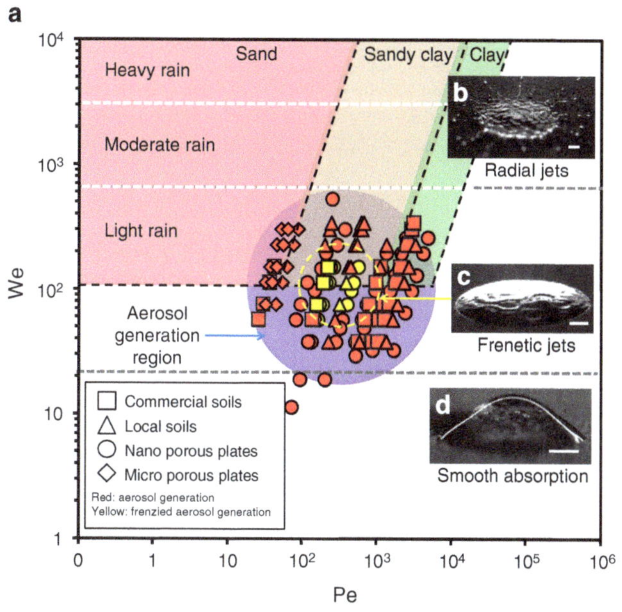

Figure 4 | Characterization of aerosol generation when droplets hit soils and porous surfaces. (**a**) The *x* axis indicates a modified Péclet number ($Pe \equiv (U_o d_o)/D_{cap}$), where U_o and d_o are the impact velocity and the diameter of the droplet. The *y* axis indicates the Weber number ($We \equiv \rho d_o U_o^2 \gamma^{-1}$), mainly varied by the impact velocity. We and Pe represent the impact condition and the surface property, respectively. The D_{cap} values of the soils were measured by the capillary rise experiments. The red symbols indicate the observation of aerosol dispersion from the droplets at the corresponding We and Pe. The blue circle indicating 'aerosol-generation region' highlights the group of the data points (the red symbols) where aerosols are generated. The yellow symbols inside the aerosol-generation region show the data points resulting in frenetic aerosol generation. Rainfalls on soils are placed on the upper part of the characterization map, and clay and sandy-clay soils have concentrated regions in the aerosol generation region. (**b-d**) The representative images of drop impingements in the regions classified with We. Scale bars, 1 mm. The horizontal dot lines and the diagonal dot lines indicate the constant We numbers and the constant D_{cap} numbers, respectively.

to understand the potential for large-scale environmental effects of aerosols generated on soil. One mechanism that could enhance the relevance of this phenomenon is wind-driven advection. Considering wind on the surface of porous media, tiny droplets or solid particles (<50 µm) can be transferred very long distances from the origin, up to a few thousand kilometres[61,62]. Furthermore, aerosols can have higher velocities than raindrops because of their small size, leading to lower inertia and drag forces[24,30]. Therefore, aerosols have the potential to migrate long distances without serious washing out or gravitational settling. In this manuscript, we focused on a novel mechanism of aerosol generation from droplets hitting porous media rather than exploring the effects of the aerosols. This discovery will lead to additional work investigating the fate of these aerosols in the environment.

This work has at least two unique contributions: first, we present visual evidence of aerosol generation from rainfall on soil; and second, we provide an effective means to generate aerosol using simple drop impingement on porous media. The results of this work indicate a strong correlation between aerosol generation, wetting properties and droplet impact velocity. Now that this phenomenon has been observed, future work could explore the environmental effects of aerosol generation on soil. In addition, results of this study could be used to develop novel chemical synthesis methods for next-generation materials employing aerosols[63]. This work could also be utilized to study the transport of viruses and bacteria impregnated in soil.

Methods

Measurements of hydraulic diffusivity. In this study, capillary rise measurements[64–66] were performed to evaluate hydraulic diffusivity on porous surfaces and soils (Table 1). For the capillary rise measurements of soils, we used a common experimental set-up used in previous work[60,64,66]. A glass tube, with 10 mm inner diameter, is filled 100 mm of soil and capped at the bottom with a glass membrane. The pore size of the membrane is in the 70- to 100-µm range. The glass tube is fixed vertically and the liquid bath height is slowly increased to initiate contact between the soil sample and the liquid surface. Capillary rise is recorded using a digital camera. For capillary rise measurements on the other surfaces, the samples, which measured 2 cm in width and 10 cm in length, were used instead of glass tubes.

Theoretical determination of hydraulic diffusivity. Hydraulic diffusivity was originally derived from Darcy's law; therefore, we need to understand how the hydraulic diffusivity correlates with the capillary rise measurement. The hydraulic diffusivity obtained from the capillary rise measurement has the same physical meaning as the hydraulic diffusivity obtained from Darcy's law, assuming a uniform porous structure. The following two sections describe the derivations of hydraulic diffusivity using both approaches.

Hydraulic diffusivity from capillary rise measurements. Capillary flows can be considered a subset of Poiseuille flow in a narrow tube with radius, r_c. The velocity within a capillary can be written as

$$v_c = \left(\frac{r_c^2}{8\mu}\right)\left(\frac{\partial P}{\partial z}\right), \tag{1}$$

where $\partial P/\partial z$ is the pressure gradient in the z direction and μ is the liquid viscosity. In the tube, the capillary pressure, P_{cap}, can be calculated from the Young–Laplace equation,

$$P_{cap} = \frac{2\gamma\cos\theta}{r_c}, \tag{2}$$

where γ is the surface tension and θ is the native contact angle on the surface.

If the capillary pressure is higher than the hydrostatic pressure in the capillary tube, the liquid rises. When a vertical capillary tube comes into contact with the free water surface, the flow velocity can be expressed by modifying equation (1). For a vertical tube with rise height, H, from the free liquid surface to the propagating liquid front, and the average velocity, v_c, can be expressed as the change in capillary height with respect to time. Washburn's equation, which is also attributed to Lucas[67], and Bell and Cameron[68], is obtained as[49]

$$v_c = \frac{dH}{dt} = \frac{r_c^2}{8\mu H}\left[\frac{2\gamma\cos\theta}{r_c} - \rho g H\right]. \tag{3}$$

If H is small, gravitational effects can be neglected and Washburn's equation can be simplified to

$$\frac{dH}{dt} = \frac{r_c\gamma\cos\theta}{4\mu H}. \tag{4}$$

Equation (4) can be integrated with respect to H and t to obtain a function relating rise height with time, resulting in the expression:

$$H^2 = \frac{r_c\gamma\cos\theta}{2\mu}t = \frac{1}{2}D_{cap}t, \tag{5}$$

where D_{cap} is the hydraulic diffusivity with the units of $m^2 s^{-1}$. Therefore, hydraulic diffusivity can be obtained from experimental measurements of the capillary rise speed.

Hydraulic diffusivity from Darcy's law. From Darcy's law, the water discharge, q (ms^{-1}), through porous media can be expressed as

$$q = -K\frac{\partial H}{\partial z}, \tag{6}$$

where K is the hydraulic conductivity (ms^{-1}) and $\partial P/\partial z$ is the hydraulic gradient in the z (or vertical) direction. When we assume the capillary head, H_{cap}, is the only driving force for the water discharge, from equations (1) and (6), we can express the hydraulic conductivity as

$$K = \left(\frac{r_c^2}{8\mu}\right)\frac{\partial P_{cap}}{\partial H_{cap}}. \tag{7}$$

Assuming water content continuity, the Richards' equation[47,69] can be derived from equation (6) as

$$\frac{\partial \Theta}{\partial t} = \frac{\partial}{\partial z}\left(K\frac{\partial H_{cap}}{\partial z}\right), \tag{8}$$

where Θ is the moisture content. Here capillary pressure is considered the only driving force; therefore, we can express equation (8) as a function of hydraulic diffusivity, D, as follows:

$$\frac{\partial \Theta}{\partial t} = \frac{\partial}{\partial z}\left(K\frac{\partial H_{cap}}{\partial z}\right) = \frac{\partial}{\partial z}\left(D\frac{\partial \Theta}{\partial z}\right), \tag{9}$$

where

$$D = K\frac{\partial H_{cap}}{\partial \Theta}. \tag{10}$$

If we assume that the porous medium is initially dry ($\Theta = 0$) and becomes fully wet ($\Theta = 1$) after capillary suction, and the capillary pressure becomes zero when $\Theta = 1$, the hydraulic diffusivity when the porous medium is dry can be expressed as

$$D|_{\Theta=0} = \left(\frac{r_c^2}{8\mu}\right)P_{cap} = \frac{r_c\gamma\cos\theta}{4\mu} = \frac{1}{2}D_{cap}, \tag{11}$$

from equations (7) and (10). Therefore, the hydraulic diffusivities obtained from Washburn's and Richard's equations have the same form.

Preparation and characterization of porous media. We used 28 different porous medium samples (see Table 1) to examine aerosol generation from drop impingement. Eight kinds of soil (from clay loam soil to peat) were purchased from World's Science (Nasco soil samples, Fort Atkinson, WI). We used the soil classification based on the product information provided by the company. Eight soil samples were collected from the Boston metropolitan area. The detailed soil sample locations are as follows; Soil A: Killian court on MIT campus, Soil B: Amherst Alley on MIT campus, Soil C and Soil D: Briggs field on MIT campus, Soil E, Soil F and Soil G: Charles River and Soil H: Nahant beach. The soils were classified considering the particle size and sampling location. In addition, we tested five kinds of commercial TLC plates purchased from Sigma-Aldrich, and seven different titania porous films produced by an electrochemical process[50].

To prepare soil substrates, first we gently broke the soils into fine grains by hand. We did not filter and grind the soils. Then, 1 cm of each soil was placed on a dish and then wetted with a water spray. The wetted soils were dried on a hot plate at 50 °C for 3 days. After cooling to room temperature, the soils were used for the drop impingement experiments. The TLC plates were used as provided from the manufacturer. The porous titania plates were rinsed with deionized water for 1 min after processing and dried at ambient conditions. From capillary rise experiments, the hydraulic diffusivity of each porous medium was evaluated with four different liquids: water, ethanol, ethylene glycol and glycerol. For the soils, the hydraulic diffusivity and aerosol generation were evaluated with water. The TLC and porous titania plates were evaluated with four different liquids: water, ethanol, ethylene glycol and glycerol. The particle sizes of the soil samples were evaluated using an image-processing and analysis software (ImageJ, US National Institutes of Health, Bethesda, MD, USA). Table 1 summarizes the surface properties and liquids used in this work.

Characterization of drop impingement on TLC plates. We investigated the behaviours of droplets hitting porous surfaces using high-speed imaging. We used the three different TLC plates (TLC-A, TLC-B and TLC-C), with wetting properties shown in Table 1, and varied the drop height to change the impact velocity. First, we compared the droplet film thickness after spreading and the bubble diameters, as shown in Fig. 3a. The droplet film thickness is measured when the droplets reach their maximum radius after impact. The bubble diameter is the average diameter of the bubbles inside the droplet when the first bubble breaks.

Delivery of materials in porous media with aerosol. In a duct with air flow, aerosol was generated from deionized water droplets hitting porous surfaces containing a fluorescent dye (Rhodamine B, Sigma-Aldrich). First, we dropped Rhodamine B suspensions (2.1×10^{-5} M) on TLC-C plates, which were then dried in a common laboratory environment for 1 day. The TLC plate was placed in a duct, with dimensions of 10 cm × 10 cm × 160 cm in width, height and length, respectively (Supplementary Fig. 1a). A transparent glass slide (Supplementary Fig. 1b) was placed at height h from the TLC plate to acquire aerosol droplets generated from the TLC plate. The transparent glass slide has a hole with the diameter of 5 mm to allow water droplets to hit the TLC plate. Using an electric fan, air flow was generated in the duct. After conducting drop impingement on the TLC surfaces (Supplementary Movie 9), we observed the transparent glass slides with a microscope to count the number of the sites where aerosol droplets were deposited. If the aerosol droplets contain traces of chemicals initially on the porous surface, we would expect to observe Rhodamine B on the glass where the aerosol droplets hit. We used a microscope (Nikon Eclipse Ti) with a UV filter, with a centre excitation wavelength of 535 nm, to distinguish the sites with Rhodamine B. Under the microscope, the regions with Rhodamine B are relatively bright and easy to distinguish (Supplementary Fig. 1c). First, we measured the number of Rhodamine B spots on the glass slides, at three different heights: 2, 3 and 5 mm from the TLC plate, when aerosol was generated in the duct without air flow. Second, we used three different air-flow speeds: 0.125, 0.25 and 0.5 ms^{-1} to verify that chemicals on the porous media convectively migrate as aerosol droplets. We counted the number of Rhodamine B spots on the downstream half of the circle.

References

1. Agranovski, I. *Aerosols: Science and Technology* (Wiley, 2011).
2. Cowling, B. J. *et al.* Aerosol transmission is an important mode of influenza A virus spread. *Nat. Commun.* **4,** 1935 (2013).
3. Seinfeld, J. H. & Pandis, S. N. *Atmospheric Chemistry and Physics: From Air Pollution to Climate Change* (Wiley, 2012).
4. Kaufman, Y. J., Tanre, D. & Boucher, O. A satellite view of aerosols in the climate system. *Nature* **419,** 215–223 (2002).
5. Woodcock, A. H. Bursting bubbles and air pollution. *Sewage Ind. Wastes* **27,** 1189–1192 (1955).
6. Chin, M. *Atmospheric Aerosol Properties and Climate Impacts* (DIANE Publishing, 2009).
7. Li, Z. *et al.* Long-term impacts of aerosols on the vertical development of clouds and precipitation. *Nat. Geosci.* **4,** 888–894 (2011).
8. Hinds, W. C. *Aerosol Technology: Properties, Behavior, and Measurement of Airborne Particles* (John Wiley & Sons, 2012).
9. Booth, B. B. B., Dunstone, N. J., Halloran, P. R., Andrews, T. & Bellouin, N. Aerosols implicated as a prime driver of twentieth-century North Atlantic climate variability. *Nature* **484,** 228–232 (2012).
10. Stevens, B. & Boucher, O. Climate science: the aerosol effect. *Nature* **490,** 40–41 (2012).
11. Carslaw, K. S. *et al.* Large contribution of natural aerosols to uncertainty in indirect forcing. *Nature* **503,** 67–71 (2013).
12. Myhre, G., Myhre, C., Samset, B. & Storelvmo, T. Aerosols and their relation to global climate and climate sensitivity. *Nat. Educ. Knowl.* **4,** 7 (2013).
13. Paasonen, P. *et al.* Warming-induced increase in aerosol number concentration likely to moderate climate change. *Nat. Geosci.* **6,** 438–442 (2013).
14. Ehn, M. *et al.* A large source of low-volatility secondary organic aerosol. *Nature* **506,** 476–479 (2014).
15. Blanchard, D. C. The size and height to which jet drops are ejected from bursting bubbles in seawater. *J. Geophys. Res. Oceans* **94,** 10999–11002 (1989).
16. Woodcock, A. H., Kientzler, C. F., Arons, A. B. & Blanchard, D. C. Giant condensation nuclei from bursting bubbles. *Nature* **172,** 1144–1145 (1953).
17. Blanchard, D. C. Sea-to-air transport of surface active material. *Science* **146,** 396–397 (1964).
18. Blanchard, D. C. The electrification of the atmosphere by particles from bubbles in the sea. *Prog. Oceanogr.* **1,** 73–202 (1963).
19. Arons, A. & Kientzler, C. Vapor pressure of sea-salt solutions. *T. Am. Geophys. Un.* **35,** 722–728 (1954).
20. Villermaux, E. & Bossa, B. Single-drop fragmentation determines size distribution of raindrops. *Nat. Phys.* **5,** 697–702 (2009).
21. Lighthart, B. & Mohr, A. J. *Atmospheric Microbial Aerosols: Theory and Applications* (Chapman & Hall, 1994).
22. Rioboo, R., Voué, M., Vaillant, A. & De Coninck, J. Drop impact on porous superhydrophobic polymer surfaces. *Langmuir* **24,** 14074–14077 (2008).
23. Feng, J. *et al.* Nanoemulsions obtained via bubble-bursting at a compound interface. *Nat. Phys.* **10,** 606–612 (2014).
24. Ruzer, L. S. & Harley, N. H. *Aerosols Handbook: Measurement, Dosimetry, and Health Effects* (Taylor & Francis, 2004).
25. Bear, I. J. & Thomas, R. G. Nature of argillaceous odour. *Nature* **201,** 993–995 (1964).
26. *Keys to Soil Taxonomy* 11th edn (USDA-Natural Resources Conservation Service, 2010).
27. Beard, K. V., Ochs, H. T. & Kubesh, R. J. Natural oscillations of small raindrops. *Nature* **342,** 408–410 (1989).
28. McDonald, J. E. The shape of raindrops. *Sci. Am.* **190,** 64–68 (1954).
29. Beard, K. V. & Chuang, C. A new model for the equilibrium shape of raindrops. *J. Atmos. Sci.* **44,** 1509–1524 (1987).
30. de Lima, J. L. M. P., Torfs, P. J. J. F. & Singh, V. P. A mathematical model for evaluating the effect of wind on downward-spraying rainfall simulators. *CATENA* **46,** 221–241 (2002).
31. Sumner, G. N. *Precipitation: Process and Analysis* (Wiley New York, 1988).
32. Hasan, O. & Prosperetti, A. Bubble entrainment by the impact of drops on liquid surfaces. *J. Fluid Mech.* **1219,** 143 (1990).
33. Prosperetti, A. & Oguz, H. N. The impact of drops on liquid surfaces and the underwater noise of rain. *Annu. Rev. Fluid Mech.* **25,** 577–602 (1993).
34. Lee, J. S. *et al.* Size limits the formation of liquid jets during bubble bursting. *Nat. Commun.* **2,** 367 (2011).
35. Bird, J. C., de Ruiter, R., Courbin, L. & Stone, H. A. Daughter bubble cascades produced by folding of ruptured thin films. *Nature* **465,** 759–762 (2010).

36. Bick, A. G., Ristenpart, W. D., van Nierop, E. A. & Stone, H. A. Bubble formation via multidrop impacts. *Phys. Fluids* **22,** 042105 (2010).
37. Boulton-Stone, J. M. & Blake, J. R. Gas bubbles bursting at a free surface. *J. Fluid Mech.* **254,** 437–466 (1993).
38. Thoroddsen, S. T., Etoh, T. G., Takehara, K., Ootsuka, N. & Hatsuki, Y. The air bubble entrapped under a drop impacting on a solid surface. *J. Fluid Mech.* **545,** 203–212 (2005).
39. Lee, J. S., Weon, B. M., Je, J. H. & Fezzaa, K. How does an air film evolve into a bubble during drop impact? *Phys. Rev. Lett.* **109,** 204501 (2012).
40. Katsuragi, H. Morphology scaling of drop impact onto a granular layer. *Phys. Rev. Lett.* **104,** 218001 (2010).
41. Marston, J. O., Thoroddsen, S. T., Ng, W. K. & Tan, R. B. H. Experimental study of liquid drop impact onto a powder surface. *Powder Technol.* **203,** 223–236 (2010).
42. Nefzaoui, E. & Skurtys, O. Impact of a liquid drop on a granular medium: Inertia, viscosity and surface tension effects on the drop deformation. *Exp. Therm. Fluid Sci.* **41,** 43–50 (2012).
43. Marston, J. O. *et al.* Drop spreading and penetration into pre-wetted powders. *Powder Technol.* **239,** 128–136 (2013).
44. Ahn, S. *et al.* Effects of hydrophobicity on splash erosion of model soil particles by a single water drop impact. *Earth Surf. Proc. Land.* **38,** 1225–1233 (2013).
45. Spiel, D. E. The number and size of jet drops produced by air bubbles bursting on a fresh water surface. *J. Geophys. Res. Oceans* **99,** 10289–10296 (1994).
46. Joung, Y. S. & Buie, C. R. Scaling laws for drop impingement on porous films and papers. *Phys. Rev. E* **89,** 013015 (2014).
47. Van Genuchten, M. T. A closed-form equation for predicting the hydraulic conductivity of unsaturated soils. *Soil Sci. Soc. Am. J.* **44,** 892–898 (1980).
48. Dullien, F. A. L., El-Sayed, M. S. & Batra, V. K. Rate of capillary rise in porous media with nonuniform pores. *J. Colloid Interf. Sci.* **60,** 497–506 (1977).
49. Washburn, E. W. The dynamics of capillary flow. *Phys. Rev.* **17,** 273 (1921).
50. Joung, Y. S. & Buie, C. R. A hybrid method employing breakdown anodization and electrophoretic deposition for superhydrophilic surfaces. *J. Phys. Chem. B* **117,** 1714–1723 (2013).
51. Mandre, S., Mani, M. & Brenner, M. P. Precursors to splashing of liquid droplets on a solid surface. *Phys. Rev. Lett.* **102,** 134502 (2009).
52. Bouwhuis, W. *et al.* Maximal air bubble entrainment at liquid-drop impact. *Phys. Rev. Lett.* **109,** 264501 (2012).
53. Latka, A., Strandburg-Peshkin, A., Driscoll, M. M., Stevens, C. S. & Nagel, S. R. Creation of prompt and thin-sheet splashing by varying surface roughness or increasing air pressure. *Phys. Rev. Lett.* **109,** 054501 (2012).
54. Martin, R. Phenomena of liquid drop impact on solid and liquid surfaces. *Fluid Dyn. Res.* **12,** 61 (1993).
55. Yarin, A. L. Drop impact dynamics: splashing, spreading, receding, bouncing.... *Annu. Rev. Fluid Mech.* **38,** 159–192 (2006).
56. Mundo, C., Sommerfeld, M. & Tropea, C. Droplet-wall collisions: experimental studies of the deformation and breakup process. *Int. J. Multiphas. Flow* **21,** 151–173 (1995).
57. Yarin, A. L. & Weiss, D. A. Impact of drops on solid surfaces: self-similar capillary waves, and splashing as a new type of kinematic discontinuity. *J. Fluid Mech.* **283,** 141–173 (1995).
58. Fredlund, D. G. & Rahardjo, H. *Soil Mechanics for Unsaturated Soils* (John Wiley & Sons, 1993).
59. Letey, J., Osborn, J. & Pelishek, R. E. Measurement of liquid-solid contact angles in soil and sand. *Soil Sci.* **93,** 149–153 (1962).
60. Goebel, M. O., Bachmann, J., Woche, S. K., Fischer, W. R. & Horton, R. Water potential and aggregate size effects on contact angle and surface energy. *Soil Sci. Soc. Am. J.* **68,** 383–393 (2004).
61. Jasper, F. K., Eric, J. R. P., Timothy, I. M. & Diana Bou, K. The physics of wind-blown sand and dust. *Rep. Prog. Phys.* **75,** 106901 (2012).
62. Parsons, A. J. & Abrahams, A. D. *Geomorphology of Desert Environments* (Springer Netherlands, 2009).
63. Lu, Y. *et al.* Aerosol-assisted self-assembly of mesostructured spherical nanoparticles. *Nature* **398,** 223–226 (1999).
64. Siebold, A., Walliser, A., Nardin, M., Oppliger, M. & Schultz, J. Capillary rise for thermodynamic characterization of solid particle surface. *J. Colloid Interf. Sci.* **186,** 60–70 (1997).
65. Lu, N. & Likos, W. J. Rate of capillary rise in soil. *J. Geotech. Geoenviron.* **130,** 646–650 (2004).
66. Carter, M. R. *Soil Sampling and Methods of Analysis* (CRC Press, 1993).
67. Lucas, R. Ueber das Zeitgesetz des kapillaren Aufstiegs von Flüssigkeiten. *Kolloid Zeitschrift* **23,** 15–22 (1918).
68. Bell, J. M. & Cameron, F. K. The flow of liquids through capillary spaces. *J. Phys. Chem.* **10,** 658–674 (1905).
69. Richards, L. A. Capillary conduction of liquids through porous mediums. *J. Appl. Phys.* **1,** 318–333 (1931).

Acknowledgements

We are grateful to Dr Jim Bales and Ms Sandra J. Lipnoski of the MIT Edgerton Center for sharing their high-speed camera. We thank Zhifei Roy Ge for help with the aerosol experiments and David Hu for introducing us to the concept of 'petrichor'. We also appreciate the reviewer comments, which bolstered the impact of this work. We thank Prof Gareth McKinley for discussions regarding the Washburn-Reynolds number.

Author contributions

Y.S.J. and C.R.B. designed the research; Y.S.J. performed the research; Y.S.J. and C.R.B. analysed the data and wrote the manuscript.

Additional information

Competing financial interests: The authors declare no competing financial interests.

7

Humidity trends imply increased sensitivity to clouds in a warming Arctic

Christopher J. Cox[1,2], Von P. Walden[3], Penny M. Rowe[4,5] & Matthew D. Shupe[1,2]

Infrared radiative processes are implicated in Arctic warming and sea-ice decline. The infrared cloud radiative effect (CRE) at the surface is modulated by cloud properties; however, CRE also depends on humidity because clouds emit at wavelengths that are semi-transparent to greenhouse gases, most notably water vapour. Here we show how temperature and humidity control CRE through competing influences between the mid- and far-infrared. At constant relative humidity, CRE does not decrease with increasing temperature/absolute humidity as expected, but rather is found to be approximately constant for temperatures characteristic of the Arctic. This stability is disrupted if relative humidity varies. Our findings explain observed seasonal and regional variability in Arctic CRE of order 10 W m^{-2}. With the physical properties of Arctic clouds held constant, we calculate recent increases in CRE of 1-5 W m^{-2} in autumn and winter, which are projected to reach 5-15 W m^{-2} by 2050, implying increased sensitivity of the surface to clouds.

[1] Cooperative Institute for Research in Environmental Sciences, University of Colorado, Boulder, Colorado 80309, USA. [2] NOAA Earth System Research Laboratory, Boulder, Colorado 80305, USA. [3] Department of Civil and Environmental Engineering, Washington State University, Pullman 99164, Washington, USA. [4] Department of Geography, University of Idaho, Moscow 83844, Idaho, USA. [5] Departamento de Física, Universidad de Santiago de Chile, Santiago 9170124, Chile. Correspondence and requests for materials should be addressed to C.J.C. (email: christopher.j.cox@noaa.gov).

Amplified warming of the Arctic and coinciding decreases in sea ice are driven in part by perturbations to the surface radiation budget[1–5]. The infrared cloud radiative effect (henceforth, 'CRE') at the surface can be 50–100 W m^{-2} (refs 6–10); however, the system is sensitive to comparatively small changes. In fact, observed decadal trends in sea ice could be forced by a perturbation of just 1 W m^{-2} (ref. 11). Variability in surface cover (due to variability in snow/ice cover) modifies albedo, which is a control on shortwave CRE that is not directly due to cloud properties[12], that is, clouds cool the surface relative to clear skies by reflecting sunlight back to space; however, this cooling is minimized over ice-covered surfaces because the albedos of the surface and cloud are similar[13]. Nevertheless, in spite of changes in surface cover in the Arctic, it is likely that cloud feedbacks at high latitudes are dominated by infrared radiation[12].

Globally, infrared emission from clouds reaches the surface primarily through the 'atmospheric window' (AW; defined here as 7–14 μm) where absorption by atmospheric gases is relatively low[14]. However, in dry regions, such as the Arctic, the far-infrared (FIR; 16–40 μm) is also semi-transparent, and variability in clouds[14], water vapour[15,16] and surface emission[16,17] all contribute substantially to energy exchanges between the surface and space. Since the infrared radiance received at the surface depends nonlinearly on absorption and emission from clouds and water vapour, both of which depend on temperature, atmospheric feedbacks and processes implicated in Arctic warming[1–5,12] are not easily disentangled. Increases in water vapour are reported in the Arctic[18,19] and projected by climate models[20], but temperature–humidity relationships in the Arctic are distinct from lower latitudes[21], making it difficult to project the effects of atmospheric change on the surface in the presence of clouds. Similar to the way that changes in surface cover influence the shortwave CRE, temperature and humidity exert an influence that is independent of cloud properties. Therefore, variability in CRE due to variability in the atmospheric state must be addressed.

Here we present an analysis of spectrally resolved CRE using surface-based measurements from three Arctic observatories. Interplay between temperature and humidity is shown to control CRE through competing influences in two semi-transparent wavelength ranges (7–14 and 16–40 μm), stabilizing CRE at constant relative humidity for temperatures characteristic of the Arctic (~230–280 K). Using reanalysis and climate model data sets, we find that, because of this mechanism, increases in temperature and precipitable water vapour (PWV) in the Arctic are likely resulting in increases in CRE that are independent of changes in cloud properties.

Results

Observations of the CRE relation with temperature and PWV. In a well-known positive feedback, atmospheric water vapour increases with temperature, and because water vapour is a strong greenhouse gas, this leads to more water vapour emission to the surface and thus further warming[22] (grey arrows in Fig. 1). Increases in water vapour also lead to increases in cloud cover and/or optical depth, both of which feedback to increased warming through additional CRE (for example, ref. 23; green and blue arrows in Fig. 1). However, absorption and emission from water vapour between the cloud and the surface mask the radiance from the cloud. In particular, cloud emission must be transmitted through the atmosphere below the cloud to reach the surface and have an impact. Thus, CRE should decrease as humidity increases concurrently with temperature. Indeed, smaller CREs are noted for the wet season in the tropics as

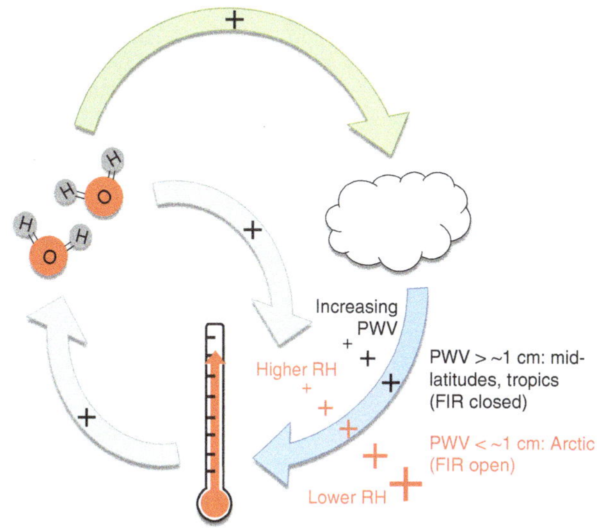

Figure 1 | Conceptual diagram of Arctic feedbacks relevant to CRE. Conceptual diagram of Arctic feedbacks involving surface temperature, atmospheric water vapour and clouds. The grey arrows represent the 'water vapour feedback'[21]. The green arrow represents the subsequent change in cloud properties; for example, an increase in the amount of clouds[23] increases the surface temperature via increased CRE (blue arrow). The magnitude of the black and red plus (+) symbols represents perturbations to the CRE feedback; note that the total change in CRE is influenced by changes in cloud properties as well. Increasing atmospheric water vapour *dampens* the increase in CRE feedback from changing cloud properties, indicated by decreasing black '+' symbols. When the FIR is open, as relative humidity (RH) decreases, the CRE feedback increases (red '+' symbols increasing), as described in this study.

compared with the dry season, due in part to differences in humidity[24]. A similar relationship likely occurs at mid-latitudes, although the effect has not been isolated from the seasonal cycle in cloud fraction[25]. This relationship plays out as decreases in the cloud feedback though each iteration of the loop in Fig. 1 (black 'plus' symbols in the lower right of the diagram signify increases in humidity with time). However, for the Arctic, we find that CRE is poorly correlated with observed surface temperature and PWV (Fig. 2a,b), suggesting that the behaviour of the cloud feedback is different in the Arctic compared with lower latitudes. Our analysis begins with an examination of the observed CRE in the Arctic with the objective of isolating the influence of the atmospheric state from the influence of variability in cloud physical properties.

CRE is defined as the difference between downwelling infrared flux from clouds and atmospheric gases together and that from gases alone,

$$\mathrm{CRE} = F\downarrow_{\text{all-sky}} - F\downarrow_{\text{clear-sky}} \quad (1)$$

In this study, CRE is calculated from the perspective of the surface. At the surface, CRE differs from cloud radiative forcing (CRF) in that CRE is for the downwelling infrared component only, rather than the net flux, but they are similar because the upward component of CRF is small. To isolate how temperature and humidity influence CRE, it is instructive to hold cloud physical properties constant, either by subsetting observations that represent samples of optically thick clouds, or by performing radiative transfer calculations for clouds with fixed optical depth. The main assumption that is made in either case is that tropospheric temperature covaries with cloud temperature.

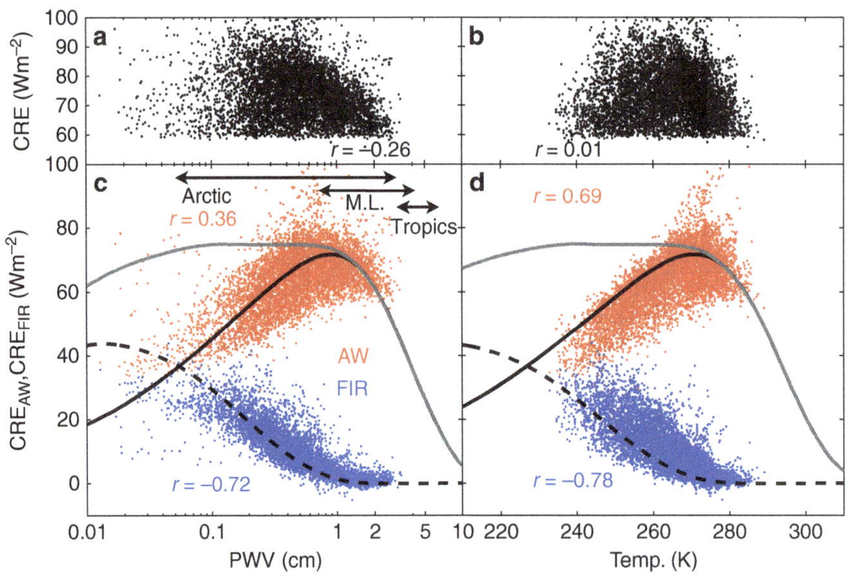

Figure 2 | CRE in the FIR and the atmospheric window. (**a**) Observations of 3-hour averages of downwelling infrared CRE from Barrow, Alaska; Eureka, Canada and Summit Station, Greenland, plotted as a function of PWV. Only CRE > 60 W m^{-2} are shown to highlight clouds that are optically thick. (**b**) Same as **a**, but plotted as a function of near-surface air temperature. (**c,d**) are similar to **a,b**, but are separated into spectral components from the atmospheric window (CRE_{AW}, red points) and the far-infrared (CRE_{FIR}, blue points). For reference, the arrows in **c** are the approximate ranges of PWV for the Arctic, mid-latitudes (M.L.) and Tropics. Curves in **c,d** are radiative transfer calculations of CRE_{FIR} (dashed black), CRE_{AW} (solid black) and their sum $CRE = CRE_{FIR} + CRE_{AW}$ (grey).

Three-hour averages of CRE at stations representative of different regions of the Arctic—Barrow, Alaska; Eureka, Canada and Summit, Greenland—are derived using observations[6,7]. The observed values of PWV at these locations span a large range from less than 0.1 cm in winter at Summit[26] to ~2 cm in summer at Barrow[6]. Like much of the far-western Arctic, Barrow is relatively moist with a high incidence of optically thick clouds[6,9,10]. Eureka is representative of the northern Canadian archipelago, a drier region with fewer optically thick clouds than Barrow[6,27]. Summit is a high-altitude station on the Greenland ice sheet that is extremely dry[26]. To assess how variability in CRE is modulated by temperature and humidity contributions in the FIR and AW, CRE is calculated here as partial-band fluxes[14,28] such that $CRE \approx CRE_{AW} + CRE_{FIR}$.

To understand the apparent lack of correlation in Fig. 2a,b, we examine observed Arctic CRE_{FIR} and CRE_{AW}, which are plotted against PWV in Fig. 2c and temperature in Fig. 2d (red and blue dots). Over the range of Arctic conditions, CRE_{AW} increases with temperature and PWV while CRE_{FIR} decreases. To put these measurements in context, we simulate CRE using a radiative transfer model for a wider range of surface temperatures and PWVs than those typical of the Arctic (black and grey curves in Fig. 2). CRE is simulated by parameterizing PWV as a function of surface temperature; this is performed by fixing the shape of the temperature profile and scaling it to fit the surface temperature and by using a fixed relative humidity profile. When CRE is plotted for the wider range of temperatures and PWVs, we see that CRE_{FIR} and CRE_{AW} exhibit similar behaviour (that is, both increase, peak and then decrease), but are shifted in temperature and PWV. This shift occurs because of differing water-vapour absorption coefficients, which are larger in the FIR relative to the AW. This is not a surprising result, but what is interesting is the compensation of the two spectral regions when summed (grey line Fig. 2c,d). This compensation obscures the dependence on temperature and humidity between ~230 and 280 K, and, thus, explains the lack of correlation in CRE shown in Fig. 2a,b. These compensating flux variations are unique to the temperature and humidity ranges observed in the Arctic. Conversely, CRE increases with temperature below ~230 K and decreases above ~280 K. Examples of individual infrared spectra that illustrate this spectral compensation are depicted in Fig. 3.

To investigate the consequences of this compensation using an idealized framework, radiative transfer calculations are performed using radiosoundings acquired at Barrow and Summit. CRE_{FIR} and CRE_{AW} are calculated based on observed temperature and humidity profiles with model clouds inserted randomly at 0, 1, 2 or 5 km. The modelled values of CRE (CRE_{FIR}, CRE_{AW} and CRE) for optically thick clouds (optical depth, $\tau = 10$) are shown in the left panels of Fig. 4; observed values (explained below) are shown in the right panels for comparison. A parameterization of the Clausius–Clapeyron relationship[29] is also plotted for reference (black line). Note that CRE is expressed here on a colour scale as a function of both surface temperature and PWV. The modelled CRE_{FIR} (Fig. 4a) varies between 0 and 40 W m^{-2} with a strong dependence on humidity and low sensitivity to temperature. Conversely, CRE_{AW} (Fig. 4b), which ranges from 40 to 85 W m^{-2}, is sensitive primarily to temperature. Owing to the compensation described previously, when summed, CRE (Fig. 4c) has values of constant flux that closely follow the Clausius–Clapeyron relationship. Thus, temporal or spatial variations in temperature and/or PWV within the Arctic temperature range do not change CRE as long as the variability is consistent with the Clausius–Clapeyron relationship. However, deviations from this relationship will either increase or decrease CRE over a range of ~40 W m^{-2} at a given temperature. This behaviour is consistent in modelled CRE for clouds with smaller optical depths, although the magnitudes of the fluxes are smaller (Supplementary Fig. 1).

Observed CREs at Barrow and Eureka, shown on the right column of Fig. 4, agree well with the modelled values in the left column and use the same subset of approximately optically thick clouds shown in Fig. 2. Both the CRE values (colours) and the variations of CRE with temperature and PWV (colour distributions) are quite similar between the model and observations, suggesting that the idealized framework is a reasonable representation of the Arctic.

The results in Fig. 4 explain some of the observed variability in Arctic CRE observed in other studies (for example, ref. 9). For example, the annual cycle in temperature and humidity at Barrow results in variability in the maximum potential CRE of $\sim$70–90 W m^{-2} with peaks in spring and autumn. As mentioned above, clouds are opaque in the simulations, whereas the true annual cycle in CRE depends also on variations in cloud properties (cloud fraction, optical depth, microphysics and so on). The results of ref. 9 include cloud-property variability and show that the actual peak in infrared CRF occurs in August at Barrow, coinciding with the annual cycle in cloud fraction. Our results are complementary to ref. 9 in that they suggest greater *sensitivity* in the infrared to clouds in the transition seasons (spring and autumn) compared with summer or winter at Barrow, but they do not include the effect of seasonal changes in cloud fraction.

Distributions of CRE for all-sky conditions in summer (JJA) at all three stations are bimodal, with one mode representing clear skies (CRE $\sim$0 W m^{-2}) and another representing optically thick clouds (CRE > 60 W m^{-2}, representative of the subset of clouds shown in Figs 2 and 4; Fig. 5a), similar to distributions reported by others[6,7,30]. The mode representing optically thick clouds is $\sim$10 W m^{-2} higher at Summit compared with the other stations. These higher CRE values were, for example, an important driver of widespread surface melting of the ice sheet in July 2012 (ref. 8). Elevated CRE at Summit is due to enhanced CRE_{FIR} (Fig. 5c), rather than CRE_{AW}, which is similar at all stations (Fig. 5b).

Observed monthly mean temperature and PWV from the stations are plotted in Fig. 6a over the idealized CRE from Fig. 4c, which is interpolated to a surface. The figure shows that differences in the peak of the upper mode of CRE in Fig. 5a are because of deviations from contours of constant CRE, that is, deviations from the Clausius–Clapeyron relationship (dashed white curve in Fig. 6a). Indeed, the conditions at Barrow and Eureka lie close to the Clausius–Clapeyron line, while the conditions at Summit are at lower PWVs relative to temperature (falling below the dashed white line in Fig. 6a). Modelling CRE as in Fig. 2, but incrementally increasing the height of the surface, indicates that station altitude can explain the lower PWV relative to temperature, and thus the higher CRE observed at Summit.

Projected changes in Arctic CRE. Output from a reanalysis product and a climate model is now used to provide a conceptual understanding of how future changes in the Arctic system might have an impact on its sensitivity to CRE. Using the same methodology to create Fig. 6a, trends in CRE are estimated using the monthly mean temperature and PWV from a reanalysis product and a climate model. First, data from the European Centre for Medium-Range Weather Forecasts (ECMWF) Interim Reanalysis[31] (ERA-Interim) are mapped on the temperature–PWV surface. The domain is north of 70°N and 70°E to 220°E, corresponding roughly to the region where interannual variability

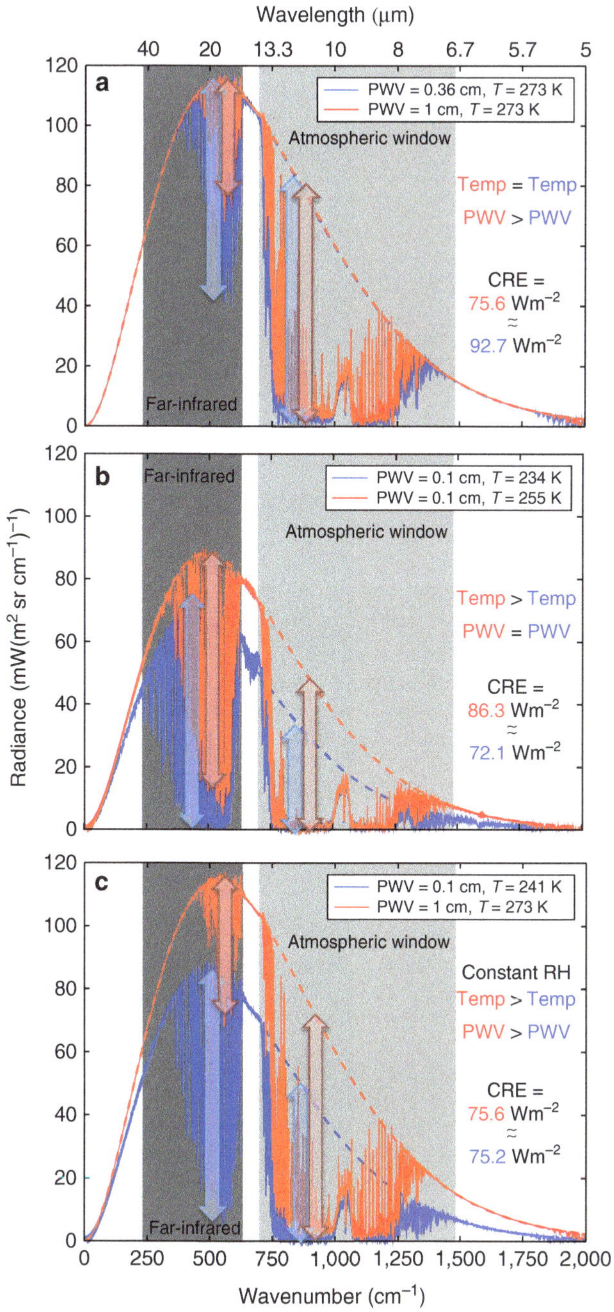

Figure 3 | Examples of infrared spectra. Examples of 'clear-sky' infrared spectra calculated from profiles of temperature and humidity measured by radiosoundings at Barrow, Alaska (red) and Summit Station, Greenland (blue). The dashed lines are Planck functions corresponding to the near-surface air temperatures for the respective cases and represent hypothetical, optically thick clouds. The FIR and AW spectral regions are shown as dark-grey and light-grey shaded regions. The downwelling infrared CRE is the spectral integral of the difference between the curves (dashed minus solid for any case). (Note that because these spectra are for zenith views, the flux (in W m^{-2}) is obtained by integrating the radiance over the hemisphere and over wavenumber (frequency) from 0 to 3,000 cm − 1 (ref. 6)). The vertical arrows indicate conceptually the magnitude of the CRE in each window. (**a**) Spectra from 12 July 2012 (Summit) and 8 May 2008 (Barrow): similar near-surface air temperatures, but different PWV. The CRE is larger at Summit because of additional CRE in the FIR associated with the low PWV. (**b**) Spectra from 7 May 2011 (Summit) and 12 March 2007 (Barrow): similar PWV, but different near-surface air temperatures. The CRE is larger at Barrow because of additional CRE in the AW associated with the higher temperature. (**c**) Spectra from 21 April 2011 (Summit) and 18 May 2008 (Barrow), where the conditions approximate the Clausius-Clapeyron parameterization in Figs 4 and 6. These cases represent both the cold/dry and warm/moist limits of the range of conditions. The total CRE ($CRE_{FIR} + CRE_{AW}$) in **c**, represented conceptually by summing the arrows, is similar between the two cases due to the compensating effects described in the main text.

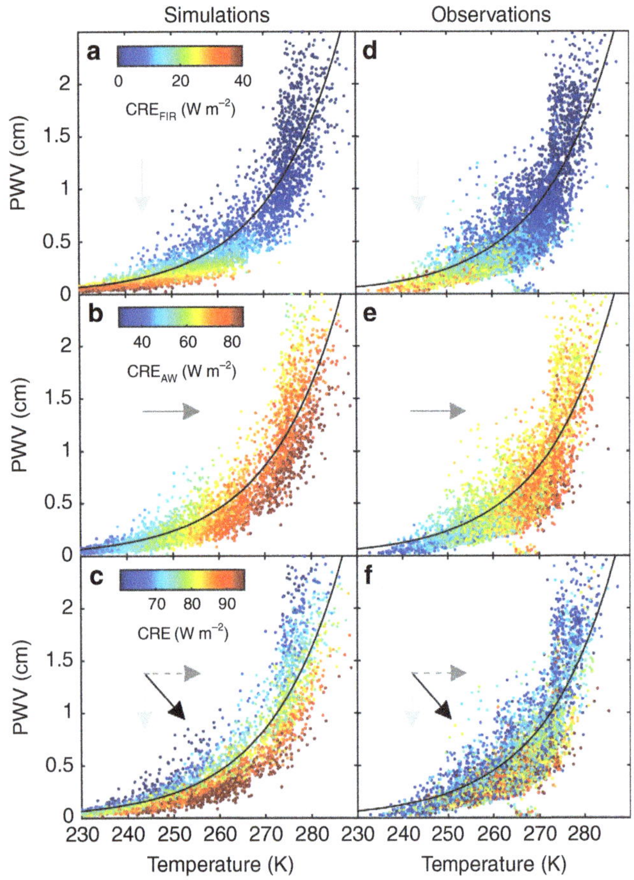

Figure 4 | Simulated and observed CRE as functions of temperature and PWV. Simulations (left panels (**a–c**)) and observations (right panels (**d–f**)) of the downwelling infrared CRE in the far-infrared (top panels (**a,d**)), atmospheric window (middle panels (**b,e**)) and the total CRE (bottom panels, (**c,f**)). The simulations are from radiative transfer calculations using observed profiles of temperature and humidity acquired by radiosoundings at Barrow, Alaska and Eureka, Canada. The black line is the Clausius-Clapeyron relationship for near-surface air temperature and PWV following ref. 29 and assuming a relative humidity (with respect to liquid) of 100% and a scale height of 3 km. The observed values are data from Barrow and Eureka, but only for clouds with large optical depths (CRE > 60 W m^{-2}). As a visual aide, arrows show the direction in the temperature-PWV space of increasing CRE, also shown as coloured symbols. Note that the colour scale changes for each row.

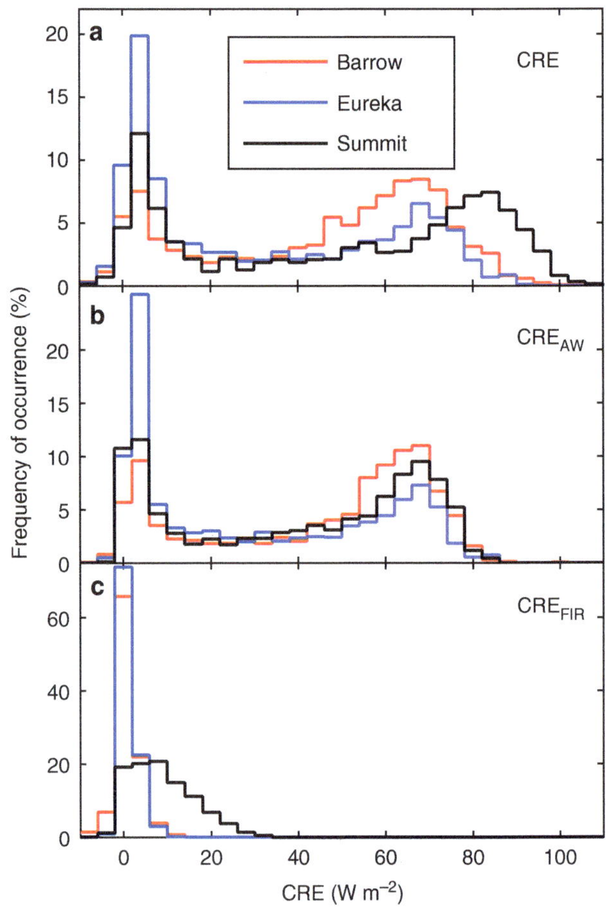

Figure 5 | Observed distributions of CRE in summer. Probability distribution functions of 3-hour averages of (**a**) CRE, (**b**) CRE$_{AW}$ and (**c**) CRE$_{FIR}$ in summer (June-July-August) from observations at Barrow, Alaska; Eureka, Canada and Summit Station, Greenland. The bin size is 4 W m^{-2}.

in sea-ice conditions is largest (Supplementary Fig. 2). For each month, the 5-year running mean CRE anomalies are calculated using 1979–2000 as a baseline. Results are then scaled to account for the annual cycle in cloud fraction and optical depth. This scaling is fixed for all years such that cloud physical properties are held constant for reasonable modern values so that the resulting projected anomalies are attributed to changes in the temperature and opacity of water vapour. The anomaly time series for each month are plotted in Fig. 6c. This same calculation is then made using the Community Earth System Model Large Ensemble (CESM-LE)[32] from 1979 through 2080, combining historical runs through 2005 and projections beginning in 2006 that use the Representative Concentration Pathway 8.5 (ref. 33; Fig. 6b,d).

Positive anomalies in CRE in both ERA-Interim and CESM-LE (Fig. 6c,d) emerge in autumn and early winter in the early 2000s. The mean anomaly for September through November from 2005 to 2012 is $+2.5\,W\,m^{-2}$ (ERA-Interim) and $+2.7 \pm 1.6\,W\,m^{-2}$ (CESM-LE; Supplementary Fig. 3). The largest anomalies are projected by CESM-LE to appear after 2040 in autumn between 5 and $15\,W\,m^{-2}$. This result is associated with temperature increases in autumn outpacing the expected water vapour increases via the Clausius–Clapeyron relationship (red versus black lines in Fig. 6b). A similar signal is observed in spring, although it is smaller, in part because of less cloud cover and generally thinner clouds in that season. These model projections provide a conceptual understanding of how future changes in the Arctic system might have an impact on its sensitivity to CRE. In particular, the results for autumn likely represent a regime shift expressed as a change in the seasonal cycle of relative humidity that is driven by the increasing amount of open water in the Arctic. The consequence of these changes is that the feedback from clouds shown in Fig. 1 likely increases despite concomitant increases in absolute humidity.

Covariability between temperature and PWV in the Arctic does not closely fit the Clausius–Clapeyron relationship because of dominance by ice saturation at low temperatures and temperature buffering from latent energy exchange[21] at 273 K. The latter results in relatively moist conditions for temperatures above 273 K, corresponding to lower CRE (that is, Figs 4 and 6b). For this reason, clouds in summer may contribute less flux in the future, by up to $3\,W\,m^{-2}$, because of increasingly moist conditions for temperatures near 273 K in summer (Fig. 6b). Therefore, a relative dampening of the cloud feedback shown in Fig. 1 might be expected in summer.

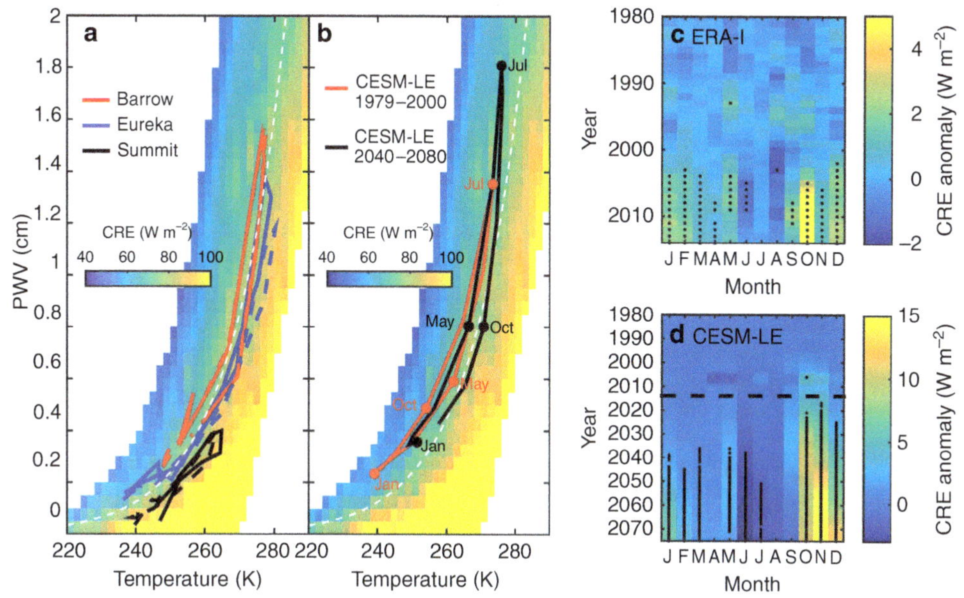

Figure 6 | Recent and projected changes in CRE associated with changes in humidity. (**a**) Downwelling infrared CRE as a function of observed near-surface air temperature and PWV for optically thick clouds at Barrow (red), Eureka (blue) and Summit (black). These observations are from the same time period as the CRE calculations in Fig. 5; the dashed lines are for averages over all times, while the solid lines are for averages over times when CRE > 60 W m^{-2} (optically thick clouds). (**b**) Similar to **a**, but the red line are CESM-LE monthly mean values from 1979 through 2000, while the black line is for 2040-2080. The CRE surfaces in **a,b** are the values from Fig. 4c interpolated to a T-PWV grid (2 K by 0.05 cm). The white dashed lines in **a,b** are the Clausius-Clapeyron relationship for near-surface air temperature and PWV from ref. 29, assuming a relative humidity (with respect to liquid) of 100% and a scale height of 3 km. (**c**) Estimated anomalies in CRE determined by mapping monthly mean ERA-Interim 2-m air temperature and PWV on the CRE surface in **b**. (**d**) is similar to **c**, but for the CESM-LE historical (1979-2005) and projected (2006-2080) data sets. Data in **c,d** are 5-year running means averaged over the far-western and Eurasian Arctic Ocean domain (north of 70N and 70-220E). Anomalies are calculated using 1979-2000 as a baseline and black dots represent differences between the mean of the 5-year sample (centred on point) and the baseline that are statistically significant ($P < 0.05$).

Discussion

This work presents the novel finding that, for a constant relative humidity profile, CRE_{FIR} and CRE_{AW} compensate each other as they change with temperature and PWV over ranges characteristic of the Arctic. The result has important implications in that the CRE remains fairly constant with warming as long as the relative humidity profile stays constant. This is in contrast to lower latitudes, where CRE is expected to decrease with increasing temperature at constant relative humidity, illustrating a unique sensitivity of the polar regions to climate change. In warmer, wetter conditions than the Arctic (for example, mid-latitudes and tropics, see Fig. 2), the FIR plays a negligible (rather than compensating) role, and CRE is thus dominated by CRE_{AW}. Because CRE_{AW} decreases with increasing temperature and PWV, warmer/wetter regions are expected to become less sensitive to clouds with warming. In very cold, dry conditions (temperatures below ~230 K), CRE_{FIR} is larger than CRE_{AW}. CRE_{AW} increases with temperature and PWV, while CRE_{FIR} is transitioning similarly to the behaviour of CRE_{AW} near 280 K. When combined, the sensitivity to temperature is dominated by CRE_{AW}, and thus CRE increases slightly with temperature and PWV. We speculate that below 230 K this may cause a positive feedback whereby increases in temperature increase CRE, which further increases temperature. Although the feedback is probably small, it is notable that it is driven by temperature alone.

In the Arctic, where changes in CRE_{AW} and CRE_{FIR} with temperature (at constant relative humidity) compensate, changes in covariability between temperature and humidity are necessary to modify CRE independently of variability in cloud properties. This work demonstrates this process both through radiative transfer calculations and through observations using spectrally resolved measurements; because similar instrumentation is available in multiple locations, we also observe how the process modulates the climate in different regions of the Arctic. Owing to the described mechanism, changes in the way temperature and moisture vary relative to each other cause shifts in the control of infrared irradiance between atmospheric gases and clouds, modifying the separation in downwelling infrared flux (W m^{-2}) between the Arctic clear and cloudy states. The cloud feedback described in Fig. 1 is amplified or dampened accordingly, and, with it, the importance of variability in cloud microphysical and macrophysical properties. Indeed, using temperature and humidity from reanalysis data and climate model projections, we show trends of increasing CRE in the Arctic in all seasons except summer. Projected changes in Arctic cloud optical depth and cloud cover in autumn correspond to +3.8 and +14 W m^{-2} (2081–2100 minus 1961–1980)[5] of infrared forcing, respectively, which is similar in magnitude to changes in CRE described here.

The variability in CRE described here is independent of shortwave CRE because the analysed perturbations are associated with atmospheric gases, which largely transmit solar radiation. This suggests a partial offset of the shortwave cooling supported by clouds and points to a reduction in the dampening of the ice-albedo feedback in the presence of increased cloud cover, a connection that is linked to interannual variability in sea ice[34]. Our findings also highlight the complexity of interpreting projections of cloud feedbacks. For example, decreases in cloud occurrence can accompany decreases in relative humidity, as was the case in summer 2007 (refs 23, 35), yet our results suggest that such changes could still result in net increases in CRE because the clouds that do form each contribute relatively more infrared flux to the surface energy budget.

ERA-Interim may be a more robust data set for climate monitoring than other reanalyses, including in the Arctic[19,36], and the CESM-LE permits accounting for natural variability[32]. Nevertheless, caution must be taken when interpreting the time series shown in Fig. 6 because potential changes in the long-term temperature–humidity covariability are poorly understood. Regardless, the results presented here demonstrate conceptually the substantial sensitivity of CRE to modest, and realistic, shifts in the Arctic atmospheric state and provides further motivation for the need to reduce uncertainties in cloud properties in climate models. More research is also needed to interpret how temperature–humidity and temperature–humidity–CRE interactions occur over long timescales in association with external forcing and low-frequency natural variability. In addition, the associated enhancement (as is likely in autumn, winter and spring) or dampening (as is likely in summer) of cloud feedbacks should be investigated.

Methods

Data. Observations from Summit (July 2010 to August 2012, 72.58°N, 38.48°W, 3,210 m) are from the Integrated Characterization of Energy, Clouds, Atmospheric state and Precipitation at Summit (ICECAPS) project[20]. Barrow observations (January 2006 to December 2008, 71.325°N, 156.625°W, 8 m) are from the US Department of Energy Atmospheric Radiation Measurement (ARM) Program obtained at the North Slope of Alaska (NSA) site. ARM and ICECAPS data are available at http://www.arm.gov. The Eureka measurements (March 2006 to December 2008, 80.053°N, 86.417°W, 10 m) were supported by the NOAA Earth System Research Laboratory (ESRL) and the Canadian Network for the Detection of Arctic Change (CANDAC). Eureka data used in this study are archived by NOAA-ESRL, available at ftp://ftp1.esrl.noaa.gov/psd3/arctic/eureka/. Observational data may also be accessed through the International Arctic Systems for Observing the Atmosphere (IASOA) data portal (http://www.esrl.noaa.gov/psd/iasoa/). CESM-LE can be acquired from https://www.earthsystemgrid.org/home.htm. ERA-Interim data are archived by ECMWF and are available at http://apps.ecmwf.int/datasets/.

Observed CRE. To determine the observed CRE, $F\downarrow_{\text{all-sky}}$ (equation (1)) is derived from radiances measured by Atmospheric Emitted Radiance Interferometers (AERI) at each station; because these measurements are at zenith and do not include the entire spectral range of interest, they are supplemented with radiative transfer calculations using collocated radiosoundings as input. $F\downarrow_{\text{clear-sky}}$ is estimated entirely from radiative transfer calculations (ref. 6 and references therein). To calculate the partial-band fluxes used in this study (CRE_{FIR} and CRE_{AW}), integrations are performed over subranges for the FIR (10–630 cm^{-1}) and AW (700–1,390 cm^{-1}), similar to ref. 28. These ranges are chosen to correspond to the nearest band edges in the Rapid Radiative Transfer Model (RRTM)[37]. A small amount of CRE occurs at frequencies outside of the partial-band spectral regions used here; the residual is $-0.6 \pm 1.2\ W\ m^{-2}$. CRE is calculated using observations and radiative transfer modelling, which account for scattering, reflection, anisotropy and cloud thickness.

Radiative transfer modelling. For deriving idealized CRE (Fig. 4a–c), radiosoundings from Barrow (2002–2012) and Summit (2010–2015) are supplemented with the sub-Arctic summer standard atmosphere[38] between the radiosonde termination height and 60 km. Carbon dioxide (CO_2) is set to 395 p.p.m. with a constant mixing ratio with height and mixing ratios of O_3, N_2O, CH_4, CO and O_2 are set to values from the standard atmosphere. Radiative transfer calculations for gaseous emission are performed using RRTM, and scattering and emission from the clouds are calculated using discrete-ordinate-method radiative transfer[39]. The surface and cloud are near thermodynamic equilibrium for most low-level overcast Arctic clouds, such that the cloud and surface temperatures are typically within ~5 K of each other (and net surface infrared radiation is typically less than 20 $W\ m^{-2}$ ref. 30). Thus, to reduce the influence of unrealistic combinations of clouds and atmospheric states, only samples when the surface temperature is within 5 K of the cloud temperature are retained ($n = 4{,}664$).

For calculations of modelled CRE (Fig. 2, curves), the radiative transfer calculations are the same as described above but only account for absorption, while scattering is ignored. These calculations are performed using the sub-Arctic summer standard atmosphere[38] with an optically thick cloud in the layer between 1 and 2 km; the temperature profile is shifted uniformly and PWV is calculated assuming constant relative humidity at all levels.

Effective cloud emissivity. CRE, calculated for optically thick clouds, can be roughly converted to time-averaged CRE by multiplying by the cloud fraction and the mean cloud emissivity as $\varepsilon_{\text{cloud,effective},\nu} = (R\downarrow_{\text{all-sky},\nu} - R\downarrow_{\text{clear-sky},\nu})/B_\nu(T_s)$, where R is the radiance at zenith and $B_\nu(T_s)$ is the Planck function using the near-surface air temperature. The calculation is made using 900 cm^{-1} for wavenumber, ν. Atmospheric transmission between the surface and the cloud base and cloud anisotropy are neglected. These approximations result in a conservative estimate for the magnitude of the anomalies displayed in Fig. 6.

References

1. Francis, J. A. & Hunter, E. Changes in the fabric of the Arctic's greenhouse blanket. *Env. Res. Lett.* **2,** 045011 (2007).
2. Screen, J. A. & Simmonds, I. The central role of diminishing sea ice in recent Arctic temperature amplification. *Nature* **464,** 1334–1337 (2010).
3. Pithan, F. & Mauritsen, T. Arctic amplification dominated by temperature feedbacks in contemporary climate models. *Nat. Geosci.* **7,** 181–184 (2014).
4. Serreze, M. C., Barrett, A. P., Stroeve, J. C., Kindig, D. N. & Holland, M. M. The emergence of surface-based Arctic amplification. *Cryosphere* **3,** 11–19 (2009).
5. Chen, Y., Miller, J. R., Francis, J. A. & Russell, G. L. Projected regime shift in Arctic cloud and water vapor feedbacks. *Environ. Res. Lett.* **6,** 044007 (2011).
6. Cox, C. J., Walden, V. P. & Rowe, P. M. A comparison of the atmospheric conditions at Eureka, Canada, and Barrow, Alaska (2006-2008). *J. Geophys. Res.* **117,** D12204 (2012).
7. Cox, C. J. *et al.* Downwelling longwave flux over Summit, Greenland, 2010-2012: Analysis of surface-based observations and evaluation of ERA-Interim using wavelets. *J. Geophys. Res.* **118,** D021975 (2014).
8. Bennartz, R. *et al.* July 2012 Greenland melt extent enhanced by low-level liquid clouds. *Nature* **496,** 83–86 (2013).
9. Dong, X. *et al.* A 10 year climatology of Arctic cloud fraction and radiative forcing at Barrow, Alaska. *J. Geophys. Res.* **115,** D17212 (2010).
10. Shupe, M. D. & Intrieri, J. M. Cloud radiative forcing of the Arctic surface: The influence of cloud properties, surface albedo, and solar zenith angle. *J. Climate* **17,** 616–628 (2004).
11. Kwok, R. & Untersteiner, N. The thinning of Arctic sea ice. *Phys. Today* **64,** 36–41 (2011).
12. Vavrus, S. The impact of cloud feedbacks on Arctic climate under greenhouse forcing. *J. Climate* **17,** 603–615 (2004).
13. Charlock, T. P. & Ramanathan, V. The albedo field and cloud radiative forcing produced by a General Circulation Model with internally generated cloud optics. *J. Atmos. Sci.* **42,** 1408–1429 (1985).
14. Huang, X. *et al.* A global climatology of outgoing longwave spectral cloud radiative effect and associated effective cloud properties. *J. Climate* **27,** 7475–7492 (2014).
15. Turner, D. D. & Mlawer, E. J. The radiative heating in underexplored bands campaigns. *Bull. Amer. Meteorol. Soc.* **91,** 911–923 (2010).
16. Harries, J. *et al.* The far-infrared Earth. *Rev. Geophys.* **46,** 34 (2008).
17. Feldman, D. R., Collins, W. D., Pincus, R., Huang, X. & Chen, X. Far-infrared surface emissivity and climate. *Proc. Natl Acad. Sci. USA* **111,** 16297–16302 (2014).
18. Wang, X. & Key, J. R. Recent trends in Arctic surface, cloud, and radiation properties from space. *Science* **299,** 1725–1728 (2003).
19. Serreze, M. C., Barrett, A. P. & Stroeve, J. Recent changes in tropospheric water vapor over the Arctic as assessed from radiosondes and atmospheric reanalyses. *J. Geophys. Res.* **117,** D10104 (2012).
20. Stephens, G. L. & Ellis, T. D. Controls of global-mean precipitation increases in Global Warming GCM experiments. *J. Climate* **21,** 6141–6155 (2008).
21. Curry, J. A., Schramm, J. L., Serreze, M. C. & Ebert, E. E. Water vapor feedback over the Arctic Ocean. *J. Geophys. Res.* **100,** 14223–14229 (1995).
22. Curry, J. A., Schramm, J. L., Rossow, W. B. & Randall, D. Overview of Arctic cloud radiation characteristics. *J. Climate* **9,** 1731–1764 (1996).
23. Dong, X. *et al.* Critical mechanisms for the formation of extreme arctic sea-ice extent in the summers of 2007 and 1996. *Clim. Dyn.* **43,** 53–70 (2013).
24. McFarlane, S. A., Long, C. N. & Flaherty, J. A climatology of surface cloud radiative effects at the ARM Tropical Western Pacific sites. *J. Appl. Meteorol. Clim.* **52,** 996–1012 (2013).
25. Dong, X., Xi, B. & Minnis, P. A climatology of midlatitude continental clouds from the ARM SGP Central Facility. Part II: cloud fraction and surface radiative forcing. *J. Climate* **19,** 1765–1783 (2006).
26. Shupe, M. D. *et al.* High and dry: new observations of tropospheric and cloud properties above the Greenland ice sheet. *Bull. Am. Meteorol. Soc.* **94,** 169–186 (2013).
27. Shupe, M. D. *et al.* Clouds at Arctic atmospheric observatories. Part I: Occurrence and macrophysical properties. *J. Appl. Meteorol. Clim.* **50,** 626–644 (2011).
28. Town, M. S., Walden, V. P. & Warren, S. G. Spectral and broadband longwave downwelling radiative fluxes, cloud radiative forcing, and fractional cloud cover over the South Pole. *J. Climate* **18,** 4235–4252 (2005).

29. Stephens, G. L. On the relationship between water vapor over the oceans and sea surface temperature. *J. Climate* **3,** 634–645 (1990).
30. Stramler, K., Del Genio, A. D. & Rossow, W. B. Synoptically driven Arctic winter states. *J. Climate* **24,** 1747–1762 (2011).
31. Dee, D. P. *et al.* The ERA-Interim reanalysis: configuration and performance of the data assimilation system. *Q. J. R. Meteorol. Soc.* **137,** 553–597 (2011).
32. Kay, J. E. *et al.* The Community Earth System Model (CESM) Large Ensemble Project: A community resource for studying climate change in the presence of internal climate variability. *Bull. Am. Meteorol. Soc.* **96,** 1333–1349 (2015).
33. Meinshausen, M. *et al.* The RCP greenhouse gas concentrations and their extensions from 1765 to 2300. *Clim. Change* **109,** 213–241 (2011).
34. Choi, Y.-S. *et al.* Connecting early summer cloud-controlled sunlight and late summer sea ice in the Arctic. *J. Geophys. Res.* **119,** 087–11,099 (2014).
35. Kay, J. E., L'Ecuyer, T., Gettelman, A., Stephens, G. & O'Dell, C. The contribution of cloud and radiation anomalies to the 2007 Arctic sea ice extent minimum. *Geophys. Res. Lett.* **35,** L08503 (2014).
36. Simmons, A. J., Willett, K. M., Jones, P. D., Thorne, P. W. & Dee, D. P. Low-frequency variations in atmospheric humidity, temperature, and precipitation: Inferences from reanalyses and monthly gridded observational data sets. *J. Geophys. Res.* **115,** D01110 (2010).
37. Mlawer, E. J., Taubman, S. J., Brown, P. D., Iacono, M. J. & Clough, S. A. RRTM, a validated correlated-k model for the longwave. *J. Geophys. Res.* **102,** 16663–16682 (1997).
38. McClatchey, R. A., Fenn, R. W., Selby, J. E. A., Volz, F. E. & Garing, J. S. *Optical Properties of the Atmosphere (3rd edn).* Tech. Rep. Report No. AFCRL-72-0497, 108 (Air Force Geophysics Laboratory, Hanscom AFB, Bedford, MA, USA, 1972).
39. Stamnes, K., Tsay, S.-C., Wiscombe, W. & Jayaweera, K. A numerically stable algorithm for discrete-ordinate-method radiative transfer in multiple scattering and emitting layered media. *Appl. Opt.* **27,** 2502–2509 (1988).

Acknowledgements

C.J.C. acknowledges funding from Arctic Research Program of the NOAA Climate Program Office and the Cooperative Institute for Research in Environmental Sciences (CIRES) Visiting Fellowship Program. V.P.W. acknowledges funding from National Science Foundation grants, ARC-0856773, PLR-1414314 and PLR-1420932. P.M.R. acknowledges funding from NSF award ARC-1108451 and from USACH-DICYT 041331CC_DAS and FONDECYT 1151034. M.D.S. acknowledges funding from NSF PLR-1314156 and US Department of Energy DE-SC0011918. We thank the anonymous reviewers for their comments, which helped to improve the manuscript. We acknowledge useful conversations with D. Feldman (LBNL), C. Kou (LBNL), C. Long (CIRES/NOAA), Z. Mariani (U. Toronto), S. Neshyba (U. Puget Sound) and K. Strong (U. Toronto).

Author contributions

C.J.C. conceived and designed the experiments; C.J.C. and P.M.R. performed the experiments; C.J.C., P.M.R., V.P.W. and M.D.S. analysed the data; C.J.C., P.M.R. and V.P.W. contributed materials/analysis tools; C.J.C., P.M.R., V.P.W. and M.D.S. co-wrote the paper.

Additional information

Competing financial interests: The authors declare no competing financial interests.

8

Quantifying the ozone and ultraviolet benefits already achieved by the Montreal Protocol

M.P. Chipperfield[1,2], S.S. Dhomse[1,2], W. Feng[1,3], R.L. McKenzie[4], G.J.M. Velders[5] & J.A. Pyle[3,6]

Chlorine- and bromine-containing ozone-depleting substances (ODSs) are controlled by the 1987 Montreal Protocol. In consequence, atmospheric equivalent chlorine peaked in 1993 and has been declining slowly since then. Consistent with this, models project a gradual increase in stratospheric ozone with the Antarctic ozone hole expected to disappear by ~2050. However, we show that by 2013 the Montreal Protocol had already achieved significant benefits for the ozone layer. Using a 3D atmospheric chemistry transport model, we demonstrate that much larger ozone depletion than observed has been avoided by the protocol, with beneficial impacts on surface ultraviolet. A deep Arctic ozone hole, with column values <120 DU, would have occurred given meteorological conditions in 2011. The Antarctic ozone hole would have grown in size by 40% by 2013, with enhanced loss at subpolar latitudes. The decline over northern hemisphere middle latitudes would have continued, more than doubling to ~15% by 2013.

[1] School of Earth and Environment, University of Leeds, Leeds LS2 9JT, UK. [2] National Centre for Earth Observation (NCEO), University of Leeds, Leeds LS2 9JT, UK. [3] National Centre for Atmospheric Science (NCAS), UK. [4] National Institute of Water and Atmospheric Research (NIWA), Lauder Private Bag 50061, New Zealand. [5] National Institute for Public Health and the Environment, PO Box 1, Bilthoven 3720 BA, The Netherlands. [6] Department of Chemistry, University of Cambridge, Cambridge CB2 1EW, UK. Correspondence and requests for materials should be addressed to M.P.C. (email: M.Chipperfield@leeds.ac.uk).

Concern over the depletion of the ozone layer started in the 1970s with the suggestion that chlorine from chlorofluorocarbons could reach the stratosphere and cause ozone loss[1]. Research activities intensified greatly in the mid-1980s following the surprise discovery of significant ozone depletion over the Antarctic[2], the so-called ozone hole. This large decrease in ozone was explained by chlorine- and bromine-catalysed loss acting in the particularly cold conditions of the Antarctic polar lower stratosphere, which allow polar stratospheric clouds to form (see, for example, Solomon[3]). Ozone depletion was subsequently observed in the Arctic springtime stratosphere, but the loss was much smaller than the Antarctic and more variable from year to year[4,5]. Cold years in the Arctic stratosphere, such as 1994/95, lead to an integrated column loss of around 120–140 DU ($\sim$35%), while warm years, without polar stratospheric cloud occurrence, produce essentially zero loss.

The discovery of the Antarctic ozone hole helped stimulate the initial signing in 1987 of the Montreal Protocol, an international treaty to limit production of chlorine- and bromine-containing ozone-depleting substances (ODSs). The Montreal Protocol has since been strengthened greatly through subsequent amendments and adjustments, supported by ongoing research, which has enhanced our understanding that ozone loss, in the polar regions and globally, is driven by chemical processes involving chlorine- and bromine-containing gases, arising mainly as breakdown products of ODSs. With global compliance with Montreal Protocol regulations, atmospheric chlorine peaked at $\sim$3.6 parts per billion by volume (p.p.b.v.) in 1993 at the surface and a few years later in the stratosphere, and then began to decline. The current tropospheric loading is 10% below the 1993 peak value.

Most of the major ODSs have long lifetimes (many decades) in the atmosphere[6], so a significant time delay is expected before the reduction in atmospheric emissions of chlorine and bromine translate into an increase in stratospheric ozone. Column ozone amounts in the northern middle latitudes reached their minimum values in the mid-1990s at about 6–8% below the 1964–1980 mean[7], due to the combination of elevated stratospheric chlorine and bromine with enhanced aerosol loading after the 1991 eruption of Mt Pinatubo. Mean northern mid-latitude ozone values at the present day are still around 4% below the long-term mean[8]. The Antarctic ozone hole continues to re-appear each spring; stratospheric chlorine is not expected to return to 1980 levels, when the Antarctic ozone hole was first detectable, until about 2050. Substantial Arctic ozone loss also continues to be observed in the late winter/early spring of some years when polar stratospheric temperatures are particularly low. The largest Arctic ozone loss observed to date occurred in the recent cold winter of 2010/11 (ref. 9). By early April 2011, about 75% of the ozone had been destroyed in a limited altitude region of the polar lower stratosphere. Despite this relatively large local loss there were substantial differences in the spatial extent and the degree of local depletion between a typical Antarctic ozone hole and the 2011 Arctic loss. Near complete removal of ozone occurs annually in the Antarctic lower stratosphere while the maximum loss in the 2011 in Arctic is estimated at just over 80% at 18–20 km within a polar vortex covering a much smaller area than in the south. By late March, around 45% of the Arctic polar vortex had column ozone values below 275 DU. However, the observed column did not fall below the threshold for the Antarctic ozone hole, taken as 220 DU. Therefore, by that definition an 'Arctic ozone hole' did not occur.

Several studies have investigated the benefits of the Montreal Protocol. Prather *et al.*[10] first discussed the possible impacts if the protocol had not come into force. Later studies have attempted to use atmospheric models to quantify the benefits. These have focused on the avoided future impacts later this century, assuming a continued strong growth in ODS emissions, and stratospheric chlorine. Morgenstern *et al.*[11] used a coupled chemistry-climate model (CCM) to compare the atmosphere in 2030 with and without the effect of the Montreal Protocol. By 2030 their assumed stratospheric chlorine loading was 9 p.p.b.v. In addition to increased stratospheric ozone loss, they found a feedback on stratospheric dynamics and an impact on surface climate. Newman *et al.*[12] used a similar CCM to investigate the impact on an uncontrolled 3% per year growth in chlorine through 2065. As expected, they found very large stratospheric ozone depletion and also large temperature changes in response to the ozone loss. Garcia *et al.*[13] used their CCM to study the chemical processes responsible for this possible collapse of the ozone layer in the mid twenty-first century. Finally, Egorova *et al.*[14] also used a CCM to study the impact of the Montreal Protocol on the ozone layer and again focused mainly on the large losses that would have occurred by the end of this century. Their analysis did extend from 1987 to 2100 but, as they were using a free-running CCM like all of the above studies, internal model variability prevented them from being able to diagnose the impact of the Montreal Protocol during specific years of the recent past.

These 'World Avoided' studies have thus mainly looked far into the future under the assumption that no action would have been taken to control ODS emissions, and compared the results with those expected with full implementation of the Montreal Protocol. In the face of the expected mounting evidence of an impact it seems quite unlikely that ODS emissions would have continued to grow at the assumed uncontrolled rates without some policy action by the middle of this century. Furthermore, these studies have partly focused on the coupled climate impact of large O_3 loss, but climate models often have temperature biases that make accurate quantitative modelling of the strongly temperature-dependent polar ozone loss extremely challenging.

In contrast, off-line chemical transport models (CTMs) have been shown[15–17] to give an accurate simulation of stratospheric ozone loss when they use meteorological analyses—our best estimate of the 'real' meteorology. CTMs have also been developed with a focus on accurate and complete representations of chemical processes, although CCMs often use computationally fast schemes in more complex climate models. Therefore, a good CTM is likely to produce a more faithful simulation of the present and past atmosphere than a good CCM.

Here we use CTM calculations, covering the past two decades during which the growth of atmospheric chlorine and bromine abundances first slowed and then started to decline, to assess the benefits already achieved by the Montreal Protocol. We show that much larger ozone depletion than observed has been avoided by the protocol. A deep Arctic ozone hole, with column values $<$120 DU, would have occurred given the meteorological conditions in 2011. The Antarctic ozone hole would have grown in size by 40% by 2013, with enhanced loss at subpolar latitudes. The decline over northern hemisphere middle latitudes would have continued, more than doubling to $\sim$15% by 2013. This ozone loss would have led to increases in surface ultraviolet of up to 8–12% in Australia and New Zealand and 14% in the United Kingdom, and consequent increases in skin cancer.

Results

Global ozone. Figure 1 shows the difference between the modelled evolution of column ozone since the year 2000 between the 'MP' model run, which includes the effects of the Montreal Protocol and is based on observed atmospheric ODS loading, and the 'NoMP' model run, which assumes continued growth of ODSs production at 3% per year after 1987 (ref. 18; Fig. 1a, plotted until 2010 which corresponds to the stratospheric loading until about 2014). MP is our best reconstruction of the recent

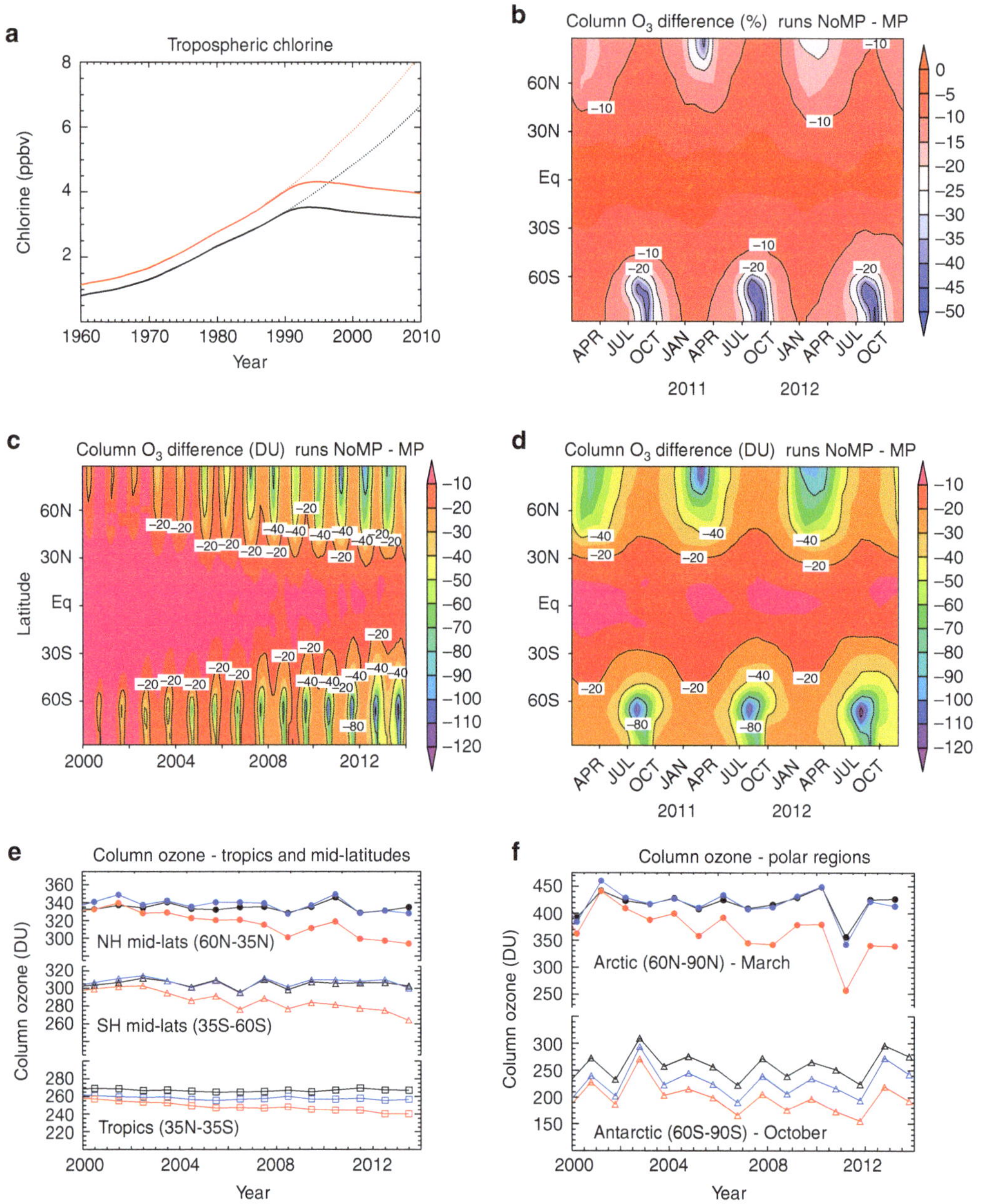

Figure 1 | Calculated impact of Montreal Protocol on atmospheric halogens and column ozone. (**a**) Time series of surface chlorine (black) and surface equivalent chlorine (chlorine + 50 × bromine, red) from observations (solid lines) and assumed 3% growth (dotted lines). (**b–d**) The difference in zonal mean monthly mean column ozone (DU) between run MP and run NoMP for (**b**) 2010–2012 (%), (**c**) 2000–2013 (DU) and (**d**) 2010–2012 (DU). (**d**) The same results as **c** on an expanded time axis for clarity. (**b**) The same results as **d** but as a percentage change. (**e**, **f**) Comparisons of SBUV observed column ozone[37] in different latitude regions (black) with model runs MP (blue) and NoMP (red).

evolution of the ozone layer, and is in excellent agreement with observations: the model simulates the global column ozone distribution[19] to within a few DU in mid-latitudes and the Arctic spring and to within 10 DU in the tropics (Fig. 1e,f). Meanwhile, the NoMP run projects what could have occurred given our specified ODS growth rate without the Montreal Protocol: it describes the 'World Avoided'.

There are significant differences between the two runs, which become greater with time. These differences are most evident in the polar regions of both hemispheres. Figure 1 also shows that a decrease of around 10 DU (~5%) has been avoided by the Montreal Protocol in the tropics, where observed ozone trends have been small. The larger differences over populated middle latitudes indicate that the Montreal Protocol has already had a very significant effect in protecting the ozone layer there. Without the Protocol ozone columns over northern middle latitudes would by 2010 have been between 5 and 10% lower than has occurred with the Protocol. Recall that the observed ozone, in response to the Protocol, is still about 4% lower in northern middle latitudes than the 1964–1980 long-term mean. So, the Protocol has by 2014 already saved us from at least a doubling in ozone depletion.

The most dramatic measure of ozone loss is the Antarctic ozone hole. Figure 1 shows that without the Montreal Protocol ozone loss at both poles would have become very much larger. In the Antarctic, the observed ozone loss is already essentially 100% in the lower stratosphere by early October. Therefore, the extent

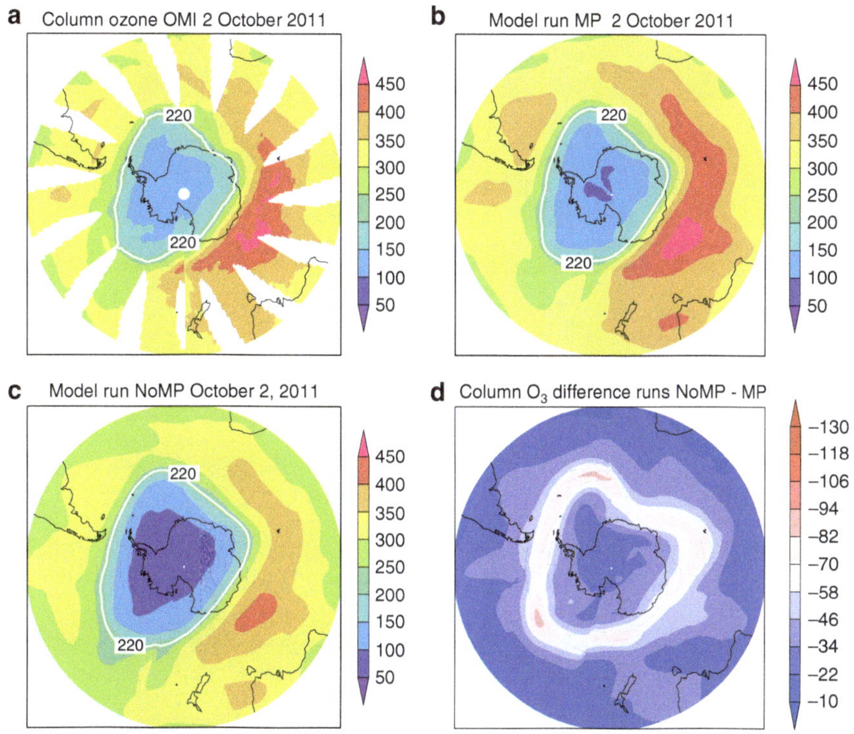

Figure 2 | Column ozone in the Antarctic from satellite observations and model for October 2011. Column ozone (DU) on 2 October 2011 (**a**) observed by OMI, (**b**) from model run MP and (**c**) from model run NoMP (with the 220 DU contour indicated in white). (**d**) Difference in column ozone between runs NoMP and MP.

of ozone depletion can only be increased by loss at higher/lower altitudes or at lower latitudes, or lasting for a longer period. Figure 2 shows a comparison of column ozone in October 2011 from runs MP and noMP. Without the Montreal Protocol our model shows around 100 DU additional loss in a substantial edge region of the polar vortex in spring by 2014 (Figs 1d and 2d).

Arctic ozone hole in 2011. We consider next a particularly dramatic polar feature, the very large additional Arctic ozone loss, which was avoided by the Montreal Protocol in 2011. Figure 3e shows the time series of minimum column ozone in the Arctic in winter 2010/11 observed by the Ozone Monitoring Instrument (OMI). The observed column has a local minimum of around 230 DU, lasting several weeks, during the main phase of chemical ozone depletion in March. Results from the run MP show excellent agreement with OMI, further demonstrating the ability of the TOMCAT CTM to reproduce the observed evolution of the ozone layer when forced by observed ODS boundary conditions and appropriate meteorology. In strong contrast to run MP and the observations, the NoMP integration indicates a minimum column below 120 DU, a value that is quite unprecedented in the observational record. Large depletion also lasts not weeks, but months. These results are further illustrated in Fig. 3a–d, which show maps of observed and modelled column ozone on 26 March 2011. The observed region of relatively low column ozone (250–275 DU) is clearly visible and is well captured by model run MP. With the increased chlorine and bromine in run NoMP there is a further dramatic decrease in column ozone by up to 130 DU over a wide region of the Arctic. The 220 DU contour in Fig. 3c clearly demonstrates that without the Montreal Protocol a deep Arctic ozone hole would have occurred in 2011.

Recent decadal variations in Antarctic and Arctic. Although the Arctic 2010/11 stands out as a major signature of additional polar ozone loss, with around 110 DU lower ozone in the monthly mean zonal mean, Fig. 1 shows that the Montreal Protocol has already had an impact in other (warmer) years in the Arctic, avoiding much larger ozone loss than was actually observed. The impact of the Montreal Protocol in recent years on the size of the Antarctic and possible Arctic ozone holes is shown in Fig. 4. In the Antarctic, the MP integration shows excellent agreement with the size of the ozone hole derived from OMI with, for example, a peak area of $25 \times 10^6\,km^2$ in 2008 and some small interannual variability. Without the Montreal Protocol, the size of the hole would have already increased substantially, by ~40% to over $37 \times 10^6\,km^2$ in 2013. Moreover, the onset of the low (<220 DU) columns would start much earlier in the season. Neither the observations nor model run MP show the occurrence of an Arctic ozone hole. As noted previously, although there was considerable chemical destruction of ozone in 2011 (ref. 9), this was offset by dynamical effects that mitigated the ozone depletion. These effects are well captured in our model. However, as discussed above, an Arctic ozone hole would clearly have occurred under the meteorological conditions of 2010/11 without the Montreal Protocol. It is also worth noting the large size of the calculated Arctic ozone hole in 2011 in run NoMP. With an area near $20 \times 10^6\,km^2$, the Arctic ozone hole would have been a similar size to the actual Antarctic hole observed in the 1990s and 2000s (see also the region contained by the 220 DU contour in Fig. 3c). Figure 4 also shows that many other recent winters (for example, 2012/13) would also have experienced periods with minimum column ozone below 220 DU, that is, the model run indicates that smaller Arctic ozone holes would have become a regular occurrence.

Discussion

Ozone absorbs radiation in the ultraviolet and infrared regions of the electromagnetic spectrum. As a result, ozone depletion affects the temperature structure of the atmosphere, which has

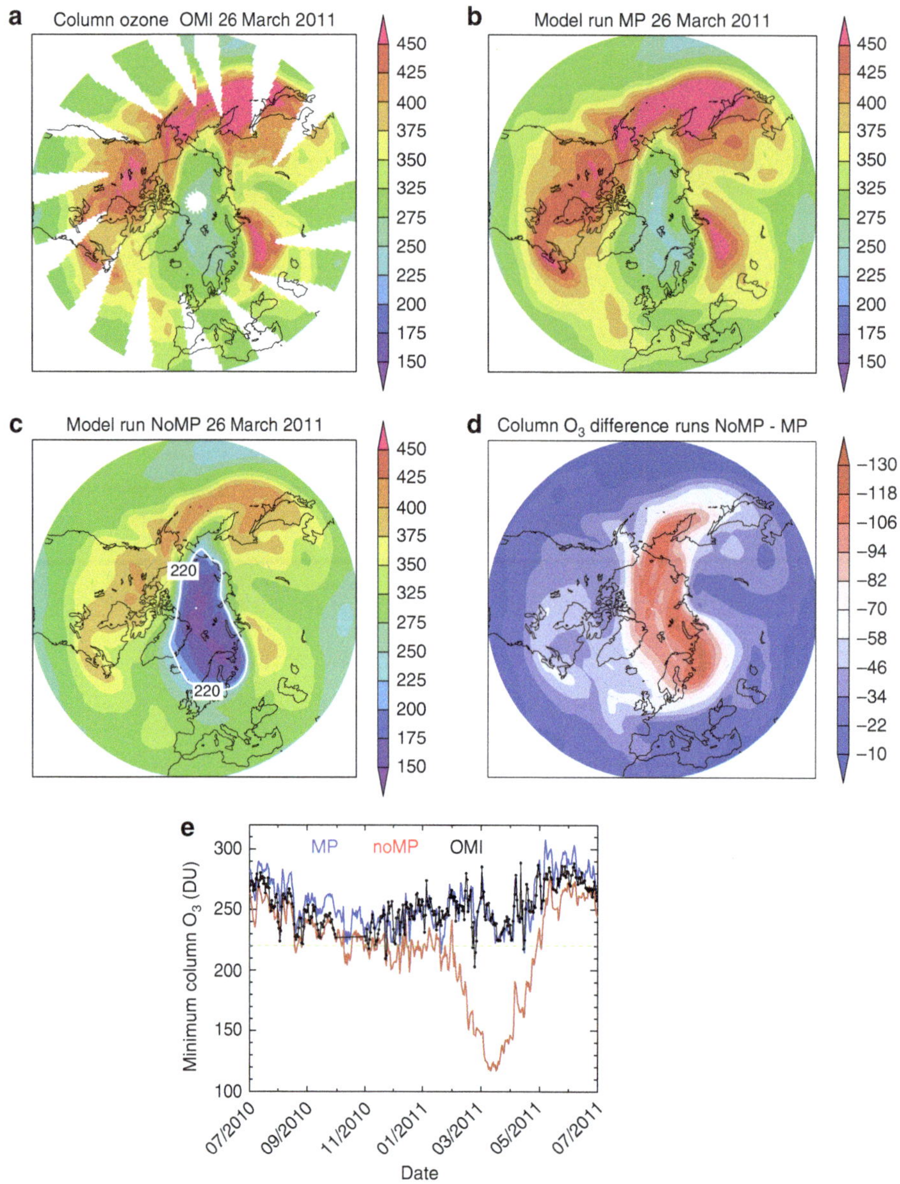

Figure 3 | Evolution of column ozone in the Arctic from satellite observations and model for winter 2010/11. Column ozone (DU) on 26 March 2011 (**a**) observed by OMI, (**b**) from model run MP and (**c**) from model run NoMP (with the 220 DU contour indicated in white). (**d**) Difference in column ozone between runs NoMP and MP. (**e**) The daily minimum O_3 column in the Arctic region (latitude >45°N) from mid-2010 to mid-2011 as observed by OMI (black points), along with equivalent model results from run MP (blue) and run NoMP (red).

important long-term consequences for circulation and climate. In addition, ozone depletion increases solar radiation at the surface thereby increasing the risk of damage to humans, flora and fauna. For example, model studies have linked the depletion of O_3 in the Antarctic to a delay in the breakdown of the polar vortex through decreased heating[7,8,20,21]. Without the Montreal Protocol, the enhanced ozone loss would have increased this effect and likely led to a further extension of the duration of the Antarctic vortex. Antarctic ozone depletion has also been linked to an observed change in the surface climate at high southern latitudes. The depletion has led to a change in the Southern Annular Mode with a poleward shift in the position of the summertime southern hemisphere jet[22,23]. Changes in high-latitude surface temperatures and in southern hemisphere rainfall have also been attributed to the ozone loss[24,25]. We do not discuss these in detail here, and these effects cannot be modelled in our offline CTM, but we simply point out that even greater loss than observed would certainly have had further important consequences.

Stratospheric ozone depletion leads to an increase in potentially harmful radiation at the surface, assuming that other absorbers, including clouds and aerosol, remain the same. Previous 'World Avoided' studies have assessed the impact of changes in surface ultraviolet into the future[26]. Based on the Morgenstern *et al.* study[11], Van Dijk *et al.*[27] estimated the number of cases of skin cancer avoided by 2030 at very roughly 2 million, or about 14% fewer skin-cancer cases per year. The much more substantial ozone declines in 2065 predicted in the study by Newman *et al.*[12] would presumably have led to very much higher skin cancer rates by 2100 (ref. 28). Many other factors can affect ultraviolet penetration to the surface and assessing the knock-on health impacts is very uncertain[29]. Nevertheless, the world

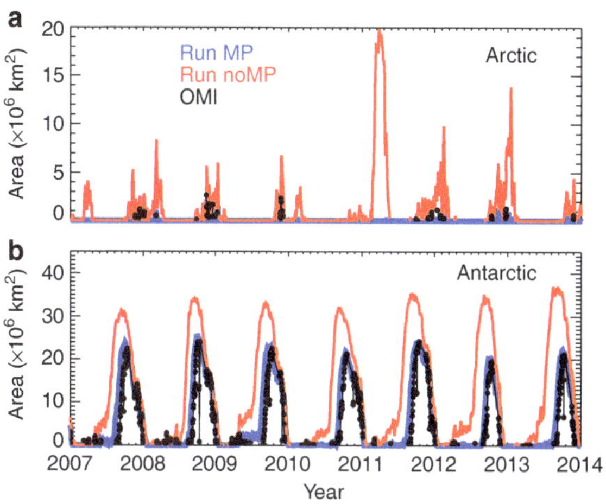

Figure 4 | Area extent of the Antarctic and Arctic ozone holes. The area of the ozone hole is defined as the area with column ozone < 220 DU. (**b**) The observed area of the Antarctic ozone hole from 2007 to 2014 estimated from OMI data (black symbols). (**a**) For the Arctic, the observed ozone hole area is near zero in all years. The figure also includes the ozone hole areas calculated from model runs MP (blue) and NoMP (red).

avoided, with large additional ozone loss, would necessarily have seen important increases in surface radiation.

We can assess the changes already avoided using our realistic calculations of ozone to estimate the ultraviolet index (UVI). The UVI describes the level of solar ultraviolet radiation at the Earth's surface, ranging from zero upward—the higher the index value, the greater the potential for damage to the skin and eye, and the less time it takes for harm to occur[30]. Figure 5a shows the noontime sea-level UVI calculated using the daily column ozone for 2011 from model run MP, showing a strong gradient between the tropics and high latitudes. The additional ozone depletion in run NoMP causes an increase in UVI at all latitudes (Fig. 5b). The largest changes occur at the edge of the Antarctic vortex in September–November, with an increase of 20–100%. Following the large ozone loss in the Arctic in 2011, the UVI increases by a mean of over 15% at 60°N in March. In the tropics, the UVI increases by about 5% all year round.

Figure 5c shows the percentage change in the 2011 annual mean noon UVI between runs NoMP and MP. This is a measure of the change in the sun-burning dose of ultraviolet. Without the Montreal Protocol, these ultraviolet changes would have been a strong function of latitude, ranging from more than a 20% increase in parts of Antarctica, to 5% or less in the tropics. At mid-

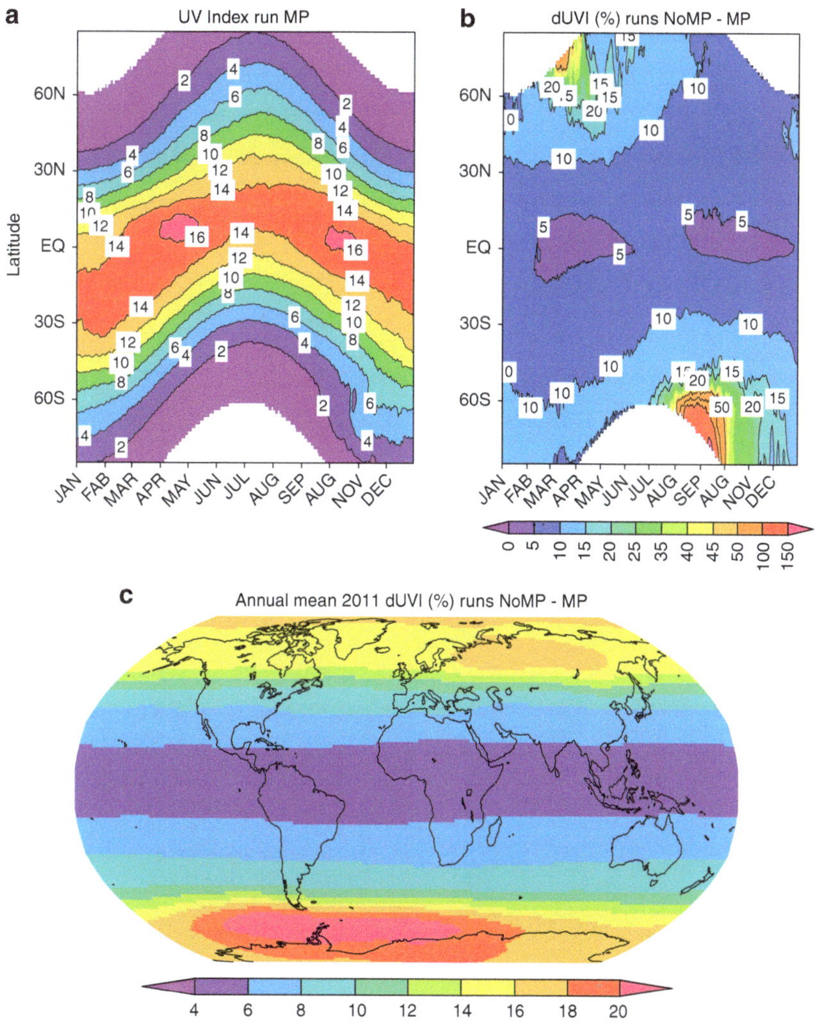

Figure 5 | The impact of the Montreal Protocol on surface ultraviolet index for 2011. (**a**) The daily zonal mean sea-level ultraviolet index at local noon calculated from the column ozone from run MP and a radiative transfer model[39]. (**b**) The percentage difference in daily zonal mean ultraviolet index at local noon between model run NoMP and run MP. (**c**) The percentage difference in annual mean ultraviolet index at local noon between model run NoMP and run MP.

latitudes, where skin types of the resident populations are typically more sensitive to ultraviolet damage, the percentage changes are potentially important. For example, in the most populated areas of Australia and New Zealand, which currently have the highest mortality rates from skin cancer[31], increases would have been 8–12%; and in Northern Europe, including the United Kingdom, increases would have exceeded 14%. Health impacts of these changes in UVI are complex and difficult to quantify. However, changes as large as these would have had potentially serious consequences in the decades that followed. In the absence of any changes in sun-exposure patterns, a 5% increase in sunburning ultraviolet is expected to lead to larger increases in the two most common forms of skin cancer: about 15% and 8% in incidence rates for squamous and basal cell carcinoma, respectively[32]. Effects on melanoma rates, with its higher mortality, are less certain because the action spectrum for melanoma in humans has not been measured. More quantitative statements would require analyses of the type described by van Dijk *et al.*[27], but focusing on a later period to take account of the large time lag between ultraviolet exposure and the development of skin cancer.

The Montreal Protocol is rightly seen as a seminal international agreement, which is successfully protecting the ozone layer. In this paper, we have shown that, just over two decades since it was ratified, the Montreal has already had major beneficial impacts, including avoiding an Arctic ozone hole.

Methods

Model configuration and experimental design. We have used the TOMCAT/SLIMCAT off-line three-dimensional (3D) CTM to calculate the impact of the Montreal Protocol on stratospheric ozone[33]. The model has been widely used in previous studies of stratospheric chemistry[34] and reproduces stratospheric ozone well. The model includes a detailed treatment of stratospheric chemistry, including a full description of processes related to polar ozone depletion. The model is forced by ERA-Interim analyses provided by the European Centre for Medium-Range Weather Forecasts[35]. The model was integrated in two experiments at a horizontal resolution of 2.8° × 2.8° with 32 levels from the surface to ~60 km. The control run (MP) was forced by observed surface mixing ratios of source gases from 1979 to 2013. Run NoMP was identical to run MP but used an alternative scenario for chlorine- and bromine-containing source gases after 1987. This scenario assumes a 3% per year growth rate in the production of ODSs.

We drive the world avoided model simulation with meteorological analyses for the period between 1979 and 2013 but allow the ODS concentrations to grow, unaffected by the Montreal Protocol. Ozone is a climate gas so, had there been no Protocol, stratospheric temperatures would have slowly evolved to a different mean state, with a consequent feedback on dynamics. The most obvious observed effect of ozone depletion has been a cooling and strengthening of the Antarctic polar vortex[20,21], which itself would exert a feedback tending to deepen and prolong the ozone hole, with subsequent impacts on tropospheric climate. We believe that our neglect of this feedback during the short time period since the implementation of the Montreal Protocol in our 'World Avoided' scenario is a justifiable approximation globally, compared with using the meteorology simulated in a climate model. However, the approximation is likely to lead to a small underestimate of the avoided ozone depletion in the Antarctic.

Satellite data. To compare to our MP model simulation, we use observations from the OMI[36] level 3 (OMTO3d) total column data. The OMTO3d are TOMS-like (Total Ozone Mapping Spectrometer) daily gridded datafiles, generated by gridding and merging only high-quality level 2 measurements (based on TOMS-like algorithm) for a given day. Data are available from 1 October 2004 at 1.0° × 1.0° resolution and is obtained from http://disc.sci.gsfc.nasa.gov/Aura/data-holdings/OMI/omto3d_v003.shtml. As OMI data are only available from 2004, Figure 1e and f shows merged Solar Backscattered Ultraviolet (SBUV) ozone data (v8.6). These are monthly mean zonal and gridded average products constructed by merging individual SBUV/SBUV/2 (total and profile ozone) satellite data sets[37] (see http://acd-ext.gsfc.nasa.gov/Data_services/merged/).

Ultraviolet Index. Skin-damaging ultraviolet radiation is usually taken to be the spectral ultraviolet irradiance weighted by the erythemal action spectrum[38]. This erythemally weighted ultraviolet (UVery) is usually measured in $W m^{-2}$. However, for public dissemination, the UVI is widely used, where UVI = 40 × UVery. UVI values were pre-calculated and tabulated as a function of ozone and solar zenith angle. The calculations were carried out with the tropospheric ultraviolet and visible (TUV) radiative transfer model[39], using the 8-stream pseudo-spherical discrete ordinate method, and apply to snow-free surfaces near sea-level under cloudless and unpolluted skies. Corrections have been applied to take account of seasonal changes in Sun–Earth separation. However, absolute values should be treated with caution, especially in polluted sites, because UVI is sensitive to factors other than ozone. For example, UVI increases by at least 5% per kilometre of altitude, and can increase by more than 20% when surfaces are snow-covered. On the other hand, extinctions by aerosols in polluted locations can lead to substantial reductions in UVI. In principle, corrections can be applied to take account of such differences. However, as we are primarily interested in differences between two scenarios for which those factors are the same, these have not been applied.

References

1. Molina, M. J. & Rowland, F. S. Stratospheric sink for chlorofluoromethanes: chlorine atom-catalysed destruction of ozone. *Nature* **249**, 810–812 (1974).
2. Farman, J. C., Gardiner, B. G. & Shanklin, J. D. Large losses of total ozone in Antarctica reveal seasonal ClO_x/NO_x interaction. *Nature* **315**, 207–210 (1985).
3. Solomon, S. Stratospheric ozone depletion: a review of concepts and history. *Rev. Geophys.* **37**, 275–316 (1999).
4. Salawitch, R. J. *et al.* Chemical loss of ozone in the Arctic polar vortex in the winter of 1991-1992. *Science* **261**, 1146–1149 (1993).
5. Rex, M. *et al.* Prolonged stratospheric ozone loss in the 1995-96 Arctic winter. *Nature* **389**, 835–838 (1997).
6. M, Ko, P., Newman, S., Reimann & S., Strahan (eds) SPARC Report on the Lifetimes of Stratospheric Ozone-Depleting Substances, Their Replacements, and Related Species, Stratospheric Processes And their Role in Climate (SPARC) Report No. 6, WCRP-15/2013 (2013).
7. Scientific Assessment of Ozone Depletion. *2010, Global Ozone Research and Monitoring Project Report No. 52* (World Meteorological Organization, 2011).
8. Scientific Assessment of Ozone Depletion. *2014, Global Ozone Research and Monitoring Project Report No. 55* (World Meteorological Organization, 2015).
9. Manney, G. L. *et al.* Unprecedented Arctic ozone loss in 2011. *Nature* **478**, 469–475 (2011).
10. Prather, M., Midgley, P., Rowland, F. S. & Stolarski, R. The ozone layer: the road not taken. *Nature* **381**, 551–554 (1996).
11. Morgenstern, O. *et al.* The world avoided by the Montreal Protocol. *Geophys. Res. Lett.* **35**, L16811 (2008).
12. Newman, P. A. *et al.* What would have happened to the ozone layer if chlorofluorocarbons (CFCs) had not been regulated? *Atmos. Chem. Phys.* **9**, 2113–2128 (2009).
13. Garcia, R. R., Kinnison, D. E. & Marsh, D. R. "World avoided" simulations with the Whole Atmosphere Community Climate Model. *J. Geophys. Res.* **117**, D23303 (2012).
14. Egorova, T., Rozanov, E., Gröbner, J., Hauser, M. & Schmutz, W. Montreal Protocol benefits simulated with CCM SOCOL. *Atmos. Chem. Phys.* **13**, 3811–3823 (2013).
15. Lefèvre, F., Brasseur, G. P., Folkins, I., Smith, A. K. & Simon, P. Chemistry of the 1991-1992 stratospheric winter: Three-dimensional model simulations. *J. Geophys. Res.* **99**, 8183–8195 (1994).
16. Chipperfield, M. P. Multiannual simulations with a three-dimensional chemical transport model. *J. Geophys. Res.* **104**, 1781–1805 (1999).
17. Aschmann, J., Sinnhuber, B.-M., Atlas, E. L. & Schauffler, S. M. Modeling the transport of very short-lived substances into the tropical upper troposphere and lower stratosphere. *Atmos. Chem. Phys.* **9**, 9237–9247 (2009).
18. Velders, G. J. M., Andersen, S. O., Daniel, J. S., Fahey, D. W. & McFarland, M. The importance of the Montreal Protocol in protecting climate. *Proc. Natl Acad. Sci. USA* **104**, 4814–4819 (2007).
19. Levelt, P. F. *et al.* The Ozone Monitoring Instrument. *IEEE Trans. Geo. Rem. Sens.* **44**, 1093–1101 (2006).
20. Kiehl, J. T., Boville, B. A. & Briegleb, B. P. Response of a general circulation model to a prescribed Antarctic ozone hole. *Nature* **332**, 501–504 (1988).
21. Mahlman, J. D., Pinto, J. P. & Umscheid, L. J. Transport, radiative and dynamical effects of the Antarctic ozone hole: A GFDL "SKYHI" model experiment. *J. Atmos. Sci.* **51**, 489–508 (1994).
22. Thompson, D. W. J. & Solomon, S. Interpretation of recent Southern Hemisphere climate change. *Science* **296**, 895–899 (2002).
23. Gillett, N. P. & Thompson, D. W. J. Simulation of recent Southern Hemisphere climate change. *Science* **302**, 273–275 (2003).
24. Marshall, G. J., Orr, A., van Lipzig, N. P. M. & King, J. C. The impact of a changing Southern Hemisphere Annular Mode on Antarctic peninsula summer temperatures. *J. Clim.* **19**, 5388–5404 (2006).
25. Son, S.-W., Tandon, N. F., Polvani, L. M. & Waugh, D. W. Ozone hole and Southern Hemisphere climate change. *Geophys. Res. Lett.* **36**, L15705 (2009).
26. Slaper, H., Velders, G. J. M., Daniel, J. S., de Gruijl, F. R. & van der Leun, J. C. Estimates of ozone depletion and skin cancer incidence to examine the Vienna Convention Achievements. *Nature* **384**, 256–258 (1996).

27. van Dijk, A. *et al.* Skin cancer risks avoided by the Montreal Protocol - worldwide modelling integrating coupled climate–chemistry models with a risk model for UV. *Photochem. Photobiol.* **89,** 234–246 (2012).
28. Newman, P. A. & McKenzie, R. L. UV impacts avoided by the Montreal Protocol. *Photochem. Photobiol. Sci.* **10,** 1152–1160 (2011).
29. Williamson, C. E. *et al.* Solar ultraviolet radiation in a changing climate. *Nat. Clim. Chang.* **4,** 434–441 (2014).
30. World Health Organisation. Global solar UV index: A practical guide, ISBN 92 4 159007 6 (2002).
31. Sneyd, M. J. Trends in melanoma incidence and mortality in New Zealand, paper presented at NIWA UV Workshop, Auckland, New Zealand, 15-17 April https://www.niwa.co.nz/sites/niwa.co.nz/files/Sneyd_UV%20Workshop_2014_Melanoma.pdf (2014).
32. Longstreth, J. *et al. Environmental Effects of Ozone Depletion: 1998 Assessment.* Ch. 2 (UNEP, 1998).
33. Chipperfield, M. P. New version of the TOMCAT/SLIMCAT off-line chemical transport model: Intercomparison of stratospheric tracer experiments. *Q. J. R. Meteorol. Soc.* **132,** 1179–1203 (2006).
34. Dhomse, S. S. *et al.* Stratospheric O_3 changes during 2001-2010: The small role of solar flux variations in a chemical transport model. *Atmos. Chem. Phys.* **13,** 10113–10123 (2013).
35. Dee, D. P. *et al.* The ERA-Interim reanalysis: configuration and performance of the data assimilation system. *Q. J. Roy. Meteorol. Soc.* **137,** 553–597 (2011).
36. McPeters, R. D. *et al.* Validation of the Aura Ozone Monitoring Instrument total column ozone product. *J. Geophys. Res.* **113** (2008).
37. McPeters, R. D., Bhartia, P. K., Haffner, D., Labow, G. J. & Flynn, L. The version 8.6 SBUV ozone data record: An overview. *J. Geophys. Res.* **118,** 8032–8039 (2013).
38. McKinlay, A. F. & Diffey, B. L. in *Human Exposure to Ultraviolet Radiation: Risks and Regulations.* (eds Passchier, W. F. & Bosnajakovic, B. F. M.) 83–87 (Elsevier, 1987).
39. Madronich, S. & Flocke, S. in *Solar Ultraviolet Radiation. NATO Series I: Advanced Study Institute* Vol. 52 (eds Zerefos, C. S. & Bais, A. F.) 23–48 (Springer, 1995).

Acknowledgements

The TOMCAT modelling work was supported by the UK Natural Environment Research Council (NERC) through the MAPLE project (NE/J008621/1) and performed on the Archer HPC machine. The TOMCAT model is a community model, which is available to NERC-funded researchers in the UK. J.A.P. thanks the ERC for support through the ACCI project (number 267760). We thank European Centre for Medium-Range Weather Forecasts for providing the ERA-Interim reanalyses. We acknowledge use of the publically available OMI and SBUV data.

Author contributions

M.P.C. conceived the idea and initiated the study in discussion with J.A.P. M.P.C., S.S.D. and W.F. performed and analysed the model runs. The figures were prepared by S.D. and M.P.C. G.J.M.V. calculated the 'no Montreal Protocol' halocarbon scenario. R.L.M. provided the scheme to calculate ultraviolet index. M.P.C. and J.A.P. wrote the paper and included the comments from all of the co-authors.

Additional information

Competing financial interests: The authors declare no competing financial interests.

9

Aerosol influence on energy balance of the middle atmosphere of Jupiter

Xi Zhang[1], Robert A. West[2], Patrick G.J. Irwin[3], Conor A. Nixon[4] & Yuk L. Yung[5]

Aerosols are ubiquitous in planetary atmospheres in the Solar System. However, radiative forcing on Jupiter has traditionally been attributed to solar heating and infrared cooling of gaseous constituents only, while the significance of aerosol radiative effects has been a long-standing controversy. Here we show, based on observations from the NASA spacecraft Voyager and Cassini, that gases alone cannot maintain the global energy balance in the middle atmosphere of Jupiter. Instead, a thick aerosol layer consisting of fluffy, fractal aggregate particles produced by photochemistry and auroral chemistry dominates the stratospheric radiative heating at middle and high latitudes, exceeding the local gas heating rate by a factor of 5-10. On a global average, aerosol heating is comparable to the gas contribution and aerosol cooling is more important than previously thought. We argue that fractal aggregate particles may also have a significant role in controlling the atmospheric radiative energy balance on other planets, as on Jupiter.

[1] Department of Earth and Planetary Sciences, University of California Santa Cruz, Santa Cruz, California 95064, USA. [2] Jet Propulsion Laboratory, California Institute of Technology, 4800 Oak Grove Drive, Pasadena, California 91109, USA. [3] Atmospheric, Oceanic and Planetary Physics, University of Oxford, Clarendon Laboratory, Parks Road, Oxford OX1 3PU, UK. [4] NASA Goddard Space Flight Center, Greenbelt, Maryland 20771, USA. [5] Division of Geological and Planetary Sciences, California Institute of Technology, Pasadena, California 91125, USA. Correspondence and requests for materials should be addressed to X.Z. (email: xiz@ucsc.edu).

As on Earth, Jupiter's atmospheric temperature profile exhibits a strong inversion above the tropopause[1], implying that its middle atmosphere, or the 'stratosphere', is convectively inhibited. Therefore, the energy budget should be dominated by radiation and the stratified middle atmosphere is in global radiative equilibrium. A first-order question is: which constituents in the atmosphere control this energy balance? About half of the incoming solar radiation on Jupiter penetrates deep into the troposphere and one third is reflected back to space (Fig. 1)[2]. The bulk constituents, hydrogen and helium, are not radiatively active except via H_2–H_2 and H_2–He collisional-induced absorption (CIA) at pressures >10 hPa (refs 3,4). The next most abundant gas, methane (CH_4), diffuses upward from the deep atmosphere and heats the stratosphere by absorbing the near-infrared solar flux[3–8]. The methane photochemical products acetylene (C_2H_2) and ethane (C_2H_6), together with H_2–H_2 and H_2–He CIA, absorb the upward mid-infrared radiation from the troposphere and re-radiate it to space, resulting in an efficient net cooling of the middle atmosphere[3–9] to compensate the solar heating.

The global maps of temperature and C_2 hydrocarbons were recently retrieved from the Jupiter flyby data from Cassini and Voyager-1 spacecraft in 2000 (refs 4,10–12) and 1979 (refs 4,12), respectively. On the basis of a state-of-the-art radiative transfer model (see Methods section), we investigate the global energy balance of Jupiter[4]. Surprisingly, the global average cooling flux by gaseous constituents in the middle atmosphere is estimated to be ~1.4 W m^{-2}, about 1.5 times larger than solar flux absorbed by the stratospheric CH_4 (~0.9 W m^{-2}; Fig. 1). Vertically, the gas solar heating rate is substantially smaller than the gas thermal cooling at pressures >10 hPa (ref. 4). The energy imbalance consistently revealed by the Voyager and Cassini data is not a seasonal effect because Jupiter has nearly zero obliquity. The Jupiter–Sun distance was different for the two flybys, varying from northern fall equinox (Voyager) to the northern summer solstice (Cassini), but the global average heating is not altered significantly. Long-term ground-based observations from 1980 to 2000 also show that the global average temperature at 20 hPa does not substantially vary with time[10], and thus neither does the thermal radiative cooling. The violation of the radiative energy equilibrium thereby suggests the presence of an additional strong heat source other than CH_4 in the middle atmosphere of Jupiter, which absorbs the missing ~0.5 W m^{-2}.

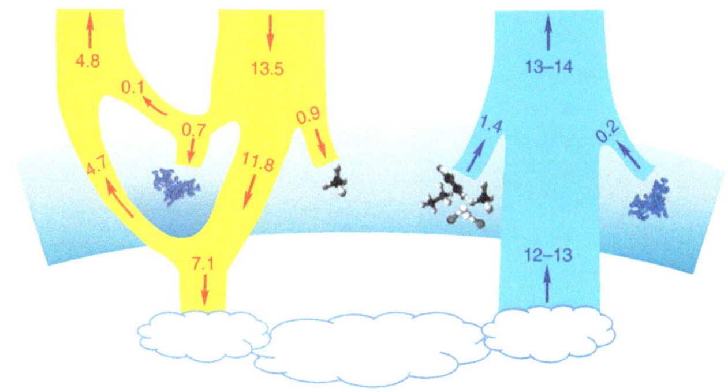

Figure 1 | Globally averaged heating and cooling fluxes on Jupiter. The heating (yellow branch) and cooling (cyan branch) fluxes are in units of W m^{-2}. The stratosphere is shaded. The heating flux is associated with the incoming solar radiation and the cooling flux is related to the outgoing thermal radiation. Of the 13.5 W m^{-2} of solar radiation incident to Jupiter's atmosphere, 0.1 W m^{-2} is reflected back to space and 11.8 W m^{-2} is transmitted to the troposphere. Tropospheric hazes and clouds absorbed 7.1 W m^{-2} and 4.7 W m^{-2} is reflected back to space[2]. The remainder of the solar energy is absorbed in the middle atmosphere by fractal haze particles (0.7 W m^{-2}) and CH_4 gas molecules (0.9 W m^{-2}). The total outgoing thermal radiation from our radiative calculation is ~13-14 W m^{-2}, consistent with that from Cassini and Voyager observations[9]. The thermal cooling flux is mainly emitted from the troposphere (12-13 W m^{-2}). In the middle atmosphere, the net cooling flux is 1.4 W m^{-2} emitted by gas molecules H_2, CH_4, C_2H_2, and C_2H_6 (black and white molecule diagrams). The upper limit of the outgoing thermal flux from the fractal aggregates (blue diagrams) is ~0.2 W m^{-2} as determined in this study.

Here we show that the missing heat source is aerosols, the end product of atmospheric chemistry on Jupiter. As a result of photochemistry, with the help of auroral chemistry, especially at high latitudes (or the 'auroral zone') where high-energy particles penetrate into the atmosphere, complex hydrocarbon compounds can form and eventually coagulate and condense as aerosols, or haze particles[13,14]. On the basis of Cassini imaging science subsystem (ISS) observations[15], here we derive the globally averaged solar flux absorbed by the stratospheric aerosols of ~0.5 to 0.7 W m^{-2}, more than half of the amount due to CH_4 (Fig. 1). The aerosol heating is predominant at middle and high latitudes, exceeding the local gas heating rate by a factor of 5–10. For the first time, we estimate the possible aerosol cooling effect, which might be as important as the cooling via hydrocarbons at high latitudes. We conclude that the aerosols maintain the atmospheric energy balance and must be partially responsible for the stratospheric temperature inversion. That the photochemistry and auroral chemistry control the atmospheric energetics via the production of aerosols suggests that Jupiter exhibits a new regime of atmospheric energy balance that is different from that of the Earth.

Results

Aerosol heating effect. The global aerosol map has been revealed from images acquired by the ISS during its Jupiter flyby[15]. At low latitudes (40° S to 25° N), an optically thin layer composed of compacted particles with radii ~0.2–0.5 μm is found to be concentrated at ~50 hPa. The rest of the atmosphere is covered by an optically thick aerosol layer at 10–20 hPa composed of fluffy, fractal particles aggregated from hundreds to thousands of ten-nanometre size monomers, similar to the haze particles on Titan[16,17]. The fractal dimension of these aggregates is assumed as 2, meaning that their geometric structure of the aggregate particles lies between a long linear chain (fractal dimension of 1) and a fully compacted cluster (fractal dimension of 3). This type of fractal aggregates is consistent with the ISS observations and the polarization observations of Jupiter, whereas spherical particles are not[16,17]. Fractal aggregates are known to be much more absorbing than spherical particles in the ultraviolet and visible wavelengths[18]. For instance, from 0.2 to 1 μm, an aggregate particle composed of a thousand ten-nanometre monomers can absorb twice as much of the solar flux as an ensemble of 0.07 μm individual spherical particles (assumed in ref. 19) with the same extinction optical depth, because the former is less scattering than the latter. Figure 2b shows that, at high latitudes, the opacity of the fractal aggregates on Jupiter is considerably larger than the CH_4 opacity in the ultraviolet and visible ranges, indicating that these particles can absorb a significant fraction of solar energy at wavelengths where the solar blackbody radiation peaks (Fig. 2a). In addition, long-term observations suggest that the seasonal variation of the Jovian north–south polarization asymmetry is only ~0.5% (ref. 20). The lack of strong temporal variation implies that the fractal aggregates constitute a steady heat source.

The radiative heating calculations demonstrate that the middle atmosphere of Jupiter is heated by two components: aerosols in

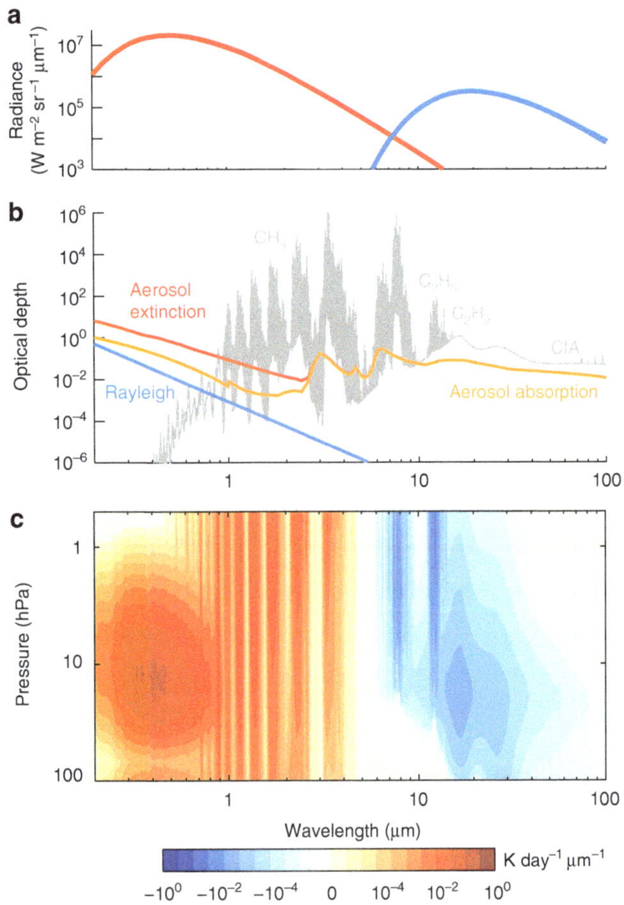

Figure 2 | Spectrally resolved heating and cooling rates and corresponding energy fluxes and opacity. (**a**) Globally averaged solar radiation received by Jupiter, approximated by a blackbody of 5,778 K (red) and Jupiter thermal radiation in the stratosphere approximated by a blackbody of 150 K (blue). (**b**) Total optical depth from the top of the atmosphere to 100 hPa as a function of wavelength at 60° S. The gas optical depth (grey) includes H_2-H_2 and H_2-He CIA and CH_4, C_2H_2 and C_2H_6 absorption. The non-gas components include Rayleigh scattering (blue), fractal aggregate aerosol extinction (red) and the aerosol absorption (orange). (**c**) Spectrally resolved zonally averaged solar heating (0.2-5 μm) and cooling (5-100 μm) map at 60° S. Absolute values of the heating/cooling rates that are $<10^{-6}$ K per day per μm are not shown. Solar heating dominates shortwards of 5 μm while Jupiter thermal cooling (shown in negative values here) dominates longwards of 5 μm. Contributions from the H_2-H_2 and H_2-He CIA and gas vibrational-rotational bands are shown. Aerosol heating is important in the ultraviolet and visible regions and aerosol cooling is important in the mid-infrared region beyond 11 μm.

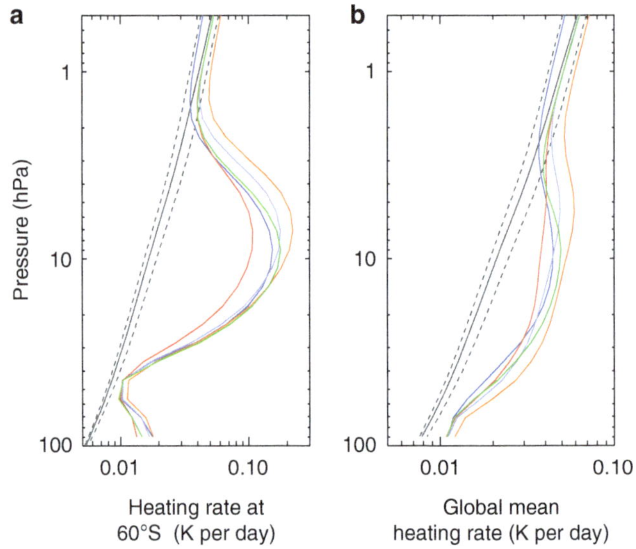

Figure 3 | Vertical heating rate profiles. (**a**) Zonally averaged heating rates at 60° S; (**b**) Globally averaged heating rates. The gas-only calculations are shown in black. The dashed black lines show the possible gas heating rates due to the uncertainty of CH_4 profiles. The coloured lines represent different aerosol retrieval solutions. Cases H1-H5 correspond to the green, red, purple, blue and orange curves, respectively. See Table 1 for detailed input information of the cases.

the ultraviolet and visible wavelengths and CH_4 in the near-infrared. Figure 2c shows the spectrally resolved zonally averaged heating rate as a function of wavelength and pressure at 60° S. At near-infrared wavelengths longer than 0.9 μm, strong CH_4 bands completely dominate the heating with minor contributions from H_2–H_2 and H_2–He CIA (Fig. 2b). At shorter wavelengths, heating by the fractal aggregates is predominant. The maximum aerosol heating occurs near the wavelength of the solar spectrum peak ($\sim$0.5 μm), but slightly shortwards owing to the increasing absorption of aerosols towards the ultraviolet. At 60° S, the integrated aerosol heating rate over all wavelengths can reach $\sim$0.2 K per day at 10 hPa, where it exceeds the CH_4 heating rate by a factor of 10 (Fig. 3a). The aerosol heating rate appears to decrease rapidly above 5 hPa and does not contribute significantly to the total heating rate in the upper stratosphere. However, this conclusion is not certain since the resolution of the near-infrared spectra[21] is not sufficient to fully characterize the upper stratosphere. On the basis of higher-resolution observations[22], another aerosol layer was found above 5 hPa at the poles that might contribute to the local heating rate in the upper stratosphere, though not to the total energy budget of the middle atmosphere due to its lower density at those levels.

On the basis of Cassini observations, the globally averaged heat flux absorbed by the stratospheric aerosols is $\sim$0.5–0.7 W m^{-2}, more than half of the amount due to CH_4 (Fig. 1). The maximum globally averaged aerosol heating rate of $\sim$0.03 K per day occurs at $\sim$10 hPa, comparable to the total near-infrared CH_4 heating rate at the same pressure level. But the globally averaged heating rate including aerosols is about twice that of the gas-only heating rate at pressures $>$20 hPa (Fig. 3b). Spatially, the aerosol heating is predominant at middle and high latitudes, which is attributable to the optically thick fractal aggregates layer (Fig. 4b). Therefore, the aerosol heating naturally, if not coincidentally, compensates for the energy deficit due to gas heating and cooling at the appropriate pressure levels (Fig. 4e).

Owing to the existence of degenerate solutions in the interpretation of ISS observations[15], we performed sensitivity tests to estimate the uncertainty ranges of the aerosol heating rate. The tropospheric haze and cloud are treated as an effective cloud layer in the troposphere[15]. Our tests show that the effective cloud albedo and phase function within the retrieved uncertainties has insignificant effect on the stratospheric heating rate. The heating rate is more sensitive to the total optical depth and single scattering albedo of the stratospheric aerosols. Given the constraints from the Cassini ISS observations, testing the sensitivity of the heating rate to each individual aerosol parameter is inappropriate. However, multiple solutions still exist in the aerosol retrieval[15]. Therefore, we adopted five typical retrieval solutions for the middle and high latitudes, namely cases

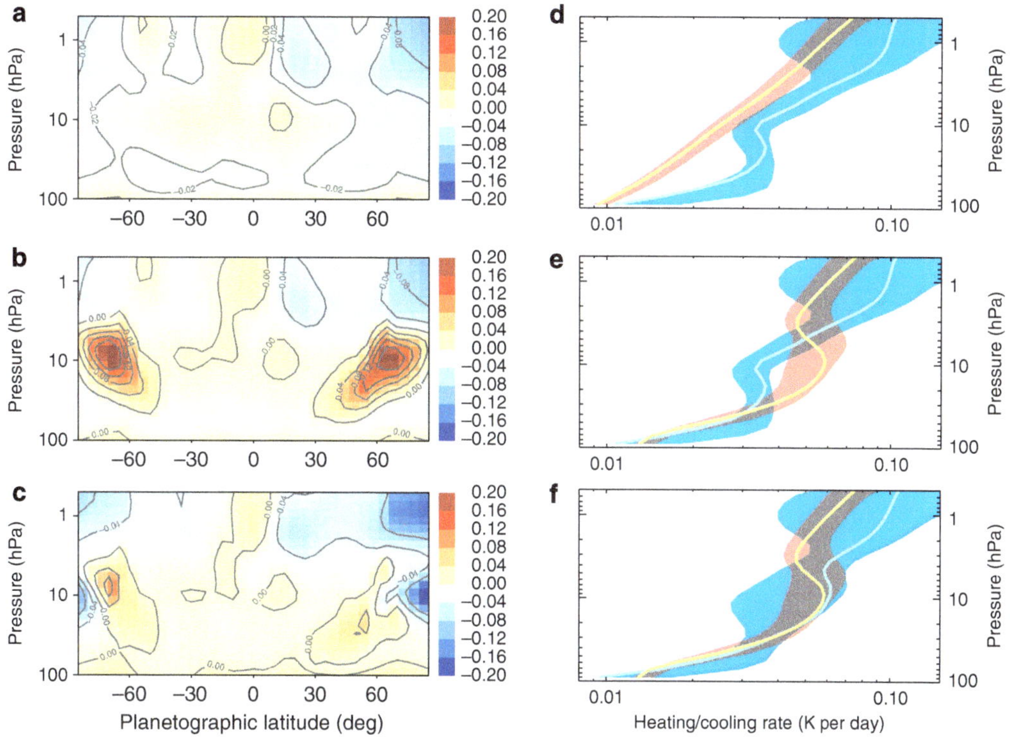

Figure 4 | Radiative balance calculation results in the middle atmosphere of Jupiter based on Cassini flyby observations. (**a**) Net radiative heating rate map (in units of K per Earth day) without aerosols; (**b**) net radiative heating rate map with aerosol heating; (**c**) net radiative heating rate map with aerosol heating and cooling; (**d**) globally averaged heating (yellow with pink shading) and cooling (cyan with blue shading) profiles without aerosols; (**e**) globally averaged heating and cooling profiles with aerosol heating; and (**f**) globally averaged heating and cooling profiles with aerosol heating and cooling. The uncertainty ranges are shaded.

Table 1 | Sensitivity cases for ISS retrieval and heating rate calculation.

Case	CH_4 mixing ratio	*k* (UV1)	*k* (CB3/MT3)	Colour
H1 (Nominal)	1.8×10^{-3}	2×10^{-2}	1×10^{-3}	Green
H2	1.8×10^{-3}	6×10^{-3}	1×10^{-4}	Red
H3	1.8×10^{-3}	8×10^{-2}	4×10^{-3}	Purple
H4	1.5×10^{-3}	2×10^{-2}	1×10^{-3}	Blue
H5	2.5×10^{-3}	2×10^{-2}	1×10^{-3}	Orange

H1–H5 in Table 1. Those cases were designed to explore the parameter space within the uncertainties of the CH_4 mixing ratio[4,23] and the imaginary part of the refractive indices (hereafter *k* values) of the UV1 and CB3/MT3 channels of Cassini ISS[15]. All cases provide good fits to the limb-darkening observations from Cassini ISS (Table 1).

The heating rate calculation at 60° S (Fig. 3a) shows that the maximum heating rate including aerosols is about a factor of 2 larger than the minimum at ~10 hPa. The maximum (H3) and minimum (H2) occurs when the *k* value of the UV1 channel reaches the upper and lower bound, respectively, outside which the UV1 limb-darkening profiles cannot be explained[15]. On a global average (Fig. 3b), the maximum heating rate including aerosols is about a factor of 1.5 larger than the minimum (shaded in red in Fig. 4e,f).

Aerosol cooling effect. Aerosols could also cool the middle atmosphere of Jupiter but this effect has not been explored in previous studies. The mid-infrared optical properties of aerosols produced in a hydrogen-dominated environment have not been measured experimentally. But at wavelengths shorter than 2.5 μm, previous laboratory experiments found that the *k* values of aerosols produced in a CH_4/H_2 gas mixture could be either larger or smaller than their counterpart in the CH_4/N_2 mixture, depending on the chemical composition and environmental pressure[24,25]. Therefore, the fractal aggregates on Jupiter might have non-negligible opacity in the mid-infrared compared with H_2–H_2 and H_2–He CIA if their optical constants behave like the aerosols on Titan, which are strongly absorbing with almost no scattering beyond 5 μm (Fig. 2b)[26,27]. To estimate the possible cooling effect from aerosols, we included aerosol absorption in a non-linear inversion model to fit the spectra from Cassini CIRS[28]. However, owing to insufficient sensitivity of CIRS observations to the Jovian aerosol opacity, a pure-gas (that is, non-detection of aerosols) model is also able to fit the spectra[4,10–12]. Future analysis on the possible C–H bending vibrational features of aerosols at 1,380 and 1,460 cm^{-1} that have been detected on Titan[27] might provide more constraints on the infrared opacity and chemical structure of aerosol particles on Jupiter. Here we aim to derive the upper limit of aerosol opacity from the CIRS spectra and estimate the upper bound of the aerosol thermal infrared cooling.

For each latitude, we included aerosols in the Non-linear optimal Estimator for MultivariatE spectral analySIS (NEMESIS) model[28] and retrieved the temperature profile and the mixing ratios of C_2H_2 and C_2H_6 following the procedure detailed in ref. 4. Owing to the lack of Jupiter-analogue aerosol measurements in mid-infrared, we tested several *k* values based on the laboratory tholin results[26] and recently derived *k* values from Cassini observations on Titan[27]. The latter shows cooling 2–3 times smaller than the former and exhibits different wavelength dependence. We gradually increased the *k* values

Table 2 | Sensitivity cases for CIRS retrieval and cooling rate calculation.

Case	*k* values in the mid-infrared	χ^2/N (600-850 cm^{-1})	χ^2/N (1,225-1,325 cm^{-1})	Colour
C1	Pure gas, no aerosol	0.544	1.094	Red
C2	Titan tholin experiment[26]	0.543	1.042	Blue
C3	Titan CIRS observations[27]	0.541	1.090	Orange
C4 (Nominal)	C2 values divided by 2	0.543	1.051	Green
C5	Five times of C2 values	0.607	1.964	Brown

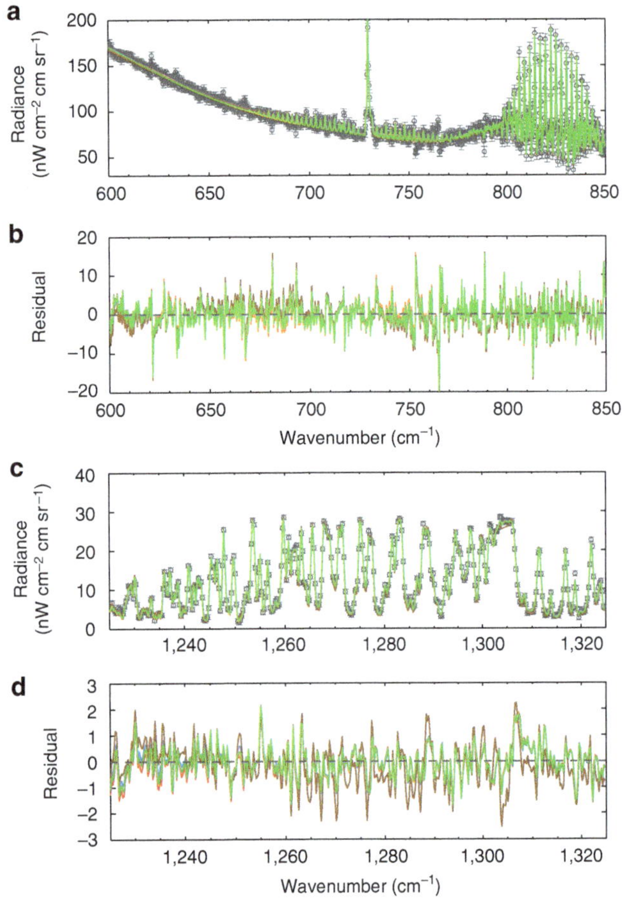

Figure 5 | Spectral inversion results at 57° S. (**a**) Spectra at 600-850 cm^{-1} region (H_2-H_2 and H_2-He CIA, C_2H_2 and C_2H_6 bands); (**b**) fitting residual at 600-850 cm^{-1} region; (**c**) spectra at 1,225-1,325 cm^{-1} region (CH_4 bands); (**d**) fitting residual at 1,225-1,325 cm^{-1} region. CIRS observations are shown as black circles. The red, blue, orange, green and brown colours represent NEMESIS retrieval cases C1-C5, respectively. The goodness of fit (χ^2/N where N is the number of measurements) in the 600-850 cm^{-1} region is ~0.5-0.6 for each case. In the CH_4 band, the goodness of fit is around unity for each case except for the brown case ($\chi^2/N = 1.96$), which does not fit the CIRS spectra. See Table 2 for detailed information of the cases.

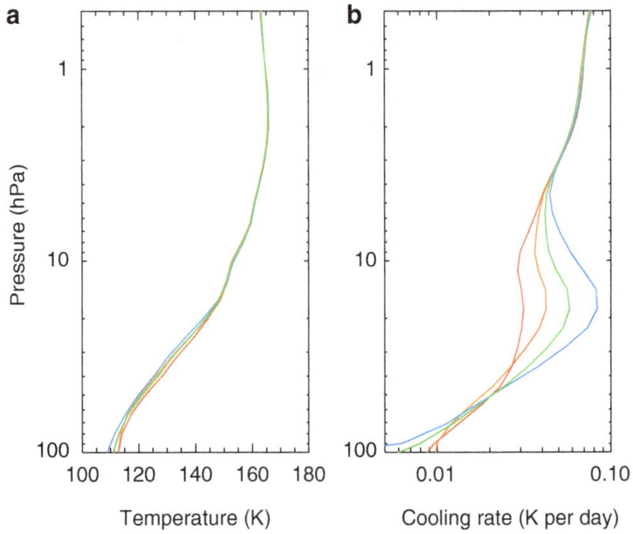

Figure 6 | Vertical temperature and cooling rate profiles at 57° S. (**a**) Retrieved temperature profiles; (**b**) corresponding zonally averaged cooling rates. The red, blue, orange and green colours represent cases C1-C4, respectively. The C5 case is not used because it cannot explain the CIRS observations.

until the CIRS spectra cannot be fitted within the measurement uncertainties. Table 2 summarized five typical tests. For each choice of optical constants, we performed an atmospheric retrieval of the Cassini CIRS spectra using NEMESIS model to constrain the aerosol opacity. Figure 5 illustrates the retrieval fitting results and residual values at 57° S and the values of goodness of fit are shown in Table 2. The retrieved temperature profiles are shown in Fig. 6a. If we enhance the optical constants from ref. 25 by a factor of 5 (case C5), we are not able to fit the CH_4 emission spectra (Fig. 5). Overall, we estimated the upper limit of the aerosol optical depth in the mid-infrared wavelengths to be ~0.1 at 100 hPa at high latitudes. This places an upper limit on the aerosol contribution to the globally averaged cooling flux of ~0.2 W m^{-2} in the stratosphere (Fig. 1).

The aerosol cooling effect could be significant at middle and high latitudes where the particles are abundant but negligible at low latitudes. When the aerosol opacity is included, the cooling rate increases in the aerosol layer (Fig. 6b). At 57° S, with a moderate choice of *k* (case C4), the zonally averaged cooling rate including aerosols is ~0.06 K per day. This is about two times larger than the gas-only cooling rate (case C1) at the pressures where the aerosol mixing ratio peaks (~10–20 hPa, Fig. 6b). The globally averaged aerosol cooling rate could be comparable to the gas cooling rate at 20 hPa and partially compensate for the aerosol heating effect (Fig. 4f). On the other hand, the cooling rates from the aerosol cases are smaller than the pure-gas case at pressure levels below the aerosol layer (Fig. 6b), a result of energy conservation as constrained by the total emission observed by CIRS. Indeed, stronger aerosol absorption leads to a colder retrieved temperature profile (Fig. 6a). For instance, the temperature profile at 57° S from the C2 case is ~3 K colder than the case without aerosols (C1 case) below ~40 hPa. This colder temperature leads to a smaller cooling effect than the pure-gas case (C1) at pressures > ~40 hPa where the cooling is dominated by H_2–H_2 and H_2–He CIA (Fig. 6b).

We estimate the possible aerosol cooling rate for each latitude and their influences on the globally averaged cooling rate by combining the 'ensemble uncertainty'[4] of the temperature and gas abundances and the aerosol cooling rate tests. The uncertainty range including the aerosol contribution (Fig. 4f) is larger than that without aerosols (Fig. 4d) because the aerosols remain elusive from the CIRS spectra. Global radiative equilibrium is achievable when both aerosol heating and cooling are included (Fig. 4f).

Discussion

Owing to the lack of sufficient observational evidence before, the importance of aerosol heating in the middle atmosphere of Jupiter has long been a controversial question. It was suggested since the 1970s that the Jovian aerosols might absorb solar radiation and heat the atmosphere[6–8]. In the 1990s, based on the International Ultraviolet Explorer and Voyager-2 data, aerosol heating was shown to have a large impact on the atmospheric circulation[29]. However, a later analysis using Hubble Space Telescope (HST) images found that the aerosols have relatively insignificant effects[19]. With better constraints on the spatial distribution and optical properties of aerosols, Cassini observations confirm a pronounced aerosol heating at high latitudes on Jupiter. Aside from the aerosol cooling effect, our conclusion is qualitatively consistent with previous estimate from the International Ultraviolet Explorer and Voyager-2 data[29] but with a much better spatial coverage. However, this work disagrees with heating rates derived solely on the basis of HST data[3,19].

The major difference between the current work and previous studies is probably attributable to the fractal nature of the aerosol and aerosol spatial distribution. Through the multi-channel–multi-phase retrieval on Cassini images, we can characterize the fractal aggregates in great detail, including the optical depth, single scattering albedo and phase function of the particles. On the basis of low-phase-angle images alone[19], the polar aerosols were assumed as tiny spherical particles of $\sim 0.07\,\mu m$ in radius instead of the submicron size aggregate particles. The latter have lower single scattering albedo than the spherical particles. Furthermore, with a larger particle size, the fractal aggregates have less backscattering than the spherical particles. Owing to the above two reasons, a larger total optical depth of the fractal aggregates is required to explain the low-phase-angle I/F observations, leading to a larger heating rate in this study, than that in ref. 19, at the south pole. Note that the k values used in ref. 19 are lower than our nominal model but still within the sensitivity test range in our study.

The vertical distributions of aerosols in previous studies[19,29] are based on microphysical simulations that are inconsistent with the near-infrared observations[21]. Previous studies adopted a haze layer located above 10 hPa at polar regions, while the near-infrared spectra reveal a main haze layer at 10–20 hPa (ref. 20). West *et al.*[29] only sampled two latitudes and estimated the other latitude information by scaling. The HST images[19] have a good latitudinal coverage, but the aerosol heating appears influential only at the south pole, not at middle latitudes or at the north pole. Moreno *et al.*[19] reported the aerosol optical depth at the north pole one order of magnitude less than that at the south pole. This result is inconsistent with other observations. For example, recent high-resolution ground-based near-infrared spectra[22] concluded that the near-infrared haze optical depth at the northern pole is comparable to that at the southern pole. The Cassini images in low and high phase angles also revealed that the haze ultraviolet optical depth at north high latitudes is not significantly less than its south counterpart[15]. Furthermore, Cassini images show that the haze optical depth in the ultraviolet channel can approach unity down to 100 hPa at middle latitudes[15]. Including the haze contribution at those latitudes would greatly enhance the aerosol heating. This will not only influence the local heat balance but also on the global energy equilibrium, especially at 10–20 hPa, as shown in our study.

Several other factors might also attribute to the difference between the current work and previous studies. For instance, we have a much better global coverage of the temperature, hydrocarbons and aerosols based on Cassini observations. The cooling rate in ref. 3 was likely to be underestimated because the temperature profile from Galileo entry probe is shown to be colder than the globally averaged temperature profile from Voyager and Cassini observations[4], albeit the cooling rate in ref. 3 is still slightly larger than the gas heating rate at pressures >10 hPa. The spectroscopic data of CH_4, C_2H_2 and C_2H_6 have been greatly improved in the last decade (see Methods section for gas opacity). The state-of-the-art line data allow us to adopt the line-by-line approach to resolve the vibrational–rotational line shape of hydrocarbons[4], the most accurate radiative transfer method to estimate the gas heating and cooling rate.

Another possible heating mechanism in the middle atmosphere is energy dissipation of upward propagating gravity waves from the troposphere. However, as per previous studies[30,31], this hypothesis has several defects. First, there is little evidence of stratospheric gravity waves at middle and high latitudes. Second, there is no direct evidence of wave breaking in the lower stratosphere of Jupiter. Third, gravity wave breaking could either heat or cool the middle atmosphere but the net effect is difficult to quantify[30].

Unlike the Earth, on which the photochemical product (ozone) only dominates the atmospheric radiative heating, Jupiter might exhibit a different regime of atmospheric energy balance where both the heating and cooling are significantly controlled by the photochemistry and auroral chemistry via the production of aerosols and C_2 hydrocarbons. Aside from the first-order global energy balance, aerosol heating and cooling on Jupiter also influence the spatial distributions of radiative forcing, which has a significant impact on the large-scale dynamical circulation in the middle atmosphere[5,19,29]. The NASA JUNO spacecraft, arriving at Jupiter in 2016, will provide more insights on the aurora processes and aerosol formation in the polar region.

Jupiter is the second planetary body and the first hydrogen-dominated planet that shows evidence of hydrocarbon aerosols playing a significant role in regulating the radiation flux and most probably the circulation of its middle atmosphere. The other one is Titan[32]. Although Jupiter's atmosphere is primarily dominated by hydrogen, Titan's is dominated by nitrogen, both of these atmospheres produce fluffy, fractal aggregate particles, suggesting that fractal aggregates might be a ubiquitous result of hydrocarbon chemistry. In view of the existence of hydrocarbon aerosols in many other atmospheres dominated by hydrogen or nitrogen, such as those of Saturn[33], Uranus[34] and Neptune[35], the early Earth[18], and possibly some exoplanets[36–40], we hypothesize that the heating and cooling from fractal aggregates could also be important for determining the radiative energy distribution and climate evolution on these planets. Owing to their strong heating effects, fractal aggregates play a significant role in creating the temperature inversion in the lower stratosphere of Jupiter. They might also be partially responsible for the temperature inversions observed on the other giant planets in the Solar System, but neglected in previous studies[41]. A typical feature of the fluffy aggregate particle is its extremely low density, which might help to explain the existence of very high and thick haze layers at pressures <0.1 hPa, as seen on the Neptune and sub-Neptune size planets GJ436b (ref. 39) and GJ1214b (ref. 40). A thorough study of fractal aggregates will shed light on how to characterize these particles in future photometry and polarization observations.

Methods

Radiative heating and cooling model. For heating rate calculations between 0.20 and 0.94 μm where the aerosol contribution is significant, we use a multiple scattering model based on the C version of the discrete ordinates radiative transfer code (DISORT Program for a multi-layered plane-parallel medium)[42]. The phase function and cross sections of low-latitude particles were calculated based on Mie theory, while that of the fractal aggregates at middle and high latitudes were computed using a parameterization method for the aggregates with a fractal dimension of two[17]. The parameterization is based on electromagnetic scattering computations using the multi-sphere method[43]. In the heating rate calculations, we use 32 streams to characterize the intensity angular distribution, which displays almost no difference from the 64-stream case. A Gaussian quadrature method with 10 zenith angles is used to average the heating rates longitudinally. The spectral resolution is 0.001 μm. The effective cloud albedo in the troposphere is interpolated between 0.20 and 0.94 μm based on the retrieved albedo from the UV1 channel and CB3/MT3 channels of Cassini ISS[15].

At longer wavelengths (0.94–200 μm), our calculations adopt the line-by-line approach, based on a state-of-the-art high-resolution radiative heating and cooling model for the stratosphere of Jupiter, which has been rigorously validated against simple but realistic analytical solutions[4]. The CH_4 heating rate calculation from 0.94 to 10 μm is performed with a spectral resolution of 0.005 cm^{-1} to resolve the CH_4 spectral line shape using the most current CH_4 line lists (see discussion for gas opacity below). The thermal cooling rate from 50 to 2,500 cm^{-1} (4–200 μm) is calculated with a spectral resolution of 0.001 cm^{-1}.

We calculate heating and cooling rates for every latitude to produce the latitude-pressure two-dimensional maps. The specific heat of Jupiter's atmosphere is taken as $1.0998 \times 10^4\ J\,kg^{-1}\,K^{-1}$ (ref. 44). The globally averaged profiles are obtained via an area-weighted mean from 90° N to 90° S. Since the data quality at latitudes north of 70° N and south of 70° S is not sufficiently good for rigorous atmospheric retrieval, we do not derive the atmospheric characteristics from the ISS images and CIRS spectra. Instead, we assume that the vertical profiles of aerosol, temperature and gases are identical to their values at 70° north and south, respectively.

This assumption might introduce some uncertainty in our estimate of the global energy balance because the heating and cooling in the polar region, especially in the infrared aurora region, are not negligible. The polar aurora is known to be highly variable both temporally and spatially[1]. According to our limited data on the polar regions from Galileo[45] and Cassini[46] spacecraft as well as ground-based observations[47,48], both regions poleward of 70° N and 70° S could be different from that ~65°–70°. For example, the Cassini CIRS instrument[46] detected a variation of thermal emission over the C_2H_2 and C_2H_6 bands at regions poleward of 65°, within about a factor of 2. However, because the surface area poleward of 70° amounts to merely 6% of the total surface area of Jupiter, increasing the polar cooling rate by a factor of 2 will probably introduce an uncertainty of only ~6% of the total cooling rate, which is still well-located within our estimated uncertainty range (Fig. 4f). On the other hand, the pressure level of infrared aurora source has not been precisely determined. If the emission originates from the upper stratosphere (for example, above 0.1 mbar level), the aurora and its variability might have little impact on the thermal cooling rate at pressures we focus here. We should also point out that the aurora might also be associated with some heating mechanisms in the polar region that have not been considered in this study. Future analysis of the Jovian polar region will provide more information on the local energetics.

The aerosol and gas opacities used in the radiative calculations are discussed below.

Aerosol opacity sources. In the visible and near-infrared wavelengths, the early laboratory studies[24–26] measured the optical properties of Titan-analogue aerosols (that is, in the nitrogen environment) and Jupiter-analogue aerosols (that is, in the hydrogen environment), respectively. Using the radio frequency glow discharge in a CH_4/H_2 gas mixture, it was found that the refractive indices of these aerosols are consistent with high-phase-angle photometry data of Uranus by Voyager-2 at 0.55 μm (ref. 24). Compared with aerosols produced in the CH_4/N_2 mixture[26], the imaginary part of the refractive index of the Jupiter-analogue aerosols could be either larger or smaller, depending on the chemical composition. Unfortunately, ref. 24 has been the only laboratory experiment of the CH_4/H_2 gas mixture to date. It should be noted that the subsequent Titan-analogue laboratory measurements show that the aerosol properties are significantly influenced by the gas composition and environmental pressure[25].

Zhang *et al.*[15] combined the ground-based infrared spectra with the Cassini ISS observations at both low and high phase angles, and determined the *k* values for the UV1 channel (0.258 μm) and the infrared channels (CB3 at 0.938 μm and MT3 at 0.889 μm). Owing to the existence of degeneracy, the *k* value at UV1 channel varies from 0.008 to 0.02 and that at 0.9 μm varies from 0.0001 to 0.004. Different choices of *k* would imply different solutions to fit the ISS data, such as the radius of the monomers from 10 to 40 nm, the number of monomers per aggregate particle from 100 to 1,000, and the abundance of particles. In all possible solutions, the total aerosol opacity only changes by ~30% among all solutions. The corresponding aerosol heating rate does not change significantly.

For the heating rate calculations, we choose the model parameters for the *k* values from ref. 15 as our nominal case (H1 in Table 1), with a careful sensitivity study in the radiatiave heating calculation. The *k* value is ~0.02 at 0.258 μm and 0.001 at ~0.9 μm. We performed an interpolation for the wavelengths between 0.20 and 0.94 μm in the coordinate of linear wavelength and logarithmic *k* value. The interpolation can be justified by the approximate same trend shown in the laboratory measurements (Fig. 7). As in ref. 15, we adopt the real part of refractive index (*n*) from ref. 26.

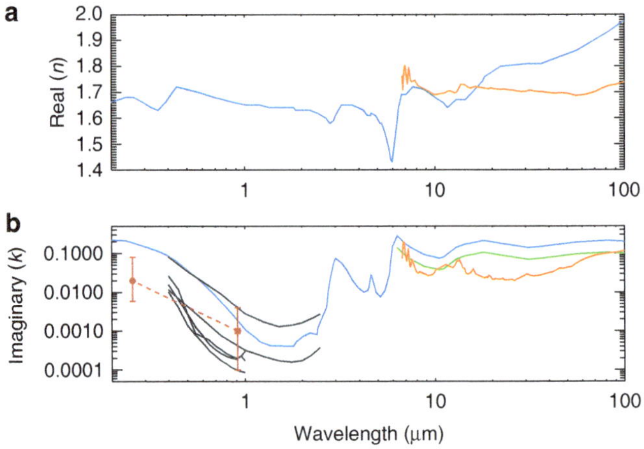

Figure 7 | Refractive indices of aerosols from 0.2 to 100 μm. (**a**) Real part of the refractive index; (**b**) imaginary part. The blue one is from the Titan tholin experiment[26]; orange is from Titan CIRS observations[27]; green is same as the blue one but reduced by a factor of 2 (case C4); black is the Jupiter-analogue aerosol experiment[24]; red is the derived from Cassini ISS observations[15], with an interpolation in the coordinate of linear wavelength and logarithmic *k* value. The red dashed line is used in the nominal case for the heating rate calculation (case H1) in this study.

Gas opacity sources. From 0.20 to 0.94 μm, the CH_4 opacity is based on ref. 49 and Rayleigh scattering optical depth is taken from ref. 14. From 0.94 to 10 μm, a line-by-line calculation is performed. We compared several CH_4 line databases, including the HITRAN2012 (refs 50,51), the database from ref. 22, and the M5 database[52]. For the CH_4 broadening width, it has been suggested that line widths in the Jovian atmosphere (H_2–He mixture) are similar to those in the Earth's atmosphere (N_2–O_2 mixture)[22]. All the above three CH_4 opacity sources result in a consistent heating rate in our calculation. However, none of the above opacity sources covers the CH_4 band between 0.94 and 1.1 μm. Recently, a '10 to 10' CH_4 line database is computed from first principles[53]. This database is shown to be roughly consistent with the HITRAN2012 CH_4 data in their overlapping near-infrared wavelengths (personal communication with J. Bailey) and therefore is helpful to fill the gap between 0.94 and 1.1 μm in our calculation. For this CH_4 band, due to the lack of laboratory measurements of line shape parameters, we adopted an average pressure-broadened half-width of 0.06 cm^{-1} for foreign-broadening and 0.077 cm^{-1} for self-broadening at 1 bar pressure, and the temperature dependence exponent is ~0.85 (refs 52,54). The contribution from this band to the total heating rate is negligible (Fig. 2c). For the heating rate calculation, we obtain the near-infrared H_2–H_2 CIA absorption from refs 55,56 and H_2–He CIA absorption from refs 57–60.

The thermal cooling rate is calculated from 50 to 2,500 cm^{-1} (4–200 μm). The opacity sources of CH_4, C_2H_2 and C_2H_6 are obtained from HITRAN2012 with hydrogen-broadening widths[4,61]. Fractal aggregates are treated as pure absorbers in the thermal wavelengths due to their negligible single scattering albedo in the mid-infrared. The mid-infrared H_2–H_2 and H_2–He CIAs are obtained from ref. 62. Figure 2c shows the spectrally resolved heating/cooling rate as a function of wavelength and pressure, in which one can see the contributions from the CIAs and different vibrational-rotational bands from gases in Fig. 2b.

Spectral inversion model. The temperature and hydrocarbon distributions are simultaneously retrieved from the Cassini CIRS spectra[4,10–12] using the NEMESIS algorithm[28]. This inversion model has been used in previous studies involving CIRS data retrieval[4,11,12]. In this study we extended the previous retrieval framework to include aerosol absorption in the mid-infrared wavelengths. No scattering calculations were needed as the fractal aggregate particles are significant absorbers and have negligible single scattering albedos at these wavelengths.

References

1. Moses, J. I. *et al.* in *Jupiter-The Planet, Satellites and Magnetosphere* (eds Bagenal, F., Dowling, T. & McKinnon, W.) 129–157 (Cambridge University Press, 2004).

2. Hanel, R., Conrath, B., Herath, L., Kunde, V. & Pirraglia, J. Albedo, internal heat, and energy balance of Jupiter: Preliminary results of the Voyager infrared investigation. *J. Geophys. Res.* **86,** 8705–8712 (1981).
3. Yelle, R. V., Griffith, C. A. & Young, L. A. Structure of the Jovian stratosphere at the Galileo probe entry site. *Icarus* **152,** 331–346 (2001).
4. Zhang, X. *et al.* Radiative forcing of the stratosphere of Jupiter, part I: atmospheric cooling rates from Voyager to Cassini. *Planet. Space Sci.* **88,** 3–25 (2013).
5. Conrath, B. J., Gierasch, P. J. & Leroy, S. S. Temperature and circulation in the stratosphere of the outer planets. *Icarus* **83,** 255–281 (1990).
6. Wallace, L., Prather, M. & Belton, M. J. S. The thermal structure of the atmosphere of Jupiter. *Astrophys. J.* **193,** 481–493 (1974).
7. Cess, R. & Chen, S. The influence of ethane and acetylene upon the thermal structure of the Jovian atmosphere. *Icarus* **26,** 444–450 (1975).
8. Appleby, J. F. & Joseph, S. H. Radiative-convective equilibrium models of Jupiter and Saturn. *Icarus* **59,** 336–366 (1984).
9. Li, L. *et al.* Emitted power of Jupiter based on Cassini CIRS and VIMS observations. *J. Geophys. Res. Planets* **117,** E11 (2012).
10. Simon-Miller, A. A. *et al.* Jupiter's atmospheric temperatures: from Voyager IRIS to Cassini CIRS. *Icarus* **180,** 98–112 (2006).
11. Nixon, C. A. *et al.* Meridional variations of C_2H_2 and C_2H_6 in Jupiter's atmosphere from Cassini CIRS infrared spectra. *Icarus* **188,** 47–71 (2007).
12. Nixon, C. A. *et al.* Abundances of Jupiter's trace hydrocarbons from Voyager and Cassini. *Planet. Space Sci.* **58,** 1667–1680 (2010).
13. Wong, A. S., Yung, Y. L. & Friedson, A. J. Benzene and haze formation in the polar atmosphere of Jupiter. *Geophys. Res. Lett.* **30,** 1–4 (2003).
14. West, R. A. *et al.* in *Jupiter-The Planet, Satellites and Magnetosphere* (eds Bagenal, F., Dowling, T. & McKinnon, W.) 79–104 (Cambridge University Press, 2004).
15. Zhang, X., West, R. A., Banfield, D. & Yung, Y. L. Stratospheric aerosols on Jupiter from Cassini observations. *Icarus* **226,** 159–171 (2013).
16. West, R. A. & Smith, P. H. Evidence for aggregate particles in the atmospheres of Titan and Jupiter. *Icarus* **90,** 330–333 (1991).
17. Tomasko, M. G. *et al.* A model of Titan's aerosols based on measurements made inside the atmosphere. *Planet. Space Sci.* **56,** 669–707 (2008).
18. Wolf, E. T. & Toon, O. B. Fractal organic Hazes provided an ultraviolet shield for early Earth. *Science* **328,** 1266–1268 (2010).
19. Moreno, F. & Sedano, J. Radiative balance and dynamics in the stratosphere of Jupiter: results from a latitude-dependent aerosol heating model. *Icarus* **130,** 36–48 (1997).
20. Starodubtseva, O. M., Akimov, L. A. & Korokhin, V. V. Seasonal variations in the north–south asymmetry of polarized light of Jupiter. *Icarus* **157,** 419–425 (2002).
21. Banfield, D., Conrath, B. J., Gierasch, P. J. & Nicholson, P. D. Near-IR spectrophotometry of Jovian aerosols—meridional and vertical distributions. *Icarus* **134,** 11–23 (1998).
22. Kedziora-Chudczer, L. & Bailey, J. Modelling the near-IR spectra of Jupiter using line-by-line methods. *Mon. Not. R. Astron. Soc.* **414,** 1483–1492 (2011).
23. Wong, M. H., Mahaffy, P. R., Atreya, S. K., Niemann, H. B. & Owen, T. C. Updated Galileo probe mass spectrometer measurements of carbon, oxygen, nitrogen, and sulfur on Jupiter. *Icarus* **171,** 153–170 (2004).
24. Khare, B. N., Sagan, C., Thompson, W., Arakawa, E. & Votaw, P. Solid hydrocarbon aerosols produced in simulated Uranian and Neptunian stratospheres. *J. Geophys. Res.* **92,** 15067–15082 (1987).
25. Imanaka, H. *et al.* Laboratory experiments of Titan tholin formed in cold plasma at various pressures: implications for nitrogen-containing polycyclic aromatic compounds in Titan haze. *Icarus* **168,** 344–366 (2004).
26. Khare, B. N. *et al.* Optical constants of organic tholins produced in a simulated Titanian atmosphere: from soft X-ray to microwave frequencies. *Icarus* **60,** 127–137 (1984).
27. Vinatier, S. *et al.* Optical constants of Titan's stratospheric aerosols in the 70–1,500 cm^{-1} range constrained by Cassini/CIRS observations. *Icarus* **219,** 5–12 (2012).
28. Irwin, P. G. J. *et al.* The NEMESIS planetary atmosphere radiative transfer and retrieval tool. *J. Quant. Spectrosc. Radiat. Transfer* **109,** 1136–1150 (2008).
29. West, R. A., Friedson, A. J. & Appleby, J. Jovian large-scale stratospheric circulation. *Icarus* **100,** 245–259 (1992).
30. Young, L. A., Yelle, R. V., Young, R., Seiff, A. & Kirk, D. B. Gravity waves in Jupiter's stratosphere, as measured by the Galileo ASI experiment. *Icarus* **173,** 185–199 (2005).
31. Watkins, C. & Cho, J. Y. K. The vertical structure of Jupiter's equatorial zonal wind above the cloud deck, derived using mesoscale gravity waves. *Geophys. Res. Lett.* **40,** 472–476 (2013).
32. Tomasko, M. G. *et al.* A model of Titan's aerosols based on measurements made inside the atmosphere. *Planet. Space Sci.* **56,** 669–707 (2008).
33. West, R. A., Baines, K. H., Karkoschka, E. & Sánchez-Lavega, A. in *Saturn from Cassini–Huygens* (eds Dougherty, M. K., Esposito, L. W. & Krimigis, S. S. M.) 161–179 (Springer, 2009).
34. Pollack, J. B. *et al.* Nature of stratospheric haze on Uranus: evidence for condensed hydrocarbons. *J. Geophys. Res.* **92,** 15037–15065 (1987).
35. Baines, K. H. & Hammel, H. B. Clouds, hazes and stratospheric methane abundance ratio in Neptune. *Icarus* **109,** 20–39 (1994).
36. Sing, D. K. *et al.* Hubble Space Telescope transmission spectroscopy of the exoplanet HD 189733b: high-altitude atmospheric haze in the optical and near-ultraviolet with STIS. *Mon. Not. R. Astron. Soc.* **416,** 1443–1455 (2011).
37. Sing, D. K. *et al.* HST hot-Jupiter transmission spectral survey: evidence for aerosols and lack of TiO in the atmosphere of WASP-12b. *Mon. Not. R. Astron. Soc* **436,** 2956–2973 (2013).
38. Deming, D. *et al.* Infrared transmission spectroscopy of the exoplanets HD 209458b and XO-1b using the wide field camera-3 on the Hubble Space Telescope. *Astrophys. J.* **774,** 95 (2013).
39. Knutson, H. A., Benneke, B., Deming, D. & Homeier, D. A featureless transmission spectrum for the Neptune-mass exoplanet GJ436b. *Nature* **505,** 66–68 (2014).
40. Kreidberg, L. *et al.* Clouds in the atmosphere of the super-Earth exoplanet GJ1214b. *Nature* **505,** 69–72 (2014).
41. Robinson, T. D. & Catling, D. C. Common 0.1-bar tropopause in thick atmospheres set by pressure-dependent infrared transparency. *Nat. Geosci.* **7,** 12–15 (2014).
42. Buras, R., Dowling, T. & Emde, C. New secondary-scattering correction in DISORT with increased efficiency for forward scattering. *J. Quant. Spectrosc. Radiat. Transfer* **112,** 2028–2034 (2011).
43. Mishchenko, M. I., Travis, L. D. & Mackowski, D. W. T-matrix computations of light scattering by nonspherical particles: A review. *J. Quant. Spectrosc. Radiat. Transfer* **55,** 535–575 (1996).
44. Irwin, P. G. J. *Giant Planets of Our Solar System: Atmospheres, Composition, and Structure* (Springer, 2009).
45. Rages, K., Beebe, R. & Senske, D. Jovian stratospheric hazes: the high phase angle view from Galileo. *Icarus* **139,** 211–226 (1999).
46. Kunde, V. G. *et al.* Jupiter's atmospheric composition from the Cassini thermal infrared spectroscopy experiment. *Science* **305,** 1582–1586 (2004).
47. Kostiuk, T., Romani, P. N., Espenak, F., Livengood, T. A. & Goldstein, J. J. Temperature and abundances in the Jovian auroral stratosphere, 2. Ethylene as a probe of the microbar region. *J. Geophys. Res. Planets* **98,** 18823–18830 (1993).
48. Livengood, T. A., Kostiuk, T., Espenak, F. & Goldstein, J. J. Temperature and abundances in the Jovian auroral stratosphere, 1. Ethane as a probe of the millibar region. *J. Geophys. Res. Planets* **98,** 18813–18822 (1993).
49. Karkoschka, E. & Tomasko, M. G. Methane absorption coefficients for the Jovian planets from laboratory, Huygens, and HST data. *Icarus* **205,** 674–694 (2010).
50. Brown, L. R. *et al.* Methane line parameters in the HITRAN2012 database. *J. Quant. Spectrosc. Radiat. Transfer* **130,** 201–219 (2013).
51. Rothman, L. S. *et al.* The HITRAN2012 molecular spectroscopic database. *J. Quant. Spectrosc. Radiat. Transfer* **130,** 4–50 (2013).
52. Sromovsky, L. A., Fry, P. M., Boudon, V., Campargue, A. & Nikitin, A. Comparison of line-by-line and band models of near-IR methane absorption applied to outer planet atmospheres. *Icarus* **218,** 1–23 (2012).
53. Yurchenko, S. N., Tennyson, J., Bailey, J., Hollis, M. D. & Tinetti, G. Spectrum of hot methane in astronomical objects using a comprehensive computed line list. *Proc. Natl Acad. Sci. USA* **111,** 9379–9383 (2014).
54. Nikitin, A. V. *et al.* GOSAT-2009 methane spectral line list in the 5,550–6,236 cm^{-1} range. *J. Quant. Spectrosc. Radiat. Transfer* **111,** 2211–2224 (2010).
55. Borysow, A. Collision-induced absorption coefficients of H_2 pairs at temperatures from 60 K to 1,000 K. *Astron. Astrophys.* **390,** 779–782 (2002).
56. Borysow, A., Trafton, L., Frommhold, L. & Birnbaum, G. Modeling of pressure-induced far-infrared absorption spectra: Molecular hydrogen pairs. *Astrophys. J.* **296,** 644–654 (1985).
57. Borysow, J., Frommhold, L. & Birnbaum, G. Collison-induced rototranslational absorption spectra of H_2-He pairs at temperatures from 40 to 3000 K. *Astrophys. J.* **326,** 509–515 (1988).
58. Borysow, A. & Frommhold, L. Collision-induced infrared spectra of H_2-He pairs at temperatures from 18 to 7,000 K. II-Overtone and hot bands. *Astrophys. J.* **341,** 549–555 (1989).
59. Borysow, A., Frommhold, L. & Moraldi, M. Collision-induced infrared spectra of H_2-He pairs involving 0-1 vibrational transitions and temperatures from 18 to 7,000 K. *Astrophys. J.* **336,** 495–503 (1989).
60. Borysow, A. New model of collision-induced infrared absorption spectra of H_2-He pairs in the 2-2.5 μm range at temperatures from 20 to 300 K: An update. *Icarus* **96,** 169–175 (1992).
61. Orton, G. S. *et al.* Semi-annual oscillations in Saturn's low-latitude stratospheric temperatures. *Nature* **453,** 196–199 (2008).
62. Orton, G. S., Gustafsson, M., Burgdorf, M. & Meadows, V. Revised ab initio models for H_2-H_2 collision-induced absorption at low temperatures. *Icarus* **189,** 544–549 (2007).

Acknowledgements

We thank E. Karkoschka, L. Brown, G. Orton, J. Bailey, T. Kostiuk, A. Showman and L. Li for useful discussions and comments. Special thanks to M. Gerstell, P. Gao, R. Hu, P. Kopparla, C. Li, M.C. Liang, S. Newman, R.L. Shia, M. Wong, X. Xi and Q. Zhang for proofreading the manuscript. The early phase of this research was supported by the Outer Planets Research program via NASA Grant JPL 1452240 to the California Institute of Technology. R.A.W. and C.A.N. are supported by the NASA Cassini project. P.G.J.I. acknowledges the support of the UK Science and Technology Facilities Council.

Author contributions

X.Z. carried out the radiative modelling and CIRS spectral retrieval; R.A.W. provided the ISS data; C.A.N. provided the CIRS data; R.A.W. and Y.L.Y. helped with radiative modelling; P.G.J.I. and C.A.N. helped with spectral inversion modelling; all authors contributed to the paper writing.

Additional information

Competing financial interests: The authors declare no competing financial interests.

10

Skilful multi-year predictions of tropical trans-basin climate variability

Yoshimitsu Chikamoto[1], Axel Timmermann[1,2], Jing-Jia Luo[3], Takashi Mochizuki[4], Masahide Kimoto[5], Masahiro Watanabe[5], Masayoshi Ishii[6], Shang-Ping Xie[7] & Fei-Fei Jin[8]

Tropical Pacific sea surface temperature anomalies influence the atmospheric circulation, impacting climate far beyond the tropics. The predictability of the corresponding atmospheric signals is typically limited to less than 1 year lead time. Here we present observational and modelling evidence for multi-year predictability of coherent trans-basin climate variations that are characterized by a zonal seesaw in tropical sea surface temperature and sea-level pressure between the Pacific and the other two ocean basins. State-of-the-art climate model forecasts initialized from a realistic ocean state show that the low-frequency trans-basin climate variability, which explains part of the El Niño Southern Oscillation flavours, can be predicted up to 3 years ahead, thus exceeding the predictive skill of current tropical climate forecasts for natural variability. This low-frequency variability emerges from the synchronization of ocean anomalies in all basins via global reorganizations of the atmospheric Walker Circulation.

[1] International Pacific Research Center, University of Hawaii at Manoa, 1680 East-West Road, Honolulu, Hawaii 96822, USA. [2] Department of Oceanography, University of Hawaii at Manoa, 1680 East-West Road, Honolulu, Hawaii 96822, USA. [3] Centre for Australian Weather and Climate Research, Bureau of Meteorology, GPO Box 1289, Melbourne, Victoria 3001, Australia. [4] Japan Agency for Marine-Earth Science and Technology, 3173-25 Showa-machi, Kanazawa-ku, Yokohama, Kanagawa 236-0001, Japan. [5] Atmosphere and Ocean Research Institute, University of Tokyo, 5-1-5 Kashiwanoha, Kashiwa, Chiba 277-8568, Japan. [6] Meteorological Research Institute, Japan Meteorological Agency, 1-1 Nagamine, Tsukuba, Ibaraki 305-0052, Japan. [7] Scripps Institution of Oceanography, University of California San Diego, 9500 Gilman Drive MC 206, La Jolla, California 92093-0206, USA. [8] Department of Meteorology, University of Hawaii at Manoa, 2525 Correa Road, Honolulu, Hawaii 96822, USA. Correspondence and requests for materials should be addressed to Y.C. (email: chika44@hawaii.edu).

The Pacific El Niño Southern Oscillation (ENSO) is a major driver for interannual atmospheric and oceanic variability, which is well documented as a global teleconnection pattern[1,2]. On decadal timescales, tropical climate anomalies can affect global warming rate[3], drought patterns[4,5], regional sea-level changes[6,7], the frequency and characteristics of El Niño events[8] and the abundance of phytoplankton and movement of fish in the oceans[9]. As an example, the recent ~20-year-cooling trend in the eastern tropical Pacific has been linked to the generation of massive droughts in Southern California[3], an intensification of Pacific trade winds and accelerated sea-level rise in the western tropical Pacific[6]. This decadal-scale cooling has also contributed to the recent hiatus of global mean surface temperature rise[3,10–12]. Forecasting the phase reversal of the recent Pacific sea surface temperature (SST) trend several years ahead—beyond the typical ENSO prediction limit of ~1 year[13]—would be of great societal benefit for instance for western Pacific island nations, which experience frequent inundations, and for the farming and energy sectors worldwide.

Climate predictions may exhibit enhanced skill on timescales of years to decades as a result of internal climate dynamics, or in response to slowly evolving boundary conditions associated with anthropogenic and natural radiative forcings, or combinations thereof. Although there is an emerging consensus that internally generated low-frequency climate variability in the extratropical North Atlantic and North Pacific[14–16] may be predictable up to half a decade[17,18], such skill is still elusive for the tropical climate system, where longer-term predictability seems to be mainly determined by external boundary forcing associated with changes in greenhouse gas (GHG)[19] and aerosol concentrations[20].

Here we revisit the subject of multi-year predictability of tropical Pacific climate, by focusing on the low-frequency modulation of trans-basin gradients in tropical SST and sea-level pressure (SLP) across the Atlantic/Indian/Pacific basins. Our study is based on a series of Coupled General Circulation Model hindcasts with the MIROC model[21] that were initialized between 1960 and 2005 from realistic assimilated ocean conditions[16] (see Methods). Every 5 years, a 10-year hindcast was performed and the time evolution of historical external forcings was prescribed during the forecast (see Methods). In addition, a second suite of uninitialized model experiments was conducted to investigate the contribution of ocean initial conditions to the multi-year predictability of the tropical climate. To evaluate the impact of ocean variability in the equatorial Pacific, the Atlantic and the Indian Ocean on multi-year predictability elsewhere, we also conducted a set of experiments for which observed ocean data were assimilated only in one of the ocean basins, whereas the other basins were allowed to evolve freely (see Methods). From these experiments, we show the multi-year predictive skills for the low-frequency tropical climate variability via global reorganizations of the atmospheric Walker Circulation.

Results

Predictability of tropical low-frequency climate variability. The model forecasts demonstrate that the external radiative forcing is still the dominant source for multi-year SST predictability over major parts of the tropical oceans (Fig. 1). However, in the central Pacific, the situation is fundamentally different: the contribution of external forcing is almost negligible, particularly for SLP anomalies. In fact, most of the multi-year predictability of SST

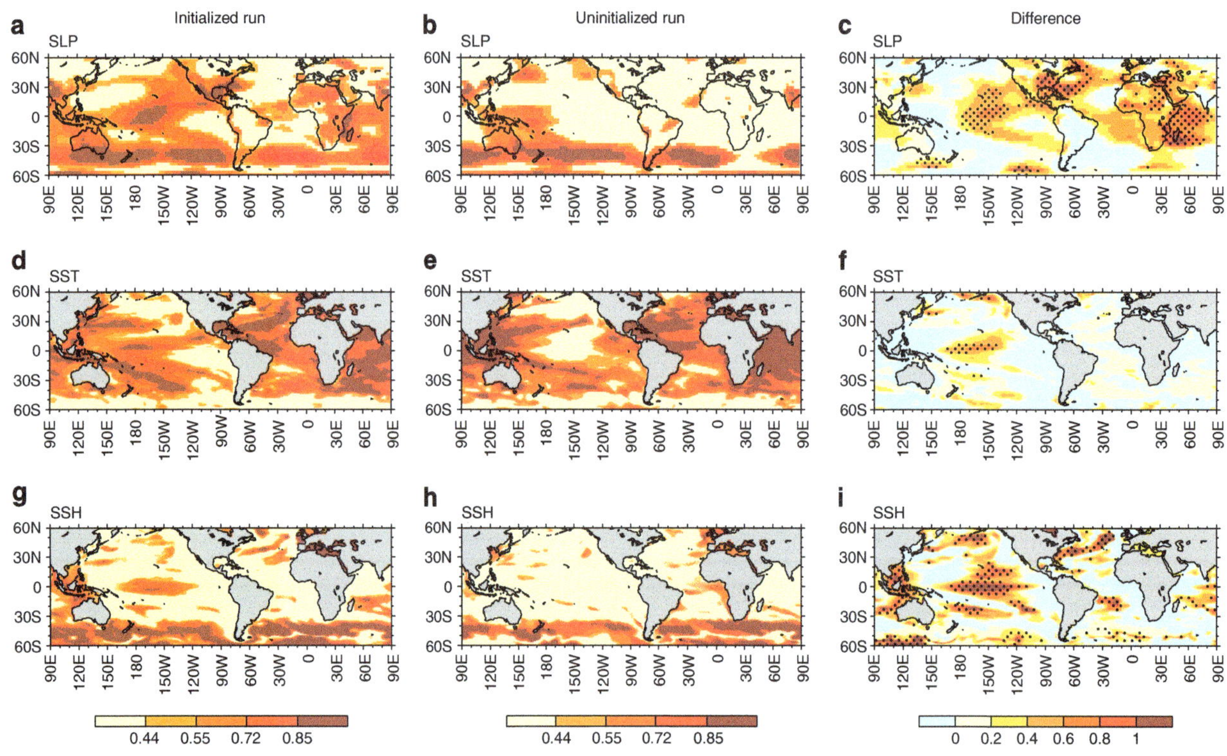

Figure 1 | Predictive skills for 2-5 years lead time. Potential predictive skills of (**a-c**) SLP, (**d-f**) SST and (**g-i**) SSH anomalies for averaged 2-5 years lead time in the initialized run (left), the uninitialized run (center) and their difference (right). Predictive skills in the initialized hindcast and uninitialized runs are measured per grid-point by the anomaly correlation coefficient and with the same number of initial cases (that is, 10 cases) by comparing with the assimilation run that is comparable to the observation. Correlation coefficients of 0.44, 0.55, 0.72 and 0.85 correspond to the statistical significance at 90, 95, 99 and 99.9% levels with eight degrees of freedom on the basis of one-side Student's *t*-test. Dotted areas indicate the statistically significant (>90%) difference in the anomaly correlation skill between the initialized and the uninitialized runs, as measured by Fishers *z*-score.

and SLP anomalies there results from the ocean initial conditions and hence the long-term memory of the ocean state (Fig. 1c,f). Consistent with these SLP and SST predictive skills, sea surface height (SSH) also shows the multi-year predictability in the central tropical Pacific and Maritime continent (Fig. 1g). This enhanced multi-year predictability in central Pacific SLP can be explained by the fact that SLP, and in particular wind variability, are more sensitive to variations in the zonal gradient of SST rather than to the nearly homogeneous SST warming[22] induced by GHGs increases. We find that, unlike in the case of SST, natural variability explains >80% of the total variance of the SLP anomalies over major parts of the tropics (Supplementary Fig. 1). In the central tropical Pacific, therefore, the natural atmosphere–ocean variability is the major contributor for multi-year predictability.

In the Atlantic and Indian basins, we also find that the initialized run outperforms the SLP predictive skill of the uninitialized run over the tropical eastern Atlantic and western Indian Ocean, as well as the northwestern Atlantic (Fig. 1c). In the initialized run, SSH exhibits the multi-year predictability in the North Atlantic and near Madagascar (Fig. 1g). On the other hand, SST predictive skills are almost comparable between the initialized and uninitialized runs because of the dominant contribution from the externally forced component (Fig. 1d–f). This spatially organized ocean/atmosphere trans-basin variability, which involves unforced and forced heat content variability in different ocean basins, is synchronized by tropical atmosphere circulation changes as shown below.

Trans-basin variability in the tropics. By determining the dominant mode of low-frequency SLP variability, we identify a pressure seesaw between the central tropical Pacific and Atlantic–Indian Ocean basins as well as its related patterns in SST, SSH and precipitation (Fig. 2). These patterns show the zonal gradients of SLP and SST anomalies between the central Pacific and tropical Atlantic/Indian Oceans. Moreover, precipitation anomalies exhibit the tropical Pacific–Atlantic contrast that is typical of a trans-basin displacement of Walker Circulation through atmosphere–ocean interactions. The trans-basin SST pattern is similar to that generated by climate models in response to specified Indian Ocean or Atlantic warming trends[5,23–25]. A large negative Atlantic/Pacific SLP gradient generates anomalous surface easterly (westerly) winds in the western Pacific (tropical Atlantic) and causes a zonal displacement of SSH anomalies in the tropical Pacific (Atlantic). We also find similar results of trans-basin displacements in the detrended anomalies or other observational re-analysis data sets for the different periods (Supplementary Figs 2 and 3). This coherent tropical low-frequency variability across the basins is referred to as tropical trans-basin variability (TBV)[5] and represented by a tropical TBV index as the difference of standardized SLP anomalies in the tropical central Pacific minus the tropical Atlantic–Indian Oceans (see Methods).

Although TBV has a similar pattern to ENSO[26,27] in the Pacific, it has very different characteristics in the Indian and Atlantic Oceans. In fact, SST anomalies in these areas oppose those of the typical ENSO teleconnection pattern[1,28,29]. The most discernable characteristic of the TBV, however, is the active role of ocean variability in the Atlantic. Obviously, SST variability in the equatorial Pacific can affect climate variability in the other oceans on interannual-to-decadal timescales[1–3]. In fact, we can capture a large fraction of tropical SLP variability[30] by partially assimilating the ocean state in the equatorial Pacific only (Supplementary Fig. 4a and see Methods). Surprisingly, however, when we partially assimilate the Atlantic Ocean state, SLP variability in the central tropical Pacific is also captured well (local maximum of correlation coefficient in Supplementary Fig. 4c is 0.6; above statistical significance at 99.9% level). This Pacific SLP multi-year variability is thus partly generated by the Atlantic SST anomalies rather than by the external radiative forcings (Supplementary Fig. 4c,g). Associated with this Atlantic-induced Pacific SLP variability, wind-driven SSH variability also shows a positive correlation in the equatorial Pacific (Supplementary Fig. 4d). The central Pacific is one of the most predictable areas for SLP on multi-year timescales demonstrated by the initialized run (Fig. 1). According to these results, the long-term memory of the TBV and the associated multi-year predictability partly originate from the Atlantic Ocean. Despite the fact that the coupled climate model still has large biases to simulate the observed variability, the atmosphere–ocean anomalies associated with the TBV, in particular the low-frequency component, are captured well by the partial assimilation over the Atlantic Ocean (Supplementary Fig. 5). Compared with the Atlantic, the Indian Ocean plays a lesser role in the central Pacific SLP variability but still contributes more than the externally forced component (Supplementary Fig. 4e,f).

Our analysis also confirms an active role of the Atlantic Ocean in the recent pause of global warming trend known as the hiatus event[3,5,12,31]. From 1992 to 2000, the first principal components associated with the TBV show a phase change from negative to positive (that is, Atlantic warming and eastern Pacific cooling in Fig. 2e), which contributes to the hiatus event after 2000 as simulated by our forecast (black circle and bars in Fig. 2e). In particular, the eastern tropical Pacific cooling trend offsets some of the greenhouse-induced Atlantic and Indian Ocean warming, thus leading to a weak global SST warming trend. Consistent with this phase change, linear trends of SST and SLP for this period show the key TBV features: trans-basin SST and SLP gradients between the Atlantic–Indian Ocean and the Pacific, La Niña-like SST pattern, and SLP increases in the North and South Pacific Oceans (Fig. 3a). Although our partial assimilation experiments indicate that the equatorial Pacific ocean variability is the major driver for those trends (Fig. 3b) as shown by previous studies[3,12], contributions from the Atlantic Ocean variability explain part of the sea surface cooling in the tropical Pacific and half of the SLP increases in the North and South Pacific Oceans (Fig. 3c). This result further supports our finding that the multi-year predictability in the tropical Pacific partly relies on the long-term memory provided by the Atlantic coupled atmosphere/ocean system. The result is also consistent with a previous study suggesting that the Atlantic SST warming contributed to improve the predictive skills of a La Niña-like SST pattern in the tropical Pacific during the late 1990s climate shift[24].

Predictability of the tropical trans-basin variability. Although ENSO-related SST prediction skill in our MIROC forecasting system quickly decreases at lead times of >1 year (Supplementary Fig. 6), in agreement with previous studies[32], the tropical TBV indices defined by the trans-basin SLP and SST gradients (see Method) show predictive skill even for lead times of up to 3 years in advance (Fig. 2e and Fig. 4, and Supplementary Fig. 7). This high persistence in TBV compared with ENSO is further illustrated by spectral analysis: ENSO exhibits much less low-frequency variability and a shorter damping timescales compared with the TBV (Supplementary Fig. 8). Because of the larger spatial scales, the TBV averages out many of the higher-frequency sub-basin-scale features, thus enhancing the low-frequency variance in contrast to ENSO. Our partial assimilation experiments also support the trans-basin climate feedbacks

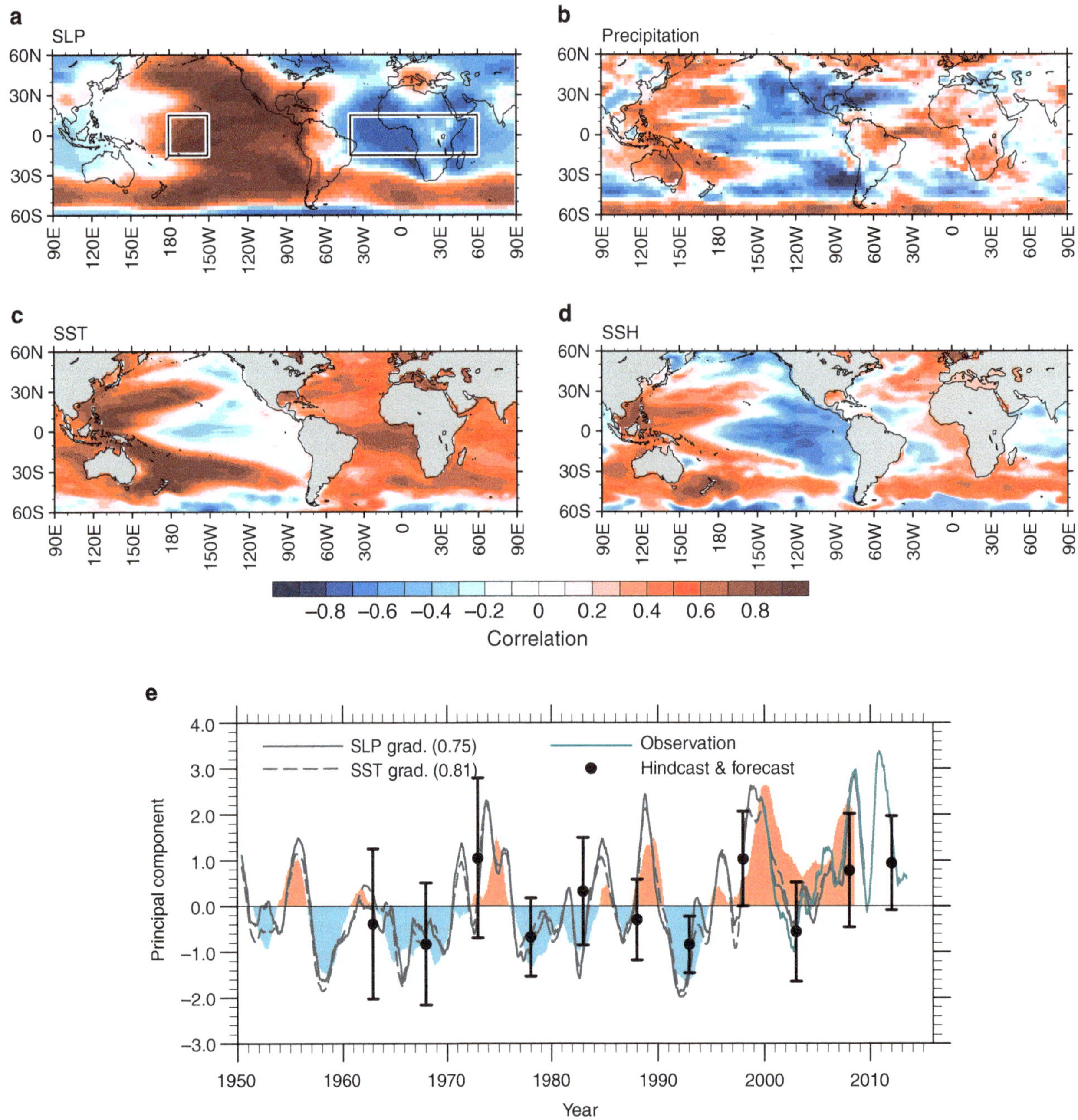

Figure 2 | Patterns associated with the trans-basin variability. Correlation maps of (**a**) SLP, (**b**) precipitation, (**c**) SST and (**d**) SSH anomalies with the principal component (shaded in **e**) of the leading correlation-matrix based EOF for global SLP anomalies (60° S–60° N) smoothed by 36-month running mean filter (explaining 31.7% of variance) in the assimilation run. Grey and turquoise solid lines (**e**) denote the 12-month running mean tropical SLP-based TBV index (Pacific minus Atlantic SLP anomalies averaged over the boxes in **a** in the assimilation run and the observation (ERA-interim[43]) from 2000 to 2013, respectively. Broken line (**e**) is the tropical SST-based TBV index (bracket; correlation coefficient with the principal component). Black circles and bars denote the hindcasting and forecasting ensemble mean of the tropical SLP-based TBV index and its spread for 2-5 years lead time.

(Supplementary Fig. 9) that are associated with an anomalous global Walker Circulation.

The multi-year predictive skill of the TBV mainly relies on the natural variability rather than on the external forcing components. This is illustrated by high prediction skill of the initialized run compared with the uninitialized run (Fig. 4 and Supplementary Fig. 7). However, predictive skill in the uninitialized runs sometimes even outperforms that of the initialized run: the SLP root-mean-squared error skill at 25–29 months (Supplementary Fig. 7a) and the SST skills at 42–49 months lead time (Fig. 4b and Supplementary Fig. 7b). Although the GHG increases contribute to the zonally uniform SST warming, some volcanic events and local aerosol forcings may modulate the trans-basin gradients owing to their zonally asymmetric structures. In fact, the external radiative forcing is the major contributor to SST predictive skill in the Atlantic and Indian Oceans (Fig. 1d–f). Although more accurate estimation of predictive skills would be achieved by the multi-model ensemble in the CMIP5 decadal prediction experiment or the enhanced intervals of the hindcast experiment (for example, initialized every year), our successful prediction at the multi-year lead time provides a new evidence for the importance of the long-term ocean memory in predicting the tropical low-frequency climate variability.

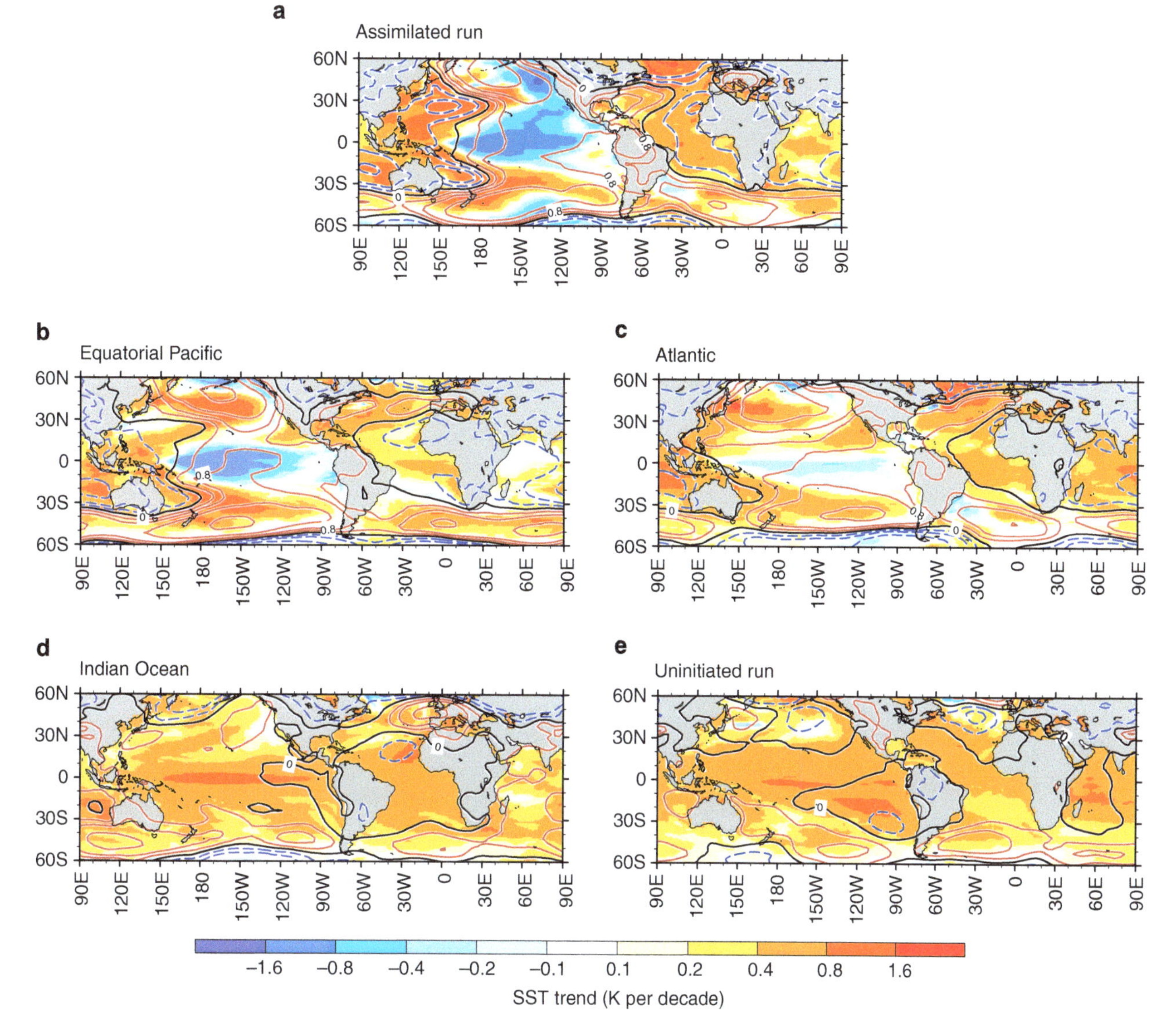

Figure 3 | Sea surface temperature and sea-level pressure trends for the 1992-2000 period. Linear trends of SST (shaded) and SLP (contoured) for the 1992-2000 period are obtained in (**a**) the assimilated run, the partial assimilations (**b**) in the equatorial Pacific, (**c**) Atlantic, (**d**) Indian Ocean and (**e**) the uninitialized externally forced twentieth century run. Contour intervals of SLP are 0, ± 0.4, ± 0.8, ± 1.6 and ± 3.2 hPa per decade. Red, blue and black lines denote positive, negative and zero contours, respectively.

Discussion

Our results suggest that the Atlantic Ocean plays an important role in low-frequency variability in the tropical Pacific via the atmospheric trans-basin coupling. Although the tropical Pacific variability is still a major driver to induce interannual-to-decadal climate variability in the other basins, the atmosphere–ocean variability in the Atlantic and Indian Oceans can feed back to the Pacific through trans-basin interactions and global displacements of the Walker Circulation. As a result, the tropical TBV predictability exceeds ENSO's predictive skill by a factor of ~3 (Fig. 4 and Supplementary Fig. 6), despite the fact that our results may be contaminated by model biases and the simple initialization technique.

In the tropical Pacific, the TBV-related SST anomaly pattern also projects onto the El Niño Modoki[33]/warm pool ENSO[34] mode and onto the Interdecadal Pacific Oscillation[35,36]. The simulated 1–12 months predictive skill for the Modoki/warm pool ENSO indices (Fig. 4c,d) is comparable to that of the Niño 3.4 ENSO SST index (Supplementary Fig. 6). This finding is in agreement with previous studies[13,19,37,38]. However, our analysis reveals a considerable recovery of skill for the central Pacific ENSO indices after the 2 or 3 years lead time (Fig. 4c,d), consistent with the idea of a remote impact from the other ocean basins. This suggests that the Modoki/warm pool ENSO mode is not solely an internal Pacific mode, but is partly controlled by Atlantic and Indian Ocean SST anomalies and related displacements of the global Walker Circulation. This conclusion is further supported by the similarity and high correlation (higher than 0.55; above statistical significance at 99.9% level) between the TBV and the Modoki/warm pool ENSO indices (Supplementary Fig. 10). The idea of predictable tropical TBV presented here provides a novel perspective for understanding global-scale climate variability and ENSO flavours. In view of the global impacts of the TBV on precipitation and sea-level anomalies (Fig. 2), operational predictions of the TBV may translate into better assessments of risks in sectors, such as coastal and water management, forestry and agriculture.

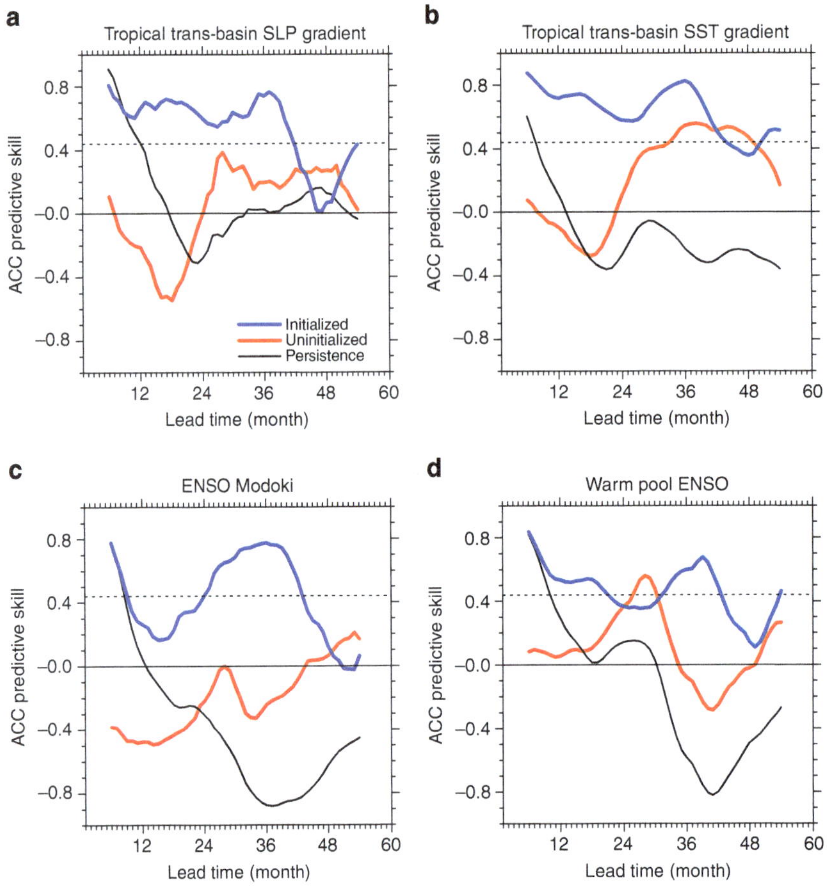

Figure 4 | Predictive skills of the tropical indices. Potential predictive skills are obtained from the tropical trans-basin (**a**) SLP and (**b**) SST gradients, (**c**) ENSO Modoki and (**d**) Warm pool ENSO indices in the initialized run (blue), the uninitialized run (red) and persistence (black), measured by anomaly correlation coefficient. A 12-month running mean is applied to indices. Black dotted line denotes the statistical significance at the 90% level with eight degrees of freedom on the basis of one-side Student's *t*-test.

Methods

Model. The coupled atmosphere–ocean general circulation model adopted here is version 3.2 of the Model for Interdisciplinary Research on Climate (MIROC)[21] with a T42 spectral resolution and 20 levels on a vertical σ-coordinate. The resolution of the ocean component is 1.4° in longitude and 0.56–1.4° in latitude with 44 vertical levels. No flux correction is applied to exchange heat, water and momentum fluxes between the atmosphere and the ocean. In all the experiments, we use historical observed data of natural and anthropogenic forcings (greenhouse gas and aerosol concentrations, solar cycle variations and major volcanic eruptions) before 2000 and the IPCC A1B-type emissions scenario after 2000.

Hindcast experiments. We conducted three main experiments using MIROC: the uninitialized, the assimilated and the initialized runs. In the uninitialized run, we prescribed natural and anthropogenic radiative forcings in the coupled model for the period of 1850–2100 with 10 ensemble members starting from different initial conditions obtained from the pre-industrial control simulation. For the 10-member assimilation runs, the model climatology defined by the 1961–1990 period from the uninitialized run was added onto the observed three dimensional oceanic temperature and salinity anomalies from surface to 700-m depth during 1945–2009 (ref. 39). The anomalies outside the sea–ice regions are assimilated into the coupled climate model using the incremental analysis update scheme[40,41]. Analysis increments estimated from a temporally and spatially invariant model-to-observation ratio in analysis errors (about one sixth) are added as forcing terms into the model's temperature and salinity tendency equation during an analysis interval of 1 day[16,42]. Due to the tight coupling between SLP and SST fields in the tropics, our assimilated run can realistically reproduce the observed trans-basin SLP variability (Supplementary Figs 3, 11 and 12), even without atmospheric assimilation. On the basis of the atmospheric and oceanic initial conditions generated by the 10-member assimilation experiments, we conducted 10-year-long ensemble hindcast experiments initialized from 1 January 1960, 1965, 1970, 1975, 1980, 1985 1990, 1995, 2000 and 2005 (the initialized runs). No post processing for removing an artificial drift is applied to our initialized run because of a negligible climate drift in the tropics during the prediction period. Details of the model experiment and assimilation procedure are described in previous studies[16,18].

Partial assimilation experiments. We have partially assimilated the observed ocean temperature and salinity anomalies in either the equatorial Pacific (10° S–10° N), the Atlantic (50° S–60° N) and Indian Ocean (30° S–30° N). These partial assimilation experiments consist of 10-member ensembles each and cover the period from 1945 to 2009. The partial assimilation experiments can capture not only the atmospheric response but also the resulting atmosphere–ocean interactions in the other ocean basins, which are allowed to vary freely in response to the atmospheric changes caused by the SST anomalies in the assimilated basins.

Tropical TBV and ENSO indices. To capture the global-scale characteristics of the TBV, the principal component time series of the 36-month running mean leading EOF mode of SLP is a suitable index. Here, for simplicity and to highlight the important role of tropics, we use the tropical TBV indices on the basis of the zonal trans-basin gradients of SLP and SST (see boxes in Fig. 2a). These surrogate indices show high correlations with the principal components of the leading EOF (Fig. 2e and Supplementary Fig. 2e). The tropical TBV indices on the basis of SLP and SST are defined as the zonal gradient of SLP and SST anomalies averaged and standardized in each region between the tropical central Pacific (15° S–15° N, 180° W–150° W) and the tropical Atlantic–Indian Ocean (15° S–15° N, 40° W–60° E). Note that these indices are more focused on the tropics compared with the global-scale TBV index introduced in a recent study[5].

The ENSO indices are defined as the zonal gradient of standardized SLP anomalies between the eastern equatorial Pacific (5° S–5° N, 80° W–130° W) and the Indonesia region (5° S–5° N, 90° E–140° E) and the SST anomalies in the Niño 3.4 region (5° S–5° N, 120° W–170° W), respectively. The ENSO index based on the SLP gradient is identical to the equatorial Southern Oscillation index.

References

1. Klein, S. A., Soden, B. J. & Lau, N. C. Remote sea surface temperature variations during ENSO: evidence for a tropical atmospheric bridge. *J. Clim.* **12**, 917–932 (1999).

2. Xie, S. P. & Carton, J. A. in *Earth Climate: Ocean-Atmosphere Interaction and Climate Variability* Vol. 147 (eds Xie, S. P. & Carton, J. A.) 121–142 (Geophysical Monograph, 2004).
3. Kosaka, Y. & Xie, S. P. Recent global-warming hiatus tied to equatorial Pacific surface cooling. *Nature* **501,** 403–407 (2013).
4. Barlow, M., Nigam, S. & Berbery, E. ENSO, Pacific decadal variability, and US summertime precipitation, drought, and stream flow. *J. Clim.* **14,** 2105–2128 (2001).
5. McGregor, S. *et al.* Recent Walker Circulation strengthening and Pacific cooling amplified by Atlantic warming. *Nat. Clim. Chang.* **4,** 888–892 (2014).
6. Timmermann, A., McGregor, S. & Jin, F.-F. Wind effects on past and future regional sea level trends in the Southern Indo-Pacific. *J. Clim.* **23,** 4429–4437 (2010).
7. Merrifield, M. A. & Maltrud, M. E. Regional sea level trends due to a Pacific trade wind intensification. *Geophys. Res. Lett.* **38,** L21605 (2011).
8. Timmermann, A., Jin, F.-F. & Abshagen, J. A nonlinear theory for El Niño bursting. *J. Atmos. Sci.* **60,** 152–165 (2003).
9. Miller, A. J., Chai, F., Chiba, S., Moisan, J. R. & Neilson, D. J. Decadal-scale climate and ecosystem interactions in the North Pacific Ocean. *J. Oceanogr.* **60,** 163–188 (2004).
10. England, M. H. *et al.* Recent intensification of wind-driven circulation in the Pacific and the ongoing warming hiatus. *Nat. Clim. Chang.* **4,** 222–227 (2014).
11. Watanabe, M. *et al.* Strengthening of ocean heat uptake efficiency associated with the recent climate hiatus. *Geophys. Res. Lett.* **40,** 3175–3179 (2013).
12. Watanabe, M. *et al.* Contribution of natural decadal variability to global-warming acceleration and hiatus. *Nat. Clim. Chang.* **4,** 893–897 (2014).
13. Meehl, G. A. *et al.* Decadal climate prediction: an update from the trenches. *Bull. Am. Meteorol. Soc.* **95,** 243–267 (2014).
14. Smith, D. M. *et al.* Improved surface temperature prediction for the coming decade from a global climate model. *Science* **317,** 796–799 (2007).
15. Keenlyside, N. S., Latif, M., Jungclaus, J., Kornblueh, L. & Roeckner, E. Advancing decadal-scale climate prediction in the North Atlantic sector. *Nature* **453,** 84–88 (2008).
16. Mochizuki, T. *et al.* Pacific decadal oscillation hindcasts relevant to near-term climate prediction. *Proc. Natl Acad. Sci. USA* **107,** 1833 (2010).
17. Matei, D. *et al.* Multiyear prediction of monthly mean Atlantic meridional overturning circulation at 26.5 N. *Science* **335,** 76–79 (2012).
18. Chikamoto, Y. *et al.* An overview of decadal climate predictability in a multi-model ensemble by climate model MIROC. *Clim. Dyn.* **40,** 1201–1222 (2013).
19. Doblas-Reyes, F. *et al.* Initialized near-term regional climate change prediction. *Nat. Commun.* **4,** 1715 (2013).
20. Booth, B. B., Dunstone, N. J., Halloran, P. R., Andrews, T. & Bellouin, N. Aerosols implicated as a prime driver of twentieth-century North Atlantic climate variability. *Nature* **484,** 228–232 (2012).
21. K-1 Model Developers. *K-1 Coupled GCM (MIROC) Description* Vol. 1Center for Climate System Research, Univ. of Tokyo, 2004).
22. Tokinaga, H., Xie, S. P., Deser, C., Kosaka, Y. & Okumura, Y. M. Slowdown of the Walker circulation driven by tropical Indo-Pacific warming. *Nature* **491,** 439–443 (2012).
23. Luo, J.-J., Sasaki, W. & Masumoto, Y. Indian Ocean warming modulates Pacific climate change. *Proc. Natl Acad. Sci.* **109,** 18701–18706 (2012).
24. Chikamoto, Y., Kimoto, M., Watanabe, M., Ishii, M. & Mochizuki, T. Relationship between the Pacific and Atlantic stepwise climate change during the 1990 s. *Geophys. Res. Lett.* **39,** L21710 (2012).
25. Ham, Y. G., Kug, J. S., Park, J. Y. & Jin, F. F. Sea surface temperature in the north tropical Atlantic as a trigger for El Niño/Southern Oscillation events. *Nat. Geosci.* **6,** 112–116 (2013).
26. Jin, F.-F. Low-frequency modes of tropical ocean dynamics. *J. Clim.* **14,** 3874–3881 (2001).
27. Wang, C. & Picaut, J. Understanding ENSO physics—A review. *Earth Climate: Ocean-Atmosphere Interaction and Climate Variability* **147,** 21–48 (2004).
28. Chiang, J. C. H. & Sobel, A. H. Tropical tropospheric temperature variations caused by ENSO and their influence on the remote tropical climate. *J. Clim.* **15,** 2616–2631 (2002).
29. Chikamoto, Y. & Tanimoto, Y. Air-sea Humidity Effects on the Generation of Tropical Atlantic SST Anomalies during the ENSO Events. *Geophys. Res. Lett.* **33,** L19702 (2006).
30. Trenberth, K. E., Fasullo, J. T., Branstator, G. & Phillips, A. S. Seasonal aspects of the recent pause in surface warming. *Nat. Clim. Chang.* **4,** 911–916 (2014).
31. Meehl, G. A., Arblaster, J. M., Fasullo, J. T., Hu, A. X. & Trenberth, K. E. Model-based evidence of deep-ocean heat uptake during surface-temperature hiatus periods. *Nat. Clim. Chang.* **1,** 360–364 (2011).
32. Jin, E. K. *et al.* Current status of ENSO prediction skill in coupled ocean—atmosphere models. *Clim. Dyn.* **31,** 647–664 (2008).
33. Ashok, K., Behera, S. K., Rao, S. A., Weng, H. & Yamagata, T. El Niño Modoki and its possible teleconnection. *J. Geophys. Res.* **112,** C11007 (2007).
34. Kug, J.-S., Jin, F.-F. & An, S.-I. Two types of El Niño events: cold tongue El Niño and warm pool El Niño. *J. Clim.* **22,** 1499–1515 (2009).
35. Power, S., Casey, T., Folland, C., Colman, A. & Mehta, V. Inter-decadal modulation of the impact of ENSO on Australia. *Clim. Dyn.* **15,** 319–324 (1999).
36. Meehl, G. A. & Hu, A. Megadroughts in the Indian monsoon region and southwest North America and a mechanism for associated multidecadal Pacific sea surface temperature anomalies. *J. Clim.* **19,** 1605–1623 (2006).
37. Hendon, H. H., Lim, E., Wang, G. M., Alves, O. & Hudson, D. Prospects for predicting two flavors of El Niño. *Geophys. Res. Lett.* **36,** L19713 (2009).
38. Jeong, H. I. *et al.* Assessment of the APCC coupled MME suite in predicting the distinctive climate impacts of two flavors of ENSO during boreal winter. *Clim. Dyn.* **39,** 475–493 (2012).
39. Ishii, M. & Kimoto, M. Reevaluation of historical ocean heat content variations with time-varying XBT and MBT depth bias corrections. *J. Oceanogr.* **65,** 287–299 (2009).
40. Bloom, S. C., Takacs, L., da Silva, A. M. & Ledvina, D. Data assimilation using Incremental Analysis Updates. *Mon. Wea. Rev.* **124,** 1256–1271 (1996).
41. Huang, B., Kinter, J. L. & Schopf, P. S. Ocean data assimilation using intermittent analyses and continuous model error correction. *Adv. Atmos. Sci.* **19,** 965–992 (2002).
42. Tatebe, H. *et al.* The initialization of the MIROC climate models with hydographic data assimilation for decadal prediction. *J. Meteorol. Soc. Japan* **90A,** 275–294 (2012).
43. Dee, D. P. *et al.* The ERA-Interim reanalysis: configuration and performance of the data assimilation system. *Q. J. R. Meteorol. Soc.* **137,** 553–597 (2011).

Acknowledgements

This study was supported by the Japanese Ministry of Education, Culture, Sports, Science and Technology, through the Program for Risk Information on Climate Change. The simulations were performed with the Earth Simulator at the Japan Agency for Marine-Earth Science and Technology and the NEC SX-8R at the National Institute for Environmental Studies. A.T. and Y.C. were supported by NSF (1049219); S.-P.X. by NSF and NOAA.

Author contributions

Y.C., M.K., J.-J.L., A.T. and F.-F.J. designed the research. Y.C., T.M. and M.I. performed the experiments. Y.C., M.K., S.-P.X., J.-J.L. and A.T. wrote the paper. All the authors discussed the results and commented on the manuscript.

Additional information

Competing financial interests: The authors declare no competing financial interests.

11

Northern-hemispheric differential warming is the key to understanding the discrepancies in the projected Sahel rainfall

Jong-Yeon Park[1,2], Jürgen Bader[1,3] & Daniela Matei[1]

Future projections of the Sahel rainfall are highly uncertain, with different climate models showing widely differing rainfall trends. Moreover, the twentieth-century cross-model consensus linking Sahel rainfall to tropical sea-surface temperatures (SSTs) is no longer applicable in the twenty-first century. Here we show that the diverse future Northern Hemisphere differential warming between extratropical and tropical SSTs can explain the discrepancy in the projected Sahel rainfall. The relationship between SST and Sahel rainfall that holds for the twentieth-century persists into the twenty-first century when the differential SST warming is taken into account. A suite of SST-sensitivity experiments confirms that strong Northern Hemisphere extratropical warming induces a significant increase in Sahel rainfall, which can predominate over the drying impact of tropical SST warming. These results indicate that a trustworthy projection of Sahel rainfall requires the estimation of the most likely future Northern-hemispheric differential warming.

[1] Max Planck Institute for Meteorology, 20146 Hamburg, Germany. [2] International Max Planck Research School on Earth System Modelling, 20146 Hamburg, Germany. [3] Uni Climate, Uni Research & the Bjerknes Centre for Climate Research, 5007 Bergen, Norway. Correspondence and requests for materials should be addressed to J.-Y.P. (email: jong-yeon.park@mpimet.mpg.de).

The semiarid Sahel, located south of the Sahara desert, has its rainy season in the boreal summer. The rainfall over the African Sahel region exhibits puzzling behaviour in twenty-first century projections[1–6]. Future Sahel rainfall projections in different climate models disagree, even on the direction of the trend. Given the robust drying in response to late twentieth-century forcing factors of the same coupled models[7], such a disagreement with respect to future Sahel rainfall is surprising. Apart from the question of whether Sahel rainfall will increase or decrease, how such disparate Sahel rainfall projections arise remains an open question.

Sea-surface temperature (SST) variations are recognized to be the main cause of the observed decadal Sahel rainfall variability[8–12], with local soil moisture or vegetation feedbacks only amplifying the SST-driven low-frequency variability[13]. However, an attempt to predict future Sahel rainfall based on its historical relationship with tropical SSTs has been unsuccessful[1]. Alternative explanations have been proposed regarding the source of the model discrepancy, such as the strength of the Sahara Low and the land surface response to radiative forcing[14,15]. Given that oceanic forcing has played a dominant role in shaping the observed decadal variations of Sahel rainfall and previous studies mainly restricted their analysis to tropical SSTs to predict future rainfall changes, it is necessary to extend the investigation to the role of global SSTs to understand the disagreement in predicted Sahel rainfall trends.

In this study, we show that SSTs are the fundamental factor in estimating future Sahel rainfall changes. The difference between future scenarios and the historical period is that the key SST area determining future Sahel rainfall changes will be shifted from the tropics to the Northern Hemisphere (NH) extratropics in the case of dominant future extratropical warming over tropical warming. A multi-model analysis reveals that future Sahel rainfall change can be substantially modified by the magnitude of extratropical warming, which is also confirmed by a series of SST-sensitivity experiments. A confident projection of future Sahel rainfall requires an understanding of the cause of the discrepancy in rainfall predictions in current climate models, and here, we contribute to such understanding by establishing NH differential warming as a crucial factor underlying the uncertainty regarding future Sahel rainfall.

Results

Statistical analysis of multi-model outputs in CMIP5. The uncertainty in future Sahel rainfall found in the third phase of the Coupled Model Intercomparison Project (CMIP3) models is present even in the new generation of climate models, that is, CMIP5 (Fig. 1a and Supplementary Fig. 1). For example, as shown in the future projection from 'gfdl-cm2.1' and 'miroc3_2_medres' in CMIP3 (refs 1,2), two CMIP5 models from the same family, GFDL-ESM2M and MIROC-ESM, still project a substantially dryer and wetter Sahel, respectively, under a medium future emission scenario (Representative Concentration Pathways 4.5; RCP4.5). A recent study using the strongest radiatively forced future scenario for CMIP5 also

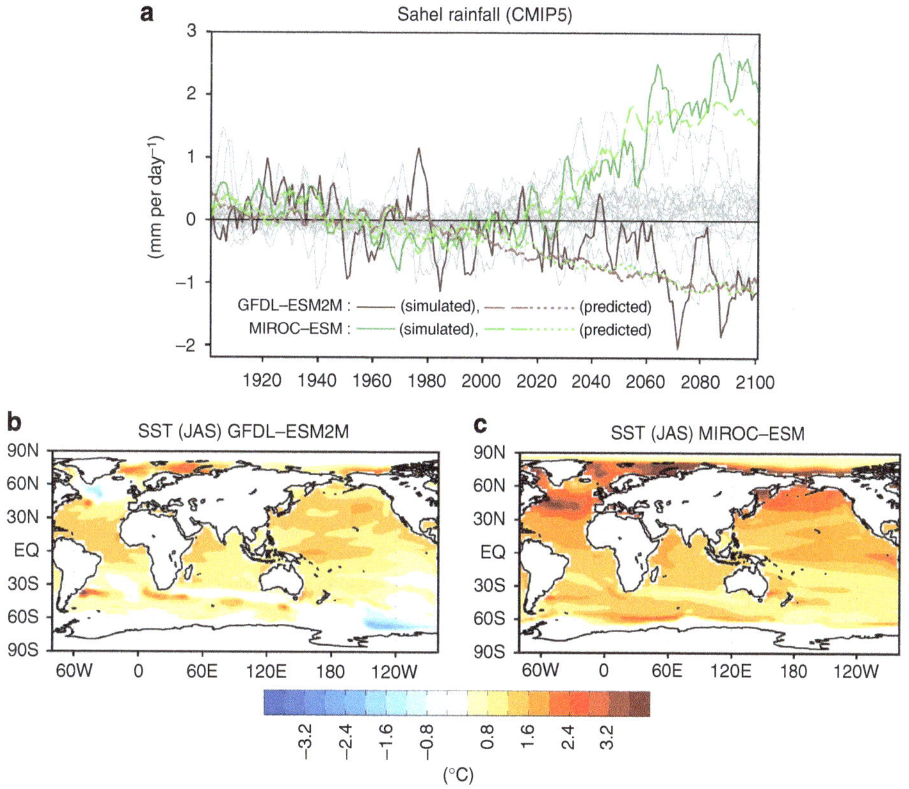

Figure 1 | Uncertainty in future Sahel rainfall and global SST warming pattern. (**a**) The JAS mean Sahel rainfall (10°W-30°E; 10°-20°N) anomalies during the twentieth and twenty-first centuries from CMIP5 models. Brown- and green-solid lines represent the Sahel rainfall anomalies simulated by GFDL-ESM2M and MIROC-ESM, which produce extreme Sahel drying and wetting, respectively. The rainfall anomalies from other CMIP5 models are indicated by grey lines. Dotted lines are the linearly predicted Sahel rainfall anomalies based on two tropical SST indices: the Atlantic SST gradient [(70°-20°W, 5°-30°N) minus (40°W-5°E, 30°S-5°N)] and Indo-Pacific SST (50°E-90°W, 20°S-20°N). Dashed lines represent predicted Sahel rainfall anomalies using an extratropical SST index (0°-360°E, 30°-75°N) in addition to the two tropical indices. The regression coefficients of the multi-linear model are calculated from the historical simulation (1901-2005). The anomalies are the differences from the historical mean and all time series data are smoothed using a 5-year running mean. (**b**,**c**) JAS SST differences between the twenty-first century and twentieth-century for GFDL-ESM2M (**b**) and for MIROC-ESM (**c**).

supports the model variability of future Sahel rainfall trends[16]. We cannot ignore those two extreme simulations because either the drying or wetting of the Sahel is a physically plausible response to increased radiative forcing, and these two models provide relatively accurate simulations of the West African summer rainfall in the historical period[3].

In addition to the opposing Sahel rainfall projections, the two models show a substantial difference in their projections of future Northern-hemispheric SST warming (Fig. 1b,c). In GFDL-ESM2M, the amplitudes of tropical and extratropical SST warming are similar, whereas in MIROC-ESM, extratropical warming is more prominent than warming in the tropics, particularly in the NH. Such association between strong NH extratropical SST warming and Sahel wetting or between weak warming and Sahel drying over the next century can also be found in other models (Supplementary Fig. 2). Therefore, we will investigate the role of NH extratropical SST warming in forcing the contradictory projections of future Sahel rainfall.

The MIROC models are key to understanding the discrepancy in the future projections because of a strong positive trend of future Sahel rainfall and a strong extratropical SST warming. Thus, the role of NH extratropical SST in predicting future Sahel rainfall is evaluated by reconstructing the simulated rainfall from MIROC-ESM based on different combinations of SST indices. First, the Sahel rainfall is linearly reconstructed using only commonly used tropical indices, namely, the interhemispheric SST gradient index in the tropical Atlantic ((70°–20°W, 5°–30°N) minus (40°W–5°E, 30°S–5°N)) and the area-averaged Indo-Pacific SST index (50°E–90°W, 20°S–20°N). The statistical model and indices are similar to those used in previous studies[1,5]. The two tropical indices reconstruct the Sahel rainfall reasonably well in the historical run, whereas they fail to reconstruct the future Sahel rainfall, exhibiting a drying trend in future Sahel rainfall (dotted green line in Fig. 1a). This shortcoming confirms that the previously used tropical SST indices are insufficient to predict a strong positive trend of future Sahel rainfall. In contrast, inclusion of the extratropical SST index (0°–360°E, 40°–75°N) in the statistical model reconstructs future Sahel rainfall considerably well with the strong positive trend similar to the simulated Sahel rainfall (dashed green line in Fig. 1a). Using a single index of NH differential SST warming ((0°–360°E, 30°–75°N) minus (0°–360°E, 20°S–20°N)) in the statistical model also successfully reproduces the strong positive trend (Supplementary Fig. 3a). Such an improved reproduction of future Sahel rainfall when considering the extratropical SST warming is not found in a similar analysis with a drying model (that is, GFDL-ESM2M, dashed brown line almost overlapped by the dotted brown line in Fig. 1a), a result that could be expected from the weak future NH extratropical warming simulated by GFDL-ESM2M.

Notably, the future development of Sahel rainfall can be reconstructed simply based on twentieth century relationships with large-scale SST indices including NH extratropical SST. That is, during the historical period Sahel rainfall is mainly affected by tropical SSTs because the extratropical SST plays a minor role due to the weak magnitude of its anomalies. However, if the NH extratropical SST warming becomes stronger under a global warming scenario compared with the warming in the tropics, then the future Sahel rainfall changes cannot be explained solely by tropical SST changes. Rather, the extratropical SST should be taken into account.

We further investigate how much the future Sahel rainfall change is related to the NH differential warming across the CMIP5 models. Future changes in Sahel rainfall and those in the NH differential warming are calculated in each CMIP5 model and displayed in a scatter plot (Fig. 2a). Models showing a stronger extratropical warming relative to the tropics

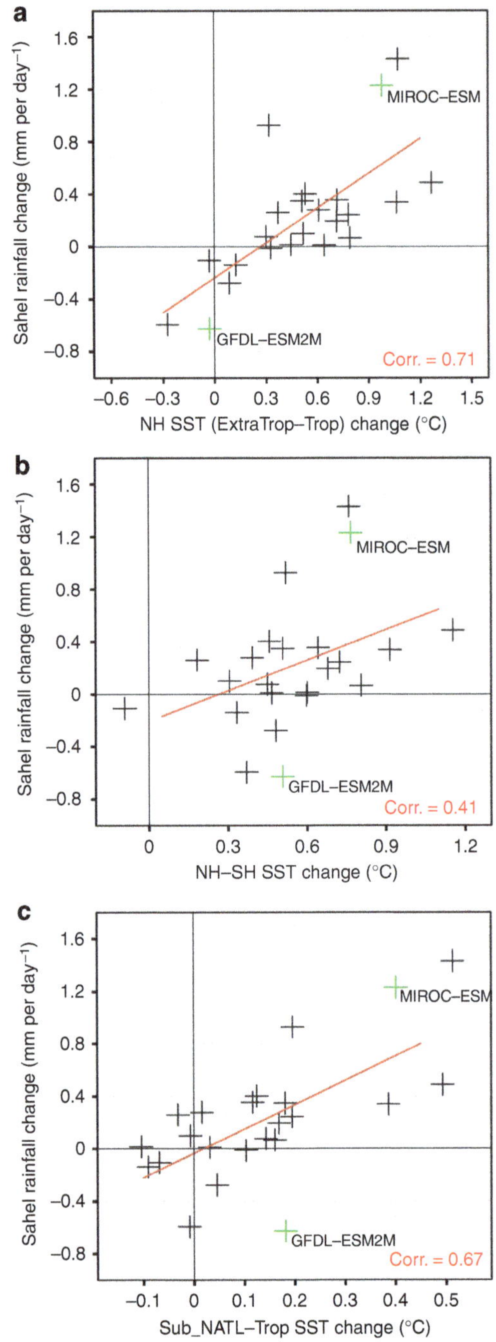

Figure 2 | Performance of various SST indices in explaining future Sahel rainfall change. (**a–c**) Scatter plots of future changes (that is, mean difference between the twenty-first century and twentieth-century) in Sahel rainfall versus those in three different SST indices across the CMIP5 models: NH extratropical SST warming relative to tropical warming ((0°-360°E, 30°-75°N) minus (0°-360°E, 20°S-20°N)) (**a**), global interhemispheric SST gradient ((0°-360°E, 0°-90°N) minus (0°-360°E, 0°-90°S)) (**b**) and difference between subtropical North Atlantic SST and global tropical SST ((75°-15°W, 10°-40°N) minus (0°-360°E, 20°S-20°N)) (**c**). Each mark represents an individual model in CMIP5. The fitted linear regression lines are shown as red lines.

tend to project a wetter Sahel. The relationship is quite robust despite other possible factors influencing future Sahel rainfall changes (for example, local land-surface or land-atmosphere feedbacks, intermodel variations in future tropical SST anomalies and different model sensitivities to SST forcing). On the basis of the fitted linear line, the same magnitude of warming over extratropical and tropical oceans leads to drying conditions over the Sahel, which is consistent with a drying Sahel rainfall response to globally uniform warming shown in a previous work[11].

A similar statistical analysis is conducted with alternative choices of previously known indices, which also include possible extratropical factors associated with Sahel rainfall, such as the global interhemispheric SST gradient[17] or the SST difference between the subtropical North Atlantic and the global tropical oceans[18]. We found that none of these indices performs comparably well to the NH intrahemispheric SST warming index (Fig. 2b,c and Supplementary Fig. 3b,c). For example, the regression model with these supplementary indices predicts a much weaker positive future rainfall trend than that with our NH differential warming SST index (see Supplementary Fig. 3a and Fig. 3b,c). The more successful reproduction of Sahel rainfall when using our NH differential warming index also holds for the historical period (Supplementary Fig. 4 and Supplementary Note 1). In addition, the two supplementary SST indices are less successful in consistently explaining future Sahel rainfall changes across the CMIP5 models (see Fig. 2a–c). Although the SST index of subtropical North Atlantic minus global tropics shows a comparable correlation with future Sahel rainfall changes, it has some limitations. For example, the relative subtropical index cannot explain the opposite future rainfall anomalies in GFDL-ESM2M and MIROC-ESM, which are key models for diverse future projections. In addition, the subtropical index shows nearly zero rainfall change corresponding to uniform global warming, which is not consistent with previous work showing a drying Sahel in response to globally uniform warming[11]. Taken together, these results highlight the distinct role of NH differential warming for future Sahel rainfall, and the model discrepancy in future Sahel rainfall projections appears to arise from different magnitudes of NH extratropical warming simulated by CMIP5 models.

AGCM experiments forced by future SST warming. Although the statistical analyses clearly emphasize the role of Northern-hemispheric differential heating as a possible source of diverse future projections of Sahel rainfall, we must verify whether the statistical link is physically consistent. Thus, a series of SST-sensitivity experiments with an atmospheric general circulation model (AGCM) has been conducted to yield more robust results and to understand the underlying processes. The model used here is the ECHAM6 (ref. 19), the atmospheric component of MPI-ESM-LR that shows nearly neutral conditions for the future Sahel rainfall under the RCP4.5 scenario (Supplementary Fig. 1). The ECHAM6 is forced by the monthly climatological SSTs (averaged over the last 30 years of the twentieth and twenty-first centuries) from both GFDL-ESM2M and MIROC-ESM models in CMIP5. Hence, the difference in precipitation responses between the experiments prescribing twentieth and twenty-first century SSTs represents the impact of future global SST warming on Sahel rainfall.

The ECHAM6 experiments forced by the global SST warming from GFDL-ESM2M and MIROC-ESM show a significant difference in Sahel rainfall, with a weak Sahel drying in the experiment using GFDL SST and a strong wetting in the MIROC SST case (Fig. 3a,b). Thus, the AGCM experiment is qualitatively consistent with the drying and wetting Sahel as simulated by GFDL-ESM2M and MIROC-ESM, respectively, in CMIP5. Note that the rainfall responses from our AGCM experiments can be further amplified if the interactive land and vegetation feedbacks, which are known to increase SST-driven Sahel rainfall changes[13], are included in the experiments. This effect highlights the importance of the projected SSTs for the future development of Sahel rainfall and shows that the difference in the projected SSTs can largely explain the divergence of future Sahel rainfall projections. Furthermore, as noted above, the AGCM used for our experiments belongs to one of the CMIP5 models showing neutral conditions for the future Sahel rainfall. This condition leads us to hypothesize that different NH future SST anomalies may be a more dominant source of uncertainty in Sahel rainfall projections than different model sensitivities to the same SST forcing, although further experiments performed with a different AGCM are required to test this hypothesis.

The individual impact of extratropical and tropical SST warming on Sahel rainfall was also evaluated by forcing ECHAM6 with the future SST warming restricted to the tropical and extratropical SST parts separately. The interesting finding here is that tropical SST warming induces less precipitation over the Sahel regardless of whether GFDL or MIROC SSTs are used (Fig. 3c,d). The difference in the magnitude of Sahel drying is presumably due to the varying patterns and magnitudes of tropical SST anomalies simulated by the two coupled models. For example, the stronger subtropical Atlantic warming in MIROC-ESM than in GFDL-ESM2M may induce less drying over the Sahel by increasing the moisture flux from the Atlantic Ocean. In contrast, both GFDL and MIROC SSTs produce an increase in Sahel rainfall when driven by the extratropical SST warming only (Fig. 3e,f). In particular, there is a strong wetting over the Sahel when the SSTs from MIROC-ESM are used, which confirms the conclusions of our previous statistical analysis using CMIP5 data. This result suggests that the precipitation decrease in GFDL-ESM2M, which shows less warming in the extratropics, is more affected by tropical warming, whereas the strong precipitation increase in MIROC-ESM is more influenced by strong extra-tropical warming. In other words, whether a model projects strong wet conditions over the Sahel depends on the dominance of extratropical warming over tropical warming.

Because we argue that the future Sahel rainfall change can be explained by the competition between the two opposing impacts of tropical and extratropical warming, the linearity of the model response is tested. The superposition of two precipitation responses from the tropical and extratropical experiments shows a remarkable similarity to the precipitation response from the global SST warming experiment, which supports the conclusion that the precipitation response to global SST warming in our model experiments can be satisfactorily obtained by the linear combination of precipitation responses to tropical and extra-tropical SST warming (Fig. 3a,b). We also examined the direct radiative impact of CO_2 increase (not mediated by SST change) on Sahel rainfall. Additional AGCM experiments forced by increased CO_2 concentrations with fixed twentieth century mean SSTs show that the Sahel rainfall slightly increases independently of the GFDL or MIROC SSTs. This result indicates that the direct radiative CO_2 impact cannot be a source of the variability in predictions of future Sahel rainfall (Supplementary Fig. 5 and Supplementary Note 2).

Mechanisms of Sahel wetting by NH differential heating. Because of the dominant role of tropical oceans on Sahel rainfall during the historical period[1,6], several mechanisms proposed by previous studies can be applied to explain the relationship between warming tropical oceans and drying Sahel. Such mechanisms include the reduced ascent at certain convective margins due to the higher moisture requirement for convective instability (the so-called 'upped-ante mechanism')[20,21],

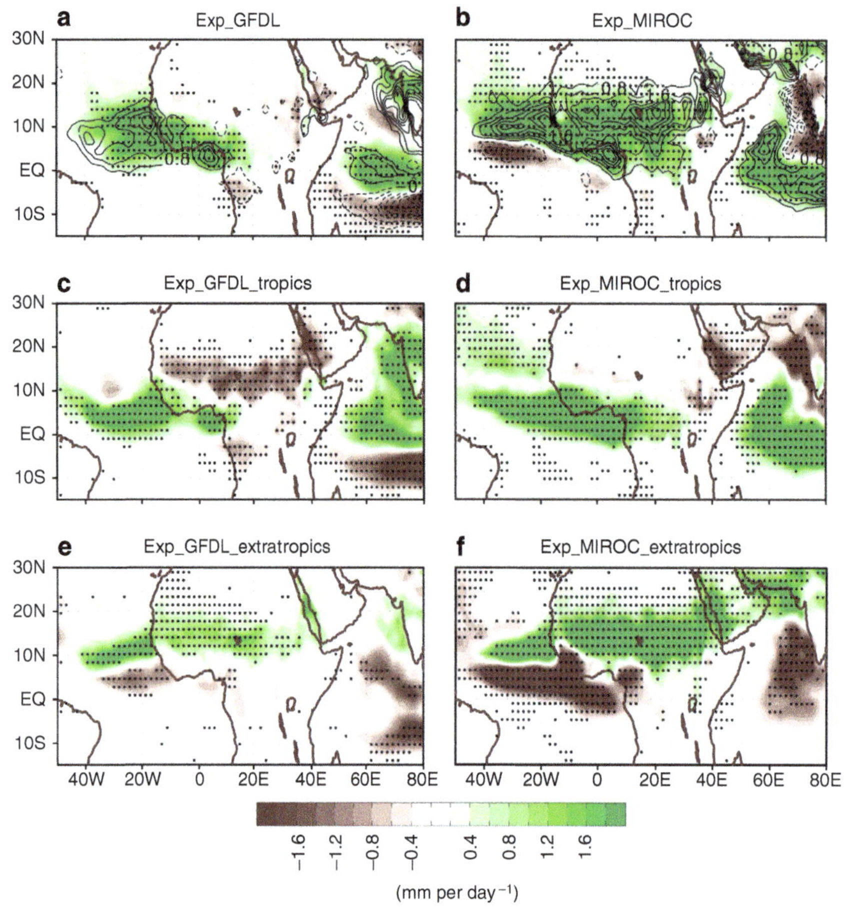

Figure 3 | Precipitation responses to SST sensitivity experiments. (**a**–**f**) Mean differences in summer (JAS) precipitation between the two experiments: the '21C run' forced by twenty-first century (2070–2099) climatological SST and the '20C run' forced by twentieth century (1970–1999) climatological SST from GFDL-ESM2M (**a**) and MIROC-ESM (**b**) in CMIP5. (**c**,**d**) are the same as (**a**,**b**) but for prescribing the twentieth century extratropical SST and the twenty-first century tropical SST for the '21C run'. Similarly, **e**,**f** are precipitation responses when considering the extratropical SST warming only. Contours in **a**,**b** represent the superposition of the precipitation responses from tropical and extratropical SST experiments. The stippled area denotes the 95% confidence region according to two-tailed *t*-test.

anomalous subsidence over the Sahel caused by tropical ocean warming[8,22,23] and weakened monsoonal flow due to the weaker contrast between ocean and continental temperatures[10]. In contrast, the role of extratropical SST warming on the Sahel rainfall has been relatively less studied. One might argue that the global interhemispheric SST gradient pattern shown in Fig. 1c is an extratropical factor inducing the Sahel rainfall increase by shifting the tropical rain belt to the north[24,25] (see also Fig. 2b and Supplementary Fig 3b). However, the zonal mean precipitation change in the extratropical warming experiment clearly indicates that the Sahel rainfall increase is only partially explained by a northward rain belt shift, and more important factors associated with the Sahel rainfall intensity change are involved given the meridionally asymmetric rainfall anomalies (Supplementary Fig. 6). Thus, we suggest potential mechanisms for how the NH differential warming is linked to the increase in Sahel rainfall by further analysing the strong extratropical warming experiment (Exp_MIROC_extratropics).

One factor is the increased moisture flux driven by anomalous subtropical low pressure over North Africa and adjacent oceans (Fig. 4a). The large-scale low-pressure anomaly can deepen the Sahara Low, a main driver of the West Africa monsoonal wind, and thus intensify the westerly flow over the tropical Atlantic Ocean and West Africa. Consequently, the anomalous westerly moisture flux enhances moisture convergence over the Sahel, and in turn increases the rainfall. Given this connection between the NH extratropical warming and Sahel rainfall increase, our results are consistent with previous works. For example, one study showed that the increased temperature and atmospheric humidity advected from the extratropics can induce a positive long-wave radiation moisture feedback and either maintain or further amplify the surface warming over North Africa[26]. This finding provides a favourable condition for the Sahel rainfall increase due to the strengthening of the land-sea thermal contrast (associated with a strong monsoon flow) and the increased moist static energy (associated with a northward shift of the climatological rain belt)[26,27]. Another study suggested the role of the Sahara Low as a possible source of the variation in future Sahel rainfall projections[14]. The disparate magnitudes of NH differential warming emphasized in the present study may be an underlying cause of the different Sahara Low intensities projected by climate models, leading to the diverse projections of Sahel rainfall.

Another possible link between the NH differential warming and the increase in Sahel rainfall can be found in upper-level jets over Africa. Two tropical jets, the Tropical Easterly Jet (TEJ) and the African Easterly Jet (AEJ), are upper-level circulation factors known to influence Sahel rainfall[28]. The intensity of the TEJ, which modulates the divergence and uplift in the upper troposphere, is associated with rainfall anomalies over much of

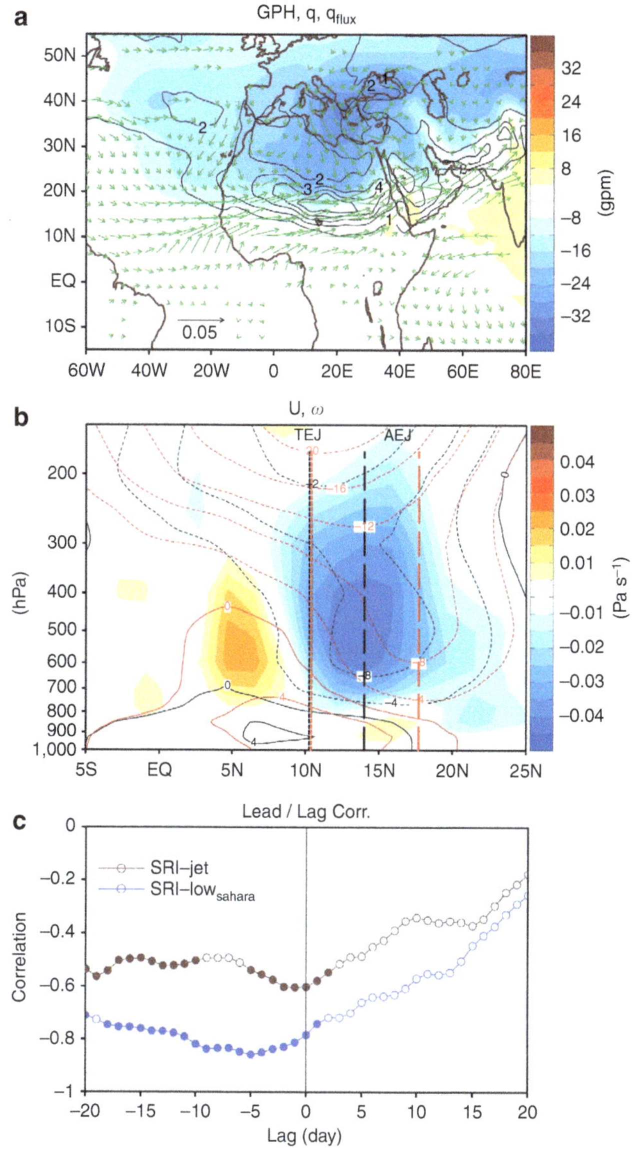

Figure 4 | Mechanisms for associating NH differential heating with Sahel rainfall increase. (**a**) The JAS mean difference in 925-hPa geopotential height (shadings, unit: m), moisture (contours, unit: 10^{-3} kg_{H_2O} kg_{air}^{-1}) and moisture flux (arrows, unit: kg_{H_2O} kg_{air}^{-1} $m s^{-1}$) between the two experiments, MIROC_21C_extratropics and MIROC_20C. The difference in moisture flux is only plotted where the mean difference is statistically significant, exceeding the 95% confidence level according to a two-tailed *t*-test. (**b**) Zonally averaged (10°W–30°E) zonal wind from MIROC_20C (black contours) and MIROC_21C_extratropics (red contours). Vertical lines represent axes of TEJ (dotted lines) and AEJ (dashed lines) for MIROC_20C (black) and for MIROC_21C_extratropics (red). Shading represents the JAS mean difference between MIROC_21C_extratropics and MIROC_20C. (**c**) Lead/Lag correlations between Sahel rainfall (10°W–30°E; 10°–20°N) and Sahara Low (10°W–30°E; 20°–30°N) and between Sahel rainfall and upper-level tropical easterly jet (10°W–30°E; 5°–15°N). The ensemble-mean daily data are used for this analysis. Negative lag indicates that Sahara Low and jets lead the Sahel rainfall. Filled circles represent correlations exceeding the 90% confidence level.

western Africa[29–31], and the location of the AEJ is linked to the latitudinal extent of the atmospheric ascent and the tropical rain belt over West Africa[28,29]. We found that the responses of these two jets in our extratropical warming experiment are in agreement with the observed relationship between the jets and Sahel rainfall. The TEJ is significantly intensified in the presence of extratropical warming, whereas the AEJ shifts northward by ~4° over the Sahel region with relatively little change in its magnitude (Fig. 4b). Such circulation features are consistent with the anomalous ascending motion over the Sahel and the corresponding rainfall increase (Fig. 4b, and Supplementary Fig. 6).

Although the aforementioned explanation is physically plausible and is consistent with previous observational results, it does not provide clear evidence of causal relationships. For example, changes in the Sahara Low or the upper-level jet might be caused by the increased precipitation. To clarify this uncertainty, both the lead–lag relationships between Sahel rainfall and the strength of Sahara Low, and between Sahel rainfall and the intensity of TEJ were analysed by using daily model outputs (Fig. 4c). We found that the changes both in the Sahara Low and jet strength generally precede Sahel rainfall change. Therefore, the change in atmospheric dynamics is likely to be a cause, rather than a consequence, of the substantial increase in Sahel rainfall.

Discussion

Similar to the atmospheric response over North Africa to the NH extratropical warming, tropical Atlantic SST warming can also result in an increase in Sahel rainfall with the intensification of the Sahara heat low and the increased low-level moisture flux[32,33]. However, the moistening impact of the specific warming pattern in the tropics is relatively weak in a warming climate, thus the drying impact by the warming of the Indian and Pacific oceans dominates the Sahel rainfall response as shown in our tropical SST experiments. Specifically, if the future extratropical warming is stronger than the tropical warming, the extratropical impact on the Sahara Low or Sahel rainfall is expected to dominate subtropical/tropical impacts. A previous modelling work based on the historical period also supports this relation, showing that tropical Atlantic warming alone cannot fully explain Sahel rainfall responses to North Atlantic warming (see Figs 3j,k and 4j,k in Sutton and Hodson[33]).

The proposed connections linking strong future extratropical warming and Sahel rainfall increase may be relevant to previous findings, particularly focused on the extratropical North Atlantic Ocean. For example, the extratropical North Atlantic cooling associated with Atlantic multi-decadal variability is known to reduce Sahel rainfall by the advection of cold, dry air to North Africa and corresponding changes in sea-level pressure and moist static energy[26]. The mid-tropospheric Rossby-wave activity originated from the North Atlantic is also a possible atmospheric factor in linking extratropical energy to the African subtropical climate[34]. An additional AGCM experiment, however, shows that only approximately one-third of the total precipitation increase shown in Fig. 3f can be explained by extratropical Atlantic warming alone, indicating that SST warming in other ocean basins, such as the Mediterranean Sea and North Pacific and Arctic Oceans, also contributes substantially to the NH extratropical warming impact on Sahel rainfall.

Over the past decade, the future projection of Sahel rainfall and the source of the discrepancies among the coupled GCMs has been an issue of great concern because of the historical climate-related calamities in this vulnerable area. Our results demonstrate that the NH differential warming is the crucial factor in consistently explaining the previously reported conundrum of future Sahel rainfall behaviour. In contrast to the observed period in which Sahel rainfall change is mainly driven by tropical SSTs, the moistening impact of strong extratropical SST warming is expected to dominate over the drying impact of tropical SST warming for the future Sahel rainfall, suggesting a fundamental shift in the pathway of SST impact in a warming climate. Therefore, a careful investigation of how the NH SST responds to greenhouse gas-induced warming is of great importance to obtain more reliable estimates of future Sahel rainfall. Given a relatively long predictability of extratropical SST[35], a better understanding and determination of the factors controlling extratropical surface warming is likely the critical issue for improving the reliability of decadal predictions and future projections of Sahel rainfall.

Methods

CMIP5 archive. The data analysed here are obtained from the CMIP5 multi-model outputs[36]. A total of 23 CMIP5 models, which provide SST and precipitation outputs during the period 1860–2100, are used in this study, focusing on the seasonal mean (July–September, JAS) values. In particular, much attention is devoted to analysing two models, 'GFDL-ESM2M' and 'MIROC-ESM', which project two opposite extreme anomalies of future Sahel rainfall. These two models originate from 'gfdl_cm2_1' and 'miroc3_2_medres' in CMIP3, which also predicted extreme drying and wetting, respectively. The two versions of models developed from the same modelling group are basically the same except for incorporating carbon dynamics and some minor tuning in physical models[37,38]. The future scenario chosen for the present study is the Representative Concentration Pathways 4.5 (RCP4.5), a mid-range mitigation scenario. All CMIP5 models used here are depicted in Supplementary Fig. 1.

Linear regression model to predict future Sahel rainfall. The statistical model used in our study is equivalent to the conventional multi-linear regression method, which is a standard technique to estimate the predictand from two or more explanatory variables. The best-fit line is calculated by the least-squares method, minimizing the sum of the squares of the deviations from the fitted line. We used the regression coefficients from the historical run for the future Sahel rainfall prediction. A cross-validation test, which is performed by calculating the coefficients from the future period and applying them to predict the historical rainfall, confirms the robustness of the statistical model.

Model and experimental design. The ECHAM6 model, developed at the Max Planck Institute for Meteorology, is used for the SST sensitivity experiments. The ECHAM models are known to reproduce the African monsoon reasonably well and have been extensively used for West African monsoon studies[39–41]. The model is run at T63 horizontal resolution (1.875° × 1.875°) with 47 vertical levels. The SST fields prescribed in the model are obtained from two CMIP5 models: GFDL-ESM2M and MIROC-ESM. Note that the corresponding sea ice fields are also used for physical consistency. Each experiment basically consists of a pair of two experiments, named '20C run' and '21C run'. In the 20C run, the model is forced by the monthly climatological mean SSTs over the last thirty years of the twentieth century (1970–1999), whereas in the 21C run the model is forced by those over the last 30 years of the twenty-first century (2070–2099). The precipitation anomalies shown in the present study represent the difference between these two experiments. In the case of isolating the impact of tropical SST, the prescribed SST in the extratropics (poleward of 30°) is fixed to the twentieth century SST (as in case in the 20C run) and only the tropical SST (30°S–30°N) from the twenty-first century is used for the 21C run (and vice versa for the case of extratropical SST). All other boundary conditions (for example, greenhouse gases and solar radiation) are identical for all integrations. A total of 20 ensemble members with different initial conditions are conducted, and only the JAS mean is shown in this study. A summary of the experimental design is provided in Supplementary Table 1.

References

1. Biasutti, M., Held, I. M., Sobel, A. H. & Giannini, A. SST forcings and Sahel rainfall variability in simulations of the twentieth and twenty-first centuries. *J. Clim.* **21,** 3471–3486 (2008).
2. Cook, K. H. Climate science: the mysteries of Sahel droughts. *Nat. Geosci.* **1,** 647–648 (2008).
3. Cook, K. H. & Vizy, E. K. Coupled model simulations of the west African monsoon system: twentieth- and Twenty-First-century simulations. *J. Clim.* **19,** 3681–3703 (2006).
4. Douville, H., Salas-Melia, D. & Tyteca, S. On the tropical origin of uncertainties in the global land precipitation response to global warming. *Clim. Dyn.* **26,** 367–385 (2006).
5. Caminade, C. & Terray, L. Twentieth century Sahel rainfall variability as simulated by the ARPEGE AGCM, and future changes. *Clim. Dyn.* **35,** 75–94 (2010).
6. Joly, M., Voldoire, A., Douville, H., Terray, P. & Royer, J.-F. African monsoon teleconnections with tropical SSTs: validation and evolution in a set of IPCC4 simulations. *Clim. Dyn.* **29,** 1–20 (2007).
7. Biasutti, M. & Giannini, A. Robust Sahel drying in response to late 20th century forcings. *Geophys. Res. Lett.* **33,** L11706 (2006).
8. Bader, J. & Latif, M. The impact of decadal-scale Indian Ocean sea surface temperature anomalies on Sahelian rainfall and the North Atlantic oscillation. *Geophys. Res. Lett.* **30,** 2169 (2003).
9. Giannini, A., Saravanan, R. & Chang, P. Oceanic forcing of Sahel rainfall on interannual to interdecadal time scales. *Science* **302,** 1027–1030 (2003).
10. Haarsma, R. J., Selten, F. M., Weber, S. L. & Kliphuis, M. Sahel rainfall variability and response to greenhouse warming. *Geophys. Res. Lett.* **32,** L17702 (2005).
11. Held, I. M., Delworth, T. L., Lu, J., Findell, K. L. & Knutson, T. R. Simulation of Sahel drought in the 20th and 21st centuries. *Proc. Natl Acad. Sci. USA* **102,** 17891–17896 (2005).
12. Hoerling, M., Hurrell, J., Eischeid, J. & Phillips, A. Detection and attribution of twentieth-century northern and southern African rainfall change. *J. Clim.* **19,** 3989–4008 (2006).
13. Zeng, N., Neelin, J. D., Lau, K. M. & Tucker, C. J. Enhancement of interdecadal climate variability in the Sahel by vegetation interaction. *Science* **286,** 1537–1540 (1999).
14. Biasutti, M., Sobel, A. H. & Camargo, S. J. The role of the Sahara low in summertime Sahel rainfall variability and change in the CMIP3 models. *J. Clim.* **22,** 5755–5771 (2009).
15. Giannini, A. Mechanisms of climate change in the semiarid African Sahel: the local view. *J. Clim.* **23,** 743–756 (2010).
16. Biasutti, M. Forced Sahel rainfall trends in the CMIP5 archive. *J. Geophys. Res.* **118,** 1613–1623 (2013).
17. Folland, C. K., Palmer, T. N. & Parker, D. E. Sahel rainfall and worldwide sea temperatures, 1901-85. *Nature* **320,** 602–607 (1986).
18. Giannini, A. *et al.* A unifying view of climate change in the Sahel linking intra-seasonal, interannual and longer time scales. *Environ. Res. Lett.* **8,** 024010 (2013).
19. Stevens, B. *et al.* Atmospheric component of the MPI-M Earth System Model: ECHAM6. *J. Adv. Model. Earth Syst.* **5,** 146–172 (2013).
20. Neelin, J. D., Chou, C. & Su, H. Tropical drought regions in global warming and El Nino teleconnections. *Geophys. Res. Lett.* **30,** 2275 (2003).
21. Chou, C. & Neelin, J. D. Mechanisms of global warming impacts on regional tropical precipitation. *J. Clim.* **17,** 2688–2701 (2004).
22. Sobel, A. H., Held, I. M. & Bretherton, C. S. The ENSO signal in tropical tropospheric temperature. *J. Clim.* **15,** 2702–2706 (2002).
23. Herceg, D., Sobel, A. H. & Sun, L. Regional modeling of decadal rainfall variability over the Sahel. *Clim. Dyn.* **29,** 89–99 (2007).
24. Broccoli, A. J., Dahl, K. A. & Stouffer, R. J. Response of the ITCZ to Northern Hemisphere cooling. *Geophys. Res. Lett.* **33,** L01702 (2006).
25. Kang, S. M., Held, I. M., Frierson, D. M. W. & Zhao, M. The response of the ITCZ to extratropical thermal forcing: idealized slab-ocean experiments with a GCM. *J. Clim.* **21,** 3521–3532 (2008).
26. Liu, Y., Chiang, J. C., Chou, C. & Patricola, C. M. Atmospheric teleconnection mechanisms of extratropical North Atlantic SST influence on Sahel rainfall. *Clim. Dyn.* **43,** 1–15 (2014).
27. Chou, C., Neelin, J. D. & Su, H. Ocean-atmosphere-land feedbacks in an idealized monsoon. *Quart. J. Roy. Meteor. Soc.* **127,** 1869–1891 (2001).
28. Nicholson, S. E. A revised picture of the structure of the 'monsoon' and land ITCZ over West Africa. *Clim. Dyn.* **32,** 1155–1171 (2009).
29. Grist, J. P. & Nicholson, S. E. A study of the dynamic factors influencing the rainfall variability in the West African Sahel. *J. Clim.* **14,** 1337–1359 (2001).
30. Nicholson, S. E. The intensity, location and structure of the tropical rainbelt over west Africa as factors in interannual variability. *Int. J. Climatol.* **28,** 1775–1785 (2008).
31. Pattanaik, D. R. & Satyan, V. Fluctuations of Tropical Easterly Jet during contrasting monsoons over India: a GCM study. *Meteorol. Atmos. Phys.* **75,** 51–60 (2000).
32. Martin, E. R., Thorncroft, C. & Booth, B. B. B. The Multidecadal Atlantic SST-Sahel Rainfall Teleconnection in CMIP5 Simulations. *J. Clim.* **27,** 784–806 (2014).
33. Sutton, R. T. & Hodson, D. L. R. Climate response to basin-scale warming and cooling of the North Atlantic Ocean. *J. Clim.* **20,** 891–907 (2007).
34. Barandiaran, D. & Wang, S.-Y. The missing teleconnection between the North Atlantic and the Sahel precipitation in CFSv2. *Atmos. Sci. Lett.* **15,** 21–28 (2014).
35. Matei, D. *et al.* Two tales of initializing decadal climate prediction experiments with the ECHAM5/MPI-OM model. *J. Clim.* **25,** 8502–8523 (2012).
36. Taylor, K. E., Stouffer, R. J. & Meehl, G. A. An overview of CMIP5 and the experiment design. *Bull. Am. Meteorol. Soc.* **93,** 485–498 (2012).
37. Dunne, J. P. *et al.* GFDL's ESM2 global coupled climate-carbon Earth System Models. Part I: physical formulation and baseline simulation characteristics. *J. Clim.* **25,** 6646–6665 (2012).
38. Watanabe, S. *et al.* MIROC-ESM 2010: model description and basic results of CMIP5-20c3m experiments. *Geosci. Model Dev.* **4,** 845–872 (2011).

39. Gaetani, M. & Mohino, E. Decadal prediction of the Sahelian precipitation in CMIP5 simulations. *J. Clim.* **26,** 7708–7719 (2013).
40. Mohino, E., Janicot, S. & Bader, J. Sahel rainfall and decadal to multi-decadal sea surface temperature variability. *Clim. Dyn.* **37,** 419–440 (2011).
41. Bader, J. & Latif, M. The 1983 drought in the West Sahel: a case study. *Clim. Dyn.* **36,** 463–472 (2011).

Acknowledgements

We acknowledge the modelling groups and the World Climate Research Programme's Working Group on Coupled Modelling for their roles in making available the CMIP5 multi-model data set, and the German Climate Computing Center (DKRZ) for providing the computer hardware for our model experiments. This work was supported by the BMBF RACE (D.M.) and MiKlip (J.B.) projects and was done in the frame of the internal MPI-M project 'Tropical VIBES'. We also thank the two reviewers and L. Silvers and D. Zanchettin for their helpful comments and feedback on the present manuscript.

Author contributions

J.B. developed the essential research idea. J.B., D.M. and J.-Y.P. contributed to developing the research. J.-Y.P. performed the analysis and SST-sensitivity experiments. All authors discussed the results and wrote the paper.

Additional information

Competing financial interests: The authors declare no competing financial interests.

12

Clouds enhance Greenland ice sheet meltwater runoff

K. Van Tricht[1], S. Lhermitte[1], J.T.M. Lenaerts[2], I.V. Gorodetskaya[1], T.S. L'Ecuyer[3], B. Noël[2], M.R. van den Broeke[2], D.D. Turner[4] & N.P.M. van Lipzig[1]

The Greenland ice sheet has become one of the main contributors to global sea level rise, predominantly through increased meltwater runoff. The main drivers of Greenland ice sheet runoff, however, remain poorly understood. Here we show that clouds enhance meltwater runoff by about one-third relative to clear skies, using a unique combination of active satellite observations, climate model data and snow model simulations. This impact results from a cloud radiative effect of 29.5 ($\pm$5.2) W m^{-2}. Contrary to conventional wisdom, however, the Greenland ice sheet responds to this energy through a new pathway by which clouds reduce meltwater refreezing as opposed to increasing surface melt directly, thereby accelerating bare-ice exposure and enhancing meltwater runoff. The high sensitivity of the Greenland ice sheet to both ice-only and liquid-bearing clouds highlights the need for accurate cloud representations in climate models, to better predict future contributions of the Greenland ice sheet to global sea level rise.

[1] KU Leuven, Department of Earth and Environmental Sciences, Celestijnenlaan 200E, Leuven 3001, Belgium. [2] Institute for Marine and Atmospheric research Utrecht—Utrecht University, Utrecht 3584CC, The Netherlands. [3] Atmospheric and Oceanic Sciences, University of Wisconsin-Madison, Madison, Wisconsin 53706, USA. [4] National Severe Storms Laboratory, NOAA, Norman, Oklahoma 73072, USA. Correspondence and requests for materials should be addressed to K.V.T. (email: Kristof.VanTricht@ees.kuleuven.be).

Regionally amplified[1] anthropogenic climate change[2] in combination with anomalous large-scale atmospheric circulations[3] have made the Greenland ice sheet (GrIS) a dominant contributor[4,5] to recent global sea level rise (0.63 ± 0.07 mm per year during 2000–2012; ref. 6). This increasing contribution is primarily driven by a reduction in GrIS surface mass balance (SMB)[6–8], related to a significant increase in GrIS surface melt extent[9] and volume[10]. Surface melt is determined by the surface energy balance (SEB)[11], the sum of incoming and outgoing energy fluxes at the surface.

Clouds are known to play a pivotal role in regulating the local SEB, with competing warming and cooling effects on the surface[12–14]. The dominating effect depends strongly on cloud properties such as vertically integrated ice and liquid water contents that determine cloud optical depth and emissivity, in addition to cloud temperature, sun position and surface albedo[13–15].

Despite their importance, the distribution of clouds over the GrIS, their phase partitioning (for example, ice-only clouds versus liquid-bearing or mixed-phase clouds) and their precise impact on the SEB and SMB are poorly understood, owing to the limited availability of measurements. Sensitivity studies based on observations have been limited to particular events or cloud types such as the role of low-level liquid-bearing clouds[16] together with the enhanced poleward transport of humid and warm air[17] in the July 2012 extreme GrIS melt event, whereas ice sheet-wide assessments of cloud SEB impact relied on climate model output to overcome the limited availability of observations[18–20].

However, analysis of the recent Coupled Model Intercomparison Project phase 5 model ensemble and a regional climate model shows a significant inter-model spread in the estimates of both ice-only and liquid-bearing cloud frequency and water paths over the GrIS (Supplementary Fig. 1 and Supplementary Methods). This highlights the need for an observation-based assessment of the cloud distribution over the GrIS.

In this study, we use a unique combination of state-of-the-art satellite remote sensing, ground-based observations and a regional climate model, to quantify the impact of clouds on the SEB over the entire GrIS. The GrIS shows a strong sensitivity to cloud presence, where both ice-only and liquid-bearing clouds are warming the ice sheet surface. Using an advanced snow model, we find that this warming enhances GrIS meltwater runoff in response to reduced refreezing rates at night, when cloud warming is highest compared with clear skies.

Results

Cloud properties from active satellite remote sensing. In this study, a total of ~6.3 million satellite cloud observations were compiled over the GrIS for the period 2007–2010 from the CloudSat and CALIPSO (Cloud-Aerosol Lidar and Infrared Pathfinder Satellite Observation) satellites. These missions carry the first complementary collocated spaceborne radar and lidar sensors[21,22], providing information on vertical distributions of cloud ice and liquid water contents, cloud phase and effective radii, supplemented by cloud optical depths derived from MODerate resolution Imaging Spectroradiometer (MODIS)[23].

These observation-based vertical profiles of cloud properties provide the main input for radiative flux calculations in the level-2 'Fluxes and Heating Rates' product (2B-FLXHR-LIDAR)[24]. An updated refined version of the 2B-FLXHR-LIDAR algorithm that includes low-level supercooled liquid-bearing clouds detected using a combination of CloudSat and CALIPSO observations[25] is used in this study. This new data set, which will form the basis of the Release 05 version of the CloudSat 2B-FLXHR-LIDAR product, provides for the first time a robust partitioning of ice-only and liquid-bearing clouds that does not suffer from errors typically present in conventional satellite products[25,26], and provides an unprecedented three-dimensional view of clouds over the GrIS and their microphysical properties[22].

Figure 1 shows that clouds occur frequently over the GrIS, where the annual mean cloud cover is 67% (Table 1). The CloudSat and CALIPSO data sets employed here suggest that liquid-bearing clouds that contain both ice and (supercooled) liquid water are present 28% of the time, consistent with other work showing that such clouds are prevalent throughout the Arctic[27]. There were almost no (< 1%) detections of liquid-only clouds over the GrIS. Whereas the average total cloud fraction is fairly stable throughout the year ranging from 63 to 70%, the satellites observe strong spatial (Fig. 1) and seasonal (Table 1) variations in ice-water path (IWP), liquid-water path (LWP) and liquid/ice partitioning. Liquid-bearing clouds occur predominantly during summer (46% of the time), associated with the presence of warmer and moister air[28], whereas winter is characterized by a much higher frequency of ice-only clouds (55%). Ice clouds occur frequently over the entire GrIS, whereas liquid-bearing clouds are much more abundant in the coastal regions. Annual mean cloud LWP in all conditions varies between 3 and 33 g m^{-2} over the entire GrIS with highest values in the coastal zones, whereas IWP is lowest (< 7 g m^{-2}) in the northern part of the interior and highest (> 100 g m^{-2}) in the high-snowfall coastal zones of the southeastern GrIS.

Cloud radiative effect. The satellite-based cloud observations allow to estimate the cloud impact on the SEB using a two-stream broadband radiative transfer model (see Methods). Cloud impact is defined here as the net cloud radiative effect (CRE)[5] at the surface, which is the difference between the all-sky and clear-sky surface radiation budgets[12].

The spatial distribution of the annual mean CRE is positive over the entire GrIS and is strongly correlated with LWP and IWP (Fig. 2a). This indicates that longwave (LW) warming dominates shortwave (SW) cooling, which can be explained by the high solar zenith angles (SZAs) and surface albedo[29].

Averaged over the GrIS, clouds reduce annual mean surface radiative heat loss by 29.5 (± 5.2) W m^{-2}, a factor of 2.7 decrease relative to clear skies. Cloud liquid water contributes only 11.2 (± 2.4) W m^{-2} to the total radiative effect. This is explained by the lower occurrence frequency and strong seasonal cycle of liquid-bearing clouds with highest frequencies in summer, when both LW warming and SW cooling constitute the CRE. Ice-only clouds occur most frequently in winter, when only LW warming contributes to the radiative effect.

Translating SEB to SMB using a snow model. The annual mean CRE of 29.5 W m^{-2} provides enough energy to melt ~90 Gt of ice in the GrIS ablation area during July and August. However, part of this excess energy is absorbed by the snowpack, warming the surface and enhancing outgoing LW radiation. Warmer snow promotes snow metamorphism, which lowers the albedo of the snowpack[30], whereas multiple reflections of SW radiation between clouds and the surface in combination with spectrally dependent cloud absorption generally increase it[31,32].

To account for these processes, which are not all included in the satellite product, the one-dimensional physical snow model SNOWPACK[33] was used. The cloud impact on the GrIS SMB is assessed by forcing SNOWPACK with different cloud inputs (all-sky, clear-sky and no-liquid). As our SNOWPACK simulations require hourly meteorological input data, which is much more frequent than the satellite overpasses (on average CloudSat

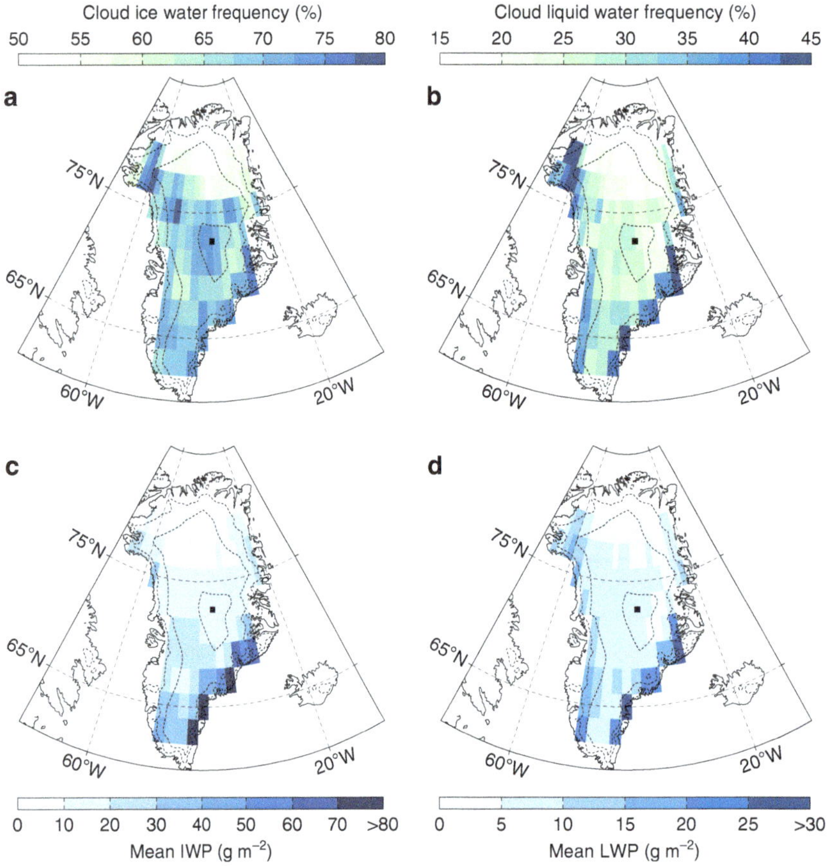

Figure 1 | Retrieved cloud properties over the GrIS by satellite remote sensing. These retrievals result from joint CloudSat, CALIPSO and MODIS observations. Both (**a**) cloud ice and (**b**) cloud liquid water occur frequently over the GrIS, although strong spatial variations exist. (**c**) Mean IWP in all conditions and (**d**) mean LWP in all conditions also show spatial variability in cloud water contents. Dashed curves indicate 1,000 m height contours and the black dot represents the location of the Summit station.

Table 1 | Seasonal and yearly cloud frequencies and mean water paths.

	All clouds (%)	Ice-only (%)	Liquid-bearing (%)	Mean IWP ($g\,m^{-2}$)	Mean LWP ($g\,m^{-2}$)
Spring (MAM)	63	41	22	20	6
Summer (JJA)	66	20	46	19	16
Autumn (SON)	68	41	27	25	8
Winter (DJF)	70	55	15	23	5
Year	67	39	28	21	9

GrIS, Greenland ice sheet; IWP, ice-water path; LWP, liquid-water path. MAM, March-April-May; JJA, June-July-August; SON, September-October-November; DJF, December-January-February.
The observed total cloud frequency (in %) over the GrIS (2007-2010) is fairly constant year round, but the partitioning and water paths of ice and liquid water exhibit strong seasonal variations. It is noteworthy that in the event of multiple detected cloud layers in a vertical profile, these are combined and reported as a single cloudy profile in these statistics.

and CALIPSO observe each 2° by 2° grid box once every 1–7 days, depending on the location), we constructed a hybrid satellite-climate model data set, combining the benefits of the observationally based estimates of CREs with the hourly temporal resolution of a regional climate model (see Methods).

The snow model simulations, which capture the evolution of the GrIS SMB from 2007 to 2010 (Fig. 3a, black dashed curve), indicate that clouds warm the GrIS surface by 1.2 ($\pm$0.1) °C on average over the entire period. The changes in the snowpack as a result of this cloud-induced warming, however, in combination with increased albedo under cloudy conditions, lower the annual mean CRE (Fig. 2a, orange line) by 8.4 $W\,m^{-2}$ ($\sim$30%) relative to the satellite estimate (Fig. 2a, purple line). The spatial distribution of this updated CRE is shown in Fig. 2b.

Enhanced meltwater runoff. Although GrIS surface melt is nearly identical under all-sky and clear-sky conditions (Fig. 3b), clouds enhance GrIS meltwater runoff by 56 $\pm$ 20 Gt per year (32 $\pm$ 12 %), with similar contributions from cloud ice (25 Gt) and cloud liquid (31 Gt) water. This suggests that the primary influence of clouds is by reducing meltwater refreezing; in clear-sky conditions, about 58% of the meltwater refreezes, but this fraction decreases to 45% in the presence of clouds.

The governing mechanisms that explain this effect are illustrated in Fig. 4, for a case study during 16–21 June 2008 at 67°N-49°E (ablation zone), during an albedo transition period. In the presence of a thick cloud (Fig. 4a), strong SW cooling and LW warming partly compensate during daytime (Fig. 4b). Surface melt occurs during and shortly after solar noon, when incoming SW radiation peaks (Fig. 4c, red curves). During the subsequent

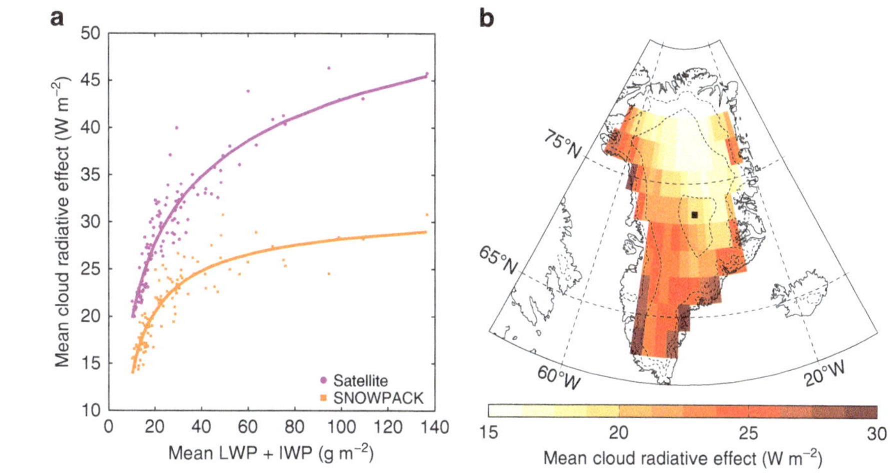

Figure 2 | CRE over the GrIS. (**a**) The average relationship between mean LWP + IWP and mean annual CRE is positive at each location, with higher radiative effects from optically thicker (more liquid and/or ice) or more frequent clouds. The purple circles were derived from satellite observations, whereas the orange squares result from the SNOWPACK runs. The satellite-retrieved CRE estimates assume fixed surface conditions (albedo and surface temperature). In reality, these conditions may change in the absence of the detected clouds. The entire snowpack changes as a result of cloud warming, reducing the initial CRE (lower values in the orange curve). (**b**) Yearly mean CRE over the GrIS (2007-2010) from the SNOWPACK output. This estimate takes into account changing surface conditions (albedo and surface temperature) and snowpack with changing energy inputs due to the presence/absense of clouds. Dashed lines indicate 1,000 m height contours and the black dot represents the location of the Summit station.

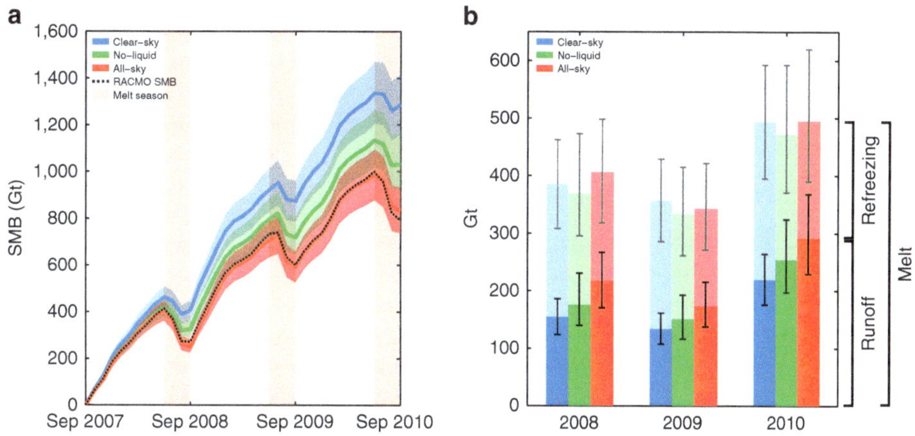

Figure 3 | The effects of clouds on the SMB during the period September 2007-September 2010. (**a**) Evolution of GrIS SMB indicates that the cloudy simulations have a lower SMB. Uncertainties are shown by the shaded areas. The black curve represents the regional climate model RACMO2.3 SMB values and show the performance of SNOWPACK in simulating the GrIS SMB. (**b**) Yearly GrIS melt, refreezing and runoff. Despite negligible differences in melt, 58% of the meltwater refreezes in clear-sky conditions, whereas only 45% refreezes in all-sky conditions. Annual meltwater runoff is therefore about one-third higher in the presence of clouds, with a slightly higher contribution of liquid-bearing clouds. The whiskers indicate an inherent SNOWPACK uncertainty and the sensitivity to the amount of LWP/IWP in the SNOWPACK simulations.

nighttime hours, when incoming SW is small (Fig. 4b, grey zones), clear-sky conditions favour strong surface radiative cooling, extracting heat from the warmer subsurface snow by conduction, (partly) refreezing the meltwater (Fig. 4c, blue curves). This radiative cooling is reduced under cloudy conditions, impeding the refreezing mechanism. The additional meltwater that does not refreeze fills up the available pore space until it runs off[34] (Fig. 4d), whereas a persistent positive CRE leads to firn warming, further enhancing melt and subsequent runoff[35]. The combined effect causes an earlier exposure of bare ice and slush under cloudy conditions, which amplifies these mechanisms through enhanced SW absorption with lower albedo values (albedo feedback, Fig. 4e). During the last 2 days of the case study, when the cloud has disappeared in the all-sky simulation, this albedo feedback is strengthened due to increased SW insolation in the absence of cloud SW blocking, enhancing melt relative to the clear-sky simulation (Fig. 4c, red curves).

Along with meltwater runoff, our results suggest that clouds enhance GrIS sublimation. In summary, liquid-bearing and ice-only clouds contribute equally to a GrIS SMB decrease during 2007–2010 of 35 ± 5% relative to clear-sky conditions by increasing meltwater runoff and sublimation, which is the average equivalent of 152 ± 20 Gt per year (Fig. 3a, red curve).

Discussion

Our results highlight the strong sensitivity of the GrIS SEB and SMB to clouds. This analysis overcomes many of the current issues that exist in observing and modelling of mixed-phase clouds by creating a hybrid satellite-climate model data set that

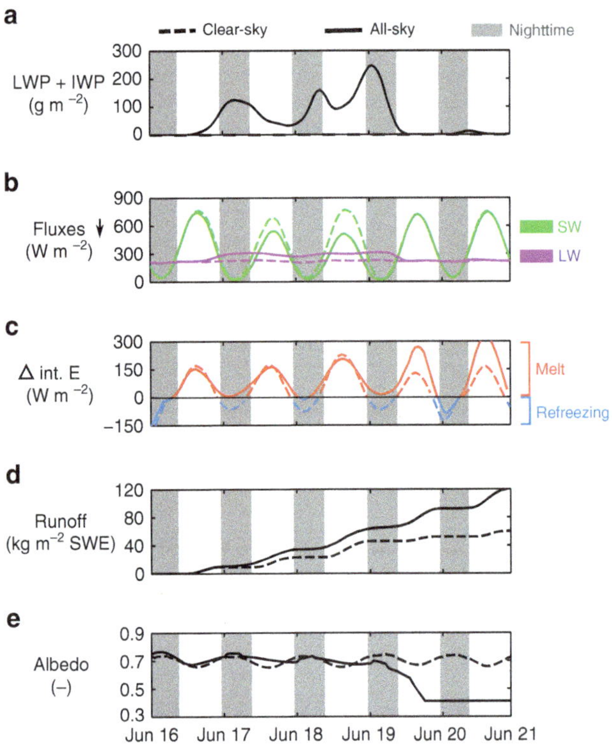

Figure 4 | Case study showing cloud impacts on melt/refreezing during 16–21 June 2008 at 67°N-49°E. Solid curves represent all-sky conditions and dashed curves represent clear-sky conditions. Grey zones are characterized by a SZA > 70°, considered to be nighttime. All variables as shown in the figure have been smoothed using a 6-hourly moving-average window. (**a**) Total (liquid + ice) water path, showing a cloud in the all-sky simulation. (**b**) Downwelling SW (green) and LW (purple) radiative fluxes. In the presence of a cloud, SW cooling (solid curve below dashed curve) and LW warming (solid curve above dashed curve) occur simultaneously during daytime, whereas LW warming dominates nighttime. (**c**) Internal energy change of the snowpack due to melting (positive) and refreezing (negative) processes. Melt rates are highest at solar noon when SW insolation peaks, whereas refreezing rates are highest at night when strong surface radiative cooling dominates. In the presence of clouds, this radiative cooling is reduced, impeding the refreezing mechanism. (**d**) Cumulative meltwater runoff in snow water equivalent (SWE) is higher under cloudy conditions, due to limited meltwater refreezing and earlier bare-ice exposure. (**e**) Surface albedo as simulated by SNOWPACK. Persistent warming by clouds enhances meltwater runoff, leading to an earlier exposure of bare ice and slush that have a much lower albedo than (fresh) snow (from 19 June onwards). At this point, the warming is amplified due to a much higher absorption of SW radiation in the all-sky simulation, as opposed to the clear-sky simulation.

closely matches satellite retrievals and ground-based observations with hourly temporal resolution. The strong positive CRE over the GrIS is found to warm the surface, enhance meltwater runoff in the GrIS ablation area and increase sublimation in the GrIS accumulation area.

However, a substantial uncertainty is associated to the sublimation term, as SNOWPACK was used in an offline configuration and does not account for any feedbacks to the atmosphere in response to changes in the snowpack. In reality, changes in surface temperature induced by clouds will modify turbulent fluxes and thus the boundary-layer air temperature and humidity. Here we did not include these second-order effects and focused exclusively on the sensitivity of the GrIS to cloud-related changes in the SEB. This constitutes a first important step in assessing the impact of clouds on the GrIS SMB.

These results further indicate that not only liquid-bearing clouds[16] but also clouds composed exclusively of ice significantly increase radiative fluxes into the surface and decrease GrIS SMB. This underscores the need for continued research into the factors that govern the formation and maintenance of these distinct cloud regimes, and their evolution in a future warmer and wetter Arctic[36]. Evidence of the large spread in cloud cover and liquid/ice partitioning over the GrIS in current state-of-the-art climate models, in combination with our limited understanding of the interaction between clouds, circulation and climate[37], suggests that improved cloud representations in climate models could significantly increase the fidelity of future projections of GrIS SMB and subsequent global sea level rise.

Methods

General approach. State-of-the-art satellite cloud observations have been used in a refined multi-stage hybrid model-observation algorithm, to estimate surface radiative fluxes. To overcome the limited temporal resolution of these satellite observations, a regional climate model has been incorporated to interpolate the observations to hourly resolution. Ground-based observations from Summit[28,38] and several automatic weather stations (AWSs) have been used to evaluate the resulting hybrid satellite-climate model data set. This new hybrid data set was finally used to force a snow model that simulates the impact of the CRE on the GrIS SMB.

CloudSat/CALIPSO satellite observations. Although radiative fluxes over the GrIS are available from the European Centre for Medium-Range Weather Forecasts (ECMWF) atmospheric re-analysis (ERA), the uncertainty in these fluxes is large, as they are calculated based on modelled clouds, excluding any cloud observational data[39]. However, with the launch of the CloudSat and CALIPSO satellites, radiative flux calculations can now be constrained by a comprehensive set of cloud observations from space[21–23]. The Cloud-Aerosol Lidar with Orthogonal Polarization (CALIOP) sensor aboard CALIPSO was designed to focus on optically thin clouds and the 94-GHz Cloud Profiling Radar (CPR) aboard CloudSat probes optically thicker clouds and precipitation. CALIOP and CPR measure the backscattered energy by cloud particles, which can be converted into vertical distributions of cloud ice and liquid water contents, and effective radii, filled in by MODIS radiance information when the retrieval algorithms of the active sensors fail to converge. Despite their limited spatiotemporal sampling along specific tracks and limited collocated availability from 2007 to 2010, these active satellite cloud observations have been proven to be of great relevance for polar cloud studies[39,40].

The 2B-FLXHR-LIDAR algorithm. The primary satellite product used in this study is the refined CloudSat 2B-FLXHR-LIDAR[24]. This product combines (i) cloud and aerosol observations from CloudSat, CALIPSO and MODIS; (ii) ancillary temperature and humidity profiles from ECMWF atmospheric re-analyses; and (iii) International Geosphere-Biosphere Programme surface albedo/emissivity data to constrain the two-stream radiative transfer model (RTM) BugsRad[24], which calculates broadband radiative fluxes. The CloudSat footprint of 1.4 km × 1.8 km provides the basis for the radiative transfer calculations, determining the horizontal resolution of the product, whereas the vertical resolution is 240 m.

We found that in the presence of low-level supercooled liquid-bearing clouds that occur frequently over the Arctic[27], the original 2B-FLXHR-LIDAR product was severely underestimating downwelling LW fluxes because of the inability of the existing algorithm to properly handle these clouds, which were falsely identified as ice-only clouds. This was attributed to the fact that the CloudSat CPR is insensitive to small supercooled liquid droplets, while an exact retrieval of water path in this case was often not available from the CALIOP sensor. In such an event, a simple linear temperature-based relationship was used in the original version, to decide on the portion of ice and liquid water to which the CALIOP backscattered energy was translated, not allowing any supercooled liquid water to occur below −20 °C, while it can physically occur down to −40 °C. In this study, we therefore refined the 2B-FLXHR-LIDAR product by introducing the use of the CloudSat level-2 'Combined Radar and Lidar Cloud Scenario Classification' product[25,41] as input for the refined 2B-FLXHR-LIDAR algorithm, which combines information from the CloudSat CPR and CALIPSO CALIOP to identify cloud phase. When low-level supercooled liquid-bearing clouds are detected and no exact retrieval is available for determining the LWP that is assigned to these clouds, we rely on a fill value based on the median LWP value retrieved by ground-based observations at Summit (see below).

A total of ~6.3 million refined 2B-FLXHR-LIDAR flux profiles were aggregated in a 2° by 2° grid, to ensure a sufficient amount of satellite profiles within each grid box. Ice and liquid water occurrence frequencies and water paths were extracted from the refined 2B-FLXHR-LIDAR product and represent the combined CloudSat/CALIPSO/MODIS observations that were used in the radiative flux retrievals.

The radiative fluxes of the original 2B-FLXHR-LIDAR product have been evaluated[24] through comparisons with Clouds and the Earth's Radiant Energy System (CERES) top-of-atmosphere observations and surface flux retrievals on a global scale. To assess the quality of the refined 2B-FLXHR-LIDAR product over the GrIS, the satellite LWP retrievals were first evaluated with ground-based LWP measurements[42], retrieved by a microwave radiometer (MWR) installed at Summit, Greenland, as part of the ICECAPS project[28], during July 2010–December 2010. Despite the limited available overlap in time of 6 months, relative frequency histograms of LWP show a strong agreement between MWR-retrieved and satellite-retrieved LWP values (Supplementary Fig. 2). Furthermore, cloud macrophysical properties such as cloud fraction agree well (not shown) with the available ground-based observations[28,38].

An additional evaluation was performed by comparing the retrieved downwelling LW/SW radiative fluxes from the satellite product with ground-based flux measurements from 11 AWSs on the GrIS (IMAU[29] and PROMICE, http://www.promice.org/). Monthly averages were computed for satellite overpasses within 100 km from the AWS locations, whereas for consistency with the original overpasses, monthly averages for the AWS-measured fluxes were based on interpolated flux measurements to the times of these overpasses. AWS stations that are located in the same 2° by 2° grid box were averaged to yield one representative flux observation for that grid box. The resulting eight comparisons (Supplementary Fig. 3) show the good performance of the 2B-FLXHR-LIDAR surface radiative fluxes, as there is in general a close agreement between satellite-retrieved and AWS-measured LW and SW fluxes. Statistics on bias and root-mean-square errors (RMSEs) are given in Supplementary Table 1. Although there is a considerable range in these statistics, the overall agreement is good, with a mean bias of -4 and 7.6 W m^{-2} for downwelling LW and SW, respectively, well below the estimated monthly mean measurement uncertainty of $\sim$5% in the radiation instruments[29,43]. RMSEs equal 13.2 W m^{-2} for LW fluxes and 24.2 W m^{-2} for SW fluxes; these RMSE values are very similar to radiative flux closure results shown at Barrow, Alaska, where the cloud properties were specified by ground-based radars, lidars and radiometers[44]. Local disagreement between satellite and AWS can be explained by heterogeneous terrain within the 100-km radius of the AWS, AWS sensor uncertainties, AWS sensor tilt affecting SW measurements and low sampling rate for some AWS locations to compare 2B-FLXHR-LIDAR and AWS observations. Despite this, our evaluation indicates that the refined 2B-FLXHR-LIDAR product provides reliable estimates of radiative fluxes over the GrIS.

Cloud radiative effect. CRE at the surface (W m^{-2}) is defined by equation (1), where $F^{\downarrow\uparrow}_{\mathrm{SW,\,all\text{-}sky}}$ and $F^{\downarrow\uparrow}_{\mathrm{LW,\,all\text{-}sky}}$ are the net surface radiative SW and LW fluxes in all-sky conditions, and $F^{\downarrow\uparrow}_{\mathrm{SW,\,clear\text{-}sky}}$ and $F^{\downarrow\uparrow}_{\mathrm{LW,\,clear\text{-}sky}}$ are the net surface radiative fluxes that would occur in the absence of clouds[12].

$$\mathrm{CRE} = F^{\downarrow\uparrow}_{\mathrm{SW,\,all\text{-}sky}} + F^{\downarrow\uparrow}_{\mathrm{LW,\,all\text{-}sky}} - \left(F^{\downarrow\uparrow}_{\mathrm{SW,\,clear\text{-}sky}} + F^{\downarrow\uparrow}_{\mathrm{LW,\,clear\text{-}sky}}\right) \tag{1}$$

A positive effect indicates net cloud warming at the surface, whereas a negative effect is indicative of net cloud cooling. The magnitude of the radiative effect is intimately connected to the amount of ice and liquid water in the cloud, its temperature, SZA and surface albedo[14].

All analyses of CRE over the GrIS were focused on three different cloud scenarios. In the 'all-sky' simulation (that is, control run), radiative fluxes were calculated based on the satellite-observed clouds and their LWP and IWP. In the 'clear-sky' simulation, all cloud ice and liquid water was removed (that is, IWP and LWP were set to zero), to simulate the radiative fluxes that would occur in the absence of clouds. In the 'no-liquid' simulation, only the retrieved cloud liquid water was removed (LWP set to zero), effectively eliminating the impact of (supercooled) liquid water in liquid-bearing clouds on the calculated radiative fluxes, whereas IWP was kept unchanged. All-sky CRE is defined as the difference in radiative fluxes between the 'all-sky' and the 'clear-sky' simulations, whereas liquid-bearing CRE is defined as the difference in radiative fluxes between the 'all-sky' and the 'no-liquid' simulations.

To determine an observationally based CRE estimate, the 2B-FLXHR-LIDAR RTM was used for the three different cloud scenarios described above. Because of the large computational requirements of running the RTM on all satellite observations between 2007 and 2010, the 'clear-sky' and 'no-liquid' simulations were limited to 4 months, distributed over the year 2010 to be able to capture seasonal variability. For this purpose, all available observations from March, June, September and December 2010 were used.

Downwelling LW radiation is described by equation (2)[45], in which $\mathrm{LW_{as}}\downarrow$ and $\mathrm{LW_{cs}}\downarrow$ are the downwelling LW radiation in all-sky and clear-sky conditions, respectively, ϵ_{cs} is the clear-sky atmospheric emissivity, T is the temperature (K) at a reference height, σ is the Stefan–Boltzmann constant and F is the cloud enhancement factor.

$$\mathrm{LW_{as}}\downarrow = \epsilon_{cs} F \sigma T^4 = F \times \mathrm{LW_{cs}}\downarrow \tag{2}$$

The cloud enhancement factor F (≥ 1) describes the increase in downwelling LW radiation relative to clear sky due to clouds[46]. In this study we estimate F for all clouds and for liquid water in liquid-bearing clouds using the previously described RTM runs, in which respectively all clouds ('clear-sky' simulation) and liquid water in liquid-bearing clouds ('no-liquid' simulation) were removed from the satellite profiles, while keeping all other parameters unchanged.

Subsequently, $F_{\mathrm{all,\,LW}}$ and $F_{\mathrm{liq,\,LW}}$, that is, the LW cloud enhancement factors F for all clouds and liquid-bearing clouds, respectively, can be calculated by equations (3) and (4).

$$F_{\mathrm{all,\,LW}} = \frac{\mathrm{LW_{as}}\downarrow}{\mathrm{LW_{cs}}\downarrow} \tag{3}$$

$$F_{\mathrm{liq,\,LW}} = \frac{\mathrm{LW_{as}}\downarrow}{\mathrm{LW_{noliq}}\downarrow} \tag{4}$$

CRE in the LW, and therefore the LW cloud enhancement factor, is primarily a function of cloud temperature (often approximated by cloud-base temperature (CBT)), cloud height and emissivity[14]. The latter is mainly determined by the cloud LWP and IWP.

In an analogous manner, we can define an SW cloud enhancement factor that describes the decrease in downwelling SW radiation due to clouds relative to clear-sky conditions. Similar to equations (3) and (4), cloud enhancement factors $F_{\mathrm{all,\,SW}}$ and $F_{\mathrm{liq,\,SW}}$ (≤ 1) that account for the decrease in SW radiation due to all clouds and liquid-bearing clouds, respectively, can be defined by equations (5) and (6).

$$F_{\mathrm{all,\,SW}} = \frac{\mathrm{SW_{as}}\downarrow}{\mathrm{SW_{cs}}\downarrow} \tag{5}$$

$$F_{\mathrm{liq,\,SW}} = \frac{\mathrm{SW_{as}}\downarrow}{\mathrm{SW_{noliq}}\downarrow} \tag{6}$$

The SW cloud enhancement factor mainly depends on cloud microphysical properties (LWP and IWP), surface albedo and SZA[13].

To extend the satellite CRE data set from the four available months to the full 2007–2010 period, we trained neural networks to predict $F_{\mathrm{all,\,LW}}$, $F_{\mathrm{all,\,SW}}$, $F_{\mathrm{liq,\,LW}}$ and $F_{\mathrm{liq,\,SW}}$ (ref. 47).

Inputs for the neural networks to predict the cloud enhancement factors were LWP, IWP and near-surface temperature for LW and LWP, IWP, SZA and surface albedo for SW. LWP, IWP, SZA and surface albedo are readily available in our data sets. However, probably the best temperature-related predictor of LW CRE is cloud temperature or CBT. Yet, this variable is not readily available from the ERA-Interim product used by 2B-FLXHR-LIDAR ('ECMWF-AUX') and it is not available in the RACMO model outputs that were used in a later stage (see below). Therefore, near-surface temperature (2 m), originating from ECMWF ERA-Interim reanalysis data, was used instead of CBT, to train the neural networks. As such, the method only requires a good correlation between near-surface temperature and CBT. We therefore investigated the relationship between both temperatures, with CBT approximated here by the temperature of the satellite bin containing the cloud base (as detected by the combined CloudSat/CALIPSO observations), for the 4 months explicitly calculated by the RTM, and found a high correlation ($r = 0.83$) for cases over the GrIS with LWP + IWP ≥ 10 g m^{-2}. From this we expect the near-surface temperature to be a good proxy to predict LW CRE.

Hence, LWP, IWP, near-surface temperature, SZA and surface albedo were used as inputs for the neural networks and the cloud enhancement factors F as targets. Of the available satellite-retrieved CRE estimates, 70% was used for training the neural networks, whereas 15% was employed to avoid overfitting and 15% was used as independent validation set to assess the neural network performance. As the range of albedo values that are used in the 2B-FLXHR-LIDAR algorithm is relatively small, we performed additional offline RTM simulations with a variable surface albedo, to extend the training data set for a wider range of albedo values. The neural networks for both SW and LW consist of one hidden layer with ten neurons. Supplementary Fig. 4 shows the error histograms of the neural networks, expressed as the difference between the F factors predicted by the neural networks, and the F factors calculated from the RTM runs. For all F factors, biases are small (0 ± 0.03 for $F_{\mathrm{liq,\,SW}}$, 0 ± 0.07 for $F_{\mathrm{liq,\,LW}}$, 0 ± 0.03 for $F_{\mathrm{all,\,SW}}$ and 0 ± 0.05 for $F_{\mathrm{all,\,LW}}$), confirming the good skill of the neural networks to predict CRE. The errors on the F factors were used to calculate the uncertainty on the estimated CRE over the GrIS (see 'Uncertainty derivations').

The final all-year CRE as estimated by the neural networks, agrees very well with ground-based observations atop the GrIS at Summit[48] (30 versus 33 W m^{-2}), whereas in the ablation zone of the GrIS this estimate is substantially higher (33 versus 23 W m^{-2}), based on AWS stations S6 and S9 (ref. 29). This disparity is almost exclusively attributed to a difference in LW CRE, most probably due to the absence of cloud observations in the ground-based analysis and related assumptions for cloud optical depth retrievals[29]. In addition, we have calculated the monthly mean CRE over 2007–2010 for Summit, to ensure that the seasonal cycle of CRE is captured well. The seasonal pattern shown in Supplementary Fig. 5 agrees well with the results of a previous study[48], where their retrieval uncertainty spans the observed differences with our results. One clear difference that stands out is a lower LW CRE in our numbers compared with theirs in wintertime. Two possible explanations are (i) the difference in years that are used for the sampling and (ii) the ground-based instruments that are more sensitive to very small amounts of LWP/IWP that are missed in the CloudSat/CALIPSO estimate. Overall, the satellite-based estimates of CRE agree well with ground-based observations.

Regional climate model RACMO2.3. As the temporal resolution of satellite observations is limited by the amount of overpasses, a hybrid approach was designed (see below) to combine the observational constraints of the satellite data with the temporal resolution of the regional climate model RACMO2.3.

RACMO version 2.3 (ref. 49) combines the atmospheric dynamics from HIRLAM with the physical processes described by ECMWF Integrated Forecast System Cycle 33R1 (ref. 50). RACMO2.3 is specifically adapted for use over polar ice sheets, as it is interactively coupled to a multi-layer snow model[51], which includes an advanced snow albedo scheme[52,53], accounts for drifting snow[54] and contains several improvements in the representation of atmospheric physics compared with RACMO2.1. In particular, changes were introduced in the cloud scheme, which now includes a new ice super-saturation parameterization, improving the representation of supercooled liquid-bearing clouds, and also in the turbulence and radiation parameterization schemes[55,56]. The original RACMO data were aggregated in the same 2° by 2° grid that was used for the satellite observations.

Part of the results in this study strongly depend on the performance of the RACMO2.3 model. As this study focuses in particular on GrIS melt, it is of utmost importance that the climate is well simulated for reproducing melt events. The RACMO2.3 regional climate model has been extensively evaluated for simulating the near-surface climate, showing a very good performance in representing air temperature and specific and relative humidity[51,57], wind speed[58,59] and surface albedo[52]. With regard to SEB, however, significant biases remain in modelled surface radiative fluxes, with, in particular, an underestimation of the downwelling LW radiation and an overestimation of the downwelling SW radiation, due to an underestimated cloud optical depth[49]. These findings are confirmed in our comparisons between the original RACMO2.3-modelled radiative fluxes and retrieved radiative fluxes by the 2B-FLXHR-LIDAR algorithm (red dots in Supplementary Fig. 6). Downwelling LW radiation is systematically underestimated with a mean bias of $-6.2\,\mathrm{W\,m^{-2}}$ and an RMSE value of $8.8\,\mathrm{W\,m^{-2}}$, whereas downwelling SW radiation is slightly overestimated with a mean bias of $2.1\,\mathrm{W\,m^{-2}}$ and an RMSE value of $5.5\,\mathrm{W\,m^{-2}}$. The earlier hypothesis[49] is supported by our IWP/LWP comparisons in the Supplementary Methods. To improve the modelled IWP/LWP values and reduce the corresponding downwelling SW/LW biases, we therefore constructed a hybrid satellite-climate model data set.

Hybrid satellite-climate model data set. The accuracy of the satellite observations and high temporal resolution of RACMO2.3 were combined by fitting 3-day moving averages through both the satellite and model LWP/IWP values and calculating an additive correction factor *CF* (that is, the difference between the satellite and model moving average value) at each model output timestep for each 2° by 2° grid box. The *CF* factor is exponentially weighted with respect to the original model values, to avoid that too much LWP/IWP is added when originally no cloud was present in the model. The resulting exponentially weighted factor was used to rescale the LWP/IWP model values, whereas the original model values were retained when no satellite retrieval was available within the moving average window. The correction procedure is mathematically shown in equation (7) for the example of LWP. The factor *P* determines the shape of the exponential function. Both the width of the moving average window and the exponential factor *P* were chosen in such a way that the resulting corrected radiative fluxes resemble the satellite-retrieved radiative fluxes as closely as possible (see below). The factor *P* ranges from 0.03 in summer to 1,000 in winter. The resulting differences between corrected and original IWP values range from -69 to $203\,\mathrm{g\,m^{-2}}$, whereas the difference in LWP values range from -13 to $82\,\mathrm{g\,m^{-2}}$ (99 percentiles).

$$\mathrm{LWP}_{\mathrm{i,corrected}} = \mathrm{LWP}_{\mathrm{i,original}} + CF \times (1 - \exp(-P \times \mathrm{LWP}_{\mathrm{i,original}})) \qquad (7)$$

There are no significant lags in time and place for RACMO2.3 simulations compared with satellite observations. RACMO2.3 is forced by ERA-Interim, giving very good correlations with observations for daily and subdaily variations in near-surface climate (temperature, wind speed and humidity)[51,57]. We also investigated the possible occurrence of lags in time by performing cross-correlation tests between RACMO2.3 LWP/IWP simulations and the satellite LWP/IWP retrievals, and found no significant lag (not shown).

Comparisons of the original RACMO and hybrid satellite-climate model LWP histograms with the ground-based retrievals by the MWR at Summit are shown in Supplementary Fig. 7. These histograms confirm that the hybrid LWP values agree significantly better with ground-based observations, although LWP values remain somewhat underestimated. Moreover, for the five available months (July to December 2010) with concurrent ground-based observations, the difference in median LWP value has decreased from an underestimation of 5.3 to $1.7\,\mathrm{g\,m^{-2}}$.

After constraining the original model LWP/IWP by satellite observations, the trained neural networks were used to scale the modelled radiative fluxes accordingly. This was done by first converting the original RACMO fluxes to clear-sky fluxes and subsequently converting these clear-sky fluxes to radiative fluxes that match the hybrid LWP/IWP data set. In this framework, clear-sky radiative fluxes were calculated by obtaining cloud enhancement factors $F_{\mathrm{all,\,LW,\,original}}$ and $F_{\mathrm{all,\,SW,\,original}}$ from the neural networks with the original LWP/IWP/T2m/SZA/albedo model values as inputs (equation (8) for the LW example).

$$\mathrm{LW_{cs}} \downarrow = \frac{\mathrm{LW_{original}} \downarrow}{F_{\mathrm{all,\,LW,\,original}}} \qquad (8)$$

Next, a new set of cloud enhancement factors $F_{\mathrm{all,\,LW,\,hybrid}}$ and $F_{\mathrm{all,\,SW,\,hybrid}}$ was calculated by using the hybrid LWP/IWP data set as inputs for the neural networks, in addition to updated albedo values based on a physical albedo parameterization[60], which match the corrected optical depths, as albedo generally increases with optical depth[32]. Finally, these adjusted cloud enhancement factors were used to calculate the hybrid all-sky $\mathrm{LW_{hybrid}}\downarrow/\mathrm{SW_{hybrid}}\downarrow$ fluxes, as shown in equation (9) for the LW.

$$\mathrm{LW_{hybrid}} \downarrow = F_{\mathrm{all,LW,hybrid}} \times \mathrm{LW_{cs}} \downarrow \qquad (9)$$

The LWP/IWP correction parameters were chosen in such a way that the corrected downwelling LW/SW fluxes over the entire GrIS match the satellite-retrieved radiative fluxes as closely as possible (Supplementary Fig. 6, blue squares). This reduced the original downwelling LW bias of $-6.2\,\mathrm{W\,m^{-2}}$ and RMSE of $8.8\,\mathrm{W\,m^{-2}}$ to $-0.9\,\mathrm{W\,m^{-2}}$ and $6.1\,\mathrm{W\,m^{-2}}$, respectively. Downwelling SW bias and RMSE were reduced from 2.1 and $5.5\,\mathrm{W\,m^{-2}}$ to -0.3 and $3.7\,\mathrm{W\,m^{-2}}$, respectively.

Finally, we independently evaluated both the original and hybrid downwelling LW/SW fluxes in RACMO by comparisons with 11 AWSs on the GrIS. The results in Supplementary Fig. 8 show that the systematic underestimation of downwelling LW fluxes at the surface in the original RACMO2.3 model has been greatly reduced in the hybrid satellite-climate model data set. Mean bias in the LW fluxes is reduced from -9.9 to $-3.2\,\mathrm{W\,m^{-2}}$, with a reduced RMSE from 10.9 to $6.8\,\mathrm{W\,m^{-2}}$. Mean bias in the SW fluxes is reduced from 4.4 to $1.8\,\mathrm{W\,m^{-2}}$, with no change in RMSE.

CRE estimates for all clouds in the hybrid satellite-climate model data set follow directly from the previous results, whereas an additional set of $F_{\mathrm{liq,\,LW,\,hybrid}}$ and $F_{\mathrm{liq,\,SW,\,hybrid}}$ factors was calculated for estimating CRE by cloud liquid water in liquid-bearing clouds. Surface albedo values from RACMO2.3 were updated for this purpose to include the effect of changing cloud optical depth on the albedo values, following a physical albedo parameterization[60].

Simulating the GrIS SMB using SNOWPACK. To couple the effect of clouds on the SEB to the SMB, SNOWPACK simulations were performed for the 'all-sky', 'no-liquid' and 'clear-sky' scenarios on the 2° by 2° grid for the three hydrological years from September 2007 to September 2010, after a 10-year spinup. SNOWPACK[33] is a one-dimensional physical snow model, which, driven by time series of standard meteorological observations, models the stratigraphy, snow microstructure, snow metamorphism, snow temperature profile and settlement, as well as surface energy exchange and mass balance of a snowpack. SNOWPACK can be run with a variety of schemes for albedo, metamorphism and water balance. The output of the model is a set of time series that describe the snow profile (albedo, temperature, grain size, density and water content) and its processes (refreezing, water retention and so on). SNOWPACK was developed for seasonal snow, but it has recently been successfully applied to Antarctica as well[61].

The all-sky simulation (that is, control run) was forced by the hybrid satellite-climate model radiative fluxes and RACMO2.3 meteorological input data (2 m air temperature, 10 m wind speed, 2 m relative humidity and total precipitation), linearly interpolated to hourly values, and represents the most realistic estimate of the actual GrIS SMB. In the 'clear-sky' simulation the LW and SW CRE of all clouds was removed to simulate the cloud impact (both ice-only and liquid-bearing) on the SMB, whereas in the 'no-liquid' simulation LW and SW CRE by cloud (supercooled) liquid water was excluded to distinguish between the ice-only and liquid-only radiative impacts. As we focus solely on the radiative effect of clouds on the SMB, their role as precipitation source is not considered.

These simulations were performed using a physically based broadband albedo parameterization[53,60], which was implemented in SNOWPACK and accounts for albedo variations in function of the specific surface area of the snow/ice profile, cloud optical depth, SZA and concentration of light-absorbing carbon in the snow/ice, but with a lower limit of albedo values that corresponds to the spatially variable ice-albedo values derived from MODIS[53]. The specific surface area of the snow/ice profile was derived from SNOWPACK geometric grain size per layer, whereas the cloud optical depth was calculated from the hybrid satellite-climate model LWP/IWP data set. Finally, the concentration of light-absorbing carbon in the snow/ice was set to a fixed value per layer together with the other SNOWPACK model parameters (Supplementary Table 2). SNOWPACK was run in an offline configuration, that is, identical atmospheric temperature, wind speed, relative humidity and precipitation were used for all three simulations. This implies that we do not allow the atmosphere to react to changes in the snowpack.

The performance of SNOWPACK to simulate GrIS surface conditions was assessed based on the comparison of SNOWPACK SMB with RACMO SMB and stake measurements, SNOWPACK melt with satellite-observed melt and SNOWPACK refreezing rates with refreezing rates from literature. First, the yearly average SMB for each grid box was calculated, showing a close correspondence between SNOWPACK and RACMO SMB (Supplementary Fig. 9). We would like to stress here that this comparison is not aimed at an absolute evaluation of GrIS SMB estimates, but merely serves as a check that SNOWPACK simulates a similar SMB when forced by the same meteorological inputs as RACMO2.3. Integrated over the

GrIS, this is also confirmed by the strong agreement between the red and black curves in Fig. 3a. The limited amount of SMB observations through yearly stake measurements hinder a Greenland-wide observation-based evaluation of SMB. Yet, we compared the SNOWPACK-simulated SMB with observations from the K-transect[62] in the same period. The spatial resolution of the SNOWPACK runs prevents individual comparisons for all stations of the K-transect. However, the S10 station is representative for an accumulation pixel, whereas the other stations cover an ablation pixel. The yearly mean recorded SMB at S10 is 203 kg m^{-2}, whereas SNOWPACK indicates 243 kg m^{-2}. The other stations have a mean yearly SMB of − 1,714 kg m^{-2}, whereas the corresponding grid box of SNOWPACK has an SMB of − 1,708 kg m^{-2}. This comparison gives a good indication that the SNOWPACK estimates are sufficiently in line with the available SMB observations in the field. In addition, melt occurrence was calculated, to assess the ability of SNOWPACK to reproduce observed melt events from spaceborne passive microwave data[63]. In this framework, melt days in the SNOWPACK simulations were defined as days during which hourly surface temperatures reached the melting point within the window of overpass of the satellite[63] to detect surface melt. The original daily melt detections[63] were aggregated to the 2° by 2° grid for comparison with SNOWPACK. The yearly amount of melt days for each grid box agree reasonably well between observations and SNOWPACK simulations (Supplementary Fig. 10). Some scatter exists due to the definition of melt days, slight timing offsets and relatively coarse resolution of the SNOWPACK simulations. Nonetheless, the general agreement with observations is good and also the spatial patterns are captured well (Supplementary Fig. 11). Based on our definition of melt days, we find a mean yearly melt index (number of days melt is observed, multiplied by the area where the melt occurs) of 3.45×10^7 km^2 days, whereas the melt index calculated from the observed melt data[63] is 3.60×10^7 km^2 days, a difference of <5%. In addition, also the timing of simulated melt events was compared with observed melt events, using the same data set[63]. The result shows a very good agreement between both simulated SNOWPACK melt events and observed melt events (Supplementary Fig. 12). Finally, we quantified the mean refreezing rate in SNOWPACK in the all-sky simulation as 45%, whereas another study[64] found 42 ± 4 %. The SMB, melt and refreezing rate comparisons illustrate the good performance of SNOWPACK for simulating the GrIS SMB.

Uncertainty derivations. Radiative fluxes retrieved from satellite observations carry an inherent uncertainty due to many different factors. In the retrieval of CRE from satellite observations, we have accounted for the following sources of uncertainty: (i) LW/SW radiative flux retrievals from 2B-FLXHR-LIDAR, (ii) cloud enhancement factors *F* predicted by the neural networks and (iii) assumed surface albedo values in 2B-FLXHR-LIDAR. To incorporate uncertainty estimates based on the LW/SW radiative flux retrievals by 2B-FLXHR-LIDAR, we used the comparisons with ground-based radiative flux measurements from AWS observations. In this framework, a general RMSE value for both LW fluxes (9.3 W m^{-2}) and SW fluxes (22.5 W m^{-2}) was derived from the individual comparisons with the stations. These RMSE values represent the mean uncertainty of the satellite-retrieved radiative fluxes over the GrIS and were subsequently used to perturb the obtained satellite LW/SW fluxes (100 iterations), by introducing random errors from a normal distribution $N(\mu = 0; \sigma = \text{RMSE})$. These were then used as inputs to the neural networks, where an additional uncertainty was added by perturbing the predicted cloud enhancement factors *F* using random errors from a normal distribution with means and s.d. based on the network performance results (Supplementary Fig. 4). Finally, a last source of uncertainty was added to the 100 sets of radiative fluxes, based on the surface albedo assumption. Comparison of the albedo values used in the 2B-FLXHR-LIDAR algorithm with the physically based albedo retrievals in the RACMO model resulted in a mean bias of 0.04 ± 0.05. A random error was introduced based on these characteristics, from which updated radiative fluxes were calculated. The final 100 sets of radiative fluxes were used to calculate corresponding sets of CRE and include the three most important sources of uncertainty. The s.d. of this data set provides the combined uncertainty estimate of CRE values that are reported in this study.

SMB terms in SNOWPACK contain a combination of uncertainties that accumulated in earlier steps of our methodology. These uncertainties were estimated using a combination of (i) perturbed CRE to force SNOWPACK and (ii) inherent SNOWPACK model uncertainty. As the SMB terms for the different simulations depend strongly on the microhpysical characteristics of clouds, we quantified the LWP/IWP uncertainty of the hybrid satellite-climate model data set by comparing these values with the satellite observations. We found a mean LWP uncertainty of 35% and IWP uncertainty of 12%. Therefore, we performed four additional SNOWPACK runs using all possible combinations of adding/removing 35% of the LWP and 12% of the IWP at each timestep, and calculating the corresponding radiative fluxes. The range of resulting outputs for the all-sky and no-liquid simulations were used as uncertainty due to the amount of cloudiness in the SNOWPACK runs. Inherent SNOWPACK uncertainty was estimated by comparing SNOWPACK and RACMO estimates of the SMB terms (Fig. 3a). Results of this comparison showed that the uncertainty on the total GrIS SMB is ~10%, whereas the uncertainty on the individual components is ~50% for sublimation and ~20% for runoff. Finally, both sources of uncertainty (inherent model uncertainty and cloudiness) on SNOWPACK SMB and individual mass terms were combined (whiskers and shaded areas in Fig. 3, and all mass-related uncertainties in the text).

Code availability. All codes that have contributed to the results reported in this study are available on request. The 2B-FLXHR-LIDAR data are available at http://www.cloudsat.cira.colostate.edu/.

References

1. Box, J. E. *et al.* Greenland ice sheet albedo feedback: thermodynamics and atmospheric drivers. *Cryosphere* **6,** 821–839 (2012).
2. Fyke, J. G., Vizcaíno, M. & Lipscomb, W. H. The pattern of anthropogenic signal emergence in Greenland Ice Sheet surface mass balance. *Geophys. Res. Lett.* **41,** 6002–6008 (2014).
3. Fettweis, X. *et al.* Brief communication 'Important role of the mid-tropospheric atmospheric circulation in the recent surface melt increase over the Greenland ice sheet'. *Cryosphere* **7,** 241–248 (2013).
4. Shepherd, A. *et al.* A reconciled estimate of ice-sheet mass balance. *Science* **338,** 1183–1189 (2012).
5. IPCC. *Climate Change 2013—The Physical Science Basis* (eds Stocker, TF. *et al.*) pp. 1535 (Cambridge Univ. Press, Cambridge, UK and New York, doi:10.1017/CBO9781107415324, 2014).
6. Enderlin, E. M. *et al.* An improved mass budget for the Greenland ice sheet. *Geophys. Res. Lett.* **41,** 866–872 (2014).
7. Hanna, E. *et al.* Increased runoff from melt from the Greenland Ice Sheet: a response to global warming. *J. Clim.* **21,** 331–341 (2008).
8. Andersen, M. *et al.* Basin-scale partitioning of Greenland ice sheet mass balance components (2007-2011). *Earth Planet Sci. Lett.* **409,** 89–95 (2015).
9. Fettweis, X., van Ypersele, J.-P., Gallée, H., Lefebre, F. & Lefebvre, W. The 1979–2005 Greenland ice sheet melt extent from passive microwave data using an improved version of the melt retrieval XPGR algorithm. *Geophys. Res. Lett.* **34,** L05502 (2007).
10. Rignot, E. & Kanagaratnam, P. Changes in the velocity structure of the Greenland Ice Sheet. *Science* **311,** 986–990 (2006).
11. van den Broeke, M. *et al.* Partitioning recent Greenland mass loss. *Science* **326,** 984–986 (2009).
12. Ramanathan, V. *et al.* Cloud-radiative forcing and climate: results from the earth radiation budget experiment. *Science* **243,** 57–63 (1989).
13. Curry, J. A., Schramm, J. L., Rossow, W. B. & Randall, D. Overview of Arctic cloud and radiation characteristics. *J. Clim.* **9,** 1731–1764 (1996).
14. Shupe, M. D. & Intrieri, J. M. Cloud radiative forcing of the Arctic surface: the influence of cloud properties, surface albedo, and solar zenith angle. *J. Clim.* **17,** 616–628 (2004).
15. Turner, D. D. *et al.* Thin Liquid water clouds: their importance and our challenge. *Bull. Am. Meteorol. Soc.* **88,** 177–190 (2007).
16. Bennartz, R. *et al.* July 2012 Greenland melt extent enhanced by low-level liquid clouds. *Nature* **496,** 83–86 (2013).
17. Neff, W., Compo, G. P., Martin Ralph, F. & Shupe, M. D. Continental heat anomalies and the extreme melting of the Greenland ice surface in 2012 and 1889. *J. Geophys. Res.-Atmos.* **119,** 6520–6536 (2014).
18. Ettema, J., van den Broeke, M. R., van Meijgaard, E. & van de Berg, W. J. Climate of the Greenland ice sheet using a high-resolution climate model—Part 2: Near-surface climate and energy balance. *Cryosphere* **4,** 529–544 (2010).
19. Franco, B., Fettweis, X. & Erpicum, M. Future projections of the Greenland ice sheet energy balance driving the surface melt. *Cryosphere* **7,** 1–18 (2013).
20. Vizcaíno, M., Lipscomb, W. H., Sacks, W. J. & van den Broeke, M. Greenland surface mass balance as simulated by the community earth system model. Part II: twenty-first-century changes. *J. Clim.* **27,** 215–226 (2014).
21. Stephens, G. L. *et al.* CloudSat mission: performance and early science after the first year of operation. *J. Geophys. Res.* **113,** D00A18 (2008).
22. L'Ecuyer, T. S. & Jiang, J. H. Touring the atmosphere aboard the A-train. *Phys. Today* **63,** 36–41 (2010).
23. Platnick, S. *et al.* The MODIS cloud products: algorithms and examples from terra. *IEEE Trans. Geosci. Remote Sens.* **41,** 459–473 (2003).
24. Henderson, D. S., L'Ecuyer, T., Stephens, G., Partain, P. & Sekiguchi, M. A multisensor perspective on the radiative impacts of clouds and aerosols. *J. Appl. Meteorol. Climatol.* **52,** 853–871 (2013).
25. Zhang, D., Wang, Z. & Liu, D. A global view of midlevel liquid-layer topped stratiform cloud distribution and phase partition from CALIPSO and CloudSat measurements. *J. Geophys. Res.* **115,** D00H13 (2010).
26. Kay, J. E., L'Ecuyer, T., Gettelman, A., Stephens, G. & O'Dell, C. The contribution of cloud and radiation anomalies to the 2007 Arctic sea ice extent minimum. *Geophys. Res. Lett.* **35,** L08503 (2008).
27. Cesana, G., Kay, J. E., Chepfer, H., English, J. M. & de Boer, G. Ubiquitous low-level liquid-containing Arctic clouds: new observations and climate model constraints from CALIPSO-GOCCP. *Geophys. Res. Lett.* **39,** L20804 (2012).
28. Shupe, M. D. *et al.* High and dry: new observations of tropospheric and cloud properties above the Greenland ice sheet. *Bull. Am. Meteorol. Soc.* **94,** 169–186 (2013).
29. van den Broeke, M., Smeets, P., Ettema, J. & Munneke, P. K. Surface radiation balance in the ablation zone of the west Greenland ice sheet. *J. Geophys. Res.* **113,** D13105 (2008).

30. Flanner, M. G. & Zender, C. S. Linking snowpack microphysics and albedo evolution. *J. Geophys. Res.* **111,** D12208 (2006).
31. Yamanouchi, T. Variations of incident solar flux and snow albedo on the solar zenith angle and cloud cover, at Mizuho Station, Antarctica. *J. Meteorol. Soc. Jpn* **61,** 879–893 (1983).
32. Niwano, M. *et al.* Numerical simulation of extreme snowmelt observed at the SIGMA-A site, northwest Greenland, during summer 2012. *Cryosphere* **9,** 971–988 (2015).
33. Lehning, M., Bartelt, P., Brown, B., Fierz, C. & Satyawali, P. A physical SNOWPACK model for the Swiss avalanche warning: part II. Snow microstructure. *Cold Regions Sci. Technol.* **35,** 147–167 (2002).
34. Harper, J., Humphrey, N., Pfeffer, W. T., Brown, J. & Fettweis, X. Greenland ice-sheet contribution to sea-level rise buffered by meltwater storage in firn. *Nature* **491,** 240–243 (2012).
35. de la Peña, S. *et al.* Changes in the firn structure of the western Greenland Ice Sheet caused by recent warming. *Cryosphere* **9,** 1203–1211 (2015).
36. Boisvert, L. N. & Stroeve, J. C. The Arctic is becoming warmer and wetter as revealed by the atmospheric infrared sounder. *Geophys. Res. Lett.* **42,** 4439–4446 (2015).
37. Bony, S. *et al.* Clouds, circulation and climate sensitivity. *Nat. Geosci.* **8,** 261–268 (2015).
38. Van Tricht, K. *et al.* An improved algorithm for polar cloud-base detection by ceilometer over the ice sheets. *Atmos. Meas. Tech.* **7,** 1153–1167 (2014).
39. Kay, J. E. & L'Ecuyer, T. Observational constraints on Arctic Ocean clouds and radiative fluxes during the early 21st century. *J. Geophys. Res.-Atmos.* **118,** 7219–7236 (2013).
40. Liu, Y., Key, J. R., Ackerman, S. A., Mace, G. G. & Zhang, Q. Arctic cloud macrophysical characteristics from CloudSat and CALIPSO. *Remote Sens. Environ.* **124,** 159–173 (2012).
41. Sassen, K., Wang, Z. & Liu, D. Global distribution of cirrus clouds from CloudSat/Cloud-Aerosol Lidar and Infrared Pathfinder Satellite Observations (CALIPSO) measurements. *J. Geophys. Res.* **113,** D00A12 (2008).
42. Turner, D. D. *et al.* Retrieving liquid water path and precipitable water vapor from the Atmospheric Radiation Measurement (ARM) microwave radiometers. *IEEE Trans. Geosci. Remote Sens.* **45,** 3680–3690 (2007).
43. van den Broeke, M., van As, D., Reijmer, C. & van de Wal, R. Assessing and improving the quality of unattended radiation observations in Antarctica. *J. Atmos. Ocean. Technol.* **21,** 1417–1431 (2004).
44. Shupe, M. D. *et al.* Deriving Arctic cloud microphysics at Barrow, Alaska: algorithms, results, and radiative closure. *J. Appl. Meteorol. Climatol.* **54,** 1675–1689 (2015).
45. Unsworth, M. H. & Monteith, J. L. Long-wave radiation at the ground I. Angular distribution of incoming radiation. *Q. J. R. Meteorol. Soc.* **101,** 13–24 (1975).
46. Sicart, J. E., Pomeroy, J. W., Essery, R. L. H. & Bewley, D. Incoming longwave radiation to melting snow: observations, sensitivity and estimation in Northern environments. *Hydrol. Process.* **20,** 3697–3708 (2006).
47. Chen, Y., Aires, F., Francis, J. A. & Miller, J. R. Observed relationships between Arctic longwave cloud forcing and cloud parameters using a neural network. *J. Clim.* **19,** 4087–4104 (2006).
48. Miller, N. B. *et al.* Cloud radiative forcing at Summit, Greenland. *J. Clim.* **28,** 6267–6280 (2015).
49. Noël, B. *et al.* Evaluation of the updated regional climate model RACMO2.3: summer snowfall impact on the Greenland Ice Sheet. *Cryosphere* **9,** 1831–1844 (2015).
50. ECMWF-IFS: Part IV : PHYSICAL PROCESSES (CY33R1). Technical Report (2008).
51. Ettema, J. *et al.* Climate of the Greenland ice sheet using a high-resolution climate model—Part 1: evaluation. *Cryosphere* **4,** 511–527 (2010).
52. Kuipers Munneke, P. *et al.* A new albedo parameterization for use in climate models over the Antarctic ice sheet. *J. Geophys. Res.* **116,** D05114 (2011).
53. van Angelen, J. H. *et al.* Sensitivity of Greenland Ice Sheet surface mass balance to surface albedo parameterization: a study with a regional climate model. *Cryosphere* **6,** 1175–1186 (2012).
54. Lenaerts, J. T. M. *et al.* Modeling drifting snow in Antarctica with a regional climate model: 1. Methods and model evaluation. *J. Geophys. Res.* **117,** D05108 (2012).
55. Van Wessem, J. M. *et al.* Updated cloud physics in a regional atmospheric climate model improves the modelled surface energy balance of Antarctica. *Cryosphere* **8,** 125–135 (2014).
56. Lenaerts, J. T. M. *et al.* Extreme precipitation and climate gradients in Patagonia revealed by high-resolution regional atmospheric climate modeling. *J. Clim.* **27,** 4607–4621 (2014).
57. Lenaerts, J. T. M. *et al.* Drifting snow measurements on the Greenland Ice Sheet and their application for model evaluation. *Cryosphere* **8,** 801–814 (2014).
58. Sanz Rodrigo, J., Buchlin, J.-M., van Beeck, J., Lenaerts, J. T. M. & van den Broeke, M. R. Evaluation of the antarctic surface wind climate from ERA reanalyses and RACMO2/ANT simulations based on automatic weather stations. *Clim. Dyn.* **40,** 353–376 (2012).
59. Gorter, W., van Angelen, J. H., Lenaerts, J. T. M. & van den Broeke, M. R. Present and future near-surface wind climate of Greenland from high resolution regional climate modelling. *Clim. Dyn.* **42,** 1595–1611 (2013).
60. Gardner, A. S. & Sharp, M. J. A review of snow and ice albedo and the development of a new physically based broadband albedo parameterization. *J. Geophys. Res.* **115,** 1–15 (2010).
61. Groot Zwaaftink, C. D. *et al.* Event-driven deposition of snow on the Antarctic Plateau: analyzing field measurements with SNOWPACK. *Cryosphere* **7,** 333–347 (2013).
62. van de Wal, R. S. W. *et al.* Twenty-one years of mass balance observations along the K-transect, West Greenland. *Earth Syst. Sci. Data* **4,** 31–35 (2012).
63. Tedesco, M. *Greenland Daily Surface Melt 25 km EASE-Grid, 2007-2010.* City College of New York, The City University of New York Digital Media (2014).
64. van Angelen, J. H., van den Broeke, M. R., Wouters, B. & Lenaerts, J. T. M. Contemporary (1960-2012) evolution of the climate and surface mass balance of the Greenland Ice Sheet. *Surv. Geophys.* **35,** 1155–1174 (2013).

Acknowledgements

K.V.T. and S.L. are funded by the Research Foundation Flanders (FWO). I.V.G. was supported via the project HYDRANT funded by the Belgian Science Policy Office under grant number EN/01/4B. The contributions of T.S.L. to this project were supported by CloudSat/CALIPSO science team grant NNX14AB35G. This study is partly funded by Utrecht University through its strategic theme Sustainability, sub-theme Water, Climate and Ecosystems, and the Netherlands Polar Programme of NWO. Part of this work was carried out under the programme of the Netherlands Earth System Science Centre (NESSC), financially supported by the Ministry of Education, Culture and Science (OCW). This work is further supported by the Belgian Federal Science Policy Office project AEROCLOUD (BR/143/A2). The Summit data were recorded in the frame of the ICECAPS project, which is supported by the US National Science Foundation under Grants ARC-0856773, 0904152 and 0856559 as part of the Arctic Observing Network (AON) programme, with additional instrumentation provided by the NOAA Earth System Research Laboratory, US Department of Energy ARM Program and Environment Canada. We thank the CloudSat Data Processing Center, in particular Phil Partain, for their help with the 2B-FLXHR-LIDAR algorithm runs.

Author contributions

K.V.T. conceived this study, constructed the hybrid satellite-climate model data set, performed most of the analyses and wrote the manuscript. S.L. implemented the SNOWPACK snow model and conducted all simulations, and participated in writing the manuscript. J.T.M.L. performed the Coupled Model Intercomparison Project phase 5 analyses and theoretical calculations of SMB impacts, provided many insights in the coupling between SEB and SMB, and participated in writing the manuscript. I.V.G. worked on the theoretical basis of the study and helped with the interpretation of results. T.S.L. was responsible for the CloudSat/CALIPSO/MODIS satellite products and has overseen the conducted 2B-FLXHR-LIDAR runs. B.N. performed RACMO2.3 runs over Greenland and supplied the necessary climate model data to construct the hybrid satellite-climate model data set. M.R.v.d.B. is responsible for the use of RACMO2.3 over Greenland and has overseen its use in this study and helped interpreting the results. D.D.T. performed the all-sky retrievals to derive LWP from the microwave radiometer at Summit and provided theoretical insights in clouds over the GrIS. N.P.M.v.L. developed the methodology for constraining the original RACMO model output by the available satellite observations and has worked on the theoretical part of the study and result interpretations. All authors commented on the manuscript.

Additional information

Competing financial interests: The authors declare no competing financial interests.

13

Subterranean atmospheres may act as daily methane sinks

Angel Fernandez-Cortes[1,2,3], Soledad Cuezva[1,3], Miriam Alvarez-Gallego[1], Elena Garcia-Anton[1], Concepcion Pla[4], David Benavente[4], Valme Jurado[5], Cesareo Saiz-Jimenez[5] & Sergio Sanchez-Moral[1]

In recent years, methane (CH_4) has received increasing scientific attention because it is the most abundant non-CO_2 atmospheric greenhouse gas (GHG) and controls numerous chemical reactions in the troposphere and stratosphere. However, there is much that is unknown about CH_4 sources and sinks and their evolution over time. Here we show that near-surface cavities in the uppermost vadose zone are now actively removing atmospheric CH_4. Through seasonal geochemical tracing of air in the atmosphere, soil and underground at diverse geographic and climatic locations in Spain, our results show that complete consumption of CH_4 is favoured in the subsurface atmosphere under near vapour-saturation conditions and without significant intervention of methanotrophic bacteria. Overall, our results indicate that subterranean atmospheres may be acting as sinks for atmospheric CH_4 on a daily scale. However, this terrestrial sink has not yet been considered in CH_4 budget balances.

[1]Department of Geology, National Museum of Natural Sciences (MNCN-CSIC), 28006 Madrid, Spain. [2]Department of Earth Sciences, Royal Holloway, University of London, Egham, Surrey TW20 0EX, UK. [3]Geomnia Natural Resources SLNE, 28003 Madrid, Spain. [4]Department of Environment and Earth Sciences, University of Alicante, San Vicente del Raspeig, 03690 Alicante, Spain. [5]Department of Agrochemistry, Environmental Microbiology and Soil Conservation, Institute of Natural Resources and Agricultural Biology (IRNAS-CSIC), 41012 Seville, Spain. Correspondence and requests for materials should be addressed to A.F-C. (email: acortes@mncn.csic.es).

Methane (CH_4) is currently the most abundant non-CO_2 greenhouse gas (GHG) in the atmosphere, reaching a global average concentration of ~1,800 p.p.b. at mid-northern latitudes[1]. Despite significant research progress in recent years, large uncertainties remain about the CH_4 budget and its evolution over time because there is a lack of knowledge about CH_4 sinks and sources[2–5]. Superimposed on the long-term trend of increasing atmospheric CH_4, there is significant interannual variability[1,6,7], and the sources of variation remain controversial[8].

Current CH_4 estimates account for ~22% of the total forcing potential of all long-lived GHGs[9]. By weight, CH_4 is 28 times more effective at trapping heat in the atmosphere than CO_2 over a 100-year period[10]. Sensible mitigation strategies also require a quantitative understanding of the CH_4 budget regarding emissions and sinks[11]. The total global emissions of CH_4 are constrained reasonably well by atmospheric observations and estimates of its lifetime, based on multiple atmospheric CH_4 inversion models (top–down studies). However, the uncertainties concerning emissions/consumption from individual sources/sinks[12] are greater, and they are poorly constrained by the current atmospheric observation network[6]. Some unaccounted sinks (or sources) could contribute the global CH_4 budget and its long-term variations.

The Earth's surface exerts its influence on the free atmosphere through the atmospheric boundary layer. This lowest portion of the atmosphere ranges from a few tens of metres to 1–2-km deep[13]. The subsurface atmosphere is usually overlooked as an important part of this boundary layer. At the top of the subsurface layer in the vadose zone (below the subsoil and above the groundwater table) highly specific biogeochemical processes occur, which may act as regulators of gas exchanges between the surface and the free atmosphere.

The uppermost part of the vadose zone may contain large amounts of underground air, that is, a CO_2-rich air reservoir permeating the unsaturated zone[14]. Underground air occupies air-filled caves, empty-water pores, cracks and voids in bedrock or unconsolidated sediment, some of which exhibit a direct connection to the open atmosphere[15]. The gas exchange between the bulk of the atmosphere and small air-filled locations, for example, fissures, cracks and voids, is minor compared with the exchange that occurs between large shallow caves and the open atmosphere. Hence, shallow caves can be thought of as small-scale representations of the uppermost part of the vadose zone, where there is intense gas exchange with the atmosphere. Shallow caves may largely represent the volume of underground air that is distributed widely across continental ecosystems.

Furthermore, shallow caves are favourable for *in situ* monitoring of gaseous exchange processes at the subsurface. They are easily accessible natural macropores that can store edaphic or atmospheric gases several tens of metres below the soil. These shallow vadose environments below the soil show significant seasonal and even daily variations in the gas composition of underground air[16–18], which involves the exchange of large amounts of other GHGs in addition to CO_2(g) with the lower troposphere[15,19]. However, there is great uncertainty regarding the behaviour and lifetime of CH_4 in these shallow vadose environments.

Here to better understand the behaviour and lifetime of the CH_4 in subsurface environments located in the vadose zone, spatiotemporal spot measurements and continuous geochemical tracing of air (in the atmosphere, soil and cave air) have been carried out by monitoring CH_4, CO_2 and stable carbon isotopic $\delta^{13}C[CO_2]$ using cavity ring-down spectroscopy (CRDS). In total, >1,000 air samples have been analysed for seven caves located in Spain. The caves exhibit a wide range of geographic, geologic and climatic features (Supplementary Table 1). For each cave, the sampling was conducted in underground air, soil air and external atmosphere. Comprehensive microclimatic monitoring of the exterior, soil and cave atmosphere was conducted simultaneously with GHG tracking. In addition, potential methanotrophy at each subsurface site was assessed by tracking the presence of methane-oxidizing bacteria (MOB) using specific DNA primers for methanotroph *pmoA* sequences[20]. The overall results show that complete consumption of CH_4 is favoured in the subsurface atmosphere under near vapour-saturation conditions without significant intervention of methanotrophic bacteria. The oxidative capacity of underground air must be related to its high density of ions, some of which are likely to react with CH_4. Overall, our results indicate that subterranean atmospheres may be acting as sinks for atmospheric CH_4 on seasonal and daily scales.

Results

GHGs in underground air. Spot air samplings were conducted during short surveys at all fieldwork sites, within a predefined network of points that were spatially distributed from the entrance to the innermost accessible areas of the cavities. Exterior and soil air were also sampled at fixed locations throughout the day, and, preferably, during maximum (midday) and minimum photosynthetic activity (early morning). Further details about the air-sampling procedures are found in Methods section. Air samples were analysed in the laboratory within 48 h of sampling for determine CO_2 and CH_4 concentrations and the isotopic $\delta^{13}CO_2$ signal using CRDS technology.

The comparison of CO_2 and CH_4 enables assessment of the influence of the exterior atmosphere and soil on the gas contents of cave air, that is, the degree of isolation of the subsurface environment. Figure 1 shows the variations in the concentration of CH_4 and CO_2 in underground air relative to their local atmospheric levels for the main four caves studied. Supplementary Figure 1 shows the extended results for the entire set of caves studied, covering a wide latitudinal transect of the Iberian Peninsula that includes different environmental features, which are summarized in Supplementary Table 1. There is an inverse relationship between the two GHGs along a subsurface–soil–atmosphere profile. Both soil and cave air exhibit a lower concentration of CH_4 compared with local atmospheric background and regional CH_4 levels (1.83–2.05 p.p.m. for the period 2011–2012, based on records for Central Spain from NOAA). An opposite pattern is found for CO_2 in cave air, which is consistent with the usual magnitude of the CO_2(g) reservoirs found in shallow caves[21,22], presenting average values that are 10–100 times the current mean atmospheric concentration (0.04% vol). The overall results indicate a higher CH_4 depletion rate (from −0.1 to −0.4 p.p.m. for each 100 p.p.m. increase in CO_2) under low-moderate contents of CO_2 (<1,000 p.p.m.). Furthermore, this CH_4 depletion rate rises in highly ventilated caves (for example, the Ojo Guareña and Sidron caves) within a narrow increment of CO_2 (500→700 p.p.m.). Shallow caves where there is intermediate and seasonal ventilation of underground air (for example, the Altamira Cave) display a lower rate of decreases in CH_4 contents (lower than −0.1 p.p.m. of CH_4 for each 100 p.p.m. increment of CO_2), especially in cave locations subjected to a stronger influence of the external atmosphere, but showing considerable CO_2 levels (1,000–3,000 p.p.m.). Most isolated caves with the lowest ventilation rates (for example, the Castañar Cave) present a steady state, with practically null CH_4 levels, thus suggesting that complete removal of CH_4 from the underground air occurs once a CO_2 concentration above 3,000 p.p.m. is reached, which is observable in the deepest

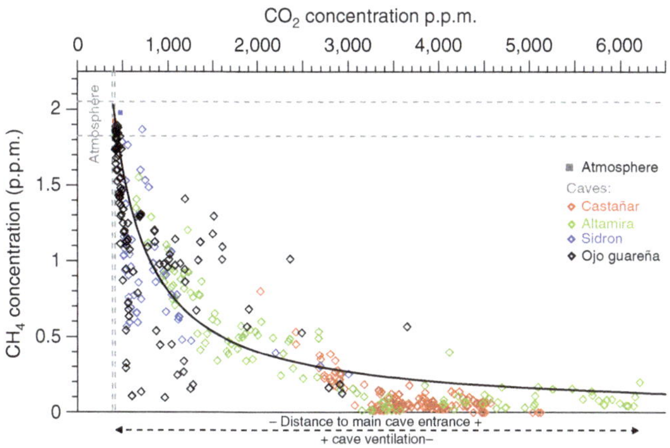

Figure 1 | Variations in the concentrations of GHGs in underground air. Caves (open diamonds), local atmosphere (solid squares). The grey dashed lines indicate the regional levels of the atmospheric CH_4 and CO_2 Dry Air Mole Fractions from the NOAA earth system research laboratory (ESRL) global monitoring division (GMD) Carbon Cycle Cooperative Global Air Sampling Network as a reference[62,63]. In particular, the grey dashed lines correspond to representative minimum and maximum data from the well-mixed troposphere from central Spain (Centro de Investigación de la Baja Atmósfera, University of Valladolid, 41.81° N, 4.93° W, 845 m a.s.l. (above sea level)) for the period 2011–2012 coeval to the cave air samplings.

remote cave locations. The dynamically ventilated caves show greater spatiotemporal variability in their CO_2 and CH_4 concentrations, which is shown by the fact that the paired data for both the parameters are more scattered compared with caves in which there is limited ventilation of underground air.

The CH_4 depletion rate and its variation relative to CO_2 content in underground air (Fig. 1) were, therefore, key parameters for assessing the ventilation of subterranean atmospheres. In highly ventilated caves, a significant volume of air circulates daily, and, consequently, there is a nearly continual contribution of air to the subterranean atmosphere, which has lower CO_2 content and atmospheric levels of CH_4. Hence, the CH_4 depletion rises with slight changes in CO_2 close to atmospheric background. Conversely, in the most isolated caves, CO_2 reaches high concentrations because air exchange with the outer atmosphere is negligible. Thus, the removed CH_4 is not rapidly replenished. As a result, the CH_4 depletion rate tends to become negligible as the CO_2 content of cave air rises.

The CO_2 concentration in cave air is determined by the mixing of background atmospheric CO_2 with soil-produced CO_2, and the mixing process fits a Keeling model with a mean y-intercept value ranging from -25 to -28‰ for the caves studied (Fig. 2). The CO_2-soil recharge by diffusion is more of an important factor at the most isolated subterranean sites, such as Castañar and Altamira caves, where the cave air usually has a higher CO_2 concentration and $\delta^{13}C[CO_2]$ below -20‰ with respect to other caves. Conversely, the lower CO_2 concentration and the $\delta^{13}C[CO_2]$ value that tends toward the local atmospheric background ($\sim 9\%$), as measured in Sidron and Ojo Guareña caves, indicates rapid air renewal that allows the air to become well-mixed with the atmosphere.

Seasonal variation of CH_4 in dynamically ventilated caves. Multi-parameter monitoring was conducted for the cave–soil–atmosphere (main climatic data and gas composition) and geochemical tracing was performed using $\delta^{13}C$–CO_2 to assess the climate-driven fluctuations of GHGs in the subsurface and to determine the spatiotemporal patterns of exchange between soil and atmosphere. A 3-km-long area of the subterranean system of Ojo Guareña (northern Spain) was chosen for field research. The Ojo Guareña cave system is one of the longest caves in Europe and is the most dynamically ventilated cave site of all the sites studied, according to our most recent studies[23].

Cave air ventilation increased from the end of the summer to early winter in the case of the Ojo Guareña subterranean system (Fig. 3; Supplementary Fig. 2). During winter, the outside air temperature is lower than the inside air temperature, and, consequently, the outside air density is higher than the cave air density. This favours advective air movement, primarily through the cave entrance and other host-rock discontinuities. Therefore, a higher proportion of CH_4-enriched air was progressively transferred from the open atmosphere to the cave during the winter months than in summer. Higher CH_4 depletion with narrow increases of CO_2 indicates a prevailing influx of outer air into the cave, and the CH_4 consumption rate consequently increases from summer to winter. The CH_4 depletion rate in underground air with an increasing CO_2 concentration is significantly higher than the rate of CH_4 consumption through soil MOB. The CH_4 depletion, coeval to increments of CO_2 concentration in underground air, is higher in cave locations that are closer to the surface (CH_4 concentration above 0.75 p.p.m.) and decreases in deeper and more isolated locations (CH_4 concentration below 0.5 p.p.m.). In the most isolated locations, CO_2 diffusion from the soil prevails over the influx of exterior air with the atmospheric CH_4 levels and lower CO_2 concentrations.

Daily variation of CH_4 consumption in ventilated caves. *In situ* and real-time measurements of CO_2, CH_4 and $\delta^{13}CO_2$ were carried out during an intensive field campaign in Altamira Cave. This intensive monitoring period was examined to assess the complete daily cycles of the GHG mixing ratios through comprehensive real-time logging using the same field-deployable CRDS analyser. The results summarized in Fig. 4 correspond to data records from the field campaign conducted in September 2011 (end of summer). In Methods section are summarized the technical procedures applied in the real-time air monitoring deployed in the field sites, including the number and frequency of records and the $\delta^{13}C$–CO_2 calculation.

The interaction of three air masses (soil–cave–exterior) is driven by soil-derived CO_2 diffusion into the cave and by the advective mixing of the cave with exterior air[18]. The diffusive process increases cave CO_2 concentration and decreases $\delta^{13}C[CO_2]$, to the extent that biological respiration in the soil causes an accumulation of CO_2 molecules with a lighter carbon isotopic signature. In contrast, advective mixing causes the CO_2 levels to decrease and causes the isotopic signal to increase because it causes the uptake of air with an atmospheric CO_2 level and a heavier $\delta^{13}C[CO_2]$ compared with both soil and cave air. The diffusive flux usually increases with soil CO_2 production, and the advective flux is driven by outer–inner density gradients. Both of these drivers are primarily regulated by the external meteorological conditions.

The monitoring results confirmed that the driving forces for the increases in CH_4 in underground air are the same as for the declines in CO_2 levels, and the daytime increase of the CH_4 content of underground air ($+0.04$ p.p.m. h^{-1} on average with the maximum reaching $+0.11$ p.p.m. h^{-1}) is, therefore, synchronous with decline in the CO_2 levels and an increase in $\delta^{13}C[CO2]$ (Fig. 4), which points to an air inlet from the outside atmosphere. Consequently, the maximum values of CH_4 begin to be reached after sunset but usually remain below the background atmospheric levels. On the contrary, during the night, the CH_4 depletion of underground air is coeval with an increase in CO_2

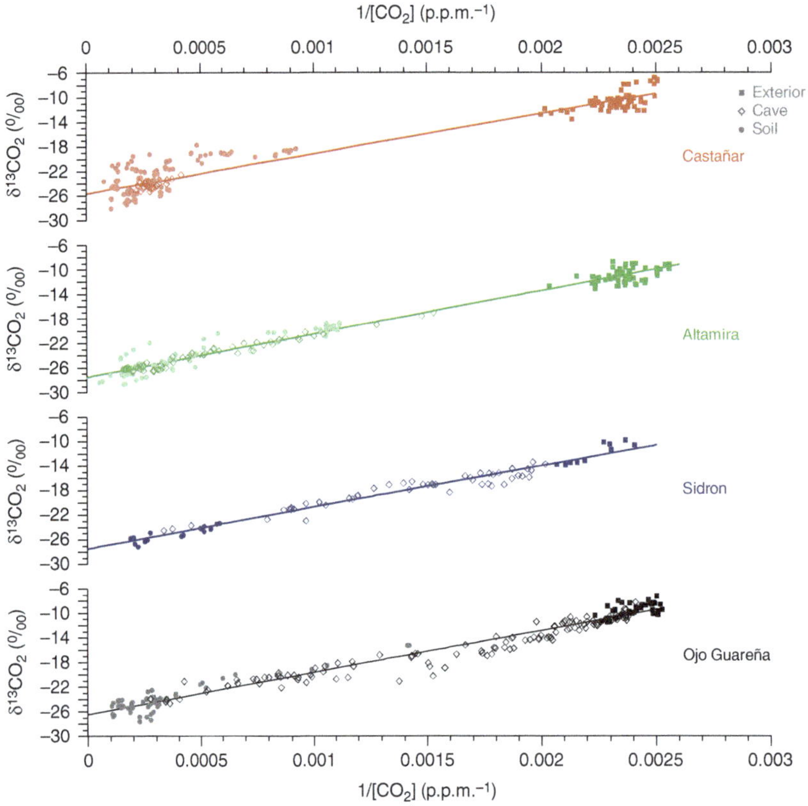

Figure 2 | Keeling models for the subterranean atmosphere at each cave site. Isotope ratios are plotted against the inverse of CO_2 concentration for all air samples (see Supplementary Table 1, and notice the number of air samples for each fieldwork site), distinguishing a three-component mixture: background atmosphere (solid squares), cave air (open diamonds) and soil air (solid circles). The y-intercept of the Keeling models corresponds to the carbon isotope composition for the integrated value of the CO_2 sources in the ecosystem: −25.71‰ for Castañar cave ($R^2 = 0.93$), −27.59‰ for Altamira cave ($R^2 = 0.98$), −27.50‰ for Sidron cave ($R^2 = 0.97$) and −26.48‰ for Ojo Guareña cave ($R^2 = 0.96$). As the distance of the cave air samples from the y-intercept of the Keeling models increases, there is greater ventilation and greater influence of the external atmosphere. Conversely, when CO_2 concentration is high and $\delta^{13}C[CO_2]$ tends toward the y-intercept, the gas enters by diffusion and has a mainly edaphic origin.

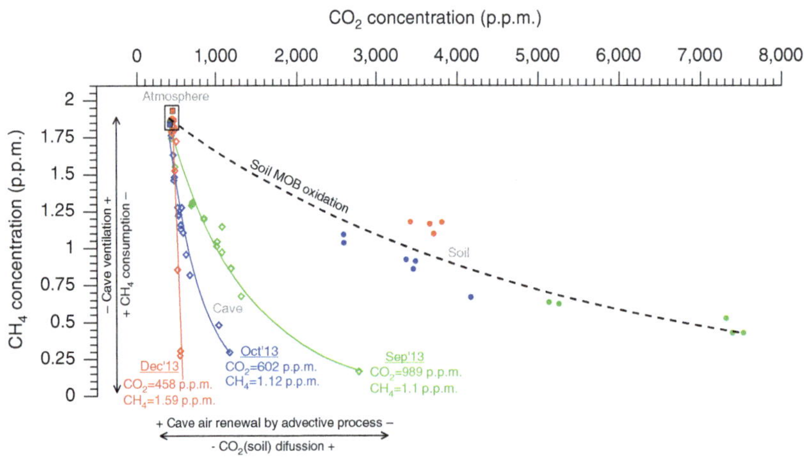

Figure 3 | Seasonal variability of GHGs in a dynamically ventilated cave. Cave (Ojo Guareña, Spain) air samples (open diamonds), soil air samples (solid circles) and atmospheric air samples (solid squares). Each colour corresponds to an air sampling date in 2013: September (green), October (blue) and December (red). The dashed black line represents CH_4 oxidation by soil MOBs as a function of CO_2 concentration in the soil air ($[CH_4, \text{p.p.m.}] = 2.04 \times \exp(-2.09 \times 10^{-4} \times [CO_2, \text{p.p.m.}])$, $R^2 = 0.96$). Coloured lines correspond to the functions for each monthly set of paired CH_4 and CO_2 data for the air from cave environments; 18 September 2013: $[CH_4, \text{p.p.m.}] = 2.72 \times \exp(-9.95 \times 10^{-4} \times [CO_2, \text{p.p.m.}])$, $R^2 = 0.98$), 29 October 2013: $[CH_4, \text{p.p.m.}] = 4.35 \times \exp(-2.3 \times 10^{-3} \times [CO_2, \text{p.p.m.}])$, $R^2 = 0.96$) and 10 December 2013: $[CH_4, \text{p.p.m.}] = -1.35 \times 10^{-2} \times [CO_2, \text{p.p.m.}] + 7.79$, $R^2 = 0.82$. The slope of these coloured lines is indicative of the consumption of atmospheric CH_4 due to the influx of outer air into the cave via advective processes. In the case of the exponential functions, an increase of the coefficient multiplying the CO_2 concentration implies a higher CH_4 depletion. The average concentrations of CH_4 and CO_2 in underground air for each month is indicated, considering the entire set of 15 cave-air-sampling points for each field campaign.

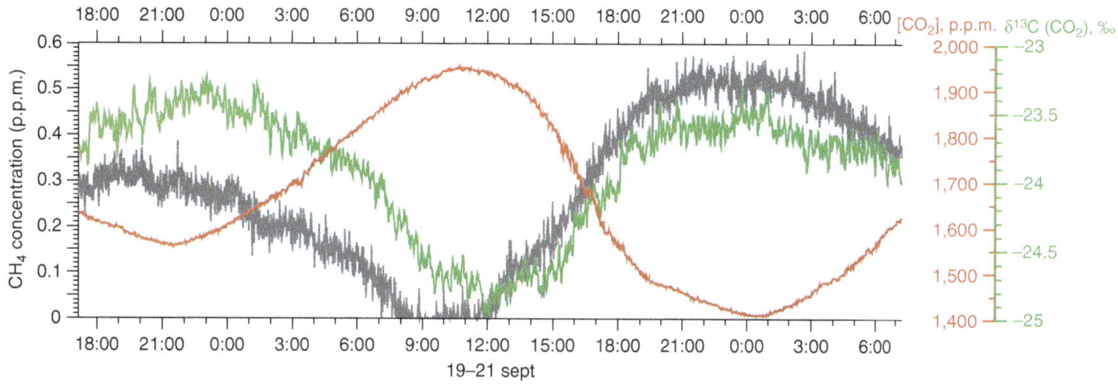

Figure 4 | Daily variations of CH_4 and CO_2 contents and the carbon isotopic signal in underground air. Data from the Altamira Cave (from day 19 to 21 September 2011). The shaded area represents the night-time period delimited by null solar radiation, in accordance with ancillary monitoring parameters (the applied methods and full time series are described by Garcia-Anton *et al.*[18]). All records are reported relative to Greenwich Mean Time.

contents and a decrease in $\delta^{13}C[CO2]$, which indicates the existence of a CO_2 gas inlet with an edaphic origin, primarily via diffusion. Conversely, the gradient of CH_4 contents between the soil and cave is not sufficient to enable an influx to underground air, due to the likely CH_4 consumption in soil through MOBs action. In the absence of nocturnal influx of atmospheric CH_4 via advective forces, CH_4 consumption mechanisms in the underground air seem to be triggered, the CH_4 content consequently decreases to an average rate of -0.03 p.p.m. h^{-1} (maximum of -0.05 p.p.m. h^{-1}), reaching 0 p.p.m. early in the morning (8:30 h Greenwich Mean Time).

Tracking the presence of MOB. We evaluated whether the CH_4 oxidation pattern by soil MOB was repeated in cave sediments. Microbiological sampling was conducted to assess the presence or absence of MOBs in the edaphic profile at the exterior and in sediments and on rock surfaces in the cave at each monitored field site. The collected samples were analysed through an ordered set of techniques, including nucleic acid extraction and PCR amplification, cloning and sequencing and sequence analyses (see Methods section for details of these analytical techniques).

The results obtained through microbiological sampling of cave sediments from the main four caves studied revealed the presence of *Methylocapsa aurea* (FN433470), *Methylomicrobium album* (EU722431), *Methylococcus capsulatus* (AE017282) and methanotrophs of the K1-1 (AF547174) and K3-16 (AF547178) strains only in the samples from the Altamira, Sidron and Ojo Guareña caves, mainly in locations where CH_4 usually reaches concentrations near to the atmospheric background levels. These analyses did not detect methanotrophs in remote subterranean locations or poorly ventilated caves, such as the Castañar Cave, where CH_4 is absent or present in minimal concentrations (below the accuracy threshold) throughout the year.

CH_4 removal in relation to other environmental parameters. Our initial measurements of the density of ions in Castañar Cave (roughly 200×10^3 ions per cm^3) revealed an extraordinary degree of air ionisation, ~100–1,000 times greater than in the local open atmosphere. The high density of ions in this cave atmosphere persists throughout the year, in the absence of CH_4. This leads us to hypothesize that the oxidative capacity of underground air is related to its high density of ions, some of which likely are able to react with CH_4.

In light of these preliminary observations, a specific and detailed multiparametric field study was conducted in the cave that had the highest ventilation rate, Ojo Guareña Cave[23]. This underground system was chosen because positive concentrations of CH_4 were always recorded throughout the year, in contrast to other isolated underground sites such as Castañar Cave. In Castañar Cave, CH_4 is hardly detected at all throughout the annual cycle. The field study included the monitoring of gas composition (air ion concentration, CH_4, CO_2 and its carbon isotopic signal) and microclimatic parameters (temperature and relative humidity). Spot air sampling was conducted during a short survey of the cave air, soil air and open atmosphere using the same procedures described above. The density of ions in underground air as measured simultaneously with environmental monitoring and air sampling using a handled air ion counter (AlphaLab Air Ion Counter, UT, USA).

Figure 5 summarizes the spatial variation of air ion concentration and CH_4 content in the underground air of Ojo Guareña Cave. The negative and positive ions have the same concentration range ($5-45 \times 10^3$ ions per cm^3) and an identical spatial distribution. Hence, Fig. 5 includes only the spatial distribution of negative ions (the positive ions show the same distribution). The spatial distributions of other ancillary parameters are displayed in Supplementary Fig. 3, including air temperature, CO_2 concentration and CO_2 carbon isotopic signal, $\delta^{13}C[CO_2]$ (the relative humidity of cave air remains close to a steady state of saturation, 100%, in the entire network of air-sampling points). Radon concentration in the air was continuously monitored in two cave locations with opposite ventilation patterns[23]. The levels of radon observed during the 24-h cycle immediately before the fieldwork campaign ranged from 693 [454–1,033] Bq m^{-3}, at the final stretch of the western gallery near the cave entrance, to 966 [760–1,386] Bq m^{-3} in the dead-end gallery located in the northeast sector of the subterranean system (Fig. 5a).

The minimum air ion concentrations, for both negative and positive ions, were measured close to the cave entrances, with values slightly higher than those observed in the exterior atmosphere. Medium values ($15-25 \times 10^3$ ions per cm^3) were measured in the cave sector between the entrances, where bidirectional air circulation is noticeable throughout the year, accordingly to our continuous monitoring studies[23]. The highest ionisation levels of cave air were found in the eastern dead-end gallery of the cave, coinciding with the highest CO_2 (>2,500 p.p.m.) concentration, the lowest levels of $\delta^{13}C[CO_2]$ (< -23‰) (Supplementary Fig. 3) and the maximum radon activity in the air, which indicates less influence of the aerodynamic processes that control cave air renewal.

The spatial pattern of CH_4 concentration is the opposite of the air ion concentration pattern (Fig. 5d). In accordance with the estimated spatial distribution of the CH_4 content in underground air, ~80% of the cave surface (~7,160 m^2) has a CH_4 concentration below the levels found in soil air (0.89–1.74 p.p.m.), with almost null values in the north-eastern galleries far from the cave entrances. The remaining 20% of the cave surface is located near entrances and has the CH_4 levels close to the local atmospheric background (1.77–1.96 p.p.m.). The atmospheric CH_4 is reduced to ~1 p.p.m. when moving from the accesses to the subterranean system to the highly ventilated sector between both cave entrances.

A linear function fits the data describing CH_4 depletion in the underground air as it becomes increasingly ionized (Fig. 5d). A CH_4 removal of 0.26 p.p.m. in the subsurface atmosphere is observed as the air ion concentrations, both positive and negative,

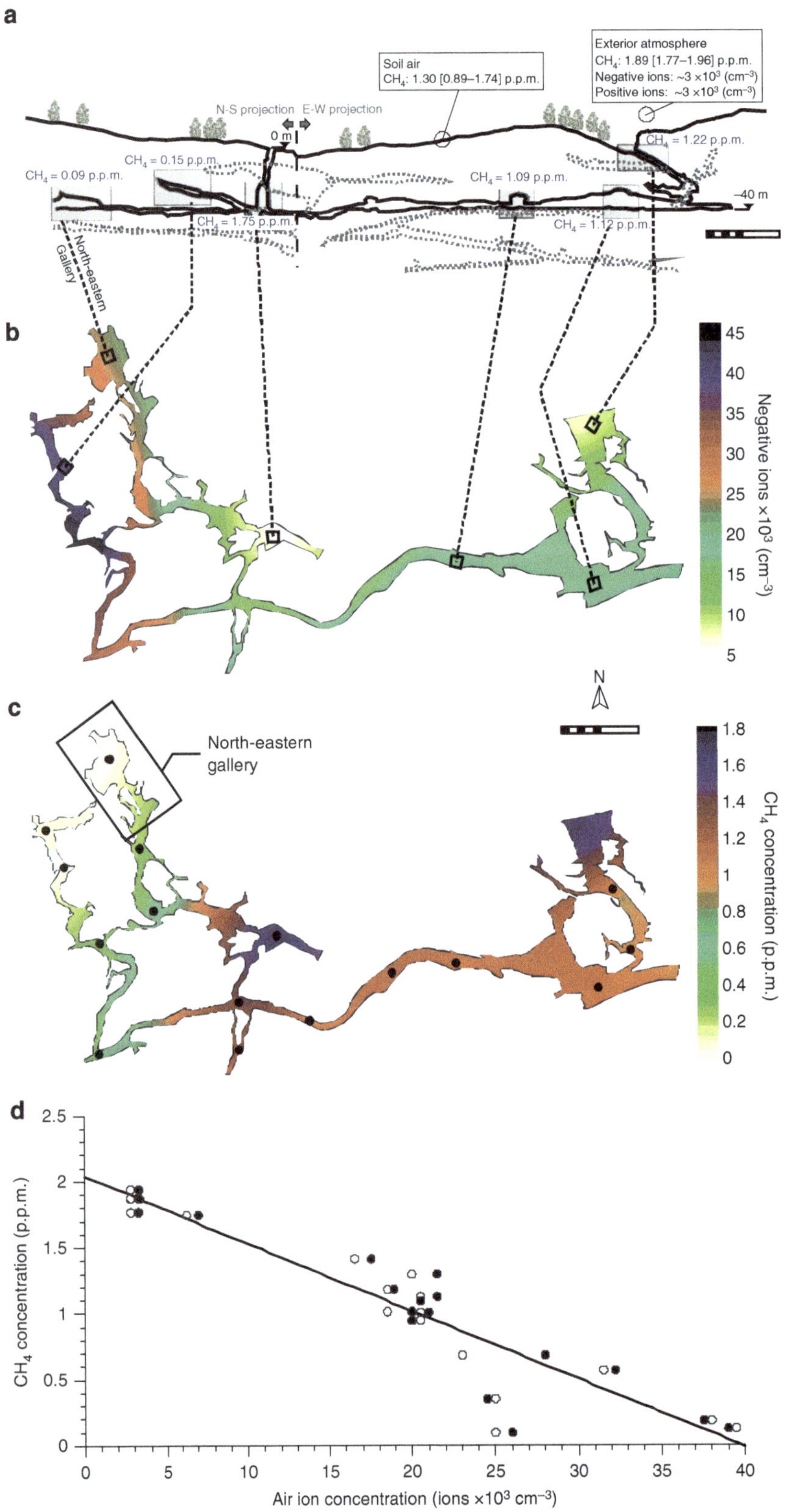

increase to 5×10^3 ions per cm^3. Depletion of local atmospheric CH_4 ranging from 1.17 to 1.36 p.p.m. is expected based on the averaged degree of ionisation of the subsurface atmosphere at the fieldwork site (~ 23.3 ions · per cm^3, for both positive and negative ions).

Some points differ to this linear trend, that is, they show a low CH_4 concentration, below 1 p.p.m., with an intermediate air ion concentration ($\sim 25 \times 10^3$ ions · per cm^3; Fig. 5d). These points correspond to air-sampling points located in a north-eastern gallery connected with the rest of the subterranean system (Fig. 5). This cave sector is under the influence of air exchange processes that cause an influx of underground air that is heavily depleted in CH_4, but less ionized than the bordering galleries with similar minimum CH_4 concentrations. Therefore, this intermediate air ion concentration ($\sim 25 \times 10^3$ ions · per cm^3) seems to be enough to cause CH_4 concentrations below 0.5 p.p.m. The values of the ancillary key parameters (lower air temperature, higher CO_2 content and a reduced $\delta^{13}C[CO_2]$ signal, as shown in Supplementary Fig. 3) confirm the distinctive higher ventilation of this cave sector relative to the surrounding galleries. The northernmost galleries with no air circulation are less influenced by the air renewal processes and that affects their maximum air ion concentration and minimum CH_4 concentration.

Discussion

The CH_4 consumption in underground air is markedly higher than CH_4 depletion in soil air due to methanotrophy. In addition, the CH_4 consumption rate in underground air varies depending on the CH_4 input ratio from the atmosphere, which is determined by the intensity of cave air ventilation represented by the CO_2 levels. Real-time measurements have revealed that consumption of atmospheric CH_4 in subsurface environments is even faster than at the soil surface, within just a few hours. Thus, the CH_4 concentration in underground air varies throughout day, showing a very similar pattern to the carbon isotope $\delta^{13}C[CO_2]$ signal, but presenting an opposite pattern to CO_2 contents. In the case of CO_2, its movement occurs through a combination of advection and diffusion processes, at both daily[17,18] and seasonal[24,25] timescale.

The outer air that enters the cave, mainly during daylight hours, is fully depleted of CH_4 within a few hours, and this underground air reservoir subsequently partially advects upwards and exhales out into the atmosphere on a daily basis. Therefore, the variation in CH_4 abundance reveals itself as a key parameter for assessing the ventilation of subterranean atmospheres, as in the case of other tracer gases (radon). This circadian pattern of both GHGs has been consistently observed over different seasons, and, preferentially, in dynamically ventilated caves during periods of the year with drier conditions. These results are fully consistent with the subsurface environments acting, at a short-term scale, as cleaning filter for atmospheric CH_4 coming from air that directly enters the subsurface, bypassing the edaphic layer via air mass advection (for example, through a cave entrance), far from the influence of biochemical oxidation by soil CH_4 and methanotrophic bacteria.

The main mechanism responsible for the complete consumption of CH_4 in the studied underground environments must be something other than the previously suggested bacterial oxidation mechanism[26,27] because the remaining null CH_4 concentration is not sufficient to satisfy energy requirements for the maintenance and growth of methanotrophic populations. In our MOB tracking, we did not detect methanotrophs in the cave sediments from poorly ventilated subterranean sites. This is consistent with the fact that we found no significant concentrations of CH_4 throughout the year. This suggests that an agent other than methanotrophs should also consume CH_4 at subsurface locations where the gas is absent for long periods. A negative correlation between CH_4 concentration and the density of ions in the cave air has been observed in highly ventilated caves (for example, Ojo Guareña). In this context, charged airborne particles with electrical mobility (primarily, inorganic ions) could play a significant oxidizing role.

The small ions exist all the time in the atmosphere, and the average concentrations of positive and negative small ions are typically $0.2–2.5 \times 10^3\,cm^{-3}$ (ref. 28). However, concentrations up to $45 \times 10^3\,cm^{-3}$ have been observed in the dynamically ventilated caves studied (for example, the Ojo Guareña Cave). Some anomalous high concentrations of ions have previously been reported in other underground locations[29]. During their lifetime of about 100 s, small ions undergo a series of ion-molecule reactions and continuously change their chemical identity[30]. The negative ions have greater affinity for the condensing vapours[31,32], consequently the ion-mediated particle formation mechanisms are expected to be important in locations with high relative humidity[33,34], close to 100% (for example, in caves). The small air ions are usually composed of the central primary ion and surrounding 4–10 molecules of water, depending on humidity, temperature, actual concentration of trace gasses[35] and polarity of the ions.

The main natural sources of small ions are air molecule ionization by radon decay, gamma and cosmic radiation, all of them exhibiting strong spatial and temporal variations[28]. The enhanced radon activity concentration causes the variation in the ionization rate, increasing the small ion concentrations observed near the ground at certain sites, especially where there is an accumulation of radon in the air[29,36–38]. The ionization rate via radon decay is highly sensitive to the elevation from the ground, so it is larger closer to the ground[39].

Figure 5 | Spatial variation of air ion concentration and CH_4 concentration in underground air. Data from a dynamically ventilated cave (Ojo Guareña, Spain). Sampling campaign of 8 October 2014. Length of scale bars is 50 m. (**a**) Detailed cross-sections of cave galleries in relation to surface geomorphology and the main entrances to the subterranean system (solid black line represents the studied galleries; the ends of these galleries continue with the rest of the subterranean system, which is not drawn; and the dotted lines show the locations of other cave levels). Data on air ion and CH_4 concentrations in the atmosphere and in soil air are displayed in relation to the average CH_4 concentration of the main cave sectors (shaded rectangles). (**b**) Spatial distribution of negative ions (both negative and positive ions have the same range of concentration, $5\text{-}45 \times 10^3$ ions per cm^3, and an identical spatial distribution. Thus, only the spatial distribution of the negative ions is displayed to avoid redundancy). (**c**) Spatial distribution of CH_4, indicating a network of points (filled black circles) at which spot sampling of cave air was conducted. CH_4 data come from the spot air sampling during short surveys and the subsequently GHG analyses on lab using the CRDS technique. The spatial maps for each parameter were constructed using geostatistical gridding based on an ordinary point kriging method, with a linear variogram model and without anisotropy constraints during interpolation (distance between the estimated points ranging from 3 to 5 m). The north-eastern gallery connected with the rest of the subterranean system is shown in both maps.
(**d**) Functional relationship between air ion concentration (open circles: negative ions and closed circles: positive ions) and CH_4 concentration in underground air. The continuous black line corresponds to a linear function that best fits the prevailing CH_4 depletion in the underground air, which occurs as the air becomes increasingly ionized ($[CH_4$, p.p.m.$] = -0.051 \times [$ions $\times 10^3$, $cm^{-3}] + 2.04$, with $R^2 = 0.87$). Air ion concentration refers to either positive ions or negative ions.

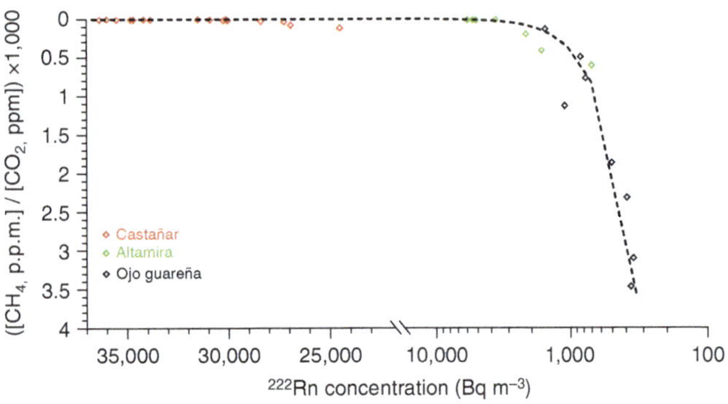

Figure 6 | Variation in the ratio of the concentrations of both greenhouse gases in underground air in relation to radon activity. The ratio of the concentrations of CO_2 and CH_4 is directly proportional to the intensity of ventilation process of the subsurface atmosphere. The coloured points correspond to paired data on the average concentrations of both parameters in the underground air for each field campaign of spot air sampling. Three main caves were considered to display this functional relationship, namely, Castañar, Altamira and Ojo Guareña. Continuous monitoring of radon activity in the underground air was conducted in these caves. The *x* axis for radon concentration in underground air is divided into two parts (with linear and logarithmic scale, respectively) to show the variations of the ratio of the concentrations of CO_2 and CH_4 in a very broad range of radon activity. The continuous black line corresponds to a second-order hyperbola that best fits the functional relationship between the ratio of the concentrations of both the greenhouse gases (CO_2 and CH_4, in p.p.m.) and its radon activity (^{222}Rn in $Bq\,m^{-3}$), $([CH_4]\cdot[CO_2]^{-1})\cdot 1{,}000 = 4.108\times10^5\cdot([^{222}Rn]\ +0.045)^{-1}$, with $R^2 = 0.96$).

Recent indoor experiments showed that determining positive air ion concentration could be an alternative method for evaluating radon activity concentrations[40]. It has been also observed that the source rate of small ions by radon activity is elevated indoors ($\sim$12 ion pairs per $cm^3\,s^{-1}$), with concentrations that follow diurnal cycles affected by ventilation. The highest concentrations are observed at night, when there is little or no ventilation[30,41]. In this regard, caves are suitable locations to storing underground air with high radon activity[42], particularly, when the underground atmosphere has a low air renewal rate (for example, Castañar Cave[16,43]). It is expected that there will be a high density of ions in such environments, and this is borne out by the measurements made in the subterranean sites that were studied. The results derived from our continuous monitoring of radon activity in underground air confirm that CH_4 depletion occurs in underground air as it is enriched in radon (Fig. 6). The ratio of the concentrations of CO_2 and CH_4 is directly proportional to the intensity of ventilation process of the subsurface atmosphere, that is, the degree of cave air renewal through the inlet of air from exterior with the atmospheric levels of both the gases (CO_2 and CH_4). The ratio of the concentrations of CO_2 and CH_4 tends to zero once the radon concentration in underground air exceeds 2,000 $Bq\,m^{-3}$ (Fig. 6).

It is widely known that the photochemical oxidation of CH_4 in the troposphere is largely dominated by its reaction with hydroxyl radicals (OH). Under the majority of environmental conditions, the hydroxyl radical lifetime is less than a second[44]. Unlike the outdoor atmosphere, however, one of the important OH sources in indoor air may be from radioactive ^{222}Rn decay. An earlier study provided basic data for the evaluation of indoor radioactive pollution using chemical amplification techniques[45], quantifying the HOx production rate in indoor air from the radiolysis of water vapour by radon decay on an average rate of $(4.31\pm0.07)\times10^5$ HOx molecules per radon decay per second (Bq) over a range of relative humidity from 3.4 to 55.0% at 22 °C.

The ionic reactions of gaseous CH_4 with its each primary ion, including ions CH_4^+, CH_3^+, CH_2^+, CH^+ and C^+, represent other mechanism of CH_4 removal. An ion-molecule reaction mechanism is implicated in the formation of these upper homologues of CH_4, according to[46,47]:

$$CH_4^+ + CH_4 \rightarrow CH_5^+ + CH_3$$

$$CH_4^+ + CH_4 \rightarrow C_2H_4^+ + 2H_2$$

$$CH_4^+ + CH_4 \rightarrow C_2H_6^+ + H_2.$$

The starting point of this ion-molecule reaction mechanism is the CH_4^+ ion, which also appears under the action of uranium–radon (and its descendants) α-attack of CH_4 (ref. 48). In the case of ion-molecule reactions of O^+ with CH_4, other product ions are generated, such as OH^- and CH_4^+ (ref. 49). Therefore, the above-mentioned reactions for CH_4 removal by oxidation with hydroxyl radicals and ion-molecule reactions provide feedback to the system.

Overall, our findings contribute to a conceptual advance in our understanding of the CH_4 budget by including the subsurface-related CH_4 sink, which has long been largely overlooked. The discovery of frequent (daily), rapid (several hours) and complete CH_4 consumption in subsurface locations makes the uppermost portion of the vadose zone a special site for a CH_4 sink. CH_4 concentrations in underground air below the atmospheric background were found at all points inside the group of caves that was studied. In addition, rapid consumption of atmospheric CH_4 was observed in the underground air of dynamically ventilated caves (for example, Altamira Cave), just after the advective mixing of the cave air with exterior air. Therefore, sub-atmospheric CH_4 concentrations have not only been observed locally but are widespread the atmospheres of all the studied caves. The caves that were studied are distributed across a range of latitudes from 36–43° N to 0–5° W (Supplementary Fig. 1; Supplementary Table 1) and they have different geomorphological properties, different locations relative to the surface and different host-rock discontinuities. Our *in situ* measurements have also documented the link between CH_4 depletion and ion concentration in the subsurface atmosphere; however, the mechanism of CH_4 consumption remains a hypothesis to be tested by laboratory experimentation.

In light of these results, a new line of research could extrapolate to a global scale and analyse the CH_4 sink that has been described here for the subterranean atmosphere. There are early observations regarding CH_4 sinks in cave atmospheres that refer to other latitudes in the southern hemisphere, which are quite different to the locations in the north temperate zone that are described here. The phenomenon described here may thus be relevant on a global scale. CH_4 consumption of 1,000 p.p.b. in only a few hours was described in the subterranean air of an Australian cave, for example[26]. There are not yet enough globally representative data on CH_4 in subsurface atmospheres, so extrapolation of the CH_4-sink model to a global scale still entails much uncertainty. Part of what is estimated at present to be soil uptake may be due to subsurface atmosphere; therefore, the CH_4 sink suggested here is probably at present embedded in the soil uptake term of forward and inverse modelling studies. However, understanding the processes that control the generation, abundance and fate of CH_4 in the atmosphere, oceans and shallow crust is critical for understanding the future of the environment, because CH_4 is a naturally and anthropogenically emitted GHG[50]. Improved knowledge about the oxidation of CH_4 in the subsurface atmosphere under temperature and pressure conditions characteristic of the troposphere near sea level (+15 °C and 103.15 kPa, respectively), could lead to the development, in the medium and long term, of technological alternatives for the removal or storage of GHGs, the identification of new cost-effective ways and methods for mitigation CH_4 emissions and the design of new strategies for atmospheric CH_4 elimination in underground environments. Therefore, different formation, transport and removal processes of CH_4 often impart distinctive isotopic fractionations and abundance of clumped isotopologues[51], which can provide information about the mechanisms involved and determine constraints on the sinks that can help guide future mitigation efforts. Finally, laboratory experiments should be conducted to attempt to reproduce the oxidation of CH_4 under different environmental conditions.

Methods

Environmental monitoring networks at fieldwork sites. Previous studies have demonstrated the availability of a wide set of facilities and devices that are especially designed to control, measure and analyse interesting parameters related to variations in cave environment conditions at each fieldwork site[16,17]. Further descriptions, technical data and results from the monitoring networks are found in these reports. These monitoring networks include sensors and data loggers continuously record local environmental parameters in cave air (pressure, temperature and relative humidity, wind speed and direction, air pressure and CO_2 and ^{222}Rn contents), soil (CO_2, temperature and moisture content) and the exterior atmosphere (rainfall, temperature, relative humidity, atmospheric pressure, solar radiation, wind speed and direction).

GHGs monitoring. A specific sampling and analysis programme for GHGs was implemented for each field site, including: (1) seasonal spot air-sampling and GHG analyses during short surveys at all fieldwork sites and (2) intensive field campaigns conducted for 38 h in the Castañar and Altamira caves, to obtain daily cycles of the GHG mixing ratio through comprehensive real-time logging using CRDS.

Spot air sampling was conducted seasonally in a predefined network of points spatially distributed inside each cave and during several short surveys (<3 h long). In total, 1,037 air samples were analysed for the seven caves studied, including 549 samples from cave air, 281 samples from soil air and 207 samples from atmosphere. Sampling was duplicated (morning/afternoon) in the most accessible and smaller caves. Exterior and soil air were sampled at fixed locations throughout the day, and, preferably, during maximum (midday) and minimum photosynthetic activity (early morning). Cave and atmosphere samples were taken 1 m above from the floor. Soil air collection was conducted at sites located vertically above each cave using a steel tube with grooved sides at its ends. The tubes were inserted to a depth of 20–50 cm near the bedrock–soil interface. Soil air was extracted using a micro-diaphragm gas pump (KNF Neuberger, Freiburg, Germany) at 3.1 l min^{-1} at atmospheric pressure, and the pump was equipped with a 0.2-μm filter to avoid contamination with airborne particles. A portable air compressor (AQUANIC s790) was run at 0.4 l min^{-1} for the cave air and atmosphere samples. Air samples were collected into 1L Tedlar bags with lock valves, especially designed to ensure the inertness and impermeability of air samples. Each air sample was purged at least one time with air from each sampling point to avoid contamination. Furthermore, the operator remained at a sufficient distance from the portable air compressor during the short interval needed for filling each 1L Tedlar bag, thereby avoiding any contamination by human breathing. The Picarro analyser works with a sample flow rate below 0.05 s.l.m. (standard litre per minute, at 760 Torr) and it is only necessary 10 min for achieving a reliable measurement of the three parameters (CO_2, CH_4 and $\delta^{13}CO_2$), therefore <50 s of pumping is enough to provide a suitable volume of air (both from cave and exterior atmospheres) for subsequent analysis using the CRDS technique. Less than 20 s of pumping was enough to provide a suitable volume of soil air for subsequent analysis using the CRDS technique. Thus, we ensured that no air was taken in from the atmosphere. The bag samples were analysed within 48 h of sampling to determine CO_2 and CH_4 concentrations and the isotopic $\delta^{13}CO_2$ signal using a laser-based analyser that employs CRDS technology[52], as described below.

In situ and real-time measurements of CO_2, CH_4 and $\delta^{13}CO_2$ were carried out during an intensive field campaign of 38 h using a field-deployable CRDS analyser model G2101-i (Picarro Inc., Santa Clara, CA, USA). This fieldwork aimed to provide detailed records of GHG variations on short-term and diurnal timescales. Continuous monitoring of GHGs was conducted in the Altamira cave, for which a previous complete environmental database regarding cave processes exists. The field campaign was deployed during the warm, dry period at the end of the summer, when the cave is most tightly coupled with the external atmosphere and ventilation is at its most effective[24,25]. A detailed time series of GHGs content was obtained (>1,18,000 records with data of $^{12}CO_2$ and $^{13}CO_2$ contents every 2 s), covering a monitoring period of 38 h (two night-time periods and one daytime period, from day 19 until 21 September 2011). The $\delta^{13}C$–CO_2 value was automatically calculated using the CRDS analyser by the raw measurements of $^{12}CO_2$ and $^{13}CO_2$ concentration and by taking a central moving average with a window of 5 min, which provides suitable precision according to the manufacturer's specifications. In addition, a comprehensive network of monitoring stations was specifically implemented to register the coeval short-term environmental changes, including parameters such as air temperature, relative humidity, air pressure and radon contents. The CRDS analyser was operated in continuous mode with a network of tubes connected to a multivalve sequencer, which enabled simultaneous air sampling from different locations (cave, soil and atmosphere) for >10 min each time by alternating the opening and closing of valves. Air from each location was pumped using a 4-mm (inner diameter) polyurethane tube from the main galleries of each cave site and using a micro-diaphragm gas pump as described above. An air extraction tube was connected to the analyser using a T-connection to avoid problems related to overpressure effects. As an alternative to continuous gas monitoring, sampling bags of exterior and soil air were also collected throughout the day.

Analyses of GHGs using both the procedures (spot-sampling bags and continuous real-time monitoring) were performed using a CRDS analyser Picarro G2101-i (Picarro Inc.). This analyser performs *in situ* and real-time measurements of atmospheric levels of carbon dioxide ($^{12}CO_2$ and $^{13}CO_2$), CH_4 and water vapour (H_2O_v). The isotopic $\delta^{13}CO_2$ relationship is automatically measured as a calculation based on $^{12}CO_2$ and $^{13}CO_2$ concentrations and considering a correction due the presence of other target species (CH_4 and H_20_v). Precisions of 200 and 10 p.p.b. are guaranteed by the manufacturer for $^{12}CO_2$ and $^{13}CO_2$, respectively, over a wide operating range, with a resulting precision of <0.3‰ for $\delta^{13}C(CO_2)$ after 5 min of analysis. The maximum drift of $\delta^{13}C(CO_2)$ is <2.0‰ over 24 h at standard air pressure (peak to peak, 1 h average). Periodic tests were conducted to check that the analyser achieved the above performance specifications (precision and drift) in accordance with the guidelines of the manufacturer (test for 24 uninterrupted hours using the analyser attached to a cylinder of compressed air with known GHG abundances). Measurements of CO_2 in air were performed relative to three reference standards whose gas mixing ratio was determined with high precision and accuracy. On each day/session of analysis using the CRDS analyser (field deployment or analysis of air bags in the lab), the internal secondary standards were run regularly, at least at the beginning and at the end of the analyses, to verify the proper functioning of the analyser. This procedure periodically confirmed that the performance specifications regarding CO_2 analyses via CRDS were met. Regarding CH_4 measurements, the instrument's response to the sample was also compared with three standards with known CH_4 contents. The measurement accuracy determined through repeated analysis of CH_4 in standard gas cylinders was <0.1 p.p.m. in the studied range of levels (0–2 p.p.m.). In addition, CH_4 measurements was corrected as a function of the presence of H_20_v through periodical calibrations of the analyser using synthetic gases with known CH_4 abundances (0, 0.5 and 1.7 p.p.m.v.) over a wide range of H_20_v levels (0–3.5%).

In addition, some initial spot measurements of air ion density were conducted in Castañar cave to test the range of values in underground air with null CH_4 concentration. A handled air ion counter was used (AlphaLab Air Ion Counter). This instrument is a true ion density metre, based on a Gerdien Tube (Gerdien Condenser) design, and it contains a fan that pulls air through the metre at a calibrated rate. The Air Ion Counter is designed to measure density of ions, that is, the number of ions per cubic centimetre in air, accomplishing the following performance specifications: range: 0–2 × 10^7 ions per m^3; resolution: 10 ions per m^3; accuracy: ±25% of the reading; and response time: 2 s. It also measures this number separately for positive and negative ions.

In light of these preliminary observations, a detailed multiparametric and spatially resolved monitoring of air ions density was conducted during a specific field survey in Ojo Guareña cave, in tandem with the analysis of gas composition (CH_4, CO_2 and $\delta^{13}C$–CO_2) and microclimatic parameters (temperature and relative humidity) in the underground air, the soil air and the external atmosphere.

For this study, continuous monitoring (every half hour) of changes in the radon concentration in the underground air was conducted at three cave sites: Castañar, Altamira and Ojo Guareña. Several Radim 5WP monitors (SMM Company, Prague, Czech Republic and distributed by GT-Analytic KG Innsbruck, Austria) were used, because these monitors are mainly designed for indoor radon measurements. These monitors determine radon concentration by measuring gross α-activity of the radon decay products ^{218}Po and ^{214}Po, which are collected electrostatically on the surface of a Si-semiconductor. The lowest activity detectable is 80 Bq m^{-3} for 1-h measurements with a statistical error equal to ±20%. The maximum detectable activity is 150 kBq m^{-3}. The instrument response is 0.4 (imp h^{-1})/(Bq m^{-3}). The monitors were periodically calibrated by the manufacturer. In addition, the monitors were checked regularly in the field (inside the caves) using a Pylon AB5 scintillator photomultiplier with a continuous passive radon detector (Pylon Electronics Inc., Ottawa, Canada) as a reference instrument. This reference instrument was periodically calibrated to provide highly accurate measurements (lowest detectable activity: 24.8 Bq m^{-3}, with a sensitivity of 0.041 c.p.m. (Bq m^{-3})$^{-1}$, accuracy of ±4% and nominal background of 0.4 c.p.m.).

MOBs tracking. Microbiological sampling was conducted to assess the presence or absence of MOBs in the edaphic profile at exterior and sediments and rock surfaces into cave from each monitored field site. Inside caves, samples were collected in the cave entrance and in the most remote but accessible locations. Sampling was carried out using sterilized materials. On collection, the samples were stored on ice and processed on arrival to the laboratory. Collected samples were analysed by an ordered set of techniques: nucleic acid extraction and PCR amplification, cloning and sequencing and sequence analyses[53].

Nucleic acid extraction and PCR amplification. Extraction of total nucleic acids from the samples was performed using the method proposed by Griffiths *et al.*[54] in conjunction with a FastPrep instrument, according to the manufacturer's instructions. Extracted nucleic acids were resuspended in 50 ml of sterile ultrapure water (Sigma-Aldrich). For RNA extraction, a 25-ml aliquot was subjected to a DNase digestion step using a RNase-Free DNase (Qiagen), and the purification of RNA using the RNeasy MinElute Cleanup kit (Qiagen), according to the manufacturer's instructions. Superscript II Reverse Transcriptase (Invitrogen) was used to synthesize complementary DNA (cDNA) using specific primers: Mb661 (5′-CCG GMG CAA CGT CYT TAC C-3′) and A682 (5′-GAA SGC NGA GAA GAA SGC-3′) for bacterial *pmoA* gene[55,56] and 1492R (5′-GGY TAC CTT GTK WCG ACT T-3′) for bacterial 16S rRNA gene[57] at a temperature of 42 °C for 50 min. Amplification of cDNA will be performed with the methanotrophs-specific primer pairs: A189F (5′-GGN GAC TGG GAC TTC TGG-3′)/Mb661, and A189F/A682 (ref 55), and a universal bacterial primer: 27F (5′-AGA GTT TGA TCC TGG CTC AG-3′)/1492R (ref 56). One or 2 ml of cDNA, diluted 1/10 was used as a template. PCR conditions were the same for DNA. PCR reactions were performed in a Bio-Rad iCycler thermal cycler (Bio-Rad). The PCR reaction mixture (1 ml) contains 100 ml of PCR buffer (Biotools), 30 ml of MgCl2 (50 mM, Biotools), 10 ml of each primer (50 mM), 100 ml of dNTPs (100 mM, Bioline), 5 ml of Taq polymerase (Biotools) and 750 ml of sterile ultrapure water. The thermal cycler for PCR amplification consisted of 94 °C for 4 min; 28 cycles of 94 °C for 1 min, 56 °C for 1 min (45 °C for 27F/1492R), 72 °C for 45 s; and a final step of 72 °C for 5 min. Positive and negative controls were included in all amplification experiments. All amplification products were purified using a JETquick PCR Purification Spin kit (Genomed) and stored at − 20 °C for further analysis.

Cloning and sequencing. DNA libraries of PCR amplified products were constructed using the pGEM-T Easy Vector (Promega) and then transformed into One Shot Max Efficiency DH5a-T1 chemically competent *Escherichia coli* (Invitrogen) according to the manufacturer' instructions. Transformants were randomly picked and transferred to other plates containing Luria–Bertani medium supplemented with 100 mg ml^{-1} ampicillin and 15% w/v glycerol and stored at − 80 °C. The amplicons that contain the correct size of *pmoA* and 16S rRNA genes fragment were re-amplified using the primer pairs as described above. On average, 100 clones from each library were sequenced at Macrogen Inc., Korea.

Sequence analyses. Low-quality (short) sequences were removed from the data set. Sequences were aligned using the software package _other[58]. Sequences were checked for chimera by chimera.slayer as implemented in _other. Putative chimeric sequences were excluded from further analysis. After this analysis, all sequences were compared with the non-redundant database of sequences deposited at the National Center for Biotechnology and EzTaxon[59] using BLASTN algorithm[60]. Aligned sequences were clustered into operational taxonomic units using _other, with a 97% sequence identity cutoff for 16S rDNA sequence and 93% for *pmoA* sequence[61]. Rarefaction curves and Venn diagrams were also obtained using _other.

References

1. Dlugokencky, E. J. *et al.* Observational constraints on recent increases in the atmospheric CH_4 burden. *Geophys. Res. Lett.* **36,** L18803 (2009).
2. Bousquet, P. *et al.* Contribution of anthropogenic and natural sources to atmospheric methane variability. *Nature* **443,** 439–443 (2006).
3. Chen, Y. H. & Prinn, R. G. Estimation of atmospheric methane emissions between 1996 and 2001 using a three-dimensional global chemical transport model. *J. Geophys. Res.* **111,** D10307 (2006).
4. Khalil, M. A. K., Butenhoff, C. L. & Rasmussen, R. A. Atmospheric methane: trends and cycles of sources and sinks. *Environ. Sci. Technol.* **41,** 2131–2137 (2007).
5. Heimann, M. Atmospheric science: enigma of the recent methane budget. *Nature* **476,** 157–158 (2011).
6. Dlugokencky, E. J., Nisbet, E. G., Fisher, R. & Lowry, D. Global atmospheric methane: budget, changes and dangers. *Philos. Trans. A Math. Phys. Eng. Sci.* **369,** 2058–2072 (2011).
7. Hartmann, D. L., Klein, A. M. G. & Rusticucci, M. in *Climate Change 2013: The Physical Science Basis.*. (eds Stocker, T. F. *et al.*) 159–254 (Cambridge Univ. Press, 2013).
8. Ciais, P. & Sabine, C. in *Climate Change 2013: The Physical Science Basis* (eds Stocker, T. F. *et al.*) 465–570 (Cambridge Univ. Press, 2013).
9. Lelieveld, J., Crutzen, P. J. & Dentener, F. J. Changing concentration, lifetime and climate forcing of atmospheric methane. *Tellus* **50B,** 128–150 (1998).
10. Myhre, G. & Shindell, D. in *Climate Change 2013: The Physical Science Basis.*. (eds Stocker, T. F. *et al.*) 659–740 (Cambridge Univ. Press, 2013).
11. Fisher, R. E. *et al.* Arctic methane sources: Isotopic evidence for atmospheric inputs. *Geophys. Res. Lett.* **38,** L21803 (2011).
12. Montzka, S. A., Dlugokencky, E. J. & Butler, J. H. Non-CO_2 greenhouse gases and climate change. *Nature* **476,** 43–50 (2011).
13. Smith, R. N. B. *Subsurface, surface and boundary layer processes.* 1–54. Unified Model Documentation Paper 24 (NCAS-CMS Climate Research Meteorological Office) (1996).
14. Atkinson, T. C. CO_2 in the atmosphere of the unsaturated zone: an important control of groundwater hardness in limestones. *J. Hydrol.* **35,** 111–123 (1977).
15. Serrano-Ortiz, P. *et al.* Hidden, abiotic CO_2 flows and gaseous reservoirs in the terrestrial carbon cycle: review and perspectives. *Agric. For. Meteorol* **150,** 321–329 (2010).
16. Fernandez-Cortes, A. *et al.* Characterization of trace gases' fluctuations on a 'low energy' cave (Castañar de Ibor, Spain) using techniques of entropy of curves. *Int. J. Climatol.* **31,** 127–143 (2011).
17. Cuezva, S. *et al.* Short-term CO_2(g) exchange between a shallow karstic cavity and the external atmosphere during summer: role of the surface soil layer. *Atmos. Environ.* **45,** 1418–1427 (2011).
18. Garcia-Anton, E., Cuezva, S., Fernandez-Cortes, A., Benavente, D. & Sanchez-Moral, S. Main drivers of diffusive and advective processes of CO_2-gas exchange between a shallow vadose zone and the atmosphere. *Int. J. Greenh. Gas Control* **21,** 113–129 (2013).
19. Bourges, F., Genthon, P., Genty, D., Mangin, A. & D'Hulst, D. Comment on "Carbon uptake by karsts in the Houzhai Basin, southwest China" by Junhua Yan *et al. J. Geophys. Res.* **117,** 1–3 (2012).
20. Holmes, A. J. *et al.* Characterization of methanotrophic bacterial populations soils showing atmospheric methane uptake. *Appl. Environ. Microbiol.* **65,** 3312–3318 (1999).
21. Ek, C. & Gewelt, M. Carbon-dioxide in cave atmospheres. New results in Belgium and comparison with some other countries. *Earth Surf. Proc. Land.* **10,** 173–187 (1985).
22. Bourges, F., Mangin, A. & D'Hulst, D. Le gaz carbonique dans la dynamique de l'atmosphère des cavités karstique: l'exemple de l'Aven d'Orgnac (Ardèche). *Earth Planet. Sci. Lett.* **333,** 685–692 (2001).
23. Fernandez-Cortes, A. *et al.*in *Hydrogeological and Environmental Investigations in Karst Systems, Environmental Earth Sciences.* (eds Andreo, B. *et al.*) 523–531 (Springer-Verlag, 2015).
24. Sanchez-Moral, S. *et al.* Inorganic deterioration affecting the Altamira Cave, N Spain: quantitative approach to wall-corrosion (solutional etching) processes induced by visitors. *Sci. Total Environ.* **243/244,** 67–84 (1999).
25. Kowalski, A. S. *et al.* Can flux tower research neglect geochemical CO2 exchange? *Agric. For. Meteorol.* **148,** 1045–1054 (2008).
26. Waring, C. L., Griffith, D. W. T., Wilson, S. & Hurry, S. Cave atmosphere: A guide to calcification and a methane sink. *Geochim. Cosmochim. Acta* **73,** A1419–A1419 (2009).
27. Mattey, D. P. *et al.* Methane in underground air in Gibraltar karst. *Earth Planet. Sci. Lett.* **374,** 71–80 (2013).
28. Hirsikko, A. *et al.* Atmospheric ions and nucleation: a review of observations. *Atmos. Chem. Phys.* **11,** 767–798 (2011).
29. Sas, D., Sladek, P. & Navratil, O. Radon and air ion balance. *Czech. J. Phys.* **56,** D141–D148 (2006).
30. Hirsikko, A. *et al.* Indoor and outdoor air ions and aerosol particles in the urban atmosphere of Helsinki: characteristics, sources and formation. *Boreal Environ. Res.* **12,** 295–310 (2007).
31. Junninen, H. *et al.* Observations on nocturnal growth of atmospheric clusters. *Tellus* **60B,** 365–371 (2008).
32. Winkler, P. M. *et al.* Heterogenous nucleation experiments bridging the scale form molecular ion cluster to nanoparticles. *Science* **7,** 1374–1377 (2008).
33. Laakso, L., Mäkelä, J. M., Pirjola, L. & Kulmala, M. Model studies on ion-induced nucleation in the atmosphere. *J. Geophys. Res.* **107,** 4427 (2002).
34. Yu, F. Ion-mediated nucleation in the atmosphere: Key controlling parameters, implications, and look-up table. *J. Geophys. Res.* **115,** D03206 (2010).
35. Luts, A. & Parts, T. Evolution of negative small air ions at two different temperatures. *J. Atmos. Sol. Terr. Phys.* **64,** 763–774 (2002).
36. Dhanorkar, S. & Kamra, A. K. Diurnal variation of ionization rate close to ground. *J. Geophys. Res.* **99,** 18523–18526 (1994).
37. Nagaraja, K., Prasad, B. S. N., Madhava, M. S. & Paramesh, L. Concentration of radon and its progeny near the surface of the earth at a continental station Pune (18_ N, 74_ E). *Ind. J. Pure Appl. Phys.* **41,** 562–569 (2003).
38. Horrak, U., Salm, J. & Tammet, H. Diurnal variation in the concentration of air ions of different mobility classes in a rural area. *J. Geophys. Res.* **108,** 4653 (2003).
39. Tammet, H., Hõrrak, U., Laakso, L. & Kulmala, M. Factors of air ion balance in a coniferous forest according to measurements in Hyytiälä, Finland. *Atmos. Chem. Phys.* **6,** 3377–3390 (2006).
40. Kolarz, P. M., Filipovic, D. M. & Marinkovic, B. P. Daily variations of indoor air-ion and radon concentrations. *Appl. Radiat. Isot.* **67,** 2062–2067 (2009).

41. Fews, A. P., Holden, N. K., Keitch, P. A. & Henshaw, D. L. A novel high-resolution small ion spectrometer to study ion nucleation of aerosols in ambient indoor and outdoor air. *Atmos. Res.* **76,** 29–48 (2005).
42. Cigna, A. A. Radon in caves. *Int. J. Speleol.* **34,** 1–18 (2005).
43. Lario, J., Sanchez-Moral, S., Cuezva, S., Taborda, M. & Soler, V. High ^{222}Rn levels in a show cave (Castañar de Ibor, Spain): proposal and application of management measures to minimize the effects on guides and visitors. *Atmos. Environ.* **40,** 7395–7400 (2006).
44. Monks, P. S. Gas-phase radical chemistry in the troposphere. *Chem. Soc. Rev.* **34,** 376–395 (2005).
45. Ding, H. & Hopke, P. K. HOx production due to radon decay in air. *J. Atmos. Chem.* **17,** 375–390 (1993).
46. Haynes, R. M. & Kebarle, P. Mass-spectrometric study of ions at near-atmospheric pressure. The alpha radiolysis of methane. *J. Chem. Phys.* **45,** 3899–3906 (1966).
47. Rasul, G., Surya Prakash, G. K. & Olah, G. A. Comparison of structures and energies of CH52 + with CH4 + and their possible role in superacidic methane activation. *Proc. Nat Acad. Sci. USA* **94,** 11159–11162 (1997).
48. Cosma, C., Suciu, I., Jantschi, L. & Bolboaca, S. Ion-molecule reactions and chemical composition of emanated from herculane spa geothermal sources. *Int. J. Mol. Sci.* **9,** 1024–1033 (2008).
49. Ottinger, C. H., Zimmerann, S. & Muschlitz, E. E. Ion-molecule reactions of O+ with CH_4. *Chem. Phys.* **81,** 211–220 (1983).
50. Wuebbles, D. J. & Hayhoe, K. Atmospheric methane and global change. *Earth Sci. Rev.* **57,** 177–210 (2002).
51. Stolper, D. A. *et al.* Combined 13C–D and D–D clumping in methane: methods and preliminary results. *Geochim. Cosmochim. Acta* **126,** 169–191 (2014).
52. Crosson, E. R. A cavity ring-down analyzer for measuring atmospheric levels of methane, carbon dioxide, and water vapor. *Appl. Phys. B* **92,** 403–408 (2008).
53. Diaz-Herraiz, M. *et al.* Deterioration of an Etruscan tomb by bacteria from the order *Rhizobiales*. *Sci. Rep.* **4,** 3610 (2014).
54. Griffiths, R. I., Whiteley, A. S., O'Donnell, A. G. & Bailey, M. J. Rapid method for coextraction of DNA and RNA from natural environments for analysis of ribosomal DNA- and rRNA-based microbial community composition. *Appl. Environ. Microbiol.* **66,** 5488–5491 (2000).
55. Costello, A. M. & Lidstrøm, M. E. Molecular characterization of functional and phylogenetic genes from natural populations of methanotrophs in lake sediments. *Appl. Environ. Microbiol.* **65,** 5066–5074 (1999).
56. Holmes, A. J., Costello, A. M., Lidstrom, M. E. & Murrell, J. C. Evidence that particulate methane monooxygenase and ammonia monooxygenase may be evolutionarily related. *FEMS Microbiol. Lett.* **132,** 203–208 (1995).
57. Pasic, L., Kovce, B., Sket, B. & Herzog-Velikonja, B. Diversity of microbial communities colonizing the walls of a Karstic cave in Slovenia. *FEMS Microbiol. Ecol.* **71,** 50–60 (2010).
58. Schloss, P. D. *et al.* Introducing mothur: Open-source, platform-independent, community-supported software for describing and comparing microbial communities. *Appl. Environ. Microbiol.* **75,** 7537–7541 (2009).
59. Kim, O. S. *et al.* Introducing EzTaxon-e: a prokaryotic 16S rRNA gene sequence database with phylotypes that represent uncultured species. *Int. J. Syst. Evol. Microbiol.* **62,** 716–721 (2012).
60. Altschul, S. F., Gish, W., Miller, W., Myers, E. W. & Lipman, D. J. Basic local alignment search tool. *J. Mol. Biol.* **215,** 403–410 (1990).
61. Degelmann, D. M., Borken, W., Drake, H. L. & Kolb, S. Different atmospheric methane-oxidizing communities in European beech and Norway spruce soils. *Appl. Environ. Microbiol.* **76,** 3228–3235 (2010).
62. Dlugokencky, E. J., Lang, P. M., Crotwell, A. M., Masarie, K. A. & Crotwell, M. J. *Atmospheric Methane Dry Air Mole Fractions from the NOAA ESRL Carbon Cycle Cooperative Global Air Sampling Network, 1983–2012* ftp://aftp.cmdl.noaa.gov/data/trace_gases/ch4/flask/surface/ (2013).
63. Dlugokencky, E. J., Lang, P. M., Masarie, K. A., Crotwell, A. M. & Crotwell, M. J. *Atmospheric Carbon Dioxide Dry Air Mole Fractions from the NOAA ESRL Carbon Cycle Cooperative Global Air Sampling Network, 1968–2012* ftp://aftp.cmdl.noaa.gov/data/trace_gases/co2/flask/surface/ (2013).

Acknowledgements

This research was funded by the Spanish Ministry of Economy and Competitiveness project CGL2013-43324-R and its programme Torres Quevedo (PTQ 13-06296 and PTQ 12-05601). Funding was also provided by the People Programme (Marie Curie Actions—Intra-European Fellowships, call 2013) of the European Union's Seventh Framework Programme (FP7/2007-2013) under REA grant agreement n°624204 and by CSIC funds (PIE project 201230E125). We thank cave managers for their collaboration throughout the entire investigation.

Author contributions

All authors contributed extensively to the work presented in this paper. A.F-C and S.C. have contributed equally to the work, oversaw all fieldwork, designed the research and edited the manuscript. M.A-G., E.G-A, C.P. and D.B. participated in the collection and analysis of samples and data. V.J. and C.S-J. performed the microbiological analyses and the interpretation of microbiological data. S.S-M. conceived and supervised the study. All authors have discussed the results and implications and commented on the manuscript at all stages.

Additional information

Competing financial interests: The authors declare no competing financial interests.

14

Bidecadal North Atlantic ocean circulation variability controlled by timing of volcanic eruptions

Didier Swingedouw[1], Pablo Ortega[2], Juliette Mignot[2,3,4], Eric Guilyardi[2,5], Valérie Masson-Delmotte[6], Paul G. Butler[7], Myriam Khodri[2] & Roland Séférian[8]

While bidecadal climate variability has been evidenced in several North Atlantic paleoclimate records, its drivers remain poorly understood. Here we show that the subset of CMIP5 historical climate simulations that produce such bidecadal variability exhibits a robust synchronization, with a maximum in Atlantic Meridional Overturning Circulation (AMOC) 15 years after the 1963 Agung eruption. The mechanisms at play involve salinity advection from the Arctic and explain the timing of Great Salinity Anomalies observed in the 1970s and the 1990s. Simulations, as well as Greenland and Iceland paleoclimate records, indicate that coherent bidecadal cycles were excited following five Agung-like volcanic eruptions of the last millennium. Climate simulations and a conceptual model reveal that destructive interference caused by the Pinatubo 1991 eruption may have damped the observed decreasing trend of the AMOC in the 2000s. Our results imply a long-lasting climatic impact and predictability following the next Agung-like eruption.

[1] Environnements et Paléoenvironnements Océaniques et Continentaux (EPOC), UMR CNRS 5805 EPOC—OASU—Université de Bordeaux, Allée Geoffroy Saint-Hilaire, Pessac 33615, France. [2] LOCEAN/IPSL Sorbonne Universités (UPMC, Univ Paris 06)-CNRS-IRD-MNHN, 4 place Jussieu, Paris F-75005, France. [3] Climate and Environmental Physics, Physics Institute, University of Bern, Falkenplatz 16, 3012 Bern, Switzerland. [4] Oeschger Centre of. Climate Change Research, University of Bern, Falkenplatz 16, Bern 3012, Switzerland. [5] NCAS-Climate, Univeristy of Reading, Reading RG6 6BB, UK. [6] Laboratoire des Sciences du Climat et de l'Environnement (Institut Pierre Simon Laplace, CEA-CNRS-UVSQ, UMR8212), 91191 Gif-sur-Yvette, France. [7] School of Ocean Sciences, Bangor University, Menai Bridge, Anglesey LL59 5AB, UK. [8] Centre National de Recherches Météorologiques-Groupe d'Etude de l'Atmosphère Météorologique/Groupe de Météorologie de Grande Echelle et Climat, Toulouse 31100, France. Correspondence and requests for materials should be addressed to D.S. (email: didier.swingedouw@u-bordeaux1.fr).

The Atlantic Meridional Overturning Circulation (AMOC) plays a key role in the meridional heat transport, and in heat and carbon storage in the ocean[1]. Changes in the AMOC affect surface oceanic conditions in the North Atlantic (salinity, temperature and sea level), with impacts on marine ecosystems and regional climate (for example, Greenland glaciers[2]). Understanding the mechanisms driving AMOC decadal variability is therefore critical for climate predictability, particularly in the Northern Hemisphere[3–5]. In response to increased greenhouse gas concentrations, climate models project a gradual AMOC slowdown during the 21st century[6,7]. AMOC strength has only recently been monitored through observation networks. Existing data sets do not reveal a strong trend in the 2000s[8,9], although a declining tendency appears from the early 2010s[10,11]. This is also supported by ocean reanalyses[12,13], which depict an AMOC maximum in the 1990s at subpolar latitudes, followed by a decrease and stabilization in the 2000s. Beyond direct information on AMOC, North Atlantic observations[14,15] and proxy records indicate a 20-year preferential variability in this region in the atmosphere[16], sea ice[17] and the ocean[18,19]. Such variability can be associated with the dynamics of subpolar gyre, whose characteristic decadal timescales are associated with advection processes and the size of the gyre[20,21].

Indeed, such advective processes have been observed during Great Salinity Anomaly (GSA) events[22–24]. For instance, a salinity anomaly was first identified in the Nordic Seas in 1968, and then detected in the Labrador Sea around 1971. This anomaly was monitored along its propagation within the subpolar gyre during the following 7 years, and it reached the eastern part of the Nordic Seas in 1978 (ref. 22).

Explosive volcanic eruptions have a short-lived but strong radiative impact through the loading of a large amount of sulphate aerosols into the stratosphere, which leads to a cooling of the Earth's surface during the 2–3 years following the onset of the eruption[25,26]. Moreover, the volcanic sulphate aerosol injection causes a significant stratospheric warming in the tropical band due to long-wave heat absorption. The subsequent strengthening of the meridional atmospheric temperature gradient leads to intensified zonal winds and jets through thermal wind balance. As a result, observations show that volcanic eruptions trigger dynamical changes[27,28], with a tendency towards a positive phase of the winter North Atlantic Oscillation (NAO, first mode of atmospheric variability in the north Atlantic sector[29]) during the few years following the eruption[27,30,31]. This dynamical mechanism is not well represented in most climate simulations, at least partly due to a too coarse vertical resolution of their atmospheric model component[27]. In addition, climate model analysis suggests that the volcanic-driven, short-lived cooling of the upper ocean can induce longer-lived changes on North Atlantic climate, notably through its influence on the AMOC[32–34]. The robustness of the mechanisms at play have, however, been challenged by differences in the simulated timing and response, suggesting a sensitivity of the results to the model used or to the prescribed volcanic forcing. Oceanographic data have not yet been used to evaluate the exact oceanic processes at play.

Here we evaluate the potential impact of moderate explosive volcanic eruptions (similar to Agung or Pinatubo) on the North Atlantic bidecadal preferential variability. For this purpose, we use available outputs from different climate models using the Coupled Model Intercomparison Project Phase 5 (CMIP5) database[35] complemented by additional simulations performed using one model, and *in situ* recent oceanic observations as well as longer paleoclimate proxy records of the last millennium. We find that moderate volcanic eruptions may reset a 20-year intrinsic variability mode in the North Atlantic both in model simulations as well as in the data analyzed, leading to interference patterns over the recent period and in the near future.

Results

Analysis of the CMIP5 database. We investigate the mechanisms involved in bidecadal North Atlantic and AMOC variations and reduced AMOC variability around the trend during the 2000s using CMIP5 simulations and sensitivity tests conducted specifically with the IPSL-CM5A-LR model[36]. Historical climate simulations (1870–2005) driven by natural and anthropogenic forcing archived in the CMIP5 database exhibit a large spread in the simulated AMOC at 48°N (Fig. 1a). Out of 19 available model simulations, we select the subset of 9 (8 + IPSL-CM5A-LR called 'Bi-Dec' ensemble) which exhibit peaks of spectral energy in the

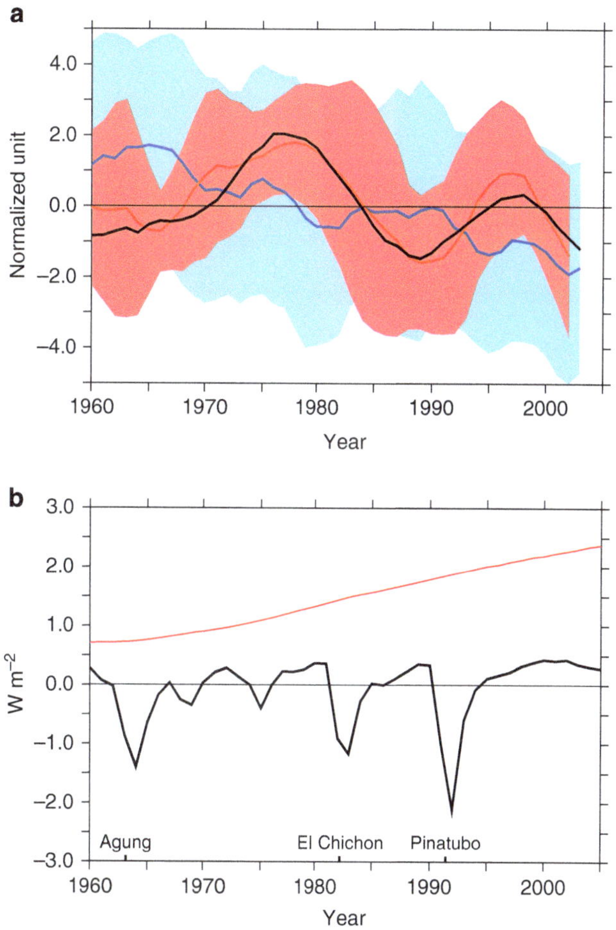

Figure 1 | Simulated AMOC changes and radiative external forcing. (**a**) Variations of the AMOC maximum at 48°N over the period 1960-2005 for the ensemble of five historical simulations performed with the IPSL-CM5A-LR model (black); for the Bi-Dec ensemble excluding the IPSL-CM5A-LR simulation (subset of eight CMIP5 models, which also exhibit variability in the 10–30 years spectral band, in red, see Methods, Supplementary Table 1 and Supplementary Fig. 1); for the ensemble mean of the 10 other CMIP5 historical simulations (in blue). The standard deviation (s.d.) of the two CMIP5 ensembles is shown with the red and blue envelopes. All the AMOC indices have been normalized with the s.d. of the detrended time series over the period 1850-2005. (**b**) External radiative forcing (in $W m^{-2}$) computed in the IPSL-CM5A-LR historical simulation. The black curve is the natural forcing including solar and volcanic eruptions and the red curve represents the anthropogenic forcing including greenhouse gas changes and the anthropogenic aerosols effects. A 5-year running mean has been applied to all time series from **a**.

10–30 years band for either the historical or the pre-industrial simulations (Supplementary Table 1, Supplementary Figs 1 and 2). Because this frequency peak is clearly evidenced in North Atlantic proxy records[16–18], we expect these selected models to potentially capture the actual variability pattern. In the IPSL-CM5A-LR model, strong 20-year variability in the North Atlantic is related to the time scale of temperature and salinity advection in the subpolar gyre, with salinity anomalies driving convection and deep ocean circulation[20]. The Bi-Dec ensemble shows coherent AMOC variability at 48°N (with an ensemble correlation coefficient of 0.64, significant at the 99% level), not seen in the ensemble of the remaining CMIP5 historical runs (Fig. 1a and Supplementary Fig. 2). More precisely, the Bi-Dec members systematically exhibit an AMOC maximum in the late 1970s. This consistent timing among them strongly suggests the response to a common external forcing. Earlier modelling studies have indeed shown intensified AMOC about a decade after strong volcanic eruptions[26,32,33], the lag reflecting the time required for the advection of cold anomalies from the subtropics to high latitudes, and for the response of the AMOC to changes in atmospheric momentum forcing. Here we suggest that the 1963 Agung eruption (Fig. 1b) has reset bidecadal variability in the eight CMIP5 simulations, leading to a first AMOC maximum about 15 years later (late 1970s). A second (weaker) simulated maximum occurs in the 1990s, corresponding to a secondary peak from the 20-year mode (Fig. 1). The same features are reproduced in a five-member ensemble of historical simulations from the IPSL-CM5A-LR model[13].

We do not expect such coherent variability to arise from the gradual long-term anthropogenic forcing. While solar forcing has been shown to influence ocean circulation variability in a climate model[37], solar forcing had a weak radiative amplitude of variability compared with volcanic eruptions, especially during the second half of the 20th century. We therefore propose that the coherent signal identified in the simulations has been triggered by the 1963 Agung volcanic eruption. Such large volcanic eruptions are indeed known to strongly cool the surface of the North Atlantic subpolar gyre and the Nordic Seas[32,34,38] (see for example, Supplementary Fig. 3). The bidecadal time scale of the subsequent dynamical adjustment to temperature anomalies involves the advection of temperature and salinity anomalies[15] (for example, GSAs) or baroclinic instabilities leading to large-scale baroclinic Rossby waves[21,39,40], strongly constrained by the size of the subpolar basin, and interactions with sea ice and atmospheric circulation. The specific processes accounting for the bidecadal variability may differ in the different climate models[21]. In the following, we will focus on one specific climate model (IPSL-CM5A-LR), for which the mechanisms underlying this variability[20,41] as well as its response to recent volcanic eruptions have been recently investigated in depth[15]. Simulations over the historical period, the last millennium, as well as additional sensitivity simulations performed with this model are discussed and compared with *in situ* oceanic data as well as proxy records.

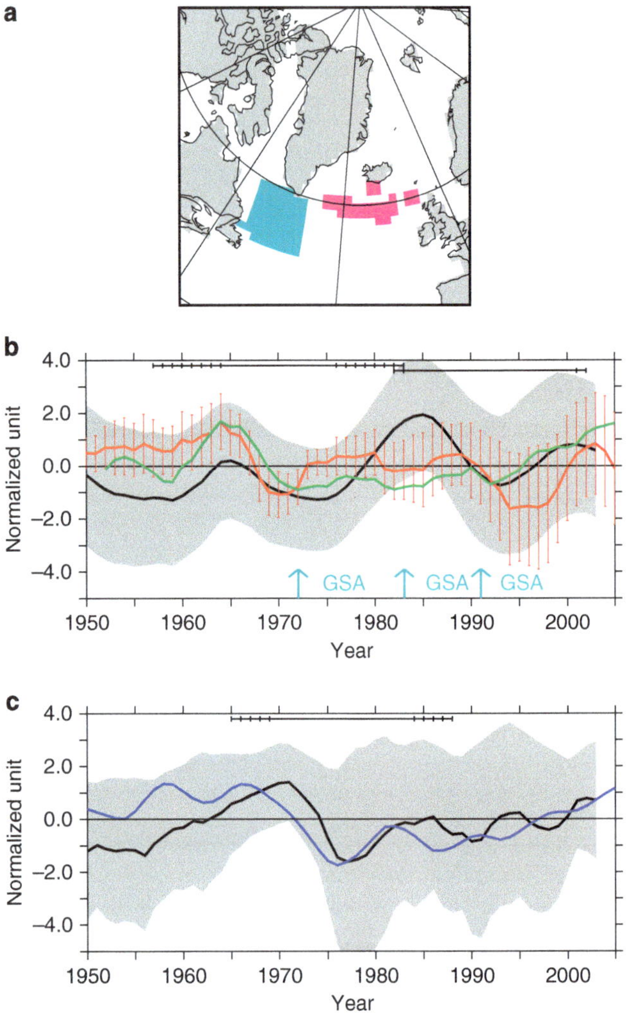

Figure 2 | Observed and simulated changes in salinity. (**a**) Location of the two key regions (Labrador Sea and subpolar gyre) studied here. (**b**) *In situ* salinity data from the Labrador Sea averaged over the cyan region down to 300 m compiled from the Bedford Institute (see for example, Methods, red line, with vertical line representing 2-s.d. errors) and from EN3 (ref. 43) (green) observational data sets. (**c**) Sea Surface Salinity from *in situ* data[42] (in blue) averaged over the eastern subpolar gyre (magenta region). In **b**,**c**, the 5-member ensemble mean outputs from historical simulations using IPSL-CM5A-LR are also displayed (black, mean value; grey shaded, s.d. envelope). The horizontal black lines at the top of **b**,**c** delimit the 20-year sliding windows for which significant correlation is detected (at the 95% confidence level) between the simulations and the observations (using the Bedford Institute data only in **b**. A 5-year running mean has been applied to all time series.

Comparison with recent oceanic data. Recent AMOC variability cannot be assessed in the short 2004-to-present observational record[8,11]. An alternative strategy is to use salinity observations, which exhibit large and well-documented variability in the North Atlantic, most notably as GSA. These GSAs occurred in the 1970, 1980 and 1990s (Fig. 2b). The current understanding is that the 1980s event was forced by the atmospheric circulation, while the 1970 and 1990s events are associated with Arctic Ocean variability via the East Greenland Current (EGC)[22,24]. Changes in salinity play a crucial role in the stability of the water column. They can drive changes in oceanic convection, which is directly related to AMOC variability. Here we use a combination of North Atlantic surface and subsurface salinity observational data sets (spanning years 1949–2010): a reconstruction of surface salinity in the eastern subpolar gyre[42], a new compilation of salinity data over a region enclosing large part of the Labrador Sea and reaching down to 300 m (compiled from the Bedford Institute of Oceanography website, see for example, Methods) and the global EN3 (ref. 43) data set (Fig. 2b). These data sets are compared with the IPSL-CM5A-LR ensemble of historical simulations.

Figure 2 shows that the IPSL-CM5A-LR simulations capture the observed 1970 and 1990s GSAs in the Labrador Sea and Eastern subpolar gyre. In this model, an increase in the EGC leads to a decreased salinity in the Labrador Sea, followed by AMOC variations about 10 years later[20]. Detailed analyses[15] have shown that changes in salinity simulated by the IPSL-CM5A-LR in the 1970s are a dynamical response to the Agung eruption. Here we conclude that the Agung eruption has excited the bidecadal mode of the North Atlantic in both observations and simulations, explaining not only the GSA of the 1970s but also the subsequent GSA of the 1990s. During both the 1970 and 1990s GSA events, salinity anomalies are associated with local temperature minima in the observations (Supplementary Fig. 4), consistently with the mechanism at play in the IPSL-CM5A-LR model[20]. This agreement between model and observations can be extended to the Bi-Dec ensemble. Indeed similar GSAs as the observed ones in the 1970 and 1990s are found in this ensemble mean as in the observations in the Labrador Sea both for salinity and temperature (Supplementary Fig. 5). The large spread in the ensemble also shows that the exact locations of the anomalies and of the processes at play are model dependent. It has been suggested that the GSA encountered in the 1980s, which is clearly weaker than the two others in the Bedford compilation (Fig. 2), and is not depicted by the IPSL-CM5A-LR historical simulations, may have a different origin[22]. This specific GSA was indeed associated with extremely severe winters of the early 1980s, with a possible contribution of the Arctic freshwater outflow via the Canadian Archipelago[22]. If this GSA was indeed driven by extreme stochastic atmospheric processes, this might explain the fact that it is not captured in the historical simulations, as well as the smaller amplitude and depth of this event as compared with the two other GSAs (Fig. 2b). We therefore hypothesize that this smaller GSA may not have strongly affected convection in the subpolar gyre, with therefore a very weak signature on the AMOC variability, in contrast with the two other GSA. This is however not fully supported by the fingerprint of the 1980s GSA in the EN3 data set, challenging this interpretation.

Last millennium perspective. To further assess the validity of the excitation of bidecadal variations by volcanic eruptions, we investigate two high-resolution proxy records and a new IPSL-CM5A-LR simulation spanning the last millennium, which includes several earlier Agung-like volcanic eruptions. This

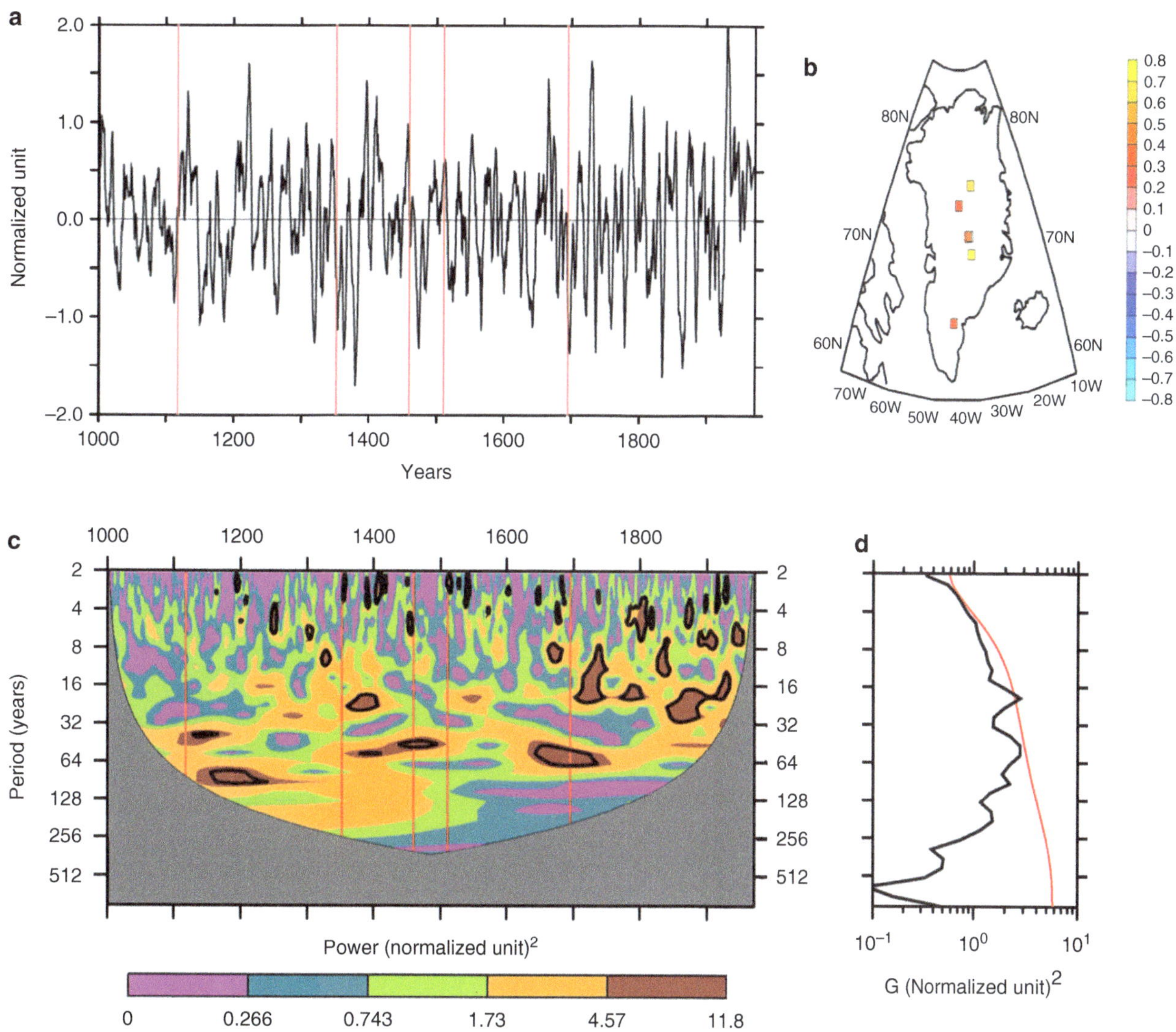

Figure 3 | Bidecadal variability in Greenland ice core data. (**a**) Principal component time series (with a 5-year running mean) associated with (**b**) the first EOF of the $\delta^{18}O$ data from six annually resolved Greenland ice cores, computed over the period 1000-1973 (see for example, Methods). (**c**) Wavelet analysis of the time series shown in **a** but without any filtering. The region significant at the 95% level are marked using bold contours. (**d**) Power spectrum of the time series in black compared with an AR1 signal shown in red. In **a**,**c**, the vertical red bars stand for the onset of the five eruptions selected over the last millennium.

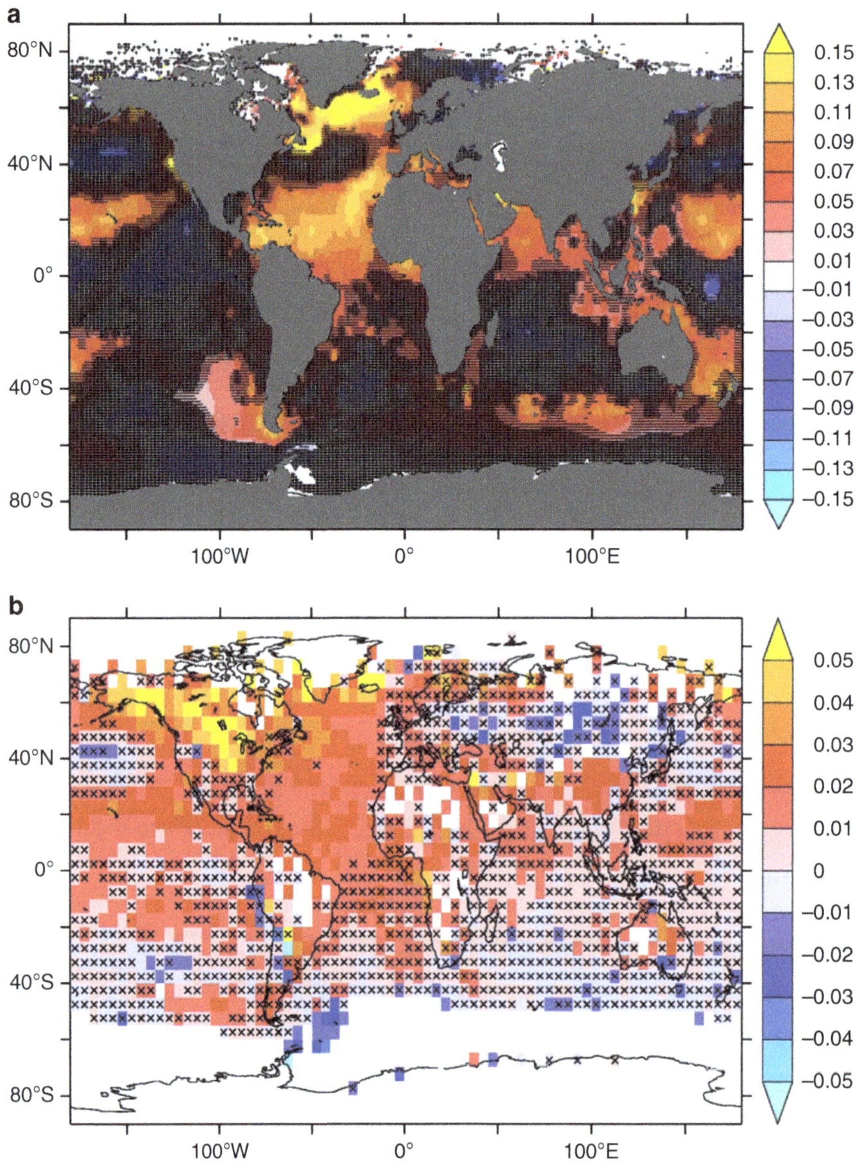

Figure 4 | Regression of the principal component of the Greenland ice core $\delta^{18}O$ with sea surface temperature data sets. (**a**) For the period 1870-1974 with SST data from HadISST[44] and (**b**) over the period 1000-1850 with gridded surface temperature reconstructions[46]. The non-significant zones for the regression at 95% level are marked with a black cross.

simulation also exhibits a clear spectral peak at the 20-year time scale for the 48°N AMOC variability (Supplementary Fig. 6).

The first proxy record is an updated compilation of six annually resolved $\delta^{18}O$ reconstructions from Greenland ice cores (see Methods and Supplementary Fig. 1). From an Empirical Orthogonal Function (EOF) analysis, their leading common signal is extracted, corresponding to a monopole that exhibits preferential variability at a 20-year time scale (Fig. 3), coherent with earlier ice core composites[3]. This data set is highly correlated with North Atlantic Sea Surface Temperature (SST) as shown by the regression of HadISST data[44] on the principal component (PC) of the first EOF from the ice cores for the period 1850–1973 (Fig. 4). Significant correlation is indeed found in the Atlantic, approximately from the equator to the Nordic Seas, a spatial pattern that is reminiscent of the Atlantic Multidecadal Oscillation[45] (AMO). The response is especially large in the North Atlantic (Fig. 4a). To test this relationship on a longer time scale, we use a gridded SST reconstruction[46] and perform the same regression analysis for the period 1000–1850. Again, significant correlation is mainly found in the Atlantic basin north of the equator (Fig. 4b). Although this second result should be taken with caution, given that the SST reconstruction is based on a multi-proxy approach, also including Greenland ice core data, we argue that the ice core compilation provides an annually resolved record of bidecadal North Atlantic surface temperature variability. Such bidecadal timescales are usually filtered out in AMO reconstructions[46,47], which indeed act as low pass filters and stress multidecadal variability. We thus interpret this first EOF of Greenland ice cores as a proxy for North Atlantic SST variations in line with a former hypothesis linking a stack of Greenland ice cores with surrounding SST[48].

The second proxy record corresponds to growth increments from the bivalve *Arctica islandica* collected north of Iceland[49], also annually resolved. These bivalve data are not an unequivocal proxy for SST as they also depend on other factors, notably the nutrient supply[49]. The latter depends itself on vertical movements that can be related to AMOC variations, notably through convection processes which can help deep ocean nutrients to come to the surface[50]. For instance, in the isopycnal layered ESM2G Earth System model, strong inter-decadal changes in surface salinity associated with changes in AMOC produce spatially heterogeneous variability in convection, nutrients supply and thus diatom biomass[44]. As the IPSL-CM5A-LR model includes a bio-geochemical component[36,51], we use it to test this hypothesis through a pseudo-proxy approach[52]. In particular, we analyze the link between nutrient concentrations (PO_4, NO_3, Si and Fe) and the 48°N AMOC index in the last millennium simulation. In this case, silicate and iron in the Nordic Seas are the main limiting nutrients for phytoplankton growth (not shown) and therefore expected to be also the limiting nutrients for bivalve growth, motivating a focus on these nutrients. Around Iceland, the simulated nutrients supply is significantly correlated with the simulated variations in 48°N AMOC, the latter leading the former by 1–3 years (Fig. 5a). We argue here that this relationship arises from changes in convection in the Nordic Seas, which precedes AMOC changes by 1–4 years[20] and changes in nutrient supplies around Iceland by 2–5 years in our model (Fig. 5c). This last lag can be explained by the upwelling of nutrients stored at depth by vertical currents associated with convection, and a subsequent period of a few years to export these nutrients from the convection sites towards north Iceland. Based on the hypothesis that the nutrient supply to Iceland is the limiting factor for the growth of the bivalve *A. islandica*, we therefore use growth increment records as a proxy of the subpolar AMOC, with a response lag of 1–3 years.

Using estimates of past volcanic radiative forcing[53], we identify volcanic eruptions with a radiative forcing ranging between one

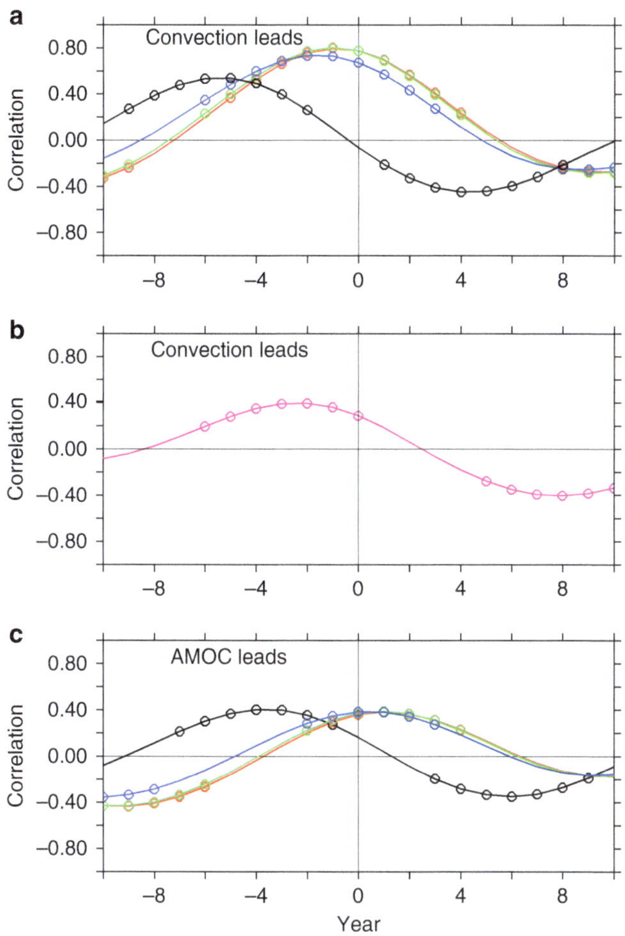

Figure 5 | Phase relationship between key variables of the last millennium IPSL-CM5A-LR simulation. Cross-correlation coefficients (*y*-axis) computed over the whole length of the last millennium simulation (850-1850) for different time lags (*x*-axis), (**a**) mixed layer depth in the Nordic Seas (an indicator of convection) and nutrient concentration north of Iceland, averaged between 50 and 150 m, where the bivalves are located. (**b**) Cross-correlation between mixed layer depth in the Nordic Seas and 48°N AMOC. (**c**) Same as **b** except that mixed layer depth in the Nordic Seas is replaced by 48°N AMOC. In **a**,**c**, different colours are used to represent different nutrients (red, NO_3; green, PO_4; blue, Si and black, Fe). Correlation coefficients which are significant using a 95% confidence level Student *t*-test (where degrees of freedom account for data autocorrelation[70]) are highlighted with circles.

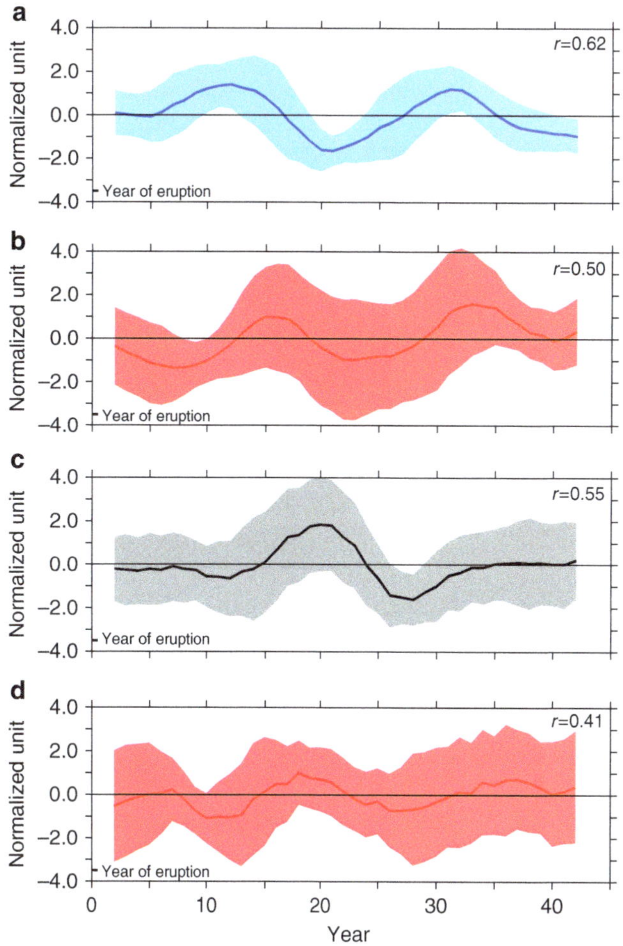

Figure 6 | Response to five Agung-like eruptions of the last millennium in reconstructions and IPSL-CM5-LR simulations. Time series of the composite formed by the ensemble of the five members for five Agung-like eruptions (taking place in 1118, 1352, 1460, 1511, 1695) over the last millennium identified through the same volcanic forcing reconstruction used in last millennium simulation[53]: (**a**) Simulated AMOC maximum at 48°N, (**b**) Growth increments in the bivalve *Arctica islandica*[49], (**c**) Simulated North Atlantic SST between 25 and 55°N, (**d**) first principal component of six yearly Greenland $\delta^{18}O$ ice core reconstructions. The ensemble correlation between the five individual time series following the respective eruptions is shown for each index in top right of each panel. A 5-year running mean has been applied to all time series. Data have been standardized and expressed in normalized units. The average of the response following the five eruptions is shown with lines, and the envelope of the response is displayed through the shading around these lines.

to one and a half times the amplitude of the 1963 Agung eruption (Pinatubo being 1.4 times Agung in terms of global radiative forcing used in IPSL-CM5A-LR, see for example, Methods). To avoid any interference from successive events (see below), we retain only those not followed within 40 years by eruptions larger than 1.5 Agung. Only five of all the eruptions recorded over the last 1,000 years meet these two criteria (onset of the eruption in 1118, 1352, 1460, 1511, 1695, see for example, Supplementary Fig. 7).

We first compare the simulated AMOC variations following these five eruptions with the Icelandic bivalve data. They exhibit a maximum both around 15 and 35 years following the eruptions (Fig. 6a,b). The slight shift that is found between the 48°N AMOC from the model and the bivalve observations[49] is consistent with the 3-year lag found in Fig. 5. Indeed the 3-year lagged correlation between the model AMOC and the bivalve data ensemble means reaches 0.66 (p-value <0.1 see for example, Supplementary Fig. 8a).

We then compare the simulated SST variations in the North Atlantic to the Greenland ice core data. They both exhibit coherent warming around 20 and 40 years after the eruptions (Fig. 6c,d and Supplementary Fig. 8b). The lag between AMOC and North Atlantic SST changes is attributed to a lagged accumulation of warm waters in the North Atlantic when the circulation accelerates[41] (Supplementary Fig. 9) and is consistent with the results of other climate models[54]. Altogether, these results for five different Agung-type eruptions of the last millennium support the mechanism by which Agung-like volcanoes reset the 20-year AMOC variability. Figure 7 summarizes the different processes proposed here as well as the time lags between them, including the proxy records analyzed.

Interference pattern. The implications of this oceanic response to moderate volcanoes like Agung or Pinatubo over the most recent 30-year period and the very near future are now investigated. Indeed, the Agung eruption was followed by the El Chichon eruption 19 years after and the Pinatubo eruption 28 years later. While the timing of El Chichon may have led to a constructive interference with the Agung-excited 20-year cycle, we make the hypothesis that the Pinatubo eruption interfered destructively; thereby explaining the damped observed and simulated AMOC variability in the 2000s.

This hypothesis is tested using a five-member ensemble of IPSL-CM5A-LR starting in 1991 and using the same forcings as in the historical simulations but without the Pinatubo eruption. This 'no Pinatubo' sensitivity ensemble deviates from the historical ensemble (Fig. 8a) and exhibits much stronger practically bidecadal variability, with a minimum around 2010, a maximum in 2015 and a new minimum in 2025–2030. By contrast, the historical ensemble does not exhibit any remarkable decadal AMOC excursions from 2005 to 2030. Although the signal-to-noise ratio remains small and leads to an overlap of the s.d. envelopes of the two ensembles (Supplementary Fig. 10), this sensitivity test confirms that the Pinatubo eruption induced a suppression of the bidecadal AMOC variability in the model. The simulated regular AMOC weakening in the 2010s after the Pinatubo eruption is consistent with the RAPID array observations[8,11] and with oceanic reanalyses[12] (Fig. 8).

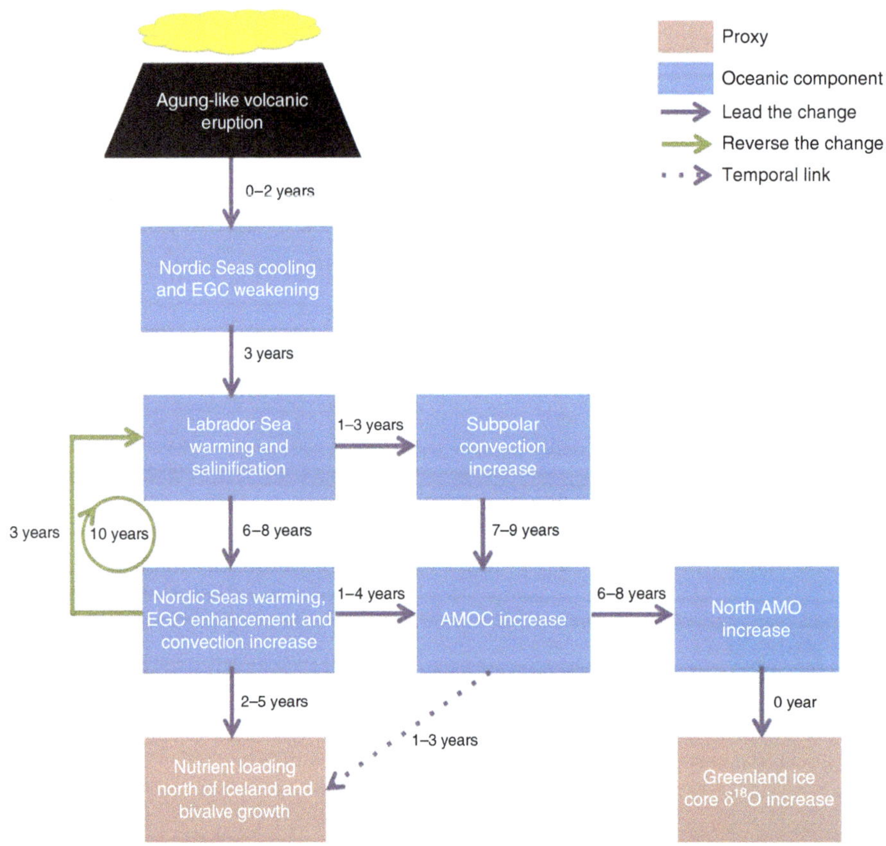

Figure 7 | Scheme summarizing the main processes at play in the North Atlantic response to volcanic eruptions. Blue boxes depict oceanic processes and red boxes depict the proxies considered here. Purple arrows represent causality links together with the associated time lags in years. Dotted arrows only depict a temporal link, with no direct interaction. The green arrow represents the delayed negative feedback that triggers the reversal phase of the 20-year cycle.

This interference pattern from successive volcanoes can be captured by a simple conceptual model (Fig. 9b), which takes into account the timing of large volcanic eruptions that excited bidecadal oscillations over the last 60 years, assuming a constant response time, together with a long-term weakening trend[6] (see Methods). This conceptual model successfully reproduces the main features of the interference theory developed here, and closely follows the IPSL-CM5A-LR simulations with and without Pinatubo (Fig. 8). However, neither the forced simulations nor this conceptual model are able to produce the rapid AMOC increase identified in the early 1990s in a few AMOC reconstructions[12,13] (Fig. 8b). It has been suggested that this specific shift is linked with internal atmospheric variability and 5-year-earlier changes in the NAO, which affected North Atlantic SST and therefore the decadal variability of the AMOC[3,55,56]. Indeed, climate simulations nudged towards observed SST anomalies performed using IPSL-CM5A-LR do produce a strong AMOC enhancement in the late 1990s that can be attributed to a NAO positive phase a few years earlier[15] (Fig. 8). Similarly, implementing additional NAO variations in the conceptual model with a 5-year lagged response of the ocean (see Methods) reproduces most of the observed variability (Fig. 8b) as well as the very recent weakening over the last few years[10].

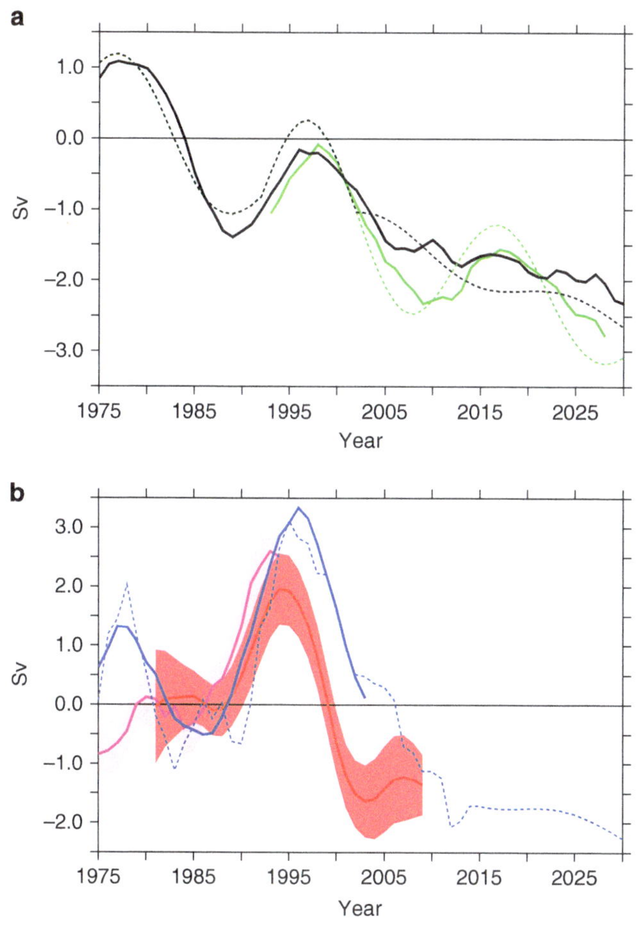

Figure 8 | Recent AMOC and projected change. (**a**) Simulated AMOC anomalies at 48°N in historical IPSL-CM5A-LR simulations (black), and in the sensitivity tests performed without taking into account the Pinatubo eruption (green). Results of the corresponding conceptual model (SI) are shown in dotted lines. (**b**) Estimates of AMOC at 45°N (1979-1988) based on hydrographic data[13] (pink curve) and ensemble mean of the AMOC maximum at 45°N from 12 ocean reanalyses[12] (red). This ensemble mean has been rescaled by a factor $\sqrt{12}$, to correct for the loss of variance caused by averaging different ocean models with different data assimilation. The five-member ensemble average of IPSL-CM5A-LR simulations nudged towards observed SST anomalies[14] is shown in blue. The dashed blue curve is the associated conceptual model including NAO variation (Methods). A 5-year running mean has been applied to all time series.

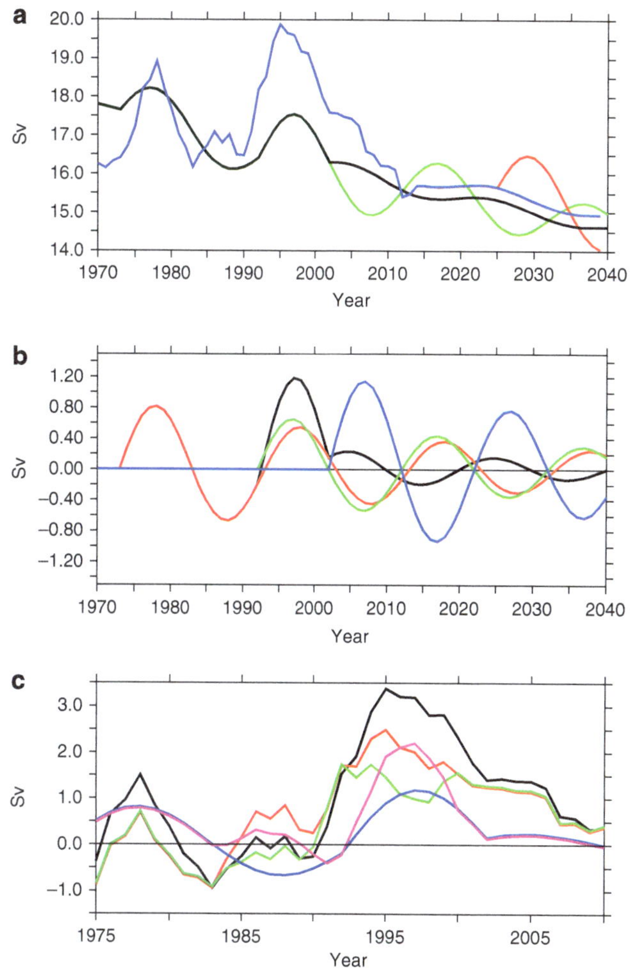

Figure 9 | Time evolution of the AMOC as computed by different conceptual models (described in Methods). (**a**) Different conceptual models with two (green) or three (black) volcanoes and with the NAO included (blue). For the future, when no NAO information is available, we take a null NAO phase. This allows illustration of the impact of an eruption in 2015 as shown in red. All the conceptual models include a long-term decreasing trend due to global warming (GW). (**b**) Decomposition of the impact of each volcano, Agung (in red), El Chichon (in green), Pinatubo (in blue) and the sum of the three (in black). (**c**) AMOC at 48°N from the different conceptual models, including or not including NAO variations following volcanic eruptions (see Methods). This decomposition allows estimation of the part of variance of the AMOC resulting from volcanic forcing and/or NAO, as compared with the total variance induced by volcanoes and NAO (black curve, 100% of the total variance). The red curve shows the calculation over the years 1975-2010 result when accounting for NAO (57% of the total variance), while the calculation accounting for NAO with the exception of the 3 years following each volcanic eruption is shown in green (44% of the total variance). The blue curve shows the bidecadal mode excitation by the volcanoes only (20% of the total variance), while the magenta curve shows the calculation when accounting for the impact of volcanoes plus the effect of the NAO in the 3 years following each eruption (30% of the total variance).

For the 1975–2010 period, the conceptual model allows the relative importance of the response to volcanoes and to the NAO to be distinguished (see Methods). The NAO-induced AMOC variations account for 57% of the total AMOC variance, while 20% of the AMOC variance is due to excitation of the AMOC 15 years after volcanic eruptions, as a damped eigen mode (see for example, Fig. 9c). The sum of the variance from this decomposition into two components does not equal 100%, indicating that there exists covariance between the components.

This simple approach does not account for the potential impact of volcanoes on the NAO. While a strong positive NAO occurred in the 2–3 years following the Pinatubo eruption[28,30], this is not a systematic response to all eruptions[57]. For instance, the winter following the Agung eruption had a negative NAO phase. This different response may arise from the intrinsic chaotic nature of the NAO, for which internal oscillations produce a large amount of noise, which can therefore mask a volcanic-forced signal. To address this issue, we repeat our analysis removing the NAO signature of the three winters following the three main volcanic eruptions considered. In the conceptual model, the fraction of AMOC variance explained by the response to volcanic forcing in 1975–2010 now increases to 30%, while NAO-only (without considering the three winters following eruptions) accounts for 44% of the variance (see for example, Fig. 9c and Methods). Once again, we note some covariance within this decomposition leading to a sum lower than 100%.

Discussion

We have shown here multiple lines of evidence supporting the long-lasting impact of moderate (Agung-size) volcanic eruptions on ocean circulation in the North Atlantic. The timing of subsequent volcanic eruptions can lead to constructive (for example, Agung and El Chichon) or destructive interferences (for example, Agung and Pinatubo) to force internal modes of ocean variability. In the IPSL-CM5A-LR model, the underlying mechanism is related to the build-up and advection of salinity anomalies near the Labrador region, and its subsequent effect on North Atlantic convection. Moderate volcanoes increase the sea-ice cover in the Nordic Seas, thus reducing the export of freshwater through the Denmark Strait and leading to anomalous positive anomalies in the Labrador region, explaining the observed GSAs of the 1970 and 1990s. While the exact processes at play may be model dependent, the coherence of the Bi-Dec model responses suggest a key role for the propagation of salinity and temperature anomalies as observed during GSA events. The exact path and pattern for this propagation may vary among the models that exhibit bidecadal AMOC variability, including, for example, the strength of the amplifying mechanisms through sea ice and atmosphere interactions, clearly depicted by the IPSL-CM5A-LR model. Further intercomparisons of the precise mechanisms at play within the different models are needed to understand this potential diversity. Nevertheless, the present results provide a good working hypothesis based on the data that is currently available. Indeed, this impact of volcanic eruptions explains several aspects of recent AMOC variations, and associated patterns (for example, GSAs). Moreover recent analysis of the observed variability mode of bidecadal sea level reveals a strong regime shift since the 1970s[14], which can be related to the ocean circulation changes described here in response to the Agung eruption.

Nevertheless, processes other than volcanic eruptions may have shaped variability in the North Atlantic over the last six decades, such as the NAO (Fig. 8) or the 1980s GSA, which is not attributed here to volcanic eruptions. We also stress the fact that the proposed mechanisms identified in the IPSL-CM5A-LR simulations are not at play for eruptions with a radiative forcing 50% larger than that of Agung (see for example, Supplementary Fig. 11). Different ranges of the sea-ice response[34] can indeed either increase or decrease convection, depending on the size of the eruption. This sensitivity to volcanic forcing is therefore expected to be model dependent and other models from our Bi-Dec ensemble may exhibit different thresholds and sensitivity to volcanic eruption size. Further analysis of this sensitivity is necessary to improve our understanding of the response of the climate system to volcanic eruptions, within a coordinated intercomparison framework. This will hopefully be implemented within the CMIP6 framework in the near future.

The present findings also imply a significant predictability of the North Atlantic dynamics if an Agung- or Pinatubo-like eruption occurs in the future (Fig. 9a). Excited bidecadal variability would be superimposed on the long-term decreasing trend, which is driven by global warming[6]. Given the potential influence of AMOC and the North Atlantic Ocean on hurricane activity[58], African Sahel drought[59], Greenland melt[2], regional sea level[60] and marine ecosystems, such decadal predictability is of key interest to a diversity of stakeholders, including the agriculture, energy, fisheries, insurance industries, water supply and management agencies.

Methods

CMIP5 analysis. We use the CMIP5 database of historical simulations for the period 1850–2005 (http://cmip-pcmdi.llnl.gov/cmip5/), performed using solar, volcanic and anthropogenic forcings, and pre-industrial control simulations, performed without changes in forcing, to explore the variability of the simulated subpolar AMOC. This analysis is restricted to the 19 models for which the meridional stream function output is available (Supplementary Table 1). Based on the information from several paleoclimate archives from the North Atlantic, which consistently show bidecadal variability (see main text), we select the subset of models for which a peak of energy significantly larger than a red noise process is present in the spectral band of 10 to 30 years. A power spectrum analysis (Supplementary Fig. 1) is used to identify this subset of models for which spectral power is expressed in the 10–30-year period band, and passes a significance test (that is, the one that refutes the red noise null hypothesis) in either the pre-industrial or historical simulations. Almost half of the CMIP5 climate models (9 among 19 models) do satisfy this criterion: CCSM4, CESM1-BGC, MRI-ESM, NorESM1-ME, NorESM1-M, CanESM2, GFDL-ESM2M, GFDL-CM3 and IPSL-CM5A-LR. Supplementary Table 1 summarizes the main findings in terms of spectral characteristics of the AMOC at 48°N in the different models and simulations. Supplementary Figure 2 shows the AMOC evolution from some of the selected models over the period 1950–2005. These models show relatively well-phased variability close to harmonic variations with a maximum in the late 1970s and another one in the late 1990s. This extends the findings of Swingedouw *et al.*[15] using the IPSL-CM5A-LR model. Supplementary Figure 5 illustrates the phasing in terms of salinity and temperature, averaged in the upper Labrador Sea, found in these models

IPSL-CM5A-LR simulations. The ocean-atmosphere coupled model mainly used in this study is the IPSL-CM5A (ref. 36) in its low-resolution (LR) version as developed for CMIP5. The atmospheric model is LMDZ5 (ref. 61) with a $96 \times 96 \times$ L39 regular grid and the oceanic model is NEMO[62] with an $182 \times 149 \times$ L31 non-regular grid, in version 3.2, including the LIM-2 sea ice model[63] and the PISCES[51] module for oceanic biogeochemistry. The internal variability of the AMOC has been described in the study by Escudier *et al.*[20] It has a 20-year preferential variability in the whole subpolar gyre. The mechanism explaining such a variability was shown to be a cycle beginning (for instance) with an intensification of the EGC that brings more cold and fresh water to the Labrador Sea where it accumulates and gives rise to negative SST and sea surface salinity (SSS) anomalies. These anomalies are advected along the subpolar gyre; they affect the convection all along their path up to the Nordics Seas, where the negative SST anomalies increase the sea-ice cover, which in turn induces a positive sea-level pressure anomaly and a localized anticyclonic atmospheric circulation. This leads to a decrease in the wind stress along the eastern coast of Greenland and thus a weakening of the EGC, leading to an opposite phase of the cycle around 10 years after its onset. This cycle impacts the AMOC variations through the contribution of SSS anomalies to the convective activity in the subpolar North Atlantic. The list of simulations performed with this model and used in this study is summarized in the Supplementary Table 2 and described below.

The five-member ensemble of historical simulations use a prescribed external radiative forcing from the observed increase in greenhouse gases and aerosol concentrations as well as the ozone changes (Fig. 1) and the land-use modifications (not shown, see the study by Dufresne *et al.*[36]). They also include estimates of solar irradiance variations and of volcanic eruptions over the historical period represented as a decrease in the total solar irradiance (depending on the intensity of the volcanic eruption, Fig. 1). These historical simulations start from year 1850. Their initial conditions come from different dates of the 1000-year control simulation under pre-industrial conditions, each separated by 10 years. This control pre-industrial simulation is itself starting after thousands of years of spin-up procedure.

To evaluate the hypothesis of a destructive interference due to the Pinatubo eruption, we prepared an additional five-member ensemble using the IPSL-CM5A-LR model. The simulations start on 1 January 1991 from each historical simulation described earlier. They use rigorously the same external forcing except for the effect of the Pinatubo eruption, which was removed from the total solar irradiance. The AMOC response with the s.d. of the ensemble is shown in Supplementary Fig. 10.

We also consider a five-member ensemble of nudged simulations, for which each simulation includes a nudging term towards observed anomalous monthly SST (ref. 15). Each simulation starts on 1 January 1949 from one of the historical simulations presented above. These nudged simulations are using the same external forcing as the historical ones (Fig. 1). The nudging technique consists in adding a heat flux term Q to the SST equation under the form $Q = -\kappa(SST'_{mod} - SST'_{obs})$ where SST'_{mod} stands for the modelled anomalous SST at each time step and grid point, and SST'_{obs} the anomalous observed SST (Reynolds *et al.* (2007)). Anomalies are computed with respect to the average SST over the period 1949–2005 in the corresponding historical simulation and in the observations. We use a restoring coefficient κ of $40\,W\,m^{-2}\,K^{-1}$ corresponding to a relaxing time scale of $\sim$60 days (for a mixed layer of 50 m depth). See the study by Swingedouw *et al.*[15] for further details.

The CMIP5 last millennium simulation starts in 850. It includes solar[64] and volcanic forcing[53]. The implementation of the volcanic forcing in this simulation is slightly different from that used for the historical simulations. In the latter, the effect of stratospheric injection of sulphate volcanic aerosols is only considered as a modulation in the total solar irradiance variations, while in this simulation the volcanic aerosols in the stratosphere are transported latitudinally in the model following Grieser and Schonwiese[65] parameterization. We consider 48 latitude bands for this transport. The model uses prescribed changes in aerosols optical depth and interactively computes the perturbed (longwave and shortwave) radiative budgets. The reconstruction of aerosols optical depth uses sulphate data from ice cores records from Greenland and Antarctica[53]. In this simulation, the AMOC at 48°N shows very strong variability in the bidecadal band (Supplementary Fig. 6), as in pre-industrial simulations analyzed in the study by Escudier *et al.*[20] and in the historical simulations (Supplementary Fig. 1).

For the analysis of the Agung-like eruptions, we consider the 40 years following each selected event as an individual sensitivity experiment, like the member of an ensemble. The selected volcanoes are shown with respect to the full time series of volcanic forcing over the last millennium in Supplementary Fig. 7. The ensemble mean and the s.d. of the five members are shown in Fig. 3. We only consider eruptions with a magnitude comparable to Agung (1963), since larger eruptions lead to very strong cooling resulting in a different regime response for the AMOC[34]. Note that although the selection of the individual volcanic events includes a condition on the absence of significant eruption during the 40 years after the event, no condition is imposed regarding the years preceding the selected eruption. It is nevertheless possible that volcanic eruptions occurring a few years earlier could also impact North Atlantic variability, and indeed it is possible that the 1460 eruption is affected in this way, although none of the other volcanoes selected. Since removing this eruption does not modify the main conclusions, we keep it in our pool.

Sensitivity of IPSL simulations to the size of eruptions. Our analyses are focused on the model response to moderate eruptions. Indeed, a composite computation of key oceanic variables in the North Atlantic for all volcanoes lower than one and a half Agung as shown in Supplementary Fig. 11 for the IPSL-CM5A-LR model clearly supports our results: this shows again that moderate-size volcanic eruptions excite the 20-year variability through a cooling of the Nordic Seas at lag 1 year, when the subpolar region is less affected. This leads in the following years to the build-up of SST positive anomalies (as well as SSS, not shown) in the Labrador Sea, which are then advected along the subpolar gyre affecting the mixed layer depth around lag 5–6 years, explaining the 15-year lag for maximum AMOC following the simple scheme from the study by Escudier *et al.*[20]

However, volcanic eruptions larger than one and half Agung have a direct negative impact on SST in the convection region south of Iceland, leading almost immediately to local negative mixed layer depth anomalies for lags 2–6 years (Supplementary Fig. 11) following the eruption onset, which will lead to AMOC weakening around year 9–13, consistent with the study by Mignot *et al.*[34]

Such sensitivity could clearly be model dependent, so that further analysis of different models will be necessary to exactly evaluate the volcanic eruption size that lead to different regimes in different models and in reality.

Salinity data. Observational salinity data have been downloaded from the Bedford Institute of Oceanography website (http://www.bio.gc.ca/science/data-donnees/base/start-commencer-eng.php). Their analysis is focused on the Labrador Sea area, which is a key for the processes involved in the 20-year variability[20]. Since changes in deep convection driving the AMOC can be triggered by salinity anomalies in the Labrador Sea as deep as 300 m, we also consider subsurface data to this depth[20]. To evaluate the observational uncertainty, we also use the version 2a of data compilation from EN3 (http://www.metoffice.gov.uk/hadobs/en3), and a SSS data compilation from ships of opportunity and oceanographic cruises[42] for the eastern part of the subpolar gyre.

***In situ* temperature anomalies.** In the IPSL-CM5A-LR climate model, the salinity anomalies are associated with temperature anomalies in the Labrador Sea and all along the subpolar gyre and Nordic Seas[20]. This is related to variations of the EGC, which brings cold and fresh water from the Arctic. We check in Supplementary Fig. 4 that, in the data analyzed, the salinity anomalies shown in Fig. 2 have a temperature counterpart (using the same source of data, that is, Bedford Institute of Oceanography and EN3). It is clearly the case in the Labrador Sea, with once again an agreement between historical simulations and temperature data, following the GSAs of the 1970 and 1990s. We note that the GSA of the 1980s is not clear in the Bedford Institute of Oceanography data set and absent in the model. We argue that this GSA is primarily driven by exceptional atmospheric forcing[22]. Given the lack of strong signals in the eastern subpolar gyre, the correspondence between temperature and salinity anomalies is weaker in this area.

Greenland ice core composite. The $\delta^{18}O$ Greenland records used herein correspond to six annually resolved millennial-long ice core records covering the period 1000–1973 (B18, Crete, DYE-3, GISP2, GRIP, NGRIP, see for example, Supplementary Table 3). During the instrumental period, the first PC associated with the leading EOF of these proxies (accounting for 25% of the variance of the six ice core records) is closely related to precipitation-weighted Greenland surface temperature changes[66], ultimately driven by changes in the four major weather regimes in the North Atlantic, namely the two phases of the NAO, the Atlantic Ridge and the Scandinavian Blocking[66]. Our PC is closely related, during the overlapping period, ($R = 0.67$, $P < 0.05$) to the average of five Arctic and Greenland $\delta^{18}O$ ice core records previously used to characterize the bidecadal SST variability in the North Atlantic during the past 700 years[16].

Bivalve proxy records. The bivalve data used are described in the study by Butler *et al.*[49] This proxy is a unique record of bivalve growth north of Iceland, annually resolved over a period of 1,357 years. Radiocarbon measurements taken on absolutely dated shell material have been used to show the co-varying influence of Atlantic and Arctic waters on the North Icelandic Shelf over the past millennium[67]. The correlation between growth and observed temperature is positive and weak (0.217, $P < 0.05$ using HadISST1 gridded data), suggesting that the bivalves are responding to a nutrient signal that is only partially temperature driven.

Link between last millennium proxies and model simulation. The AMO[45] represents the low-frequency variations of SST averaged over the Atlantic from the equator to 60°N. These multidecadal variations influence surrounding regions. In particular, Greenland climate is considered to be very sensitive to these variations. For instance, surface air temperature at Nuuk, West Greenland and the AMO Index are highly correlated ($r = 0.49$, $P = 0.05$) over the last 150 years[68].

Last, we show that in IPSL-CM5A-LR (and other models) the AMOC is leading the AMO by a few years[41]. Here we focus on the link between the AMOC and the SST in the North Atlantic and investigate the phase lag in Supplementary Fig. 8 using 1,000 years of the last millennium simulation. We show that AMOC index at 48°N is leading the SST in the Atlantic between 25 and 55°N by around 5–10 years. This phasing also clearly appears in the 40 years following selected volcanic eruptions from Fig. 3. Strong correlations are also identified for the control simulation under pre-industrial conditions, showing that this behaviour is not triggered by external forcing but is linked to the internal mechanisms relating AMOC and North Atlantic SST (not shown).

Conceptual models. We build here a family of simple conceptual models to capture the essence of the main mechanisms at play in our climate model and hopefully in reality.

Our first assumption, suggested by our results from the different CMIP5 screened models and data sets, is that Pinatubo-like eruptions lead to an AMOC maximum around 15 years after the onset of the volcanic eruption. This delay is related to the characteristic time required for the North Atlantic to adjust to the volcanic signal, through induced changes in convection. It accounts for the lag between cooling of Nordic Seas, reductions in EGC, positive anomalies advected along the subpolar gyre, changes in convection sites and, a few years later, changes in the AMOC[15]. This AMOC response corresponds to the excitation of damped oscillatory variations with a typical time scale of 20 years[69].

We call τ the time delay for a particular volcanic eruption to have an impact on the AMOC; it is expected to depend on the typical 15-year delay as well as on the exact timing of the eruption. The subsequent damped oscillation response excited by the volcano can then be described through the following function of time $f(t)$:

$$f(t) = H(t-\tau)\sin\left(\frac{2\pi}{20}t\right)e^{\frac{t-\tau}{\Delta}} \quad (1)$$

where H stands for the Heaviside function, t is the time and Δ is a characteristic damping time scale.

Over the time period 1960–2010, there have been three main eruptions (Agung, El Chichon and Pinatubo) whose radiative forcing in the IPSL-CM5A-LR historical simulations are 1.5 W m^{-2}, 1.2 W m^{-2} and 2.1 W m^{-2}, respectively (Fig. 1). To account for the size of the eruption, we introduce a scaling factor λ_i for each volcano (equal to its radiative forcing). We finally use a factor α to scale globally the effect from all the volcanoes.

For comparison with historical simulations and projections, we must also take into account the AMOC decreasing trend produced in response to anthropogenic forcing by CMIP5 models[6]. This effect of global warming is implemented through the GW(t) function, which corresponds to the anthropogenic radiative forcing multiplied by a scaling factor β.

$$f(t) = \alpha\sum_{i=1}^{3}\lambda_i H(t-\tau_i)\sin\left(\frac{2\pi}{20}t\right)e^{\frac{t-\tau_i}{\Delta}} - \beta\times\mathrm{GW}(t) \quad (2)$$

This conceptual model can be tested with and without the Pinatubo eruption. When excluding the Pinatubo eruption, the sum in equation (2) only accounts for the two other volcanoes, and the rest stays the same.

In reality, AMOC is also affected by the internal variability of the atmospheric circulation and is particularly sensitive to the NAO[3,55,56] with a 5-year lag. To include this aspect in the conceptual model, an additional term is introduced. It corresponds to the winter NAO index from NOAA (http://www.cpc.ncep.noaa.gov/products/precip/CWlink/pna/JFM_season_nao_index.shtml), with a lag of 5 years, due to oceanic adjustment to changes in surface fluxes and scaled by a coefficient γ.

$$f(t) = \alpha\sum_{i=1}^{3}\lambda_i H(t-\tau_i)\sin\left(\frac{2\pi}{20}t\right)e^{\frac{t-\tau_i}{\Delta}} - \beta\times\mathrm{GW}(t) + \gamma\times\mathrm{NAO}(t-5) \quad (3)$$

The tuning of the adjustable parameters of this conceptual model (α, β, Δ and γ) is performed with a minimization technique using IPSL-CM5A-LR simulations as targets. The global warming scaling β is first adjusted using all IPSL-CM5A-LR available projections. A linear regression performed between anthropogenic forcing and the simulated 48°N AMOC index leads to a scaling factor β of 1.2 Sv m^2 W^{-1}. We then define a cost function to be minimized, as the mean squared error between the conceptual model results and the reference simulation (historical and sensitivity experiments without Pinatubo eruption) over the period 1975–2030 (Supplementary Fig. 12). The best choice for the scaling factor α for the volcanoes and the damping factor Δ for the oscillations is unequivocally 0.5 Sv and 60 years, respectively. Finally, the scaling factor γ for the NAO-forced AMOC variations is adjusted using the IPSL-CM5A-LR nudged simulations towards observed SST and optimized for a value of 2.5 Sv.

We compare the results of two different conceptual models, which account for volcanic-induced variability and NAO forcing. Indeed, between 1975 and 2010, the volcanic excitations of the bidecadal mode acts on the timing of decadal variations; however, NAO is the main driver of the magnitude of AMOC variations in the conceptual model scaled on IPSL-CM5A-LR simulations (Fig. 9c). Indeed, if we compute the ratio of variance between the different conceptual models (Fig. 9c), including different pieces of forcing of the AMOC, we find that NAO alone accounts for 57% of the variance of the model including all processes, while the excitation of the bidecadal model accounts for 20%. Given that for any X and Y random variables, we have: $\mathrm{Var}(X+Y) = \mathrm{Var}(X) + \mathrm{Var}(Y) + Cov(X,Y)$ where Var() is the variance operator and Cov() the covariance one. The sum of the variance explained by NAO and by volcanic-induced variability is not 100%, due to covariance.

To avoid accounting twice for the role of volcanic eruption (through the radiative forcing and through the potential consequence on the subsequent winter NAO phase[27]), we consider an alternative conceptual model, in which we remove the NAO forcing in the three winters following each eruption. In this case, the variance ratio between the variance of AMOC volcanically induced and the variance of AMOC, which is purely induced by the NAO (removing the three years following volcanic eruptions), is 30% and only 44% for the pure NAO-forced AMOC with volcanic years removed, respectively (Fig. 9c).

The conceptual model is finally used to explore the impact of a hypothetical Pinatubo-like eruption starting in 2015 (Fig. 9a). Without such an event, and without large NAO changes, the conceptual model simulates only the gradual decreasing trend driven by global warming. If a large eruption were to occur in 2015, then it would excite large AMOC decadal oscillations with maximum strength in 2030 and minimum strength in 2040, potentially complicating the detection of long-term trends associated with global warming.

References

1. Wunsch, C. Oceanography. What is the thermohaline circulation? *Science* **298,** 1179–1181 (2002).
2. Straneo, F. & Heimbach, P. North Atlantic warming and the retreat of Greenland's outlet glaciers. *Nature* **504,** 36–43 (2013).
3. Latif, M., Collins, M., Pohlmann, H. & Keenlyside, N. A review of predictability studies of Atlantic sector climate on decadal time scales. *J. Clim.* **19,** 5971–5987 (2006).
4. Smith, D. M. *et al.* Improved surface temperature prediction for the coming decade from a global climate model. *Science* **317,** 796–799 (2007).
5. Doblas-Reyes, F. J. *et al.* Initialized near-term regional climate change prediction. *Nat. Commun.* **4,** 1715 (2013).
6. Weaver, A. J. *et al.* Stability of the Atlantic meridional overturning circulation: a model intercomparison. *Geophys. Res. Lett.* **39,** L70209. doi: 10.1029/2012GL053763 (2012).
7. Swingedouw, D. *et al.* On the reduced sensitivity of the Atlantic overturning to Greenland ice sheet melting in projections: a multi-model assessment. *Clim. Dyn.* doi: 10.1007/s00382-014-2184-7 (2015).
8. McCarthy, G. *et al.* Observed interannual variability of the Atlantic meridional overturning circulation at 26.5°N. *Geophys. Res. Lett.* **39,** L19609. doi: 10.1029/2012GL052933 (2012).
9. Mercier, H. *et al.* Variability of the meridional overturning circulation at the Greenland–Portugal OVIDE section from 1993 to 2010. *Prog. Oceanogr.* (in the press).
10. Robson, J., Hodson, D., Hawkins, E. & Sutton, R. Atlantic overturning in decline? *Nat. Geosci.* **7,** 2–3 (2013).
11. Smeed, D. a. *et al.* Observed decline of the Atlantic Meridional Overturning Circulation 2004 to 2012. *Ocean Sci.* **10,** 1619–1645 (2013).
12. Reichler, T., Kim, J., Manzini, E. & Kröger, J. A stratospheric connection to Atlantic climate variability. *Nat. Geosci.* **5,** 783–787 (2012).
13. Huck, T., de Verdière, A. C., Estrade, P. & Schopp, R. Low-frequency variations of the large-scale ocean circulation and heat transport in the North Atlantic from 1955–1998 in situ temperature and salinity data. *Geophys. Res. Lett.* **35,** 1–5 (2008).
14. Vianna, M. L. & Menezes, V. V. Bidecadal sea level modes in the North and South Atlantic Oceans. *Geophys. Res. Lett.* **40,** 5926–5931 (2013).
15. Swingedouw, D., Mignot, J., Labetoulle, S., Guilyardi, E. & Madec, G. Initialisation and predictability of the AMOC over the last 50 years in a climate model. *Clim. Dyn.* **40,** 2381–2399 (2013).
16. Chylek, P., Folland, C. K., Dijkstra, H. A., Lesins, G. & Dubey, M. K. Ice-core data evidence for a prominent near 20 year time-scale of the Atlantic Multidecadal Oscillation. *Geophys. Res. Lett.* **38,** L13704. doi 10.1029/2011GL047501 (2011).
17. Divine, D. V. & Dick, C. Historical variability of sea ice edge position in the Nordic Seas. *J. Geophys. Res.* **111,** C01001 (2006).
18. Sicre, M.-A. *et al.* Decadal variability of sea surface temperatures off North Iceland over the last 2000 years. *Earth Planet. Sci. Lett.* **268,** 137–142 (2008).
19. Cronin, T., Farmer, J. & Marzen, R. Late holocene sea-level variability and Atlantic meridional overturning circulation. *Paleoceanography* **29,** 765–777 (2014).
20. Escudier, R., Mignot, J. & Swingedouw, D. A 20-year coupled ocean-sea ice-atmosphere variability mode in the North Atlantic in an AOGCM. *Clim. Dyn.* **40,** 619–636 (2013).
21. Frankcombe, L. M., von der Heydt, A. & Dijkstra, H. A. North Atlantic multidecadal climate variability: an investigation of dominant time scales and processes. *J. Clim.* **23,** 3626–3638 (2010).
22. Belkin, I. M., Levitus, S., Antonov, J. & Malmberg, S.-A. 'Great Salinity Anomalies' in the North Atlantic. *Prog. Oceanogr.* **41,** 1–68 (1998).
23. Belkin, I. M. Propagation of the 'Great Salinity Anomaly' of the 1990s around the northern North Atlantic. *Geophys. Res. Lett.* **31,** 5–8 (2004).
24. Sundby, S. & Drinkwater, K. On the mechanisms behind salinity anomaly signals of the northern North Atlantic. *Prog. Oceanogr.* **73,** 190–202 (2007).
25. Robock, A. Volcanic eruptions and climate. *Rev. Geophys.* **38,** 191–219 (2000).
26. Timmreck, C. Modeling the climatic effects of large explosive volcanic eruptions. *Wiley Interdiscip. Rev. Clim. Change* **3,** 545–564 (2012).
27. Driscoll, S., Bozzo, A., Gray, L. J., Robock, A. & Stenchikov, G. Coupled Model Intercomparison Project 5 (CMIP5) simulations of climate following volcanic eruptions. *J. Geophys. Res. Atmos.* **117,** D17105 doi 10.1029/2012JD017607 (2012).
28. Stenchikov, G. Arctic Oscillation response to the 1991 Mount Pinatubo eruption: effects of volcanic aerosols and ozone depletion. *J. Geophys. Res.* **107,** 4803 (2002).
29. Hurrell, J. W. Decadal trends in the north atlantic oscillation: regional temperatures and precipitation. *Science* **269,** 676–679 (1995).
30. Robock, A. & Mao, J. Winter warming from large volcanic eruptions. *Geophys. Res. Lett.* **19,** 2405–2408 (1992).
31. Otterå, O. H. Simulating the effects of the 1991 Mount Pinatubo volcanic eruption using the ARPEGE atmosphere general circulation model. *Adv. Atmos. Sci.* **25,** 213–226 (2008).
32. Zanchettin, D. *et al.* Bi-decadal variability excited in the coupled ocean–atmosphere system by strong tropical volcanic eruptions. *Clim. Dyn.* **39,** 419–444 (2012).

33. Otterå, O. H., Bentsen, M., Drange, H. & Suo, L. External forcing as a metronome for Atlantic multidecadal variability. *Nat. Geosci.* **3,** 688–694 (2010).
34. Mignot, J., Khodri, M., Frankignoul, C. & Servonnat, J. Volcanic impact on the Atlantic Ocean over the last millennium. *Clim. Past* **7,** 1439–1455 (2011).
35. Taylor, K. E., Stouffer, R. J. & Meehl, G. A. A. Summary of the CMIP5 experiment design. *Bull. Am. Meteorol. Soc.* **93,** 485–498 (2012).
36. Dufresne, J.-L. *et al.* Climate change projections using the IPSL-CM5 Earth System Model: from CMIP3 to CMIP5. *Clim. Dyn.* **40,** 2123–2165 (2013).
37. Menary, M. B. & Scaife, A. A. Naturally forced multidecadal variability of the Atlantic meridional overturning circulation. *Clim. Dyn.* **42,** 1347–1362 (2014).
38. Stenchikov, G. *et al.* Arctic Oscillation response to volcanic eruptions in the IPCC AR4 climate models. *J. Geophys. Res. D Atmos.* **107,** D244803 doi: 10.1029/2002JD002090 (2002).
39. Colin de Verdière, A. & Huck, T. Baroclinic instability: an oceanic wavemaker for interdecadal variability. *J. Phys. Oceanogr.* **29,** 893–910 (1999).
40. Dijkstra, H. A. & Ghil, M. Low-frequency variability of the large-scale ocean circulation: a dynamical systems approach. *Rev. Geophys.* **43,** 1–38 (2005).
41. Persechino, A., Mignot, J., Swingedouw, D., Labetoulle, S. & Guilyardi, E. Decadal predictability of the Atlantic meridional overturning circulation and climate in the IPSL-CM5A-LR model. *Clim. Dyn.* **40,** 2359–2380 (2013).
42. Reverdin, G. North Atlantic subpolar gyre surface variability (1895–2009). *J. Clim.* **23,** 4571–4584 (2010).
43. Ingleby, B. & Huddleston, M. Quality control of ocean temperature and salinity profiles—Historical and real-time data. *J. Mar. Syst.* **65,** 158–175 (2007).
44. Rayner, N. a. Global analyses of sea surface temperature, sea ice, and night marine air temperature since the late nineteenth century. *J. Geophys. Res.* **108,** D14 4407. doi: 10.1029/2002JD002670 (2003).
45. Kerr, R. A. A north atlantic climate pacemaker for the centuries. *Science* **288,** 1984–1985 (2000).
46. Mann, M. E. *et al.* Global signatures and dynamical origins of the Little Ice Age and Medieval Climate Anomaly. *Science* **326,** 1256–1260 (2009).
47. Gray, S. T., Graumlich, L. J., Betancourt, J. L. & Pederson, G. A tree-ring based reconstruction of the Atlantic Multidecadal Oscillation since 1567A.D. *Geophys. Res. Lett.* **31,** 2–5 (2004).
48. Chylek, P. *et al.* Greenland ice core evidence for spatial and temporal variability of the Atlantic Multidecadal Oscillation. *Geophys. Res. Lett.* **39,** L09705. doi: 10.1029/2012GL051241 (2012).
49. Butler, P. G., Wanamaker, A. D., Scourse, J. D., Richardson, C. a. & Reynolds, D. J. Variability of marine climate on the North Icelandic Shelf in a 1357-year proxy archive based on growth increments in the bivalve *Arctica islandica. Palaeogeogr. Palaeoclimatol. Palaeoecol.* **373,** 141–151 (2013).
50. Auger, P. a. *et al.* Interannual control of plankton communities by deep winter mixing and prey/predator interactions in the NW Mediterranean: results from a 30-year 3D modeling study. *Prog. Oceanogr.* **124,** 12–27 (2014).
51. Aumont, O. & Bopp, L. Globalizing results from ocean in situ iron fertilization studies. *Glob. Biogeochem. Cycles* **20,** GB2017. doi: 10.1029/2005GB002591 (2006).
52. Lehner, F., Raible, C. C. & Stocker, T. F. Testing the robustness of a precipitation proxy-based North Atlantic Oscillation reconstruction. *Quat. Sci. Rev.* **45,** 85–94 (2012).
53. Ammann, C. M., Joos, F., Schimel, D. S., Otto-Bliesner, B. L. & Tomas, R. A. Solar influence on climate during the past millennium: results from transient simulations with the NCAR Climate System Model. *Proc. Natl Acad. Sci. USA* **104,** 3713–3718 (2007).
54. Knight, J. R. A signature of persistent natural thermohaline circulation cycles in observed climate. *Geophys. Res. Lett.* **32,** L20728. doi: 10.1029/2005GL024233 (2005).
55. Eden, C. & Willebrand, J. Mechanism of interannual to decadal variability of the North Atlantic circulation. *J. Clim.* **14,** 2266–2280 (2001).
56. Böning, C. W., Scheinert, M., Dengg, J., Biastoch, A. & Funk, A. Decadal variability of subpolar gyre transport and its reverberation in the North Atlantic overturning. *Geophys. Res. Lett.* **33,** L21S01 (2006).
57. Graf, H.-F., Zanchettin, D., Timmreck, C. & Bittner, M. Observational constraints on the tropospheric and near-surface winter signature of the Northern Hemisphere stratospheric polar vortex. *Clim. Dyn.* **43,** 3245–3266 (2014).
58. Goldenberg, S. B., Landsea, C. W., Mestas-Nunez, A. M. & Gray, W. M. The recent increase in Atlantic hurricane activity: causes and implications. *Science* **293,** 474–479 (2001).
59. Hoerling, M., Hurrell, J., Eischeid, J. & Phillips, A. Detection and attribution of twentieth-century Northern and Southern African rainfall change. *J. Clim.* **19,** 3989–4008 (2006).
60. Levermann, A., Griesel, A., Hofmann, M., Montoya, M. & Rahmstorf, S. Dynamic sea level changes following changes in the thermohaline circulation. *Clim. Dyn.* **24,** 347–354 (2005).
61. Hourdin, F. *et al.* The LMDZ4 general circulation model: climate performance and sensitivity to parametrized physics with emphasis on tropical convection. *Clim. Dyn.* **27,** 787–813 (2006).
62. Madec, G. *NEMO Ocean Engine* (Institut Pierre-Simon Laplace (IPSL), 2008).
63. Fichefet, T. & Morales Maqueda, A. M. Sensitivity of a global sea ice model to the treatment of ice thermodynamics and dynamics. *J. Geophys. Res.* **102,** 2609–2612 (1997).
64. Vieira, L. E. A., Solanki, S. K., Krivova, N. A. & Usoskin, I. Evolution of the solar irradiance during the Holocene. *Astron. Astrophys.* **531,** A6 doi: 10.1051/0004-6361/201015843 (2011).
65. Grieser, J. & Schonwiese, C. D. Parameterization of spatio-temporal patterns of volcanic aerosol induced stratospheric optical depth and its climate radiative forcing. *Atmosfera* **12,** 111–133 (1999).
66. Ortega, P. *et al.* Characterizing atmospheric circulation signals in Greenland ice cores: insights from a weather regime approach. *Clim. Dyn.* (2014).
67. Wanamaker, A. D. *et al.* Surface changes in the North Atlantic meridional overturning circulation during the last millennium. *Nat. Commun.* **3,** 899 (2012).
68. Drinkwater, K. F. *et al.* The Atlantic Multidecadal Oscillation: its manifestations and impacts with special emphasis on the Atlantic region north of 60°N. *J. Mar. Syst.* **133,** 117–130 (2013).
69. Sévellec, F. & Fedorov, A. V. The leading, interdecadal Eigenmode of the Atlantic meridional overturning circulation in a realistic ocean model. *J. Clim.* **26,** 2160–2183 (2013).
70. Bretherton, C., Widmann, M., Dymnikov, V., Wallace, J. & Blade, I. The effective number of spatial degrees of freedom of a time-varying field. *J. Clim.* **12,** 1990–2009 (1999).

Acknowledgements

This research was partly funded by the ANR CEPS Green Greenland project and by ANR MORDICUS project (ANR-13-SENV-0002-02). It is also funded by the SPECS project funded by the European Commission's Seventh Framework Research Programme under the grant agreement 308378 and by the EMBRACE project with Research Number 282672. To analyze the CMIP5 data, this study benefited from the IPSL Prodiguer-Ciclad facility, which is supported by CNRS, UPMC, Labex L-IPSL, which is funded by the ANR (Grant #ANR-10-LABX-0018) and by the European FP7 IS-ENES2 project (Grant #312979). We are grateful to the Bedford Institute of Oceanography, EN3 and Gilles Reverdin for freely providing their ocean data. We also thank Laurent Terray and Aurélien Ribes for interesting discussions around this paper and Vincent Hanquiez and Patrick Brockmann for help with the figures.

Author contributions

D.S. conceived the study and led the writing of the paper. P.O. performed the Greenland ice core analysis. All authors contributed to data interpretation and writing of the manuscript.

Additional information

Competing financial interests: The authors declare no competing financial interests.

15

Enhanced light absorption by mixed source black and brown carbon particles in UK winter

Shang Liu[1,2], Allison C. Aiken[1], Kyle Gorkowski[1,3], Manvendra K. Dubey[1], Christopher D. Cappa[4], Leah R. Williams[5], Scott C. Herndon[5], Paola Massoli[5], Edward C. Fortner[5], Puneet S. Chhabra[5,6], William A. Brooks[5], Timothy B. Onasch[5,7], John T. Jayne[5], Douglas R. Worsnop[5], Swarup China[8], Noopur Sharma[8], Claudio Mazzoleni[8], Lu Xu[9], Nga L. Ng[9,10], Dantong Liu[11], James D. Allan[11,12], James D. Lee[13], Zoë L. Fleming[14], Claudia Mohr[15,16], Peter Zotter[17,18], Sönke Szidat[19] & André S.H. Prévôt[17]

Black carbon (BC) and light-absorbing organic carbon (brown carbon, BrC) play key roles in warming the atmosphere, but the magnitude of their effects remains highly uncertain. Theoretical modelling and laboratory experiments demonstrate that coatings on BC can enhance BC's light absorption, therefore many climate models simply assume enhanced BC absorption by a factor of ~1.5. However, recent field observations show negligible absorption enhancement, implying models may overestimate BC's warming. Here we report direct evidence of substantial field-measured BC absorption enhancement, with the magnitude strongly depending on BC coating amount. Increases in BC coating result from a combination of changing sources and photochemical aging processes. When the influence of BrC is accounted for, observationally constrained model calculations of the BC absorption enhancement can be reconciled with the observations. We conclude that the influence of coatings on BC absorption should be treated as a source and regionally specific parameter in climate models.

[1] Earth and Environmental Sciences Division, Los Alamos National Laboratory, Los Alamos, New Mexico 87545, USA. [2] Cooperative Institute for Research in the Environmental Sciences and Department of Chemistry and Biochemistry, University of Colorado, Boulder, Colorado 80309, USA. [3] Department of Civil and Environmental Engineering, Carnegie Mellon University, Pittsburgh, Pennsylvania 15213, USA. [4] Department of Civil and Environmental Engineering, University of California, Davis, California 95616, USA. [5] Aerodyne Research, Inc. Billerica, Massachusetts 01821, USA. [6] Department of Chemical Engineering, University of Texas at Austin, Austin, Texas 78712, USA. [7] Department of Chemistry, Boston College, Boston, Massachusetts 02467, USA. [8] Physics Department and Atmospheric Sciences Program, Michigan Technological University, Houghton, Michigan 49931, USA. [9] School of Chemical and Biomolecular Engineering, Georgia Institute of Technology, Atlanta, Georgia 30332, USA. [10] School of Earth and Atmospheric Sciences, Georgia Institute of Technology, Atlanta, Georgia 30332, USA. [11] School of Earth, Atmospheric and Environmental Science, University of Manchester, Manchester M13 9PL, UK. [12] National Centre for Atmospheric Science, University of Manchester, Manchester M13 9PL, UK. [13] Wolfson Atmospheric Chemistry Laboratory and National Centre for Atmospheric Science, University of York, York YO10 5DD, UK. [14] National Centre for Atmospheric Science, Department of Chemistry, University of Leicester, Leicester LE1 7RH, UK. [15] Department of Atmospheric Sciences, University of Washington, Seattle, Washington 98195, USA. [16] Institute for Meteorology and Climate Research, Karlsruhe Institute of Technology, Eggenstein-Leopoldshafen 76344, Germany. [17] Laboratory of Atmospheric Chemistry, Paul Scherrer Institute, Villigen 5232, Switzerland. [18] Lucerne School of Engineering and Architecture, Bioenergy Research, Lucerne University of Applied Sciences and Arts, Horw 6048, Switzerland. [19] Department of Chemistry and Biochemistry and Oeschger Centre for Climate Change Research, University of Bern, Bern 3012, Switzerland. Correspondence and requests for materials should be addressed to S.L. (email: shang.liu@colorado.edu) or to M.K.D. (email: dubey@lanl.gov).

Quantitative prediction of the direct warming effect of absorbing carbonaceous particles is challenging but crucial to assess their climate forcing[1–4]. The major atmospheric absorbing carbonaceous components are black carbon (BC) and brown carbon (BrC). BC strongly absorbs across the solar spectrum, whereas BrC preferentially absorbs at short wavelengths[2,5]. Uncertainties in the absorptive properties of BC and BrC contribute substantially to the total uncertainty estimates of their direct warming effects, in addition to other uncertainties associated with emission estimates, atmospheric transport simulations, and removal rates. Bond *et al.*[6] suggest that the mass absorption cross-section (*MAC*) of uncoated BC has an uncertainty of 16% at 550 nm. Additional uncertainties in absorption can be introduced as the optical features of BC are continuously modified during atmospheric aging processes. These aging processes mix BC with non-BC species, including BrC. Such mixing, and more specifically formation of coatings on BC particles, can occur over a few hours after BC emission and continue for days[7]. The coatings can enhance the absorption above that of uncoated BC particles through the so-called 'lensing effect.' The lensing effect has been confirmed by theoretical calculations[8–11] and laboratory measurements[12–16] but is not in line with field observations, discussed further below. The inconsistency suggests that using a lensing effect calculated from Mie theory may introduce ~50% uncertainty in BC's absorption, which can translate to ~50% uncertainty of modelled BC's direct warming effect. An additional uncertainty of ~20% may be introduced when including the absorption of BrC[4], which is ignored by most current climate models. Accordingly, it is important to constrain the dynamic absorptive properties of BC and BrC to improve model predictions.

The influence of BC coatings on BC's absorption has been studied theoretically and experimentally in the laboratory and in the field. Mie theory calculations that assume spherical BC-containing particles with a core–shell configuration, such as those employed by many current climate models, indicate that enhancement factors of up to 3 are plausible[8–11]. Laboratory studies demonstrate that such absorption enhancement can occur[12–16]. Therefore, current climate models typically either calculate an enhancement factor based on simplified mixing-state assumptions[17,18] or assume a constant enhancement value of ~1.5 for calculating BC absorption[19,20]. In contrast, field observations of BC absorption enhancement in urban and coastal California (USA) and Japan[14,21,22] yield small values (on average no more than 1.1 at 405, 532, and 781 nm), even when substantial coatings are present[14,21]. If true globally, this result questions BC's role as the second most important anthropogenic climate warmer after carbon dioxide[1,3]. The difference may be due to a preponderance of ambient BC being only partially encapsulated or attached to the host particle surface[23]. However, larger absorption enhancements of 1.4 (at 532 nm) have been observed in air masses strongly impacted by biomass burning[24]. Because of the paucity of comprehensive ambient measurements, the mechanisms responsible for the model-observation discrepancies remain elusive, underscoring the need for field studies in other representative environments, in particular those with mixed BC and BrC sources.

We perform an in-depth field study of light absorption enhancement factors (E_{abs}) for BC particles emitted from fossil fuel and residential burning sources in wintertime UK. The measurements are part of the 2012 Clean Air for London (ClearfLo) project[25] at Detling, a rural site 45 km southeast of central London. We demonstrate substantial light absorption enhancement of BC from our measurements. Through detailed chemical and microphysical analyses, we find that the magnitude of the observed E_{abs} strongly depends on the amount of coatings on BC, the BC sources, and the extent of particle aging. We also show that the E_{abs} at 405 nm is affected by low-volatility BrC. In addition to the observations, we systematically examine the theoretically calculated E_{abs} and determine the refractive indices for BC and BrC that result in agreement between calculated and observed E_{abs}. Furthermore, single-particle morphological analysis provides mechanistic insight of E_{abs} in comparison to previous studies. In light of these analyses, we conclude that the E_{abs} is source and regionally dependent. Therefore, parameters used for Mie modelling should be calibrated with observations to account for this effect.

Results

Direct observation of E_{abs}. Observed E_{abs} values were calculated using two independent and complementary methods. In one, E_{abs} was taken as the ratio between the *MAC* observed for ambient particles and for particles heated at 250 °C. BC is typically considered refractory at these temperatures and the heating is therefore used to remove the part of the coating material that is volatile. Particle heating was achieved by passing particles through a heated thermodenuder (TD). *MAC* is defined as the ratio of the absorption coefficient (b_{abs}) to refractory BC (rBC) concentration ([rBC]): $MAC = b_{abs}/[rBC]$. b_{abs} at 405 and 781 nm were quantified using photoacoustic spectroscopy[26] and [rBC] was measured using a single-particle soot photometer (SP2)[27,28]. Normalizing b_{abs} by [rBC] accounted for particle losses in the TD and temporal variations of b_{abs} during each ambient and TD measurement cycle. The observed E_{abs} calculated using this method is termed $E^{\lambda}_{abs-obs-TD}$ (λ denotes wavelength). The campaign average $E^{405}_{abs-obs-TD}$ and $E^{781}_{abs-obs-TD}$ values were 1.3 and 1.4, respectively, with their time series and frequency distributions shown in Supplementary Fig. 1. The second method was based on the absolute *MAC* values for ambient particles and the estimated enhancement is termed $E^{\lambda}_{abs-obs-MAC}$. $E^{\lambda}_{abs-obs-MAC}$ was calculated as $MAC^{\lambda}_{obs}/MAC^{\lambda}_{ref}$, where MAC_{obs} is the absolute *MAC* for ambient particles (shown in Supplementary Fig. 2.), and MAC_{ref} is the literature reference value for uncoated BC ($7.5 \pm 1.2\,m^2\,g^{-1}$ at 550 nm)[6]. Using the inverse wavelength dependence relationship for uncoated BC[6], we derived MAC_{ref} values of 10.2 ± 1.6 and $5.3 \pm 0.8\,m^2\,g^{-1}$ at 405 and 781 nm, respectively.

Since BrC absorbs at 405 nm but not at 781 nm, the observed E_{abs} at 405 nm (using both methods) includes the lensing-driven enhancement and the enhancement induced by semi-volatile BrC, whereas the observed E_{abs} at 781 nm represents the lensing-driven enhancement only. A glossary of the abbreviations used in this study is listed in Supplementary Table 1.

Effects of internal mixing on E_{abs}. We measured the composition of total non-refractory organic mass (OM) in submicron particles using a high-resolution time-of-flight aerosol mass spectrometer (HR-ToF-AMS)[29,30] and the non-refractory OM internally mixed with BC with a soot particle aerosol mass spectrometer (SP-AMS)[31]. The SP-AMS was operated in laser-only mode, which allows for detection of only rBC-containing particles (Methods). The ratio of non-rBC mass to rBC mass in rBC-containing particles is termed R_{BC}. Assuming a core–shell particle configuration, R_{BC} is a measure of relative coating thickness.

The observed E_{abs} at Detling increased substantially with increasing R_{BC} (Fig. 1a,b) unlike the previous studies in California[14,21], even though the maximum R_{BC} values observed here were three times smaller. In our study, there was good agreement between $E^{781}_{abs-obs-MAC}$ and $E^{781}_{abs-obs-TD}$, where BrC is thought to absorb negligibly. This result indicates that a

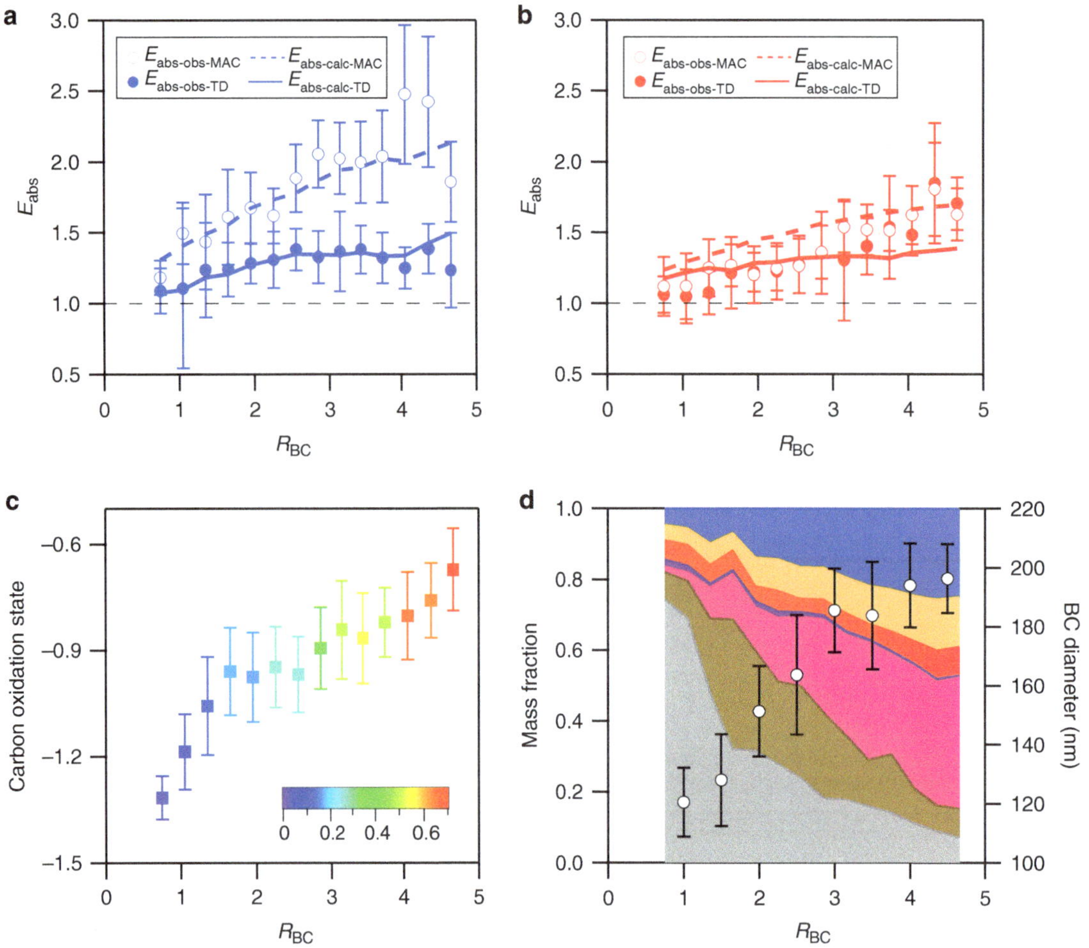

Figure 1 | Variation of E_{abs} and particle composition as a function of R_{BC}. (**a**,**b**) show measured and calculated absorption enhancement values versus R_{BC} at 405 and 781 nm, respectively. k_{BrC} of 0.004 and 0.032 were used for modelling absorption of ambient and thermodenuded organic mass at 405 nm, respectively. (**c**) Particle average oxidation state versus R_{BC}. The points are coloured by the organic mass fraction of the oxygenated organic aerosol factor, which was derived from factor analysis of the BC-associated OM. (**d**) Mass fraction of the non-refractory components internally mixed with BC (shaded areas) and BC core median volume-weighted diameter (open circles) as a function of R_{BC}. The colours represent nitrate (blue), ammonium (orange), sulfate (red), chloride (purple), oxygenated organic aerosol factor (pink), solid fuel organic aerosol factor (brown) and hydrocarbon-like organic aerosol factor (grey). The error bars in (**a**-**d**) represent s.d. of the values for each R_{BC} interval.

lensing-driven absorption enhancement was, at times, substantial in Detling. It also suggests that residual non-rBC materials did not have a notable impact on absorption by the thermodenuded particles. In contrast, $E^{405}_{abs-obs-MAC}$ was notably higher than $E^{405}_{abs-obs-TD}$. Furthermore, $E^{405}_{abs-obs-MAC}$ was larger than $E^{781}_{abs-obs-MAC}$ at a given R_{BC}. These features resulted from the direct contribution of BrC on the absorption at 405 nm.

As BrC contributes to the absorption at 405 nm, it is necessary to separate the absorption by BrC and evaluate the effect of BrC on E_{abs} at 405 nm. b_{abs} at 405 nm $\left(b^{405}_{abs}\right)$ was apportioned into absorption by residual BC (BC and low-volatility coatings on BC that remain on the particles after thermodenuding at 250 °C), low-volatility BrC that did not fully evaporate in the TD (BrC_{LV}), semi-volatile BrC that evaporated in the TD (BrC_{SV}), and the increment of BC absorption due to the lensing effect, with their respective fractions of 53, 24, 7 and 16% (Methods). The apportioned values were similar to the apportionment results of Lack *et al.*[24] at 404 nm inside a wildfire plume (Supplementary Table 2), even though we sampled a mixture of regional fossil fuel and residential burning sources with much lower absorption and OM concentrations. Given that the absorption of BrC_{LV} was 3.4 times the absorption of BrC_{SV} and that the mass of BrC_{LV} was only 9.3% of the mass of BrC_{SV} (measured by the HR-ToF-AMS assuming all OM is BrC), we estimated that the *MAC* of the BrC_{LV} was ~36 times the *MAC* of BrC_{SV}. The much higher *MAC* of BrC_{LV} was likely because BrC_{LV} contained larger molecules with more conjugated carbon bonds (compared with BrC_{SV}) that have lower volatility. This result is consistent with a recent study that shows BrC absorption from biomass burning is mostly associated with compounds of extremely low volatility[32]. Since the total non-refractory OM was substantially greater than the non-refractory OM that was associated with rBC, it is likely that much of the BrC was externally mixed with rBC.

The observed E_{abs} at 405 nm using the TD method was affected by BrC_{LV}. The increase of BrC_{LV} with R_{BC} (Supplementary Fig. 3a) suggests a larger contribution of BrC_{LV} to the absorption of thermodenuded particles as the rBC coating thickness increased, resulting in lower $E^{405}_{abs-obs-TD}$ at high R_{BC}. As a result, $E^{405}_{abs-obs-TD}$ leveled off for R_{BC} values greater than ~3 (Fig. 1a). In contrast, $E^{781}_{abs-obs-TD}$ continuously increased with R_{BC} (Fig. 1b) as the absorption of BrC was negligible at 781 nm.

Dependence of E_{abs} on sources and aging. To elucidate the mechanisms causing the enhanced BC absorption and formation of BrC_{LV}, the BC-associated OM (measured by the SP-AMS) and total OM (measured by the HR-ToF-AMS) were apportioned into linearly independent components using positive matrix factorization[33,34]. The analysis resulted in three factors for both BC-associated OM and total OM, namely oxygenated organic aerosol, solid fuel organic aerosol, and hydrocarbon-like organic aerosol. Based on the mass spectra, correlations with source tracers, and correlations with air mass sources from back trajectories, the oxygenated organic aerosol components were associated with long-range transport and aging of particles, the solid fuel organic aerosol components were associated with solid fuel burning for residential heating including biomass and coal, and the hydrocarbon-like organic aerosol components were associated with traffic emissions. The organic carbon in the total oxygenated organic aerosol component (oxygenated organic carbon) was further separated into fossil and non-fossil fractions using a combination of radiocarbon (^{14}C), factor analysis, and BC source analyses (Methods). The apportionment showed that total oxygenated organic carbon (and likely BC-associated oxygenated organic carbon) was largely (73–90%) derived from the non-fossil sources. In addition, we find that the absorption by BrC_{LV} correlated with the total oxygenated organic aerosol component (Supplementary Fig. 3b), indicating that BrC_{LV} may be secondary, as previous laboratory studies proposed[35,36]. In contrast, such correlation was not found for BrC_{SV}.

The variation of E_{abs} with R_{BC} was associated with changing sources as well as photochemical processing (Fig. 1c,d). The overall composition of rBC-containing particles varied with R_{BC} (Fig. 1d). The relative abundance of secondary components (inorganics and oxygenated organic aerosol component) increased with R_{BC}, while the relative abundance of primary components (hydrocarbon-like organic aerosol and solid fuel organic aerosol components) decreased. At lower R_{BC}, the BC-associated OM was dominated by traffic sources and the E_{abs} value (~ 1.1) was similar to the value observed at much higher R_{BC} by Cappa *et al.*[14], in which fossil fuel combustion is the dominant source for BC. Since the total oxygenated organic carbon was largely from non-fossil sources, the increase in R_{BC} was correlated with changing sources that could produce BC particles in distinct size regimes. Given that BC particles from biomass burning are typically larger than those from fossil fuel combustion[37], the correlation of BC core median volume-weighted diameter (measured by the SP2) with R_{BC} (Fig. 1d) was consistent with the oxygenated organic aerosol components being increasingly produced by biomass combustion related sources as the oxygenated organic aerosol concentration increased. The increase of BC diameter with R_{BC} may also indicate that coagulation during transport increased the BC size. However, since coagulation rates scale with the square of particle number (which decreases away from sources), and only coagulation between two BC-containing particles can lead to an increase in the BC core size, this hypothesis seems unlikely.

The increase in R_{BC} was correlated with an increasing average carbon oxidation state of BC-containing particles (Fig. 1c), which was calculated from the SP-AMS measured atomic O:C and H:C ratios (oxidation state $= 2 \times$ O:C—H:C)[38]. The atomic O:C and H:C ratios were calculated using the parameterization in Canagaratna *et al.*[39] Particle oxidation state increases with atmospheric oxidation and was used as a proxy for particle photochemical age here (we do not have the necessary measurements to calculate photochemical age). The correlation of R_{BC} with oxidation state suggests that photochemical processes over long timescales were likely important for producing both organic and inorganic coatings on BC.

Comparison of calculated and observed E_{abs}. An important consideration is the extent to which the observations are consistent with Mie theory calculations of BC absorption, as core-shell Mie theory is used in many current climate models. The E_{abs} values were calculated using observationally constrained core-shell Mie theory (Methods) by either assuming that no ($E_{abs\text{-}calc\text{-}MAC}$) or some ($E_{abs\text{-}calc\text{-}TD}$) residual material remained both on the BC particles and as externally mixed particles after heating[21]. The $E_{abs\text{-}calc\text{-}MAC}$ and $E_{abs\text{-}calc\text{-}TD}$ were conceptually comparable to $E_{abs\text{-}obs\text{-}MAC}$ and $E_{abs\text{-}obs\text{-}TD}$, respectively, assuming core-shell configuration for both ambient and thermodenuded BC-containing particles. An additional set of calculations were performed under the Rayleigh–Debye–Gans (RDG) approximation: The absorption by both uncoated and coated BC was determined by the properties of the individual spherules that made up a single BC particle[40]. The total absorption by a single particle is approximated as the sum of the absorption from individual spherules comprising a single particle, as calculated from spherical-particle Mie theory. As the size of the spherule was not explicitly known, and further might vary with the BC source, calculations were performed assuming spherules of 40 and 70 nm (diameter) to assess the sensitivity of the results to the assumed spherule size. To account for BrC contributions at 405 nm, calculations were performed using a range of imaginary refractive indices for the OM to find solutions that match the observed values. The base case assumed that OM internally mixed with BC had the same refractive index as that externally mixed from BC, and that the refractive index of the ambient OM was the same as the residual OM after heating. Alternative cases were also considered, and results of the calculations are shown in Fig. 2.

Recalling that $E_{abs\text{-}obs\text{-}TD} \sim E_{abs\text{-}obs\text{-}MAC}$ at 781 nm, the calculated E_{abs} generally reproduced the observed trend with R_{BC} (Fig. 1b). Comparison with the observed values at 781 nm showed that: (1) $E_{abs\text{-}calc\text{-}MAC}$ overestimated observations by 14% when R_{BC} was smaller than 2.9; (2) $E_{abs\text{-}calc\text{-}TD}$ overestimated observations by 14% for $R_{BC} < 1.4$ and underestimated observations by 17% for $R_{BC} > 3.8$; and (3) $E_{abs\text{-}calc\text{-}TD}$ was less dependent on R_{BC} compared to $E_{abs\text{-}obs\text{-}TD}$. However, the differences between calculated and observed absorption enhancement were comparable to the measurement uncertainty (15%) of $E_{abs\text{-}obs\text{-}TD}$. The difference between modelled $E_{abs\text{-}calc\text{-}MAC}$ and $E_{abs\text{-}calc\text{-}TD}$ at 781 nm resulted from (theoretical) contributions of residual coatings to the absorption by thermodenuded particles. As the observations suggested negligible contributions of residual coatings, this indicates that the $E_{abs\text{-}calc\text{-}MAC}$ overestimated the absorption enhancement due to the lensing effect for thermodenuded particles at a given R_{BC} compared with the observations.

The $E_{abs\text{-}calc\text{-}MAC}$ at 405 nm were lower than the observations when OM was assumed to be non-absorbing for both the Mie and RDG cases (Fig. 2), suggesting a role for BrC absorption. Direct absorption by the internally mixed BrC was very small compared with externally mixed BrC, which was primarily a reflection of the much larger mass concentration of externally mixed OM compared with BC-associated OM. This was assessed by comparing two cases, one with a bulk, campaign average imaginary refractive index (k_{BrC}) of 0.008 and the other with $k_{BrC} = 0$ for the internally mixed OM while leaving $k_{BrC} = 0.008$ for the externally mixed OM. The difference in observed and calculated $E_{abs\text{-}MAC}$ can be reconciled if the ambient OM was assumed to be slightly absorbing, with a bulk, campaign average $k_{BrC} = 0.004$ for the Mie case, $k_{BrC} \sim 0.008$ for RDG(70 nm) case, and $k_{BrC} \sim 0.012$ for the RDG(40 nm) case at 405 nm. The larger k_{BrC} required for RDG (40 nm) and RDG (70 nm) compared with Mie resulted from the larger initial difference between the $E_{abs\text{-}MAC}$ values for RDG (40 nm) and RDG (70 nm) when $k_{BrC} = 0$.

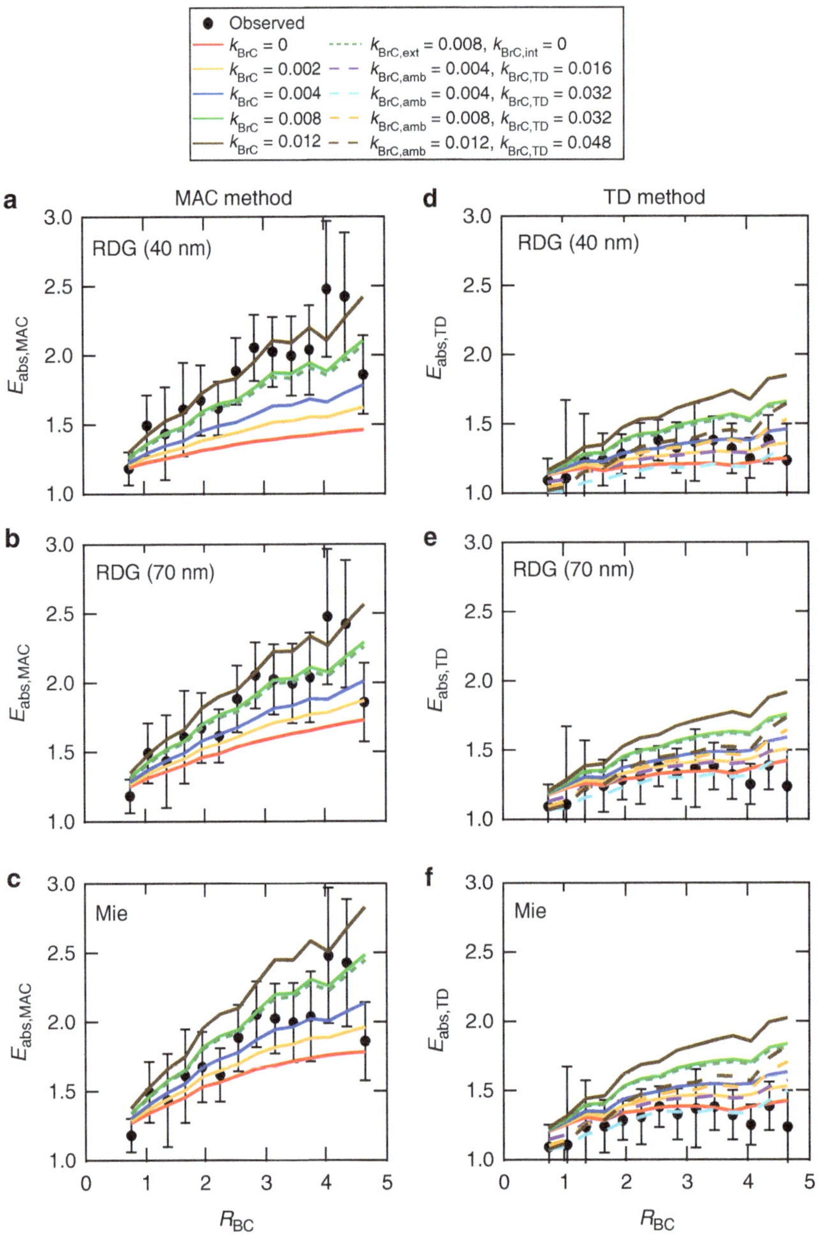

Figure 2 | Comparison of observed and modelled $E_{abs,MAC}$ and $E_{abs,TD}$ at 405 nm. Mie and RDG calculations (assuming BC spherule sizes of 40 and 70 nm) were performed using a range of k_{BrC} values that are shown in the legend. k_{BrC} indicates the cases in which a single k_{BrC} value was assumed for OM; $k_{BrC,ext}$ and $k_{BrC,int}$ represent the k_{BrC} values for OM that was externally and internally mixed with BC, respectively; $k_{BrC,amb}$ and $k_{BrC,TD}$ represent the k_{BrC} values for ambient and thermodenuded OM, respectively. The error bars in (**a**–**f**) represent s.d. of the measured absorption enhancement values for each R_{BC} interval.

The $E_{abs\text{-}TD}$ at 405 nm was well-simulated when the OM was assumed non-absorbing, which is inconsistent with the above finding that the OM needed to be absorbing at 405 nm to match the $E_{abs\text{-}MAC}$ measurements. Alternatively, reasonable quantitative agreement between the observed and calculated $E_{abs\text{-}TD}$ at 405 nm was obtained when the small amount of residual, low-volatility OM after thermodenuding was assumed to be more absorbing than the ambient OM. Specifically, good agreement (on average within 5%) was obtained for the Mie calculation when k_{BrC} of thermodenuded OM ($k_{BrC,TD}$) equals 0.032 at 405 nm, while $k_{BrC} = 0.004$. The *MAC* calculated for the BrC scales approximately linearly with k_{BrC}, and thus these results indicate the residual OM was approximately 8 times as absorbing per mass as the ambient OM. This finding indicates that the OM components that evaporated were comparably much less absorbing, with an estimated imaginary refractive index of 0.001 at 405 nm assuming volume mixing and that the average residual OM was 8.5% of the total (measured by the HR-ToF-AMS). This result gives a ratio between the $k_{BrC,TD}$ and the k_{BrC} for the evaporated components of ~ 32, consistent with the results from the apportionment of b_{abs}^{405} above. For the RDG calculation, good observation-model agreement can be achieved when the residual OM was six times as absorbing as the ambient OM, that is, $k_{BrC,TD} = 0.048$ for RDG(70 nm) and $k_{BrC,TD} = 0.072$ for RDG(40 nm). The overall results indicate that there was a non-negligible lensing-driven absorption enhancement of BC in Detling, and BrC contributed substantially to the observed particle absorption, and thus as well to the observed E_{abs} at 405 nm.

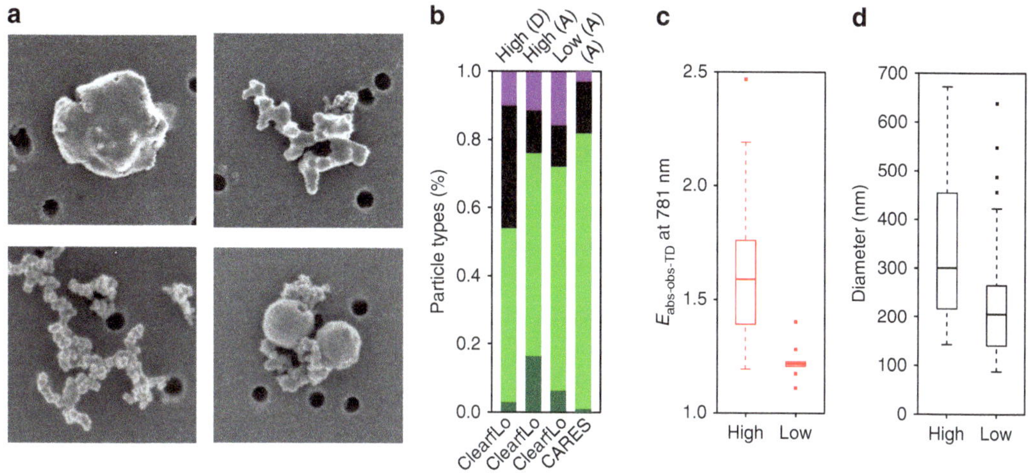

Figure 3 | Morphology and statistics of BC-containing particles. (**a**) Representative electron microscopy images of BC-containing particles collected at the Detling site for embedded (top left), partly coated (top right), thinly coated (bottom left) and partially encapsulated and/or surface attached (bottom right) BC particle types. The size of each panel is 1 µm by 1 µm. (**b**) Average number fraction of particle types with the colours indicating embedded (dark green), partly coated (green), thinly coated (black) and partially encapsulated and/or surface attached (purple) BC particle types for the high- and low-$E_{abs\text{-}obs\text{-}TD}$ samples at Detling during ClearfLo and at Sacramento during CARES. The 'D' and 'A' in the parenthesis indicate thermodenuded and ambient samples, respectively. (**c**) Box plot of $E^{781}_{abs-obs-TD}$ for the high- and low-$E_{abs\text{-}obs\text{-}TD}$ single particle sampling periods in Detling. (**d**) Box plot of area equivalent diameter of the embedded BC particles for the high-$E_{abs\text{-}obs\text{-}TD}$ samples (#1 and #2 in Supplementary Table 3) and low-$E_{abs\text{-}obs\text{-}TD}$ samples (#3 and #4 in Supplementary Table 3) for the Detling measurements. In the box plots (**c**,**d**), the bounds of each box represent quartiles, and the whiskers extend to the most extreme data points that are within 1.5 interquartile range of the box. The horizontal bars inside the boxes represent the median values.

Mechanistic insight of E_{abs} from particle morphology. To develop a mechanistic understanding of the notable dependence of E_{abs} on R_{BC} at Detling, single-particle images from scanning electron microscopy were analysed. Individual BC particles were classified visually based on their coating and morphologies as 'embedded', 'partly coated', 'thinly coated', and 'partially encapsulated and/or surface attached' (Fig. 3a), similarly to China *et al.*[41,42] The number fraction of each type was determined (Supplementary Table 3) and the averages are shown in Fig. 3b. (The electron microscopy analysis may exhibit some composition-dependent biases due to possible losses of semi-volatile materials; Methods.)

Two samples were selected for analysis when the $E_{abs\text{-}obs\text{-}TD}$ at 781 nm was high and another two samples were selected during low-$E_{abs\text{-}obs\text{-}TD}$ periods. The $E^{781}_{abs-obs-TD}$ values for the high- and low-$E_{abs\text{-}obs\text{-}TD}$ samples were 1.67 ± 0.42 (s.d.) and 1.23 ± 0.07 (s.d.), respectively (Fig. 3c). The embedded, partly coated, thinly coated and partially encapsulated and/or surface attached particle types had number fractions of 17%, 60%, 12% and 11% (out of 2187 particles) for the high-$E_{abs\text{-}obs\text{-}TD}$ and 6%, 66%, 12% and 16% (out of 2017 particles) for the low-$E_{abs\text{-}obs\text{-}TD}$ samples. Therefore, the largest difference between the high- and low-$E_{abs\text{-}obs\text{-}TD}$ samples was that the high-$E_{abs\text{-}obs\text{-}TD}$ samples contained ~ 3 times more embedded BC particles. The morphological configuration of embedded BC particles is closest to the core–shell structure commonly used in models and it is theoretically expected to show the largest absorption enhancement. In addition, the embedded particles for the high-$E_{abs\text{-}obs\text{-}TD}$ samples had significantly larger area equivalent diameter than those for the low-$E_{abs\text{-}obs\text{-}TD}$ samples (Fig. 3d) with the difference ranging between 58 and 159 nm at the 95% confidence level, suggesting that the BC particles associated with high-$E_{abs\text{-}obs\text{-}TD}$ underwent substantial coating. Moreover, the thermodenuded particles during the high-$E_{abs\text{-}obs\text{-}TD}$ period (Supplementary Table 3) showed a substantially larger fraction (36%) of thinly coated BC particles and a smaller fraction (3%) of embedded BC particles compared with the simultaneously collected ambient sample (Fig. 3b), consistent with the presence of semi-volatile components in the coatings on BC particles.

A similar analysis for particles collected at the urban T0 site (14 km northeast of the Sacramento downtown, CA, USA) during the Carbonaceous Aerosols and Radiative Effects Study (CARES) campaign is also shown in Supplementary Table 3 and the fractions of each type are shown in Fig. 3b. The embedded fraction was substantially higher at Detling (12%) than in Sacramento during the CARES campaign (1%), where only small E_{abs} is observed even at relatively large photochemical age (up to ~ 15 h)[14]. In addition, the convexity[42], which is associated with the compactness of the BC-containing particles, was higher for Detling (0.82 ± 0.005) than CARES (0.76 ± 0.005), suggesting different particle morphology at the two sites. This result indicates that mixing state may play an important role in determining E_{abs}. However, comparable electron microscopy measurements are not available for the CalNex campaign[14], where small E_{abs} values are observed even though similar SP-AMS measurements show substantially larger R_{BC} compared to the Detling measurements. Therefore, the hypothesis that the morphological details of the internal mixing-state significantly influence the absorption enhancement needs additional evidence with further field studies covering a range of different environments.

Discussion

Our field results demonstrate clearly that coatings can substantially enhance light absorption by BC emitted from mixed fossil fuel and residential solid fuel combustion sources. Photochemical aging can play an important role in BC's absorption enhancement by coating BC particles with condensable materials. Comparison with results from other field campaigns indicates that the influence of coatings on BC absorption may be source and regionally specific. We hypothesize that mixed combustion sources and atmospheric transformations can produce favourable conditions of enhanced BC absorption by coatings, including

BrC. Our analysis of absorption enhancement as a function of morphology, mixing state, coating thickness and the amount of low-volatility and semi-volatile BrC at wintertime Detling supports our hypothesis. We highlight that the effects of BrC, in particular the low-volatility BrC, on BC's absorption enhancement at short wavelengths should be considered in future studies. Our findings should be evaluated by future measurements of atmospherically processed BC particles from mixed combustion sources at other locations. We recommend that models treat the variability of BC's absorption to account for the dynamic aging processes and regional variations in sources by calibrating Mie models to measurements. Continued assessment of the absorption enhancement by BC coatings (including BrC) through comparing observations and observationally constrained calculations is necessary to develop robust numerical models, in order to accurately quantify the past, present and future contributions of absorbing particles to climate change.

Methods

Thermodenuder. A thermodenuder (TD, Aerodyne Research, Inc., Billerica, MA, USA) was installed upstream of all the instruments used in this study to evaporate semi-volatile particulate components. The flowrate through the TD was 2.3 l.p.m., resulting in a residence time of 5.3 s, which was shorter than the 8.5 s residence time in the study by Cappa *et al.*[14]

Ambient particles and heated particles were alternately sampled during 5–15 February 2012 with TD temperatures of 120 and 250 °C. Each ambient-TD cycle included 10-min ambient measurements and 10-min TD measurements. Unless stated otherwise, the E_{abs} values derived from 250 °C thermodenuding were reported.

Photoacoustic soot spectrometer. A three-wavelength photoacoustic soot spectrometer (PASS-3, Droplet Measurement Technologies, Inc., Boulder, CO, USA) was used to continuously measure the absorption and scattering coefficients of fine particles at 405, 532 and 781 nm. In the PASS-3, the particle laden sample is illuminated with a modulated laser. The energy absorbed from the incident light is thermally transferred to the air surrounding the particle and the subsequent air expansion produces a sound wave[26]. The sound wave is measured by a microphone that was calibrated at 532 nm at Droplet Measurement Technologies, Inc. using NO_2 (ref. 43). The microphone calibration is independent of wavelength and the lasers at 405, 532 and 781 nm were calibrated using a laser power meter. The calibration was checked using size selected fullerene soot particles (Alpha Aesar, Ward Hill, Massachusetts) following the procedures described by Flowers *et al.*[44] An NO_2 scrubber was installed upstream of the PASS-3 to remove the interference of NO_2 absorption at 405 nm. Background measurements of filtered air were made every 2.5 min to allow for correction of changes in gas composition. During the Detling measurements, low 532 nm laser power and drifts caused large variations of the signal in this channel. Therefore, b_{abs}^{532} was used as a qualitative measurement only and is not included here. Allan variance plots were made before the measurements to evaluate the precision and stability of the performance of the PASS-3 (refs 45,46). The precision at 2.5 min integration time was 0.5 M m^{-1} for b_{abs}^{405} and 0.3 M m^{-1} for b_{abs}^{781}. Zero tests using high-efficiency particulate arrestance filters suggested that the gas-phase absorption was within the precision of the PASS-3 measurements. The upper limit of the gas-phase contribution to absorption was estimated to be 4 and 6% at 405 and 781 nm, respectively. The relative humidity inside the PASS-3 instrument was smaller than 30%, suggesting that the particles were dry by the time they were measured. The low relative humidity inside the PASS-3 instrument was due to warm up of the cold air brought into the warmer container and the relatively high temperature (20–30 °C) in the PASS-3 measurement cell induced by the electronics.

To verify the PASS-3 measurements, the absorption measured by the PASS-3 was compared with the absorption calculated from a 7-wavelength Aethalometer (MAGEE Scientific, model AE31) that operated simultaneously at the same site[47]. The PASS-3 measured absorption correlated well with the Aethalometer-derived absorption with Pearson's $r \geq 0.94$ at 405 and 781 nm. Major axis line-fitting suggests that the PASS-3 and Aethalometer measurements agreed well, with slopes close to 1 (0.934 ± 0.004 at 405 and 0.918 ± 0.004 at 781 nm) and intercepts slightly negative (-2.724 ± 0.085 and -1.847 ± 0.037 M m^{-1}).

SP2. The refractory BC (rBC) mass concentration was quantified using an SP2 (Droplet Measurement Technologies, Inc., Boulder, CO, USA). Measurement principles of the SP2 have been described in detail in previous publications[27,28]. In brief, the SP2 selectively measures the rBC-containing particles using a diode-pumped 1064 nm Nd:YAG laser on a single-particle basis. The rBC-containing particles absorb the energy emitted by the laser and are heated to their vaporization temperature and incandesce. The peak intensity of the incandescence signal is proportional to the mass of rBC, allowing for the quantification of rBC mass concentration. The SP2 measures single rBC-containing particles with an rBC mass range of ~3–300 fg (ref. 27). The size of the rBC was calculated assuming a density of 1.8 g cm^{-3} and assuming spherical geometry with no internal voids[48]. Size-selected fullerene soot (Alfa Aesar, Ward Hill, Massachusetts; Stock# 40971, Lot# L18U002) was used for the SP2 mass calibration[49]. Contributions of rBC particles outside the detection window—which can vary with rBC source—were quantitatively estimated by fitting time-dependent rBC size distributions assuming a bimodal lognormal shape. The good correlation between absorption and [rBC] (Supplementary Fig. 4) suggests that the lognormal correction was sufficient to account for the rBC mass outside the SP2 detection limit.

HR-ToF-AMS. An Aerodyne HR-ToF-AMS (Aerodyne, Research Inc., Billerica, MA, USA) was deployed to measure the non-refractory organics, sulfate, nitrate, ammonium and chloride in submicron particles with 2-min time resolution. The HR-ToF-AMS configuration and principle of operation have been described in detail previously[29,30]. Briefly, particles that are focused and accelerated by an aerodynamic lens impact a heated vaporizer (600 °C). The non-refractory components in the particles are flash vapourized and ionized, and the ionized fragments are measured by a high-resolution time-of-flight mass spectrometer.

SP-AMS. The SP-AMS (Aerodyne Research, Inc., Billerica, MA, USA) incorporates a 1064 nm Nd:YAG laser vaporizer (the same laser as used in the SP2 instrument) into the standard Aerodyne HR-ToF-AMS to detect rBC via laser-induced heating and vaporization[31]. During 5–15 February 2012, the tungsten thermal vaporizer was removed from the SP-AMS so that only rBC and the rBC-associated non-refractory particulate matter (including organics, sulfate, nitrate, ammonium and chloride) were detected. By comparing the ambient and thermodenuded OM that was associated with BC (normalized to [rBC] to account for particle losses in the heated TD), we find that on average 24% of OM that was internally mixed with rBC remained associated with the rBC particles after heating to 250 °C. This result is consistent with a previous study[50] that shows large contribution (~11%) of organic components to particulate residuals after particle heating at 300 °C.

Electron microscopy analysis. Single particles were collected on Nuclepore filters (100 nm pores) using an aspiration sampling technique. The sample collection time was 3–6 h depending on the particle concentration. For each collection time period, two filter samples, one representing ambient and the other representing heated (at 120 °C through a TD) particles, were collected simultaneously. This TD was different from the TD used for the PASS-3, HR-ToF-AMS and SP-AMS, and therefore it may remove a different amount of material. After collection, the samples were stored in the dark before analysis. The particle images were acquired using a field emission scanning electron microscope (Hitachi S-4700). We note that only particles larger than 100 nm were analysed to reduce uncertainties in particle classification. In addition, the number of embedded BC-containing particles was likely underestimated because (1) electron microscopy images only the surface features of the particles, thereby the BC-containing particles that were entirely engulfed by other materials and completely lost their original BC morphology could not be identified and (2) the electron microscopy analysis was conducted under vacuum, so evaporation of semi-volatile coatings on BC particles could not be avoided. Therefore, the number fraction for each particle type derived from the electron microscopy analysis was semi-quantitative.

Calculation of $E_{abs\text{-}obs\text{-}TD}$. The TD method is described by the following equation

$$E_{abs-obs-TD}^{\lambda} = \frac{\left(b_{abs,amb}^{\lambda}(t-1)/[rBC](t-1) + b_{abs,amb}^{\lambda}(t+1)/[rBC](t+1)\right)/2}{b_{abs,TD}^{\lambda}(t)/[rBC](t)} \quad (1)$$

where $b_{abs,amb}^{\lambda}$ and $b_{abs,TD}^{\lambda}$ represent the absorption coefficients of ambient and thermodenuded particles, respectively; [rBC] is the mass concentration of rBC measured by the SP2; t represents the TD measurement, and $t-1$ and $t+1$ represent the times corresponding to ambient measurements before and after the TD measurement, respectively. Over 95% of the $E_{abs\text{-}obs\text{-}TD}$ values were >1, with the values smaller than 1 resulting from temporal atmospheric variability during the sequential ambient and TD measurements.

Calculation of $E_{abs\text{-}obs\text{-}MAC}$. In the calculation of $E_{abs\text{-}obs\text{-}MAC}$, the measurements were binned by the R_{BC} intervals, and the *MAC* of BC was determined as the slope of the line-fitting of b_{abs}^{λ} as a function of [rBC] for each measurement section (Supplementary Fig. 4). We examined the linear regression, major axis (also called orthogonal regression), and standardized major axis (also called reduced major axis) least square line-fitting methods. The linear regression method only accounts for the errors in the Y direction, while the other two methods take into account errors in both X and Y directions. The standardized major axis is the major axis calculated on standardized data. The standardized major axis is typically used when

X and Y data are not measured on comparable scales[51], which is appropriate for the b_{abs} versus [rBC] line-fitting in this study. In addition, comparison of the three fitting methods results showed that the slopes derived from standardized major axis were in between those derived from the linear regression and major axis fitting methods (Supplementary Fig. 2). An additional method was to take the average point-to-point b_{abs}/[rBC] as the MAC value for each R_{BC} interval, defined as the 'ratio method.' The ratio method did not give the same results as those derived from the slope methods. We suspect that this difference was due to a systematic offset in the PASS-3 measurements that affected the ratio method but not the slope methods. For these reasons, we used the slopes derived from the standardized major axis fitting as the MAC values in this study. $E_{abs\text{-}obs\text{-}MAC}$ was calculated as the ratio of the calculated MAC to the reference MAC of uncoated BC.

Apportionment of the absorption at 405 nm. The measured b_{abs}^{405} was the sum of the absorption of BC, BrC (both internally and externally mixed with BC), and the enhanced absorption due to the lensing effect of coatings on BC. Upon heating at 250 °C, the semi-volatile OM and semi-volatile inorganic species evaporated, leaving only BC and low-volatility non-BC materials. Unless stated otherwise, the temperature threshold for separating semi-volatile and low-volatility components was defined as 250 °C in this study. b_{abs}^{405} can be apportioned using $b_{abs,amb}^{405}$ and $b_{abs,TD}^{405}$ (at 250 °C) based on the following assumptions: (1) BrC absorbs at 405 nm but not at 781 nm (ref. 24); (2) the absorption of the residual BC particles (after heating) follows the inverse wavelength dependence; and (3) to a first-order approximation the lensing-driven enhancement for BC particles is wavelength independent, that is, $E_{abs}^{405}(\text{Lens}_{SV}) = E_{abs}^{781}(\text{Lens}_{SV})$ for the same BC-containing particle[52]. Note that the measured ambient $E_{abs-obs-TD}^{405}$ included the lensing-driven enhancement and the enhancement induced by evaporation of BrC_{SV}, and further $E_{abs-obs-TD}^{405}$ was affected by BrC_{LV} as discussed in the main text. Therefore $E_{abs-obs-TD}^{405}$ is different from $E_{abs}^{405}(\text{Lens}_{SV})$ used in the equations below: $E_{abs}^{405}(\text{Lens}_{SV})$ excludes the effects of BrC (BrC_{LV} and BrC_{SV}) and represents the lensing-driven enhancement only. In comparison with $E_{abs-obs-TD}^{405}$, the measured $E_{abs-obs-TD}^{781}$ was due to the lensing-driven enhancement only, that is, $E_{abs-obs-TD}^{781} = E_{abs}^{781}(\text{Lens}_{SV})$. The apportionment is described by the following equations with the variables described in Supplementary Table 1:

$$b_{abs}^{405}(T_{250}) = b_{abs}^{405}(\text{BC}_{residual}) + b_{abs}^{405}(\text{BrC}_{LV}) \quad (2)$$

$$b_{abs}^{781}(T_{250}) = b_{abs}^{781}(\text{BC}_{residual}) \quad (3)$$

$$\frac{b_{abs}^{405}(\text{BC}_{residual})}{b_{abs}^{781}(\text{BC}_{residual})} = \left(\frac{405}{781}\right)^{-1} \quad (4)$$

$$b_{abs}^{405}(T_{amb}) = b_{abs}^{405}(\text{BC}_{residual}) \cdot E_{abs}^{405}(\text{Lens}_{SV}) + b_{abs}^{405}(\text{BrC}_{LV}) + b_{abs}^{405}(\text{BrC}_{SV}) \quad (5)$$

$$E_{abs}^{405}(\text{Lens}_{SV}) = E_{abs}^{781}(\text{Lens}_{SV}) = E_{abs-obs-TD}^{781} \quad (6)$$

From equations (2–4), $b_{abs}^{405}(\text{BrC}_{LV})$ and $b_{abs}^{405}(\text{BC}_{residual})$ were solved; $b_{abs}^{405}(\text{BrC}_{SV})$ was then derived using equations (5) and (6).

Similar apportionment of b_{abs}^{405} was performed using the b_{abs}^{405} for particles that were thermodenuded at 120 °C. In this case, the BrC_{LV} was referred to as the fraction of BrC that remained in the particles after thermodenuding at 120 °C. The apportioned $b_{abs}^{405}(\text{BrC}_{LV})$ at 120 °C and 250 °C correlated in time with a Pearson's r of 0.8, and the latter was 19% lower on average, which was likely because of the absorption of BrC that evaporated between 120 and 250 °C. Since charring of OM is unlikely to take place at 120 °C, the correlation of apportioned $b_{abs}^{405}(\text{BrC}_{LV})$ at 120 and 250 °C suggests that $b_{abs}^{405}(\text{BrC}_{LV})$ at 250 °C was a real signal rather than an artefact resulting from pyrolyzed OM. This result is consistent with previous studies that confirm that the interference of pyrolyzed OM on the absorption measurements is $<1.7\%$ at 300 °C (ref. 53).

Non-fossil fractions of oxygenated organic carbon. Radiocarbon (^{14}C) analysis was used to distinguish the fossil and non-fossil carbon fractions of the oxygenated organic aerosols retrieved from the HR-ToF-AMS analysis. Non-fossil carbon fraction (f_{NF}) was calculated as the ratio of $^{14}C/^{12}C$ for the sample to $^{14}C/^{12}C$ in 1950 after correcting for field blanks and nuclear weapon tests in the 1950s and 1960s (ref. 54). The f_{NF} for total carbon ($f_{NF,TC}$) in Detling has been reported by Crilley *et al.*[55], in which the total carbon is the sum of organic carbon and elemental carbon measured with a Sunset thermal-optical carbon analyser. f_{NF} of BC ($f_{NF,BC}$) in Detling was assumed to be the wood burning fraction of BC, apportioned using the Aethalometer measurements and a source apportionment model[56] as reported in Crilley *et al.* The average $f_{NF,TC}$ and $f_{NF,BC}$ at Detling during ClearfLo are $64 \pm 7\%$ and $30 \pm 13\%$ (ref. 55), respectively.

Here, we expanded the Crilley *et al.* analysis to analyse the f_{NF} for oxygenated organic aerosols using the method described in Zotter *et al.*[54] The equation we used for the f_{NF} calculation is as follows

$$f_{NF,OOC} \times \text{OOC} = f_{NF,TC} \times \text{TC} - f_{NF,BC} \times \text{rBC} - f_{NF,HOC} \times \text{HOC} - f_{NF,SFOC} \times \text{SFOC} \quad (7)$$

where HOC, SFOC and OOC are the organic carbon concentrations for hydrocarbon-like organic aerosol, solid fuel organic aerosol, and oxygenated organic aerosol factors, respectively. The organic carbon concentration of the factor was converted from the corresponding factor OM concentration, using the OM to organic carbon ratios derived from the factor spectra. rBC is the ambient refractory BC concentration measured by the SP2. TC is the total carbon concentration, equivalent to the sum of HOC, SFOC, OOC, and BC concentrations. $f_{NF,TC}$ and $f_{NF,BC}$ values are the same as reported in Crilley *et al.*[55] We used the $f_{NF,TC}$ and $f_{NF,BC}$ values for 5–12 February 2012 that overlapped the E_{abs} measurements. $f_{NF,HOC}$ and $f_{NF,SFOC}$ are the non-fossil carbon fractions for HOC and SOFC, respectively. Since $f_{NF,HOC}$ and $f_{NF,SFOC}$ were not explicitly known, we performed sensitivity analysis using a pseudo Monte Carlo method. In this method, $f_{NF,HOC}$ was varied between 0 and 0.3, representing its substantial fossil emissions and accounting for the potential contribution of non-fossil cooking and biofuel combustion (for example, biodiesel) emissions that were not separated from HOC; $f_{NF,SFOC}$ was varied between 0.9 and 1.0 to consider the dominance of wood burning and account for the contribution of residential (fossil) coal burning to SFOC.

The apportionment result (Supplementary Fig. 5) showed that the $f_{NF,OOC}$ ranged from 0.75 to 0.93, suggesting that the majority of oxygenated organic aerosols was from non-fossil sources in Detling during ClearfLo.

Mie and RDG model calculations. The Mie calculations were performed using the same Igor code (Wavemetrics Inc) as in Cappa *et al.*[14], which was adapted from the Fortran code of Bohren and Huffman[57]. The input size distributions for the calculation were constrained by observations using the same set of measurements (from the same types of instruments) and methods as detailed in Cappa *et al.*, that is, the core BC size distribution was constrained by the SP2 measurements, the coated BC-containing particle size distribution was determined by R_{BC} and the core BC size distribution assuming that the coated components were evenly distributed around BC particles, and the non-BC particle size distribution was derived by subtracting the SP2-measured BC size distribution from the total particle size distribution measured by a scanning mobility particle sizer.

In the RDG calculation, the number of spherules per particle was calculated by the assumed spherule size and the SP2-measured spherical equivalent volumes of the overall BC core. It was assumed that the amount of coating on each spherule was proportionally the same as for the total particle, that is, that R_{BC} was the same for the spherule as for the total particle. This assumption is equivalent to having the ratio between the coated diameter and the core diameter conserved between the standard Mie (total particle) formulation and the RDG formulation. The absorption enhancements for the RDG cases were calculated analogously to the Mie case.

The refractive index of BC was assumed to be $n = 1.88 + 0.8i$ (ref. 14). Since the imaginary refractive index of BrC (k_{BrC}) is highly uncertain and can vary greatly, we performed calculations to estimate the k_{BrC} that resulted in good agreement between predicted and measured $E_{abs\text{-}MAC}$ and $E_{abs\text{-}TD}$ at 405 nm. It was assumed that the refractive index was a bulk average for OM.

References

1. Ramanathan, V. & Carmichael, G. Global and regional climate changes due to black carbon. *Nat. Geosci.* **1,** 221–227 (2008).
2. Andreae, M. O. & Gelencsér, A. Black carbon or brown carbon? The nature of light-absorbing carbonaceous aerosols. *Atmos. Chem. Phys.* **6,** 3131–3148 (2006).
3. Bond, T. *et al.* Bounding the role of black carbon in the climate system: a scientific assessment. *J. Geophys. Res.* **118,** 5380–5552 (2013).
4. Feng, Y., Ramanathan, V. & Kotamarthi, V. Brown carbon: a significant atmospheric absorber of solar radiation? *Atmos. Chem. Phys.* **13,** 8607–8621 (2013).
5. Kirchstetter, T. W. & Novakov, T. Evidence that the spectral dependence of light absorption by aerosols is affected by organic carbon. *J. Geophys. Res.* **109,** D21208 (2004).
6. Bond, T. C. & Bergstrom, R. W. Light absorption by carbonaceous particles: an investigative review. *Aerosol Sci. Technol.* **40,** 27–67 (2006).
7. Moffet, R. C. & Prather, K. A. *In-situ* measurements of the mixing state and optical properties of soot with implications for radiative forcing estimates. *Proc. Natl Acad. Sci. USA* **106,** 11872–11877 (2009).
8. Fuller, K. A., Malm, W. C. & Kreidenweis, S. M. Effects of mixing on extinction by carbonaceous particles. *J. Geophys. Res.* **104,** 15941–15954 (1999).
9. Jacobson, M. Z. Strong radiative heating due to the mixing state of black carbon in atmospheric aerosols. *Nature* **409,** 695–697 (2001).
10. Bond, T. C., Habib, G. & Bergstrom, R. W. Limitations in the enhancement of visible light absorption due to mixing state. *J. Geophys. Res.* **111,** D20211 (2006).
11. Schwarz, J. *et al.* Coatings and their enhancement of black carbon light absorption in the tropical atmosphere. *J. Geophys. Res.* **113,** D03203 (2008).
12. Cross, E. S. *et al.* Soot particle studies—instrument inter-comparison—project overview. *Aerosol Sci. Technol* **44,** 592–611 (2010).

13. Schnaiter, M. *et al.* Absorption amplification of black carbon internally mixed with secondary organic aerosol. *J. Geophys. Res.* **110,** D19204 (2005).
14. Cappa, C. D. *et al.* Radiative absorption enhancements due to the mixing state of atmospheric black carbon. *Science* **337,** 1078–1081 (2012).
15. Shiraiwa, M., Kondo, Y., Iwamoto, T. & Kita, K. Amplification of light absorption of black carbon by organic coating. *Aerosol Sci. Technol.* **44,** 46–54 (2010).
16. Zhang, R. *et al.* Variability in morphology, hygroscopicity, and optical properties of soot aerosols during atmospheric processing. *Proc. Natl Acad. Sci. USA* **105,** 10291–10296 (2008).
17. Chung, S. H. & Seinfeld, J. H. Climate response of direct radiative forcing of anthropogenic black carbon. *J. Geophys. Res.* **110,** D11102 (2005).
18. Penner, J. E., C. Chuang, C. C. & Grant, K. Climate forcing by carbonaceous and sulfate aerosols. *Clim. Dyn.* **14,** 839–851 (1998).
19. Flanner, M. G., Zender, C. S., Randerson, J. T. & Rasch, P. J. Present-day climate forcing and response from black carbon in snow. *J. Geophys. Res.* **112,** D11202 (2007).
20. Wang, Q. *et al.* Global budget and radiative forcing of black carbon aerosol: constraints from pole-to-pole (HIPPO) observations across the Pacific. *J. Geophys. Res.* **119,** 195–206 (2014).
21. Cappa, C. D. *et al.* Response to comment on 'radiative absorption enhancements due to the mixing state of atmospheric black carbon'. *Science* **339,** 393–393 (2013).
22. Nakayama, T. *et al.* Properties of light-absorbing aerosols in the Nagoya urban area, Japan, in August 2011 and January 2012: contributions of brown carbon and lensing effect. *J. Geophys. Res.* **119,** 12721–12739 (2014).
23. Adachi, K. & Buseck, P. R. Changes of ns-soot mixing states and shapes in an urban area during CalNex. *J. Geophys. Res.* **118,** 3723–3730 (2013).
24. Lack, D. A. *et al.* Brown carbon and internal mixing in biomass burning particles. *Proc. Natl Acad. Sci. USA* **109,** 14802–14807 (2012).
25. Bohnenstengel, S. *et al.* Meteorology, air quality, and health in London: The ClearfLo project. *Bull. Am. Meteor. Soc.* **96,** 779–804 (2015).
26. Arnott, W. P., Moosmüller, H., Rogers, C. F., Jin, T. F. & Bruch, R. Photoacoustic spectrometer for measuring light absorption by aerosol: instrument description. *Atmos. Environ.* **33,** 2845–2852 (1999).
27. Schwarz, J. P. *et al.* Single-particle measurements of midlatitude black carbon and light-scattering aerosols from the boundary layer to the lower stratosphere. *J. Geophys. Res.* **111,** D16207 (2006).
28. Slowik, J. G. *et al.* An inter-comparison of instruments measuring black carbon content of soot particles. *Aerosol Sci. Technol.* **41,** 295–314 (2007).
29. Canagaratna, M. *et al.* Chemical and microphysical characterization of ambient aerosols with the Aerodyne aerosol mass spectrometer. *Mass Spectrom. Rev.* **26,** 185–222 (2007).
30. DeCarlo, P. F. *et al.* Field-deployable, high-resolution, time-of-flight aerosol mass spectrometer. *Anal. Chem.* **78,** 8281–8289 (2006).
31. Onasch, T. B. *et al.* Soot particle aerosol mass spectrometer: development, validation, and initial application. *Aerosol Sci. Technol.* **46,** 804–817 (2012).
32. Saleh, R. *et al.* Brownness of organics in aerosols from biomass burning linked to their black carbon content. *Nature Geosci.* **7,** 647–650 (2014).
33. Paatero, P. & Tapper, U. Positive matrix factorization: A non-negative factor model with optimal utilization of error estimates of data values. *Environmetrics* **5,** 111–126 (1994).
34. Zhang, Q. *et al.* Understanding atmospheric organic aerosols via factor analysis of aerosol mass spectrometry: a review. *Anal. Bioanal. Chem.* **401,** 3045–3067 (2011).
35. Hoffer, A., Kiss, G., Blazso, M. & Gelencsér, A. Chemical characterization of humic-like substances (HULIS) formed from a lignin-type precursor in model cloud water. *Geophys. Res. Lett.* **31,** L06115 (2004).
36. Limbeck, A., Kulmala, M. & Puxbaum, H. Secondary organic aerosol formation in the atmosphere via heterogeneous reaction of gaseous isoprene on acidic particles. *Geophys. Res. Lett.* **30** (2003).
37. Schwarz, J. P. *et al.* Measurement of the mixing state, mass, and optical size of individual black carbon particles in urban and biomass burning emissions. *Geophys. Res. Lett.* **35,** L13810 (2008).
38. Kroll, J. H. *et al.* Carbon oxidation state as a metric for describing the chemistry of atmospheric organic aerosol. *Nature Chem.* **3,** 133–139 (2011).
39. Canagaratna, M. R. *et al.* Elemental ratio measurements of organic compounds using aerosol mass spectrometry: characterization, improved calibration, and implications. *Atmos. Chem. Phys.* **15,** 253–272 (2015).
40. Liu, D. *et al.* The effect of complex black carbon microphysics on the determination of the optical properties of brown carbon. *Geophys. Res. Lett.* **42,** 613–619 (2015).
41. China, S., Mazzoleni, C., Gorkowski, K., Aiken, A. C. & Dubey, M. K. Morphology and mixing state of individual freshly emitted wildfire carbonaceous particles. *Nat. Commun.* **4,** 2122 (2013).
42. China, S. *et al.* Morphology and mixing state of aged soot particles at a remote marine free troposphere site: implications for optical properties. *Geophys. Res. Lett.* **42,** 1243–1250 (2015).
43. Arnott, W. P., Moosmüller, H. & Walker, J. W. Nitrogen dioxide and kerosene-flame soot calibration of photoacoustic instruments for measurement of light absorption by aerosols. *Rev. Sci. Instrum.* **71,** 4545–4552 (2000).
44. Flowers, B. A. *et al.* Optical-chemical-microphysical relationships and closure studies for mixed carbonaceous aerosols observed at Jeju Island; 3-laser photoacoustic spectrometer, particle sizing, and filter analysis. *Atmos. Chem. Phys.* **10,** 10387–10398 (2010).
45. Flowers, B. A., Powers, H. H., Dubey, M. K. & McDowell, N. G. Inter-comparison of two high-accuracy fast-response spectroscopic sensors of carbon dioxide: a case study. *Atmos. Meas. Tech.* **5,** 991–997 (2012).
46. Werle, P., Mucke, R. & Slemr, F. The limits of signal averaging in atmospheric trace-gas monitoring by tunable diode-laser absorption spectroscopy (TDLAS). *Appl. Phys. B* **57,** 131–139 (1993).
47. Mohr, C. *et al.* Contribution of nitrated phenols to wood burning brown carbon light absorption in Detling, UK during winter time. *Environ. Sci. Technol.* **47,** 6316–6324 (2013).
48. Moteki, N., Kondo, Y. & Nakamura, S. Method to measure refractive indices of small nonspherical particles: Application to black carbon particles. *J. Aerosol Sci.* **41,** 513–521 (2010).
49. Baumgardner, D. *et al.* Soot reference materials for instrument calibration and intercomparisons: a workshop summary with recommendations. *Atmos. Meas. Tech* **5,** 1869–1887 (2012).
50. Thornberry, T. *et al.* Persistence of organic carbon in heated aerosol residuals measured during tropical composition cloud and climate coupling (TC4). *J. Geophys. Res.* **115,** D00J02 (2010).
51. Warton, D. I., Wright, I. J., Falster, D. S. & Westoby, M. Bivariate line-fitting methods for allometry. *Biol. Rev.* **81,** 259–291 (2006).
52. Lack, D. A. & Cappa, C. D. Impact of brown and clear carbon on light absorption enhancement, single scatter albedo and absorption wavelength dependence of black carbon. *Atmos. Chem. Phys.* **10,** 4207–4220 (2010).
53. Kondo, Y. *et al.* Consistency and traceability of black carbon measurements made by laser-induced incandescence, thermal-optical transmittance, and filter-based photo-absorption techniques. *Aerosol Sci. Technol.* **45,** 295–312 (2011).
54. Zotter, P. *et al.* Diurnal cycle of fossil and nonfossil carbon using radiocarbon analyses during CalNex. *J. Geophys. Res.* **119,** 6818–6835 (2014).
55. Crilley, L. R. *et al.* Sources and contributions of wood smoke during winter in London: assessing local and regional influences. *Atmos. Chem. Phys.* **15,** 3149–3171 (2015).
56. Sandradewi, J. *et al.* Using aerosol light absorption measurements for the quantitative determination of wood burning and traffic emission contributions to particulate matter. *Environ. Sci. Technol.* **42,** 3316–3323 (2008).
57. Bohren, C. F. & Huffman, D. R. in *Absorption and Scattering of Light by Small Particles* 475–497 (Wiley, 1998).

Acknowledgements

This research was supported by the US Department of Energy Office of Science Atmospheric System Research (ASR) program grant F265 to LANL (PI M.K.D.). A.C.A. acknowledges Director's postdoctoral funding from LANL's LDRD program. C.D.C. was supported by US DOE Award No. DE-SC0008937. The Aerodyne and Georgia Institute of Technology participants were supported by US DOE Award No. DE-SC0006002. The electron microscopy analysis was supported by US DOE grant No. DE-SC0010019. The authors are grateful to the UK Met Office for the use of their NAME dispersion model. We would like to thank David Green of Kings' College London for his assistance in setting up the Detling site. Detling site logistics were supported by the NERC ClearfLo project (Grant ref. NE/H00324X/1).

Author contributions

The LANL team (S.L., M.K.D., A.C.A, K.G.) led the experimental design, synthesis and analysis of this study. A.C.A., K.G., M.K.D., L.R.W., S.C.H., P.M., E.C.F., P.S.C., W.A.B., J.T.J., L.X. and N.L.N. conducted the measurements. S.L., A.C.A., L.R.W., S.C.H., C.D.C., T.B.O., S.C., N.S., C.M., L.X., N.L.N., S.S. and P.Z. analysed the data and interpreted the results. C.D.C. performed the Mie and RDG modelling. Z.L.F. performed the back trajectory analysis. J.D.A., D.L., A.S.H.P., J.D.L. and D.R.W. contributed to interpretation of the results. S.L., M.K.D., C.D.C. and L.R.W. wrote the manuscript. All authors discussed the results and commented on the paper.

Additional information

Competing financial interests: The authors declare no competing financial interests.

16

Tropical cyclone rainfall area controlled by relative sea surface temperature

Yanluan Lin[1], Ming Zhao[2] & Minghua Zhang[3]

Tropical cyclone rainfall rates have been projected to increase in a warmer climate. The area coverage of tropical cyclones influences their impact on human lives, yet little is known about how tropical cyclone rainfall area will change in the future. Here, using satellite data and global atmospheric model simulations, we show that tropical cyclone rainfall area is controlled primarily by its environmental sea surface temperature (SST) relative to the tropical mean SST (that is, the relative SST), while rainfall rate increases with increasing absolute SST. Our result is consistent with previous numerical simulations that indicated tight relationships between tropical cyclone size and mid-tropospheric relative humidity. Global statistics of tropical cyclone rainfall area are not expected to change markedly under a warmer climate provided that SST change is relatively uniform, implying that increases in total rainfall will be confined to similar size domains with higher rainfall rates.

[1] Ministry of Education Key Laboratory for Earth System Modeling, Center for Earth System Science, Tsinghua University, Beijing 100084, China. [2] University Corporation for Atmospheric Research/Geophysical Fluid Dynamics Laboratory, Princeton, New Jersey 08540-6649, USA. [3] School of Marine and Atmospheric Sciences, Stony Brook University, Stony Brook, New York 11794-5000, USA. Correspondence and requests for materials should be addressed to Y.L. (email: yanluan@tsinghua.edu.cn).

Tropical cyclones (TCs) induce heavy precipitation and flooding in addition to damages associated with high winds and storm surges[1]. Precipitation associated with TCs accounts for ~6–9% of total precipitation over the Tropics on average, but can account for as much as 50% of total precipitation over large portions of ocean basins[2]. The cumulative rainfall and potential damages associated with a TC depend in large part on the rainfall rate, rainfall area coverage and translation speed of the storm. Past studies have investigated potential changes in the frequency and rainfall intensity of TCs under a warmer climate. These studies project a decrease in the total number of TCs but an increase in the occurrence frequency of stronger TCs[3–6]. One of these studies also projects an ~20% increase in TC precipitation within 100 km of the storm centre by the late 21st century[6]. Few studies have examined possible changes in TC rainfall area. TC rainfall area is closely related to the TC wind field, with heavy precipitation generally confined within the outermost closed isobar[7,8]. TC rainfall area is therefore a gross measure of TC size, particularly with respect to the outer radius[8]. Note that this definition of TC size can differ from definitions based on the TC wind field, such as the radius of maximum wind or gale force wind. Rainfall area or size can be influenced by a number of factors, such as environmental humidity, low-level vorticity, vertical wind shear, TC latitude and TC intensity[9–19]. It varies greatly from one TC to another. Theoretical studies and idealized radiative convective equilibrium simulations have suggested that TC potential intensity (PI) divided by the Coriolis parameter may be a good measure of TC size[14–17]; however, the extent to which this scaling is realized in the real world remains uncertain[18,19]. Key questions remain regarding the mechanisms that control TC size and its changes in a warming climate.

Using currently available satellite measurements and global atmospheric model simulations, we examine the sea surface temperature (SST) dependence of rainfall area and rainfall rate in TCs. We find that TC rainfall area is controlled primarily by the relative magnitude of SST in the TC local environment with respect to the tropical mean SST (that is, the relative SST), while TC rainfall rate increases with absolute SST. TC rainfall area is therefore not likely to change markedly in a warmer climate provided that SST changes are relatively uniform throughout the tropics. Increases in TC rainfall are then mostly attributable to increases in TC rainfall rate. The strong dependence of TC rainfall area on relative SST is consistent with previous numerical simulations that suggested a tight relationship between TC size and mid-tropospheric relative humidity (RH) in the TC environment. In the tropics, relative SST strongly regulates the spatial distribution of mid-tropospheric humidity. TCs tend to expand as they move into regions where the mid-tropospheric humidity is high, which closely correspond to regions of high relative SST. Although the impact of relative SST can be partly understood using the PI framework, our analyses of global atmospheric model simulations suggest that relative SST is a better predictor of TC rainfall area than PI.

Results

TC rainfall characteristics. We determine TC rainfall area using two objective methods (see Methods) based on 11 years of satellite precipitation retrievals (the Tropical Rainfall Measuring Mission, or TRMM, 3B42 data set)[20] and 21 years of cloud brightness temperatures (Cloud Archive User Services, or CLAUS)[21] in tandem with observed TC tracks from the International Best Track Archive for Climate Stewardship (IBTrACS)[22]. We use these data to study the climatic controls on TC rainfall area. We use TC precipitation statistics from Geophysical Fluid Dynamics Laboratory (GFDL) high-resolution atmospheric model (HIRAM)[23] simulations at a 50-km resolution both to corroborate our observational findings and to examine the potential impacts of global warming induced by increasing concentrations of anthropogenic greenhouse gases on TC rainfall and size.

TC rainfall distributions and areas (see Methods) are stratified according to different categories of TC intensity and relative SST, where relative SST is defined as the SST in the TC environment minus the tropical (30°S-30°N) mean SST (Fig. 1). The relationship between TC intensity and TC size (the horizontal direction in Fig. 1) is weak, consistent with previous results[9]. By contrast, there is a robust increase in the average TC rainfall area with increasing relative SST in all intensity categories (the vertical direction in Fig. 1). The dependence of rainfall rate on intensity and relative SST is further illustrated in Fig. 2, which shows the radial distributions of azimuthally averaged rainfall rate for storms of different intensities within the [3 °C,4 °C] relative SST bin (row 5 in Fig. 1) and for tropical storms in different relative SST bins (column 1 in Fig. 1). As expected, TC rainfall rate increases markedly with TC intensity, especially towards the storm centre, but rainfall radius (defined as the radius where the azimuthally averaged rainfall rate reaches a threshold rainfall rate, see Methods) changes little with TC intensity (Fig. 2a). By contrast, high rainfall rates expand outwards to larger radii as relative SST increases (Fig. 2b). Increases in rainfall area with relative SST mean that total TC rainfall also increases with relative SST.

Dependence of TC rainfall area and rate on SST. TC rainfall area determined using both objective methods is typically largest in the Western North Pacific and smallest in the Eastern Pacific. This result is consistent with geographic variations in TC size metrics based on wind fields[9,18,19]. The mean TC rainfall radius increases by ~100 km as relative SST increases from the [0 °C,1 °C] bin to the [3 °C,4 °C] bin (Fig. 3a). This result is qualitatively robust for all categories of TCs. Analysis of rainfall radii determined using an alternative objective method based on brightness temperatures also shows an increase in rainfall radius with relative SST (Fig. 3b), although thc mean radii are smaller because of differences in the methodology (See Methods). Furthermore, this dependence of TC rainfall size on relative SST is reproduced in an Atmospheric Model Intercomparison Project (AMIP)-type (that is, using historical SSTs and sea ice as boundary conditions) simulation using HIRAM (Fig. 3c). The success of the model in simulating the TC size dependence suggests that the underlying mechanism may be relatively simple and controlled by large-scale atmospheric conditions rather than cumulus-scale dynamics. The strong dependence of TC size on relative SST can also be identified in an analysis of absolute SST, although the strength of the regression is slightly degraded (Supplementary Fig. 1). Temporal fluctuations in tropical mean SST are relatively small, so that changes in relative and absolute SST are largely coherent. Both relative and absolute SST reflect regional SST variations in the present climate. We are therefore unable to differentiate between relative and absolute SST impacts on TC rainfall area based on observations alone.

Two pairs of HIRAM experiments were conducted to examine the dependence of TC rainfall area and rate on relative and absolute SST. The first pair contrasts a 25-year control simulation with prescribed climatological (seasonally-varying) SSTs and sea ice with a perturbed simulation in which a uniform 2 K warming (relative to the control simulation) is imposed (P2K). The second pair contrasts an AMIP-type simulation that uses historical SSTs and sea ice from 1981 to 2009 as the lower boundary conditions with a simulation in which a uniform 4 K warming (relative to the AMIP simulation) is imposed (P4K). Global warming simulations

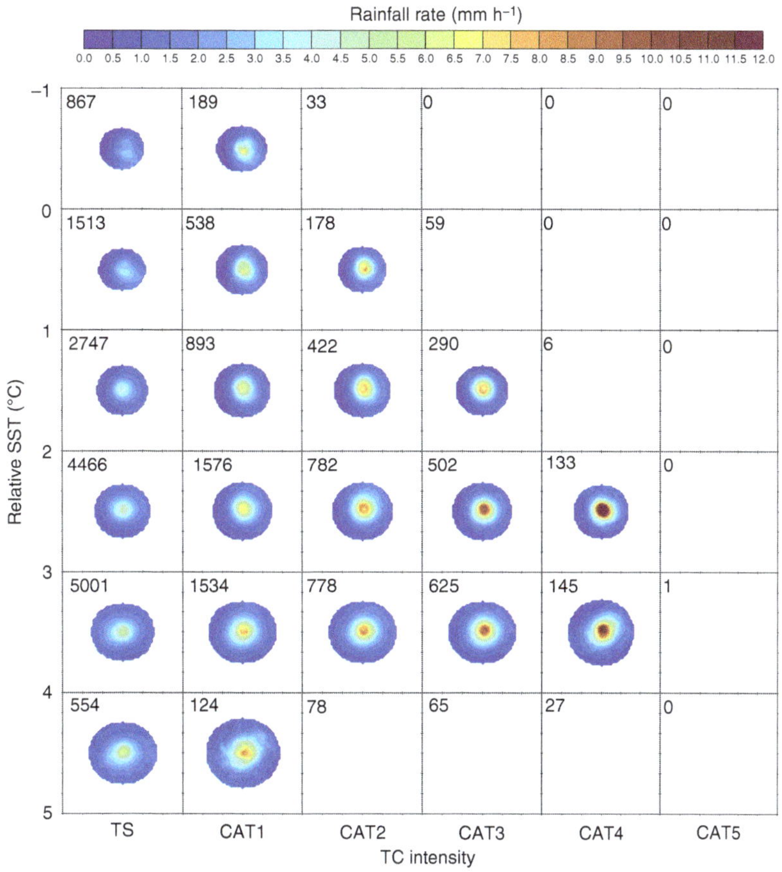

Figure 1 | Average rainfall distribution composited by relative SST and intensity. Rainfall rate (colour shaded in mm h^{-1}) composited by relative SST (*y*-axis) and tropical cyclone intensity (*x*-axis, from tropical storm (TS) to category 5 (CAT5)). The number in the upper left in each box denotes the number of samples in each composite. Only composites with sample sizes >100 are shown. For each composite, pixels with occurrence frequencies <50% are omitted.

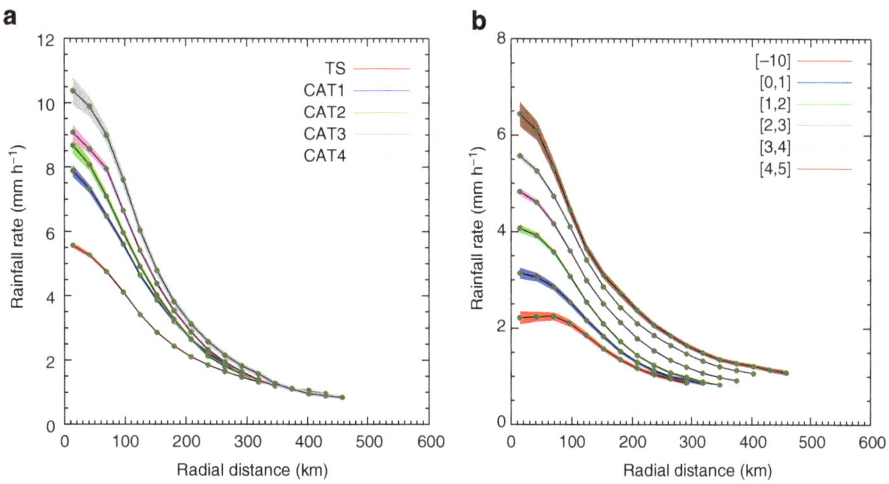

Figure 2 | Radial distribution of azimuthally averaged rainfall rate. (**a**) Radial profiles of rainfall rate for tropical cyclones of different intensities (from tropical storm (TS) to category 4 (CAT4)) in the [3 °C, 4 °C] relative SST bin. (**b**) Radial profiles of rainfall rate for tropical storms within different relative SST bins. Shading indicates one s.e.m.

use the same CO_2 concentration as the control to isolate the SST impacts from CO_2 impacts[24]. Absolute SSTs increase by 2 and 4 K in the P2K and P4K simulations, but relative SSTs stay the same as the control and AMIP simulations. TC genesis locations and tracks are broadly similar in all of these simulations (Supplementary Fig. 2).

The frequency distributions of TC rainfall radius based on all of these simulations are very similar (Fig. 4a). The overall increase in rainfall radius is only around 3 km for the P2K simulation (relative to the control simulation) and around 10 km for the P4K simulation (relative to the AMIP simulation). These small increases in rainfall radius are mainly due to increases in rainfall

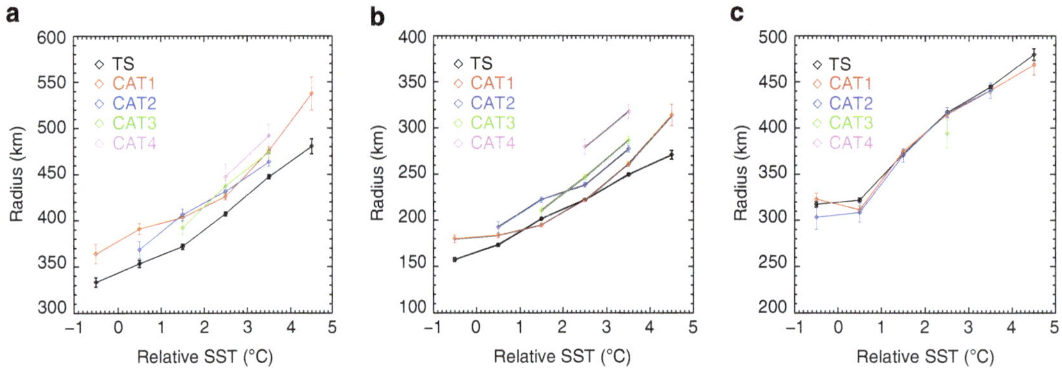

Figure 3 | Variation of tropical cyclone rainfall radius with relative SST. (**a**) Average tropical cyclone rainfall radii in the specified one-degree relative SST bins based on TRMM data using the first method (see Methods for details). Different colours indicate sets of tropical cyclones of different intensities, from tropical storm (TS) to category 4 (CAT4). (**b**) Same as (**a**) but based on CLAUS data using the second method. (**c**) Same as (**a**) but based on the HIRAM AMIP simulation using the first method. Error bars indicate one s.e.m.

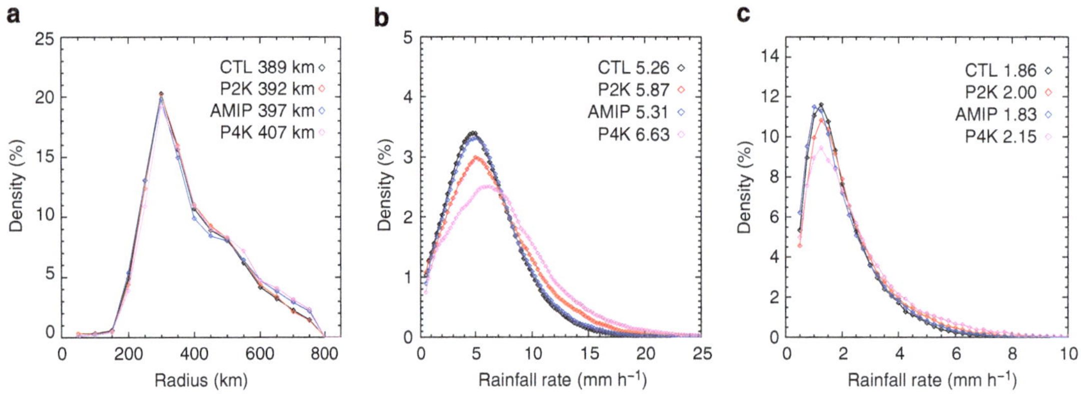

Figure 4 | Frequency distributions of rainfall radius and rainfall rate from HIRAM simulations. (**a**) Frequency distributions of TC rainfall radius from the four HIRAM simulations (CTL = the control simulation; AMIP = the AMIP simulation; P2K = the P2K simulation and P4K = the P4K simulation; see text for details). (**b**) Frequency distributions of TC rainfall rate within a 100-km radius of the TC centre. (**c**) Same as (**b**) but for TC rainfall rate within the objectively determined rainfall radius. Mean values of rainfall radius (in km) and rainfall rate (in mm h^{-1}) are labelled in each panel.

rate with increasing absolute SST (the rainfall rate threshold used in the method is fixed at 0.5 mm h^{-1} (see Fig. 5 and Methods). The negligible change in TC rainfall area under global warming scenarios is in sharp contrast to the strong dependence of TC size on absolute SST, and is consistent with the hypothesis that rainfall area depends primarily on relative SST. This suggests that, to the first order, TC rainfall area is primarily controlled by relative SST instead of absolute SST. Relative SST has also been identified as better than absolute SST for explaining variations in Atlantic TC frequency[5,25,26] and tropical convection[27].

In contrast to the negligible change in rainfall area, rainfall rate changes significantly among these experiments, both within 100 km of the TC centre (Fig. 4b) and within the TC rainfall area (Fig. 4c). Rainfall rate within 100 km of the TC centre increases by ~12% for the P2K experiment (relative to the control simulation) and ~25% for the P4K simulation (relative to the AMIP simulation). These increases decrease slightly to ~8% and ~18%, respectively, within the TC rainfall area. These increases in rainfall rate with absolute SST are consistent with previous studies[6,12,28–30]. Overall, the results suggest that TC rainfall rate is strongly impacted by absolute SST, while rainfall area is mainly controlled by relative SST.

Discussion

The rainfall rate within a TC approximately equals the vertically integrated condensation of water vapour in the atmosphere, which is in turn proportional to the upward transport of water vapour out of the boundary layer[31]. Given a radial profile of vertical velocity $w(r)$ at the top of the boundary layer, the rainfall rate is related to absolute SST as

$$P(r) \approx w(r)(\rho \cdot RH \cdot q_s(SST)) \quad (1)$$

in which P is the rainfall rate, r is the distance from the centre of an axisymmetric hurricane, $q_s(SST)$ is saturation water vapour specific humidity at the given SST, RH is the average boundary layer RH and ρ is the air density. We find that we can reproduce the shape of the observed radial distribution of rainfall rate using a fit that exponentially decreases with the square of the radius near the centre and exponentially decreases with the radius beyond a certain radius (~ 150 km here) (Fig. 5). This approach is consistent with previous findings that the structure of a TC consists of an inner core and an outer region. The exponential decrease of vertical motion with the radius in the outer region is similar to previous work[14–16,32]. As absolute SST increases, if the TC wind field (and thus $w(r)$) does not change, the TC rainfall rate P will increase with SST following the Clausius–Clapeyron relationship. A 3 K increase in absolute SST would then increase the TC rainfall rate by ~22.5% and the TC size by ~20 km (assuming a fixed threshold of 0.5 mm h^{-1}); the rainfall distribution would then follow the brown dashed line in Fig. 5. This change in TC size is much less than the ~100 km increase in TC size that results from a relative SST increase of 3 K (grey line in Fig. 5), and is consistent with the idea that TC size is largely insensitive to changes in absolute SST. The rainfall distribution will expand outwards as the relative SST increases (as shown by the red line in Fig. 5), so that the significant increase in rainfall area with increasing relative SST must be associated with an expansion of the TC wind field. This result is consistent with previous numerical simulations showing TC size increases as the environment becomes more humid[10,11].

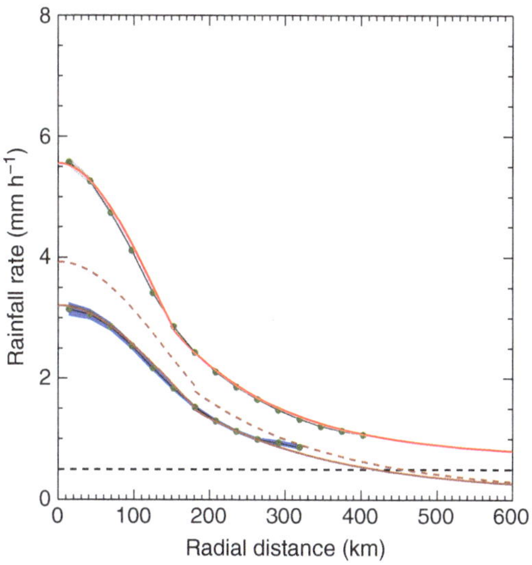

Figure 5 | Expansion of TC rainfall radius with increasing relative SST. The solid brown and red lines indicate fits to the radial profile of rainfall rate of tropical storms in the [0 °C,1 °C] and [3 °C,4 °C] relative SST bins (original data shown in blue and grey). The brown dashed line shows the estimated increase in rainfall rate associated with a 3 °C increase in absolute SST from the [0 °C, 1 °C] distribution (assuming an increase of $\sim$7.5% °n^{-n} and a fixed profile of vertical velocity). The black horizontal dotted line indicates the 0.5 mm h^{-1} rainfall rate threshold used to determine the rainfall radius.

Previous modelling studies have suggested that TC size increases with increasing TC environmental RH due to the production of potential vorticity by active convection located at greater distance from the storm centre[10,11,33]. Free tropospheric RH in the tropics is large over regions of warmer relative SSTs due to the prevalent convective moistening in these regions[34]. Relative SST is a good indicator of mid-tropospheric RH at longer time scales in the tropics. As a result, a TC moving into an area of warmer relative SST tends to encounter higher mid-tropospheric RH, which in turn tends to increase condensational heating and reduce evaporative cooling. Both changes promote the expansion of TC tangential wind, size and rainfall area[10,11]. This result can explain why TCs tend to be the largest in the western North Pacific, where the largest values of relative SST and mid-tropospheric RH in the tropics are located. A uniform increase of SST does not necessarily raise the mid-tropospheric RH. Mid-tropospheric RH is expected to remain roughly constant with warming[35], so that TC size is largely insensitive to absolute SST. These inferences are consistent with our results regarding the dependence of TC size on relative SST.

Theoretical studies and idealized simulations suggest that PI can affect TC size[14–17]. PI is closely related to relative SST[25].The strong dependence of TC rainfall area on relative SST might therefore indicate a dependence on PI. We attempt to distinguish the effects of relative SST from the effects of PI using the HIRAM simulations. In these simulations only the absolute SST is increasing while the relative SST is fixed, and thus they can be used to isolate the effects of absolute SST alone. As expected,

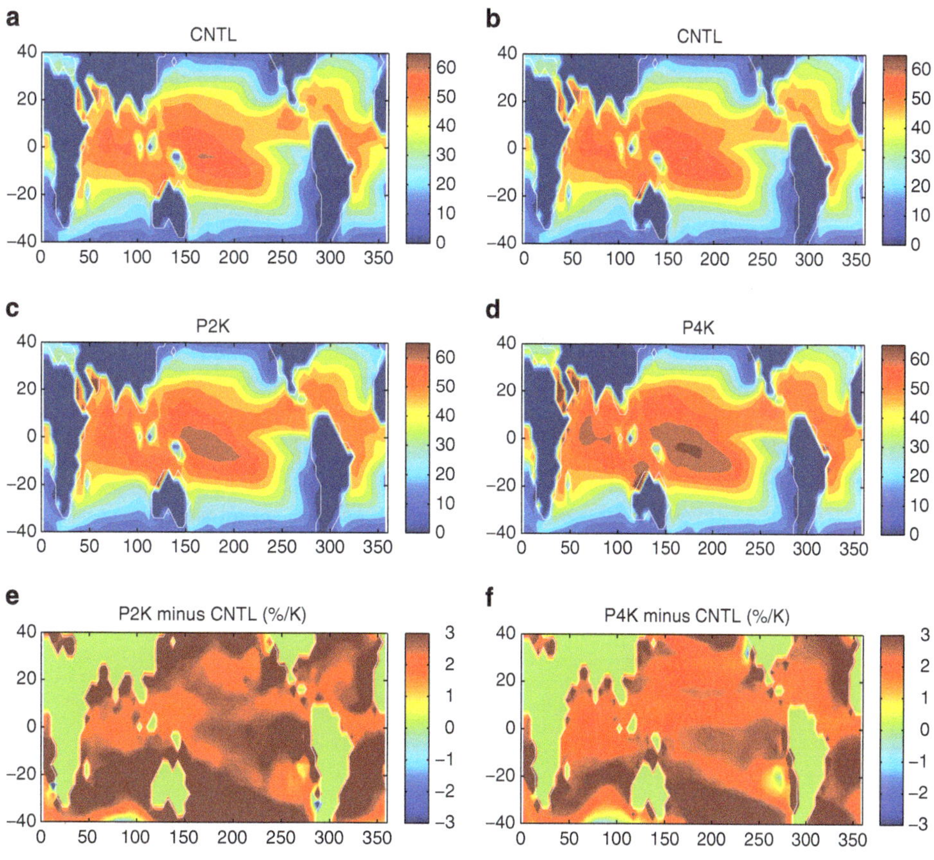

Figure 6 | Potential intensity from HIRAM simulations. Geographical distribution of mean PI (m s^{-1}) for the (**a**) control, (**b**) AMIP, (**c**) P2K and (**d**) P4K simulations, along with relative changes in PI (% K^{-1}) between (**e**) the P2K and control simulations, and (**f**) the P4K and AMIP simulations.

PI increases in the warmer climate simulations (Fig. 6), with the global PI weighted by TC track frequency increasing by roughly $1.2\,m\,s^{-1}K^{-1}$ in the P2K simulation (relative to the control simulation) and $1.0\,m\,s^{-1}K^{-1}$ in the P4K simulation (relative to the AMIP simulation; Supplementary Fig. 3). This amounts to an increase of roughly 2.3–2.7% K^{-1} (global PI weighted by TC track frequency is 43 $m\,s^{-1}$), four times larger than the simulated increase in rainfall area (~0.4–0.6% K^{-1}) (Fig. 4). This result suggests that TC rainfall area may increase at a much slower rate than PI as the climate warms.

In contrast, if the increase in TC rainfall area in the warmer climate simulation is completely determined by changes in PI, then TC size should increase by 1.25–2.5 km for each 1 $m\,s^{-1}$ increase in PI. The increase in PI with increasing relative SST reported by a previous study (~8 $m\,s^{-1}K^{-1}$)[36] suggests an increase in TC size of 10–20 km K^{-1}, significantly less than the estimated using the results shown in Fig. 3 (30–40 km K^{-1}). Relative SST therefore appears to be a better predictor of TC rainfall size than PI, although further investigations are needed to systematically quantify the roles of relative SST and PI on changes in TC size under global warming.

The above analyses lead us to conclude that changes in flooding induced by TCs under global warming will be caused primarily by increases in rainfall rate, rather than by increases in rainfall area. A significant increase in TC rainfall area is not expected unless the genesis locations and tracks of TCs shift to regions of higher relative SST or the spatial pattern of SST in the tropics changes.

Methods

Data. We use 11 years (1998–2008) of 3-hourly TRMM 3B42 precipitation data at 0.25° resolution[20]. TC track and intensities are taken from the 6-h International Best Track Archive for Climate Stewardship[22] (IBTrACS) data, which combines data from multiple sources to provide a complete global climatology of TCs. A total of 1142 TCs occurred during the 11-year analysis period. We also use the global Cloud Archive User Services[21] (CLAUS) satellite data set, which is based on ISCCP B3 data with additional quality control and satellite inter-calibration. Brightness temperatures at a resolution of one-third of a degree are used for the period 1985–2005. A total of 2,319 TCs occurred during this 21-year period. The absolute and relative SSTs at the TC positions are determined via linear interpolations from HadISST[37] data at 1° resolution. The use of monthly SST data (rather than higher frequency SST data, such as daily SSTs) minimizes the effects of transient cooling of SSTs by TC induced mixing of the ocean surface layer. Environmental RH, which is relevant for TC rainfall area, also varies relatively slowly over the tropics.

Model simulations. Four HIRAM simulations are used. HIRAM produces realistic simulations of TC statistics, including the global climatology and variations at seasonal and interannual time scales. The first pair of HIRAM simulations consists of a 25-year (1981–2005) control simulation using climatological SSTs and sea ice (seasonal variations are included but interannual variations are not) and a 25-year companion simulation in which a uniform 2 K warming is imposed (P2K). The second pair of HIRAM simulations consists of a 29-year (1981–2009) Atmospheric Model Intercomparison Project (AMIP) type simulation using historical monthly SSTs and sea ice and a companion simulation in which a uniform 4 K warming is imposed (P4K).

TC rainfall size determination. The calculation of TC rainfall area requires objective methods to determine the region that contains most of the TC rainfall. TCs generally possess an annular structure in their inner cores, although rain bands can sometimes distort this axisymmetry. The first method, which is applied to the TRMM observations and the model simulations, assumes that TC rainfall is axi-symmetric and a varying radius covering most of the TC rainfall is determined. To increase the sample size and fully utilize the 3-hourly TRMM data, we determine the 3-h position of each TC by linearly interpolating the 6-h TC positions from the IBTrACS. At each TC instance, TRMM data within a 30° × 30° region centred at the TC center (121 × 121 grid cells) are prepared first. The 2-dimensional data are then converted to a polar projection and azimuthally averaged rainfall is computed in 0.25° bins. The rainfall radius is determined by the following three criteria: average rainfall greater than a threshold value (here, 0.5 $mm\,h^{-1}$); fractional s.d. < 1.0 (large variance indicates strong asymmetry in the precipitation field, possibly due to mesoscale convective systems near the TC); and a radial gradient of average rainfall < 0.2 (average rainfall generally decreases quickly away from the storm centre, so that this criterion can be used to separate TC rainfall from nearby rainfall not directly associated with the TC). Supplementary Figure 4a shows an example of this methodology for TC Heta in the South Pacific. The method is able to capture the overall rainfall area, specifically the TC inner core and the primary and secondary rainbands. The exact value of the radius is sensitive to the parameters used in the criteria, but this sensitivity does not affect our results so long as all TCs are analysed using the same objective method. This method is also used to determine the rainfall areas and radii of TCs simulated by the HIRAM model. As shown in Fig. 5, rainfall radius can increase simply because of an increase of rainfall rate with increasing absolute SST without a change in the TC wind field. To evaluate the impacts of changes in absolute SST on the rainfall radius, we reduce TC rainfall rates in the P2K simulation by 15% and in the P4K simulation by 30% (approximately following the Clausius–Clapeyron scaling) before applying the method. The mean rainfall size in the P2K and P4K simulations after this scaling is slightly smaller (by ~3 km) than the mean rainfall size in the control and AMIP simulations.

Another objective method is used to identify TC rainfall area based on CLAUS brightness temperatures. This method accounts for persistence and asymmetries in the distribution of TC rainfall. The method is based on a region of interest (ROI) image processing algorithm, which automatically determines the closed blob of rainfall associated with a TC. First, brightness temperature data within a 30° × 30° area centred on the TC position is prepared. Second, all ROIs within this area are identified and labelled, using a threshold brightness temperature of 219 K (which corresponds well to deep raining convective clouds)[38]. ROIs with populations > 15 pixels (~19,000 km^2) located within a radius of 2° from the TC centre are assigned to be the TC rainfall field. Only one ROI is identified in ~98% of the cases and this ROI is then used to estimate TC rainfall area. In the other ~2% of the cases, either zero or two (or more) ROIs are identified and a missing value is recorded for these cases. The area equivalent radius is computed by summing the pixel areas (converted from latitude-longitude to distance) within the ROI. The rainfall area of TC Heta determined using this method covers the inner core and the primary rainband (Supplementary Fig. 4b). This method generally gives a smaller radius than the first method. The brightness temperature field is spatially more continuous, and therefore more suitable for the ROI method. The TRMM rainfall field is spatially heterogeneous, so that the ROI method easily fails.

We composite the radius, rainfall radial distribution and 2-D rainfall field based on the intensity and the absolute and relative SSTs associated with each TC (where the latter two are linearly interpolated in both time and space from monthly mean HadISST data to the TC centre). Using average SSTs within 100 km of the TC centre gives similar results. Only storms over ocean pixels within the tropics (30°S–30°N) are included in the composite to exclude complications with landfall or interactions with mid-latitude synoptic systems.

References

1. Rappaport, E. N. Loss of life in the United States associated with recent Atlantic tropical cyclones. *Bull. Am.. Meteor. Soc.* **81** 9 2065–2074 (2000).
2. Jiang, H. & Zipser, E. J. Contribution of tropical cyclones to the global precipitation from eight seasons of TRMM data: Regional, seasonal, and interannual variations. *J. Climate* **23,** 1526–1543 (2010).
3. Sugi, M., Murakami, H. & Yoshimura, J. A reduction in global tropical cyclone frequency due to global warming. *SOLA* **5,** 164–167 (2009).
4. Bender, M. A. *et al.* Modeled impact of anthropogenic warming of the frequency of intense Atlantic hurricanes. *Science* **327,** 454–458 (2010).
5. Zhao, M. & Held, I. M. TC-Permitting GCM simulations of hurricane frequency response to sea surface temperature anomalies projected for the late-twenty-first century. *J. Climate* **25,** 2995–3009 (2012).
6. Knutson, T. R. *et al.* Tropical cyclones and climate change. *Nat. Geosci.* **3,** 157–163 (2010).
7. Weatherford, C. L. & Gray, W. M. Typhoon structure as revealed by aircraft reconnaissance. Part I: Data analysis and climatology. *Mon. Wea. Rev.* **116,** 1032–1043 (1988).
8. Matyas, C. J. Association between the size of hurricane rain fields at landfall and their surrounding environments. *Meteorol. Atmos. Phys.* **106,** 135–148 (2010).
9. Merrill, R. T. A comparison of large and small tropical cyclones. *Mon. Wea. Rev.* **112,** 1408–1418 (1984).
10. Wang, Y. How do outer spiral rainbands affect tropical cyclone structure and intensity? *J. Atmos. Sci.* **66,** 1250–1273 (2009).
11. Hill, K. A. & Lackmann, G. M. Influence of environmental humidity on tropical cyclone size. *Mon. Wea. Rev.* **137,** 3294–3315 (2009).
12. Hill, K. A. & Lackmann, G. M. The impact of future climate change on TC intensity and structure: A downscaling approach. *J. Climate* **24,** 4644–4661 (2011).
13. Quiring, S., Schumacher, A. C., Labosier, C. & Zhu, L. Variations in mean annual tropical cyclone size in the Atlantic. *J. Geophys. Res.* **116,** D09114 (2011).
14. Emanuel, K. A. An air-sea interaction theory for tropical cyclones. Part I: Steady state maintenance. *J. Atmos. Sci.* **43,** 585–604 (1986).
15. Emanuel, K. A. & Rotunno, R. Self-stratification of tropical cyclone outflow. Part I: Implications for storm structure. *J. Atmos. Sci.* **68,** 2236–2249 (2011).

16. Chavas, D. R. & Emanuel, K. A. Equilibrium tropical cyclone size in an idealized state of axisymmetric radiative-convective equilibrium. *J. Atmos. Sci.* **71,** 1663–1680 (2014).
17. Khairoutdinov, M. & Emanuel, K. A. Rotating radiative-convective equilibrium simulated by a cloud-resolving model. *J. Adv. Model. Earth Sys.* **5,** 816–825 (2013).
18. Chavas, D. R. & Emanuel, K. A. A QuickSCAT climatology of tropical cyclone size. *Geophys. Res. Lett.* **37,** L18816 (2010).
19. Knaff, J. A., Longmore, S. P. & Molenar, D. A. An objective satellite-based tropical cyclone size climatology. *J. Climate* **27,** 455–476 (2014).
20. Huffman, G. J. *et al.* The TRMM multi-satellite precipitation analysis: quasi-global, multi-year, combined-sensor precipitation estimates at fine scale. *J. Hydrometeor.* **8,** 38–55 (2007).
21. Hodges, K., Chappell, D. W., Robinson, G. J. & Yang, G. An improved algorithm for generating global window brightness temperatures from multiple satellite infra-red imagery. *J. Atmos.Ocean Technol.* **17,** 1296–1312 (2000).
22. Knapp, K. R., Kruk, M. C., Levinson, D. H., Diamond, H. J. & Neumann, C. J. The International Best Track Archive for Climate Stewardship (IBTrACS): unifying tropical cyclone best track data. *Bull. Am. Meteor. Soc.* **91,** 363–376 (2010).
23. Zhao, M., Held, I. M., Lin, S.-J. & Vecchi, G. A. Simulations of global hurricane climatology, interannual variability, and response to global warming using a 50 km resolution GCM. *J. Climate* **22,** 6653–6678 (2009).
24. Held, I. M. & Zhao, M. The response of tropical cyclone statistics to an increase in co2 with fixed sea surface temperatures. *J. Clim.* **24,** 5353–5364 (2011).
25. Vecchi, G. A. & Soden, B. J. Effect of remote sea surface temperature change on tropical cyclone potential intensity. *Nature* **450,** 1066–1070 (2007).
26. Vecchi, G. A., Swanson, K. L. & Soden, B. J. Whither hurricane activity. *Science* **322,** 687–689 (2008).
27. Johnson, N. C. & Xie, S. P. Changes in the sea surface temperature threshold for tropical convection. *Nat. Geosci.* **3,** 842–845 (2010).
28. Emanuel, K. A. & Sobel, A. Response of tropical sea surface temperature, precipitation, and tropical cyclone-related variables to changes in global and local forcing. *J. Adv. Model. Earth Sys.* **5.2,** 447–458 (2013).
29. Knutson, T. R. *et al.* Dynamical downscaling projections of twenty-first-century atlantic hurricane activity: CMIP3 and CMIP5 model-based scenarios. *J. Clim.* **26,** 6591–6617 (2013).
30. Villarini, G. *et al.* Sensitivity of tropical cyclone rainfall to idealized global scale forcings. *J. Climate* **27,** 4622–4641 (2014).
31. Langousis, A. & Veniele, V. Theoretical model of rainfall in tropical cyclones for the assessment of long-term risk. *J. Geophys. Res.* **114,** D02106 (2008).
32. Emanuel, K. A tropical cyclone energetics and structure. in *Atmos. Turbulence Mesoscale Meteorol.* (eds Fedorovich, E., Rotunno, R. & Stevens, B.) (Cambridge University Press, 280, 2004).
33. Sun, Y., Zhong, Z., Yi, L., Ha, Y. & Sun, Y. The opposite effects of inner and outer sea surface temperature on tropical cyclone intensity. *J. Geophys. Res.* 2193–2208 (2014).
34. Bretherton, C. S., Peters, M. E. & Back, L. E. Relationships between water vapor path and precipitation over the tropical oceans. *J. Climate* **17,** 1517–1528 (2004).
35. Rappin, E. D., Nolan, D. S. & Emanuel, K. A. Thermodynamic control of tropical cyclogenesis in environments of radiative-convective equilibrium with shear. *Quart. J. R. Meteor. Soc.* **136,** 1954–1971 (2010).
36. Ramsay, H. A. & Sobel, A. H. Effects of relative and absolute sea surface temperature on tropical cyclone potential intensity using a single-column model. *J. Climate* **24,** 183–193 (2011).
37. Rayner, N. A. *et al.* Global analyses of sea surface temperature, sea ice, and night marine air temperature since the late nineteenth century. *J. Geophys. Res.* **108,** 4407 (2003).
38. Evans, J. L. & Shemo, R. E. A procedure for automated satellite-based identification and climatology development of various classes of organized convection. *J. Appl. Meteor.* **35,** 638–652 (1996).

Acknowledgements

The research is supported by Tsinghua University Initiative Scientific Research Program (no. 20131089356) and the Ministry of Science and Technology of China (Grant 2013CBA01805, 2010CB951800, 2012BAC19B08). It was additionally supported by the Office of Science of the US Department of Energy and the National Science Foundation to the Stony Brook University. The CLAUS archive is held at the British Atmospheric Data Centre, produced using ISCCP source data distributed by the NASA Langley Data Center.

Author contributions

All authors contributed to the central ideas presented in the paper. Y. L. performed the analysis and wrote the paper. M. Z. conducted model simulations. All authors discussed the results and edited the manuscript.

Additional information

Competing financial interests: The authors declare no competing financial interests.

17

A pervasive role for biomass burning in tropical high ozone/low water structures

Daniel C. Anderson[1], Julie M. Nicely[2], Ross J. Salawitch[1,2,3], Timothy P. Canty[1], Russell R. Dickerson[1], Thomas F. Hanisco[4], Glenn M. Wolfe[4,5], Eric C. Apel[6], Elliot Atlas[7], Thomas Bannan[8], Stephane Bauguitte[9], Nicola J. Blake[10], James F. Bresch[11], Teresa L. Campos[6], Lucy J. Carpenter[12], Mark D. Cohen[13], Mathew Evans[12,14], Rafael P. Fernandez[15,16], Brian H. Kahn[17], Douglas E. Kinnison[6], Samuel R. Hall[6], Neil R.P. Harris[18], Rebecca S. Hornbrook[6], Jean-Francois Lamarque[6,19], Michael Le Breton[8], James D. Lee[14], Carl Percival[8], Leonhard Pfister[20], R. Bradley Pierce[21], Daniel D. Riemer[7], Alfonso Saiz-Lopez[15], Barbara J.B. Stunder[13], Anne M. Thompson[4], Kirk Ullmann[6], Adam Vaughan[14] & Andrew J. Weinheimer[6]

Air parcels with mixing ratios of high O_3 and low H_2O (HOLW) are common features in the tropical western Pacific (TWP) mid-troposphere (300-700 hPa). Here, using data collected during aircraft sampling of the TWP in winter 2014, we find strong, positive correlations of O_3 with multiple biomass burning tracers in these HOLW structures. Ozone levels in these structures are about a factor of three larger than background. Models, satellite data and aircraft observations are used to show fires in tropical Africa and Southeast Asia are the dominant source of high O_3 and that low H_2O results from large-scale descent within the tropical troposphere. Previous explanations that attribute HOLW structures to transport from the stratosphere or mid-latitude troposphere are inconsistent with our observations. This study suggest a larger role for biomass burning in the radiative forcing of climate in the remote TWP than is commonly appreciated.

[1] Department of Atmospheric and Oceanic Science, University of Maryland, College Park, Maryland 20742, USA. [2] Department of Chemistry and Biochemistry, University of Maryland, College Park, Maryland 20742, USA. [3] Earth System Science Interdisciplinary Center, University of Maryland, College Park, Maryland 20742, USA. [4] NASA Goddard Space Flight Center, Greenbelt, Maryland 20771, USA. [5] Joint Center for Earth Systems Technology, University of Maryland Baltimore County, Baltimore, Maryland 21250, USA. [6] Atmospheric Chemistry Observation and Modeling Laboratory, National Center for Atmospheric Research, Boulder, Colorado 80305, USA. [7] Department of Atmospheric Sciences, Rosenstiel School of Marine and Atmospheric Science, University of Miami, Miami, Florida 33149, USA. [8] Centre for Atmospheric Science, School of Earth, Atmospheric, and Environmental Science, The University of Manchester, Manchester M13 9PL, UK. [9] Facility for Airborne Atmospheric Measurements, Cranfield MK43 0JR, UK. [10] Deparment of Chemistry, University of California, Irvine, California 92697, USA. [11] Mesoscale and Microscale Meteorology Laboratory, National Center for Atmospheric Research, Boulder, Colorado 80305, USA. [12] Wolfson Atmospheric Chemistry Laboratories, Department of Chemistry, University of York, York YO10 5DD, UK. [13] NOAA Air Resources Laboratory, College Park, Maryland 20740, USA. [14] National Centre for Atmospheric Science, Department of Chemistry, University of York, York YO10 5DD, UK. [15] Department of Atmospheric Chemistry and Climate, Institute of Physical Chemistry Rocasolano, CSIC, Madrid 28006, Spain. [16] Department of Natural Science, National Research Council (CONICET), FCEN-UNCuyo, Mendoza 5501, Argentina. [17] Jet Propulsion Laboratory, California Institute of Technology, Pasadena, California 91109, USA. [18] Department of Chemistry, Cambridge University, Cambridge CB2 1EW, UK. [19] Climate and Global Dynamics Laboratory, National Center for Atmospheric Research, Boulder, Colorado 80305, USA. [20] Earth Sciences Division, NASA Ames Research Center, Moffett Field, California 94035, USA. [21] NOAA/NESDIS Center for Satellite Applications and Research, Madison, Wisconsin 53706, USA. Correspondence and requests for materials should be addressed to D.C.A. (email: danderson@atmos.umd.edu).

Tropospheric O_3 is an important greenhouse gas. Ozone has exerted an increase in the global radiative forcing of climate of $\sim 0.4\,Wm^{-2}$ between 1750 and 2011, almost equal to that of CH_4 over the same time period[1]. The largest contribution to the climatic influence of O_3 is due to enhancements over background in the tropical troposphere[2,3]. Elevated surface O_3 adversely affects human health and agriculture[4,5]. Legislation enacted to protect public health has significantly reduced emissions of O_3 precursors from automobiles, factories and power plants throughout the industrialized world, particularly in the Northern Hemisphere mid-latitudes[6,7]. Surface O_3 levels in the industrialized extra-tropics have plateaued or fallen dramatically in response to these actions[8,9]. It is unclear whether the measures taken to reduce surface O_3 in the extra-tropics will reduce the climatic impact of O_3, since its largest radiative influence is in the tropics.

In the marine boundary layer (MBL) of the tropical western Pacific (TWP), O_3 is removed by photochemical reactions involving halogen radicals of marine biogenic origin, resulting in O_3 abundances of $\sim$20 p.p.b.v. or lower[10,11]. Local convection can transmit this low O_3 air throughout the tropospheric column[12], resulting at times in O_3 profiles that have mixing ratios of $\sim$20 p.p.b.v. over an extended altitude range[13,14]. Air masses with elevated O_3 are frequently accompanied by water vapour mixing ratios depressed with respect to the local background, particularly in the mid-troposphere[15]. These high ozone/low water (HOLW) structures inhibit mixing and convection[16], alter the local radiative heating profile[17], and affect the atmosphere's oxidative capacity[13].

Large increases in O_3 relative to these low background values have frequently been observed in the TWP mid-troposphere[12,14,18–25]. Many of these studies note that water vapour tends to be depressed, with respect to the local background, for these high O_3 air parcels[12,14,18–22,24], often attributing the HOLW structures to sources outside the tropical troposphere. Stoller *et al.*[19] conclude that stratospheric air is the dominant source of HOLW air parcels observed over the TWP during multiple aircraft campaigns conducted in September and October 1991, February and March 1994 and September and October 1996. This conclusion was mainly based on the co-location of high potential vorticity (PV) with a few HOLW structures. These early campaigns lacked the instrumentation to measure the suite of chemical compounds sampled during modern campaigns, particularly the biomass burning tracer HCN. Newell *et al.*[18] reach a similar conclusion, analysing the same data as Stoller *et al.* in addition to observations from the MOZAIC campaign from between 1994 and 1997. They note elevated CO and CH_4 in some of their observed HOLW structures, citing entrainment of biomass burning emissions into stratospheric air. Using back trajectory analysis in addition to sonde measurements of O_3 and H_2O from 6 years of Southern Hemisphere ADditional OZonesonde Experiment (SHADOZ) data at three Pacific sites, Hayashi *et al.*[20] argue that transport from the mid-latitude upper troposphere (mlUT) is the dominant source of high mid-tropospheric O_3 in the TWP, with biomass burning only a minor contributor in some months. Kley *et al.*[12] likewise hypothesize a mlUT origin for HOLW features near 700 hPa measured by ozonesondes in the equatorial western Pacific during the CEPEX and TOGA-COARE campaigns conducted during fall 1992. Finally, Ridder *et al.*[25] attribute enhancements in O_3 and CO observed remotely in the Northern Hemisphere TWP during October and November 2009 to fossil fuel combustion in Asia, Europe and North America (that is, the mid-latitudes) and in the Southern Hemisphere (SH) to both fossil fuel combustion and biomass burning.

Numerous other studies claim that low H_2O in the TWP mid-troposphere, defined here as 300–700 hPa, is dominated by the transport of dry air from the mid-latitudes. Yoneyama and Parsons[26] conclude horizontal advection from the mid-latitudes, likely through Rossby wave breaking, creates the observed distribution of dry air during the TOGA-COARE experiment. Combining satellite data and trajectory simulations, Waugh[27] finds that high PV intrusions in the subtropics lead to low relative humidity (RH) in the intrusion itself and high RH ahead of the intrusion, consistent with transport from the lower stratosphere and the deep tropical troposphere, respectively. Cau *et al.*[28] conclude that a dominant dry air source in the TWP is dehydration of air parcels via poleward movement of tropical, upper tropospheric air into the jet followed by slow descent driven by radiative cooling. This study relied on 24-day back trajectories and did not account for the presence of convective precipitation, which is known to greatly alter air mass composition. Galewsky *et al.*[29] conclude, using a tagged tracer method and reanalysis data, that mid-latitude eddies and isentropic transport are the dominant source of low H_2O in the subtropics in December–February 2001/2002. This analysis focused on the zonal mean distribution of H_2O. The subtropical location of their water minimum is coincident with earth's major deserts, which means this region is decoupled from convective precipitation that rehydrates other regions of the tropics. Conversely, Dessler and Minschwaner[30] concluded that deep convective outflow associated with the Hadley circulation is the largest source of low H_2O in the tropical eastern Pacific. Our analysis, which focuses on the tropics (20° N–20° S) of the western Pacific, similarly suggests low water is controlled by outflow of the Hadley circulation.

Other studies cite the importance of biomass burning in controlling tropical O_3, particularly over the Atlantic Ocean and in the SH, but fail to account for the origin of low H_2O in HOLW air parcels. Jacob *et al.*[31] find that tropical processes alone, including *in situ* photochemical production from biomass burning emissions and lightning NO_x ($NO_x = NO + NO_2$), can reproduce observed tropical O_3 distributions using measurements over the South Atlantic during the TRACE-A campaign in September and October 1992. Other studies have attributed high O_3 observed in the TWP to photochemical production in biomass burning plumes. Oltmans *et al.*[14] investigated the O_3 climatology over three locations (Fiji, Samoa and Tahiti) in the TWP from the SHADOZ network. Elevated, mid-tropospheric O_3 was found at these sites, particularly in September and October. Back trajectory analysis connected some of these O_3 enhancements in Fiji and Samoa to biomass burning in Indonesia, and it was noted that air parcels descended $\sim$3 km in transit. In the PEM-Tropics A mission conducted in the Eastern and Central Pacific of the SH in August and September 1996, both Blake *et al.*[21] and Singh *et al.*[22] attribute elevated O_3 in HOLW parcels to photochemical production from biomass burning in Africa and South America based on elevated mixing ratios of biomass burning tracers (C_2H_2, NO_x, CO, CH_3Cl, PAN and C_2H_6) in the structures. Similar to Oltmans *et al.*, Blake *et al.* and Singh *et al.* posit large-scale subsidence as a possible cause of the low H_2O but offer no quantitative support for this supposition. Folkins *et al.*[24] observed HOLW air over Fiji (18° S) at $\sim$120 hPa during the ASHOE/MAESA campaign in October 1994. These air parcels were enhanced in reactive nitrogen compounds and had high N_2O mixing ratios, indicative of a tropospheric origin. They attributed the low H_2O to transport from the mlUT using 10-day isentropic trajectories but did not provide any quantitative analysis of the effect of thermodynamics or descent on observed H_2O. Finally, Kondo *et al.*[23] attribute increases in O_3 of 26 p.p.b.v. over a 31 p.p.b.v. background in the western

Pacific to biomass burning during the TRACE-P campaign (February–April 2001). These studies either ignore the low H_2O (ref. 23), offer limited explanation[14,21,22], attribute low H_2O to extra-tropical transport[24], or do not describe H_2O (ref. 31). The source of high O_3 as well as the dynamical processes controlling the low H_2O in HOLW structures prevalent in the TWP is clearly unresolved.

Here we analyse aircraft and ozonesonde observations, model output, and satellite data to understand the origin of the HOLW air parcels observed in the TWP during winter 2014. The Coordinated Airborne Studies in the Tropics (CAST) and CONvective TRansport of Active Species in the Tropics (CONTRAST) aircraft campaigns based in Guam (13.5° N, 144.8° E) provide a comprehensive suite of chemical measurements. We examine the 17 CAST and 11 CONTRAST flights that provided unprecedented sampling of the TWP mid-troposphere during January and February 2014 (Fig. 1). These *in situ* observations are supplemented by satellite measurements of water vapour and outgoing longwave radiation (OLR) from the daily Atmospheric Infrared Sounder (AIRS) product as well as fire counts from the Moderate Resolution Imaging Spectroradiometer (MODIS) instrument. Ten day, kinematic back trajectories were initialized along the CAST and CONTRAST flight tracks using the Hybrid Single Particle Lagrangian Integrated Trajectory (HYSPLIT) model to connect observed air parcels to their source regions. The CAM-Chem chemical transport model was run with tagged biomass burning tracers to further elucidate ozone sources and was evaluated with ozonesonde observations from the SHADOZ network. Models, satellite data and aircraft observations are used to show that anthropogenic fires in tropical Africa and Southeast Asia are the dominant source of the high O_3 and that the low H_2O results from large-scale descent within the tropical troposphere, after detrainment of biomass burning plumes.

Results

Source of high ozone in the TWP. Representative O_3, CO and H_2O profiles from the CAST and CONTRAST campaigns are shown in Fig. 2. Unless otherwise specified, all references to H_2O in this paper refer to water vapour observations from an open-path laser hygrometer with complete discrimination against condensed water[32]. Mixing ratios of O_3 and H_2O in these profiles exhibit strong anti-correlation; that is, elevated O_3 is closely associated with low RH[15]. These HOLW structures were a pervasive feature seen throughout both campaigns. Here we define a HOLW structure as an air parcel satisfying the simultaneous criteria of $O_3 > 40$ p.p.b.v. and RH $< 20\%$, where RH in this study is with respect to water for $T > 273$ K and to ice for $T < 273$ K. Relying solely on composition to define HOLW structures allows for the inclusion of both sharply defined, thin features (for example, Fig. 2a) as well as those that occupy a large fraction of the tropospheric column (for example, Fig. 2d). HOLW features were primarily located between 300 and 700 hPa and observed during 45 of 104 independent profiles (that is, air parcels were not intentionally resampled). For air parcels with RH $> 20\%$, median O_3 was nearly constant with altitude at ~20 p.p.b.v. (Fig. 3a). For air with RH $< 20\%$, median O_3 peaked at 60 p.p.b.v. near 450 hPa, three times greater than background, and decreased to ~40 p.p.b.v. at 750 and 300 hPa.

Kinematic trajectories connect nearly all of these HOLW structures to tropical regions with active biomass burning. Figure 4a,b shows 10-day back trajectories found using the HYSPLIT model[33]. Trajectories are coloured by observed aircraft O_3 (invariant along each trajectory) and RH along the trajectory as output by HYSPLIT (see Methods section). Each trajectory stops at our estimate of the point of last precipitating convection, derived from a combination of cloud top height and precipitation

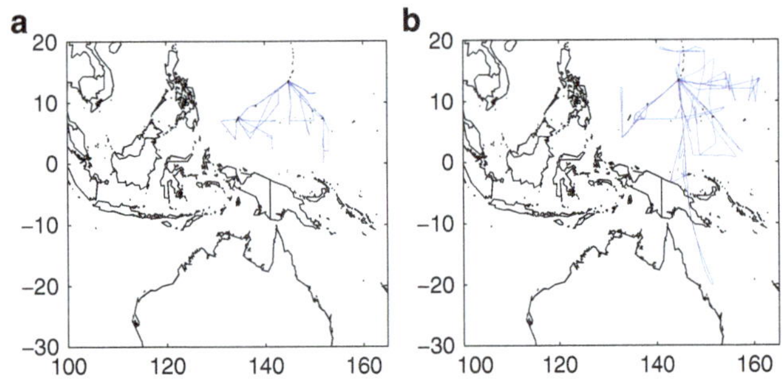

Figure 1 | Flight tracks for the CAST and CONTRAST campaigns. Tracks for (**a**) CAST and (**b**) CONTRAST flights analysed in this study.

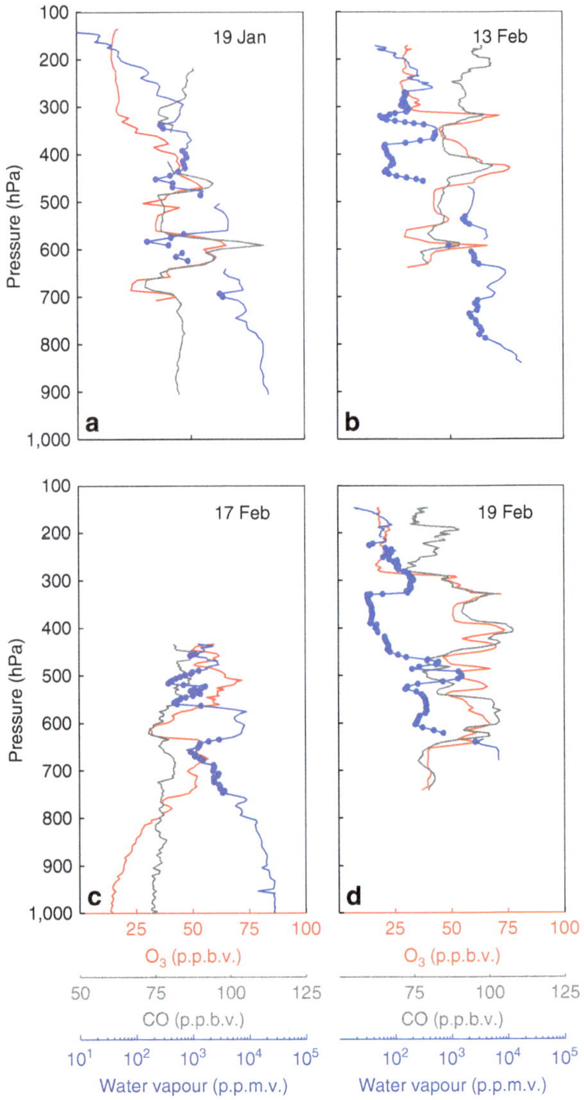

Figure 2 | Ozone, water vapour and CO profiles in the TWP. Sample profiles of 10 s averaged O_3 (red), CO (grey) and H_2O (blue) from four flights during CONTRAST (**a**,**b**,**d**) and CAST (**c**). Blue circles indicate measurements of H_2O mixing ratios for which RH $< 20\%$. RH is with respect to water and with respect to ice for temperatures above and below 273 K, respectively. Vertical profiles have a characteristic horizontal length of ~300 km.

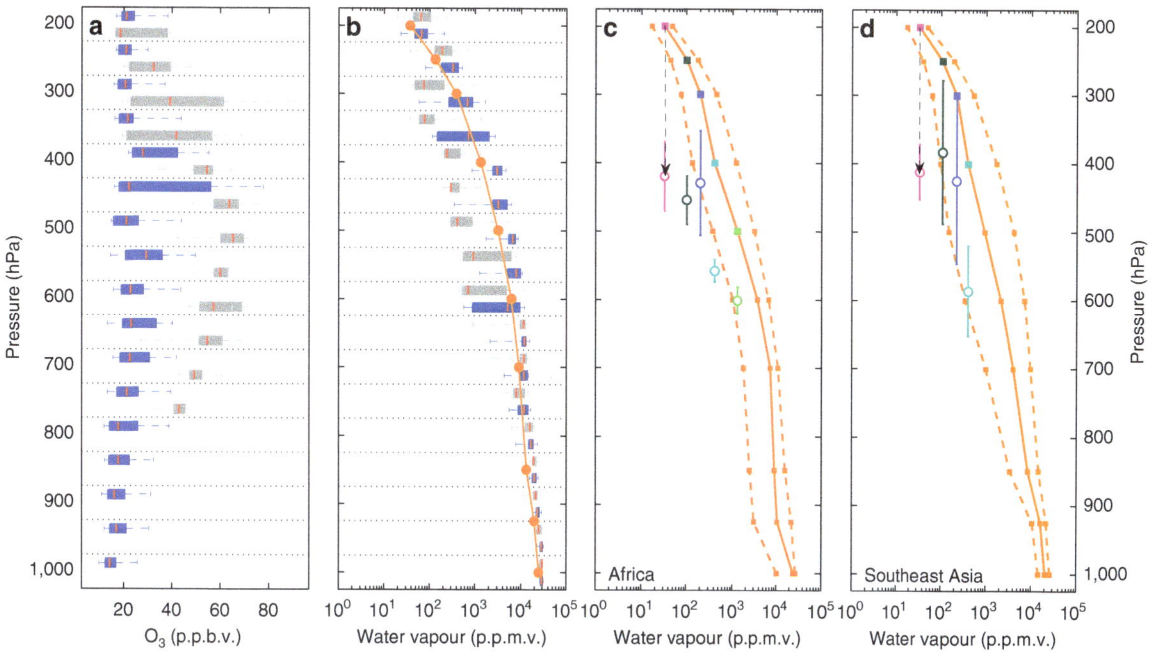

Figure 3 | Structure of ozone and water vapour in the TWP. (**a**) Profiles of *in situ* O_3 during CAST and CONTRAST for two modes of RH (blue, RH >20%; grey, RH <20%). Box and whisker plots show 5th, 25th, 50th, 75th and 95th percentiles for 50 hPa pressure bins. Each bin is delimited by a dotted line, and the two modes (grey and blue boxes) for a given pressure bin are offset for clarity. Between 400 and 700 hPa, the average (minimum) number of observations per blue and grey box is 650 (414) and 330 (43), respectively. Observations between 300 and 400 hPa were more limited with a minimum of 20 and a maximum of 56 observations. (**b**) Distribution of H_2O for two modes of HCN (grey, HCN >150 p.p.t.v.; blue, HCN <150 p.p.t.v.). Median AIRS H_2O over the ascending branch of the Hadley Cell (orange). Between 300 and 700 hPa, the average (minimum) number of observations per blue and grey box is 129 (28) and 111 (19), respectively. (**c**) Median (solid), 5th and 95th percentiles (dashed) of AIRS H_2O over the African biomass burning region (orange). Open circles represent the mean end point of descent $\pm 1\sigma$ of trajectories starting over the African biomass burning region and arriving over the TWP, for various initial pressures (closed squares). **d** Same as **c** but for trajectories starting over Southeast Asia.

data from geostationary satellites and the Tropical Rainfall Measuring Mission satellite. We stop the trajectories at the point of last precipitating convection because convection lofts MBL air into the upper troposphere, where it detrains throughout the column, creating a nearly uniform O_3 profile and a water vapour profile controlled by the local saturation vapour pressure[15]. This convection effectively resets air parcel composition, removing any influence from outside the TWP. Further, global transport models are unable to reproduce the small-scale wind fields associated with deep convection, creating large uncertainty for back trajectories beyond this point. Air parcels originating from the clean SH and eastern Pacific had unimodal O_3 and RH distributions, with medians of 22 p.p.b.v. and 63%, respectively. Trajectories originating west of the study region, primarily over Africa and Southeast Asia, exhibit a bimodal distribution[15]. Assuming a mixture of two Gaussian distributions, the two modes are described by one with peak O_3 and RH of 21 p.p.b.v. and 60% and another with peaks of 55 p.p.b.v. and 4.5%. The low O_3 trajectories are indicative of air parcels under local influence (that is, convectively controlled in the TWP). The high O_3 trajectories originated over regions of active biomass burning in the tropics and reached the TWP along the upper branch of the Walker circulation (Fig. 4a). The rather substantial contribution of the HOLW structures to the overall atmospheric state of the TWP troposphere is quantified by Pan *et al.*[15].

Simultaneous elevation of CO and O_3 in the HOLW structures, in addition to regression analysis, suggests a tropospheric origin. Local maxima in O_3 and CO profiles closely track one another, with r^2 values between these two species ranging from 0.46 to 0.72 (Fig. 5) for the profiles shown in Fig. 2. A campaign-wide regression of CO and O_3 for air parcels that trace back to continental regions yields $r^2 = 0.61$. Photochemical enhancement ratios of O_3 with respect to CO ($\Delta O_3/\Delta CO$) for the profiles shown in Fig. 2 range between 0.97 and 2.85 mol mol^{-1}. Mauzerall *et al.*[34] found fresh biomass burning plumes had enhancement ratios of $\sim$0.15 mol mol^{-1} while plumes older than 6 days had values on the order of 0.75 mol mol^{-1}. These results are consistent with Parrington *et al.*[35] who found mean enhancement ratios of 0.81 mol mol^{-1} with a maximum ratio of 2.55 for boreal biomass burning plumes older than 5 days. Both comparisons suggest the air parcels analysed here are significantly aged and the enhancement ratios lie at the upper end of previous observations. This agrees with the back trajectories and photochemical aging analysis, discussed below, both of which yield air parcel ages of $\sim$10 days. NO is also elevated in these air parcels (Supplementary Fig. 1), though the correlation with O_3 is not as prominent as for other species. This is likely due to the conversion of NO to other nitrogen-containing species, such as alkyl nitrates, peroxyacetyl nitrate and HNO_3, none of which were measured during CAST and CONTRAST.

Regression of CO against O_3 for all observations (Supplementary Fig. 2) reveals four distinct air mass types observed during CAST and CONTRAST: stratospheric, marine and two distinct polluted regimes, one with elevated CO, NO and O_3 and the other with only elevated CO. Known stratospheric air, characterized by high NO and a strong anti-correlation between CO and O_3, was encountered on two flights into the subtropics. Data collected during these two flights are excluded from the majority of the HOLW analysis, since our focus is on the tropical troposphere, but are used below to argue against a stratospheric origin of these structures. Parcels with marine characteristics (low CO, NO and O_3) all originated from the SH or eastern

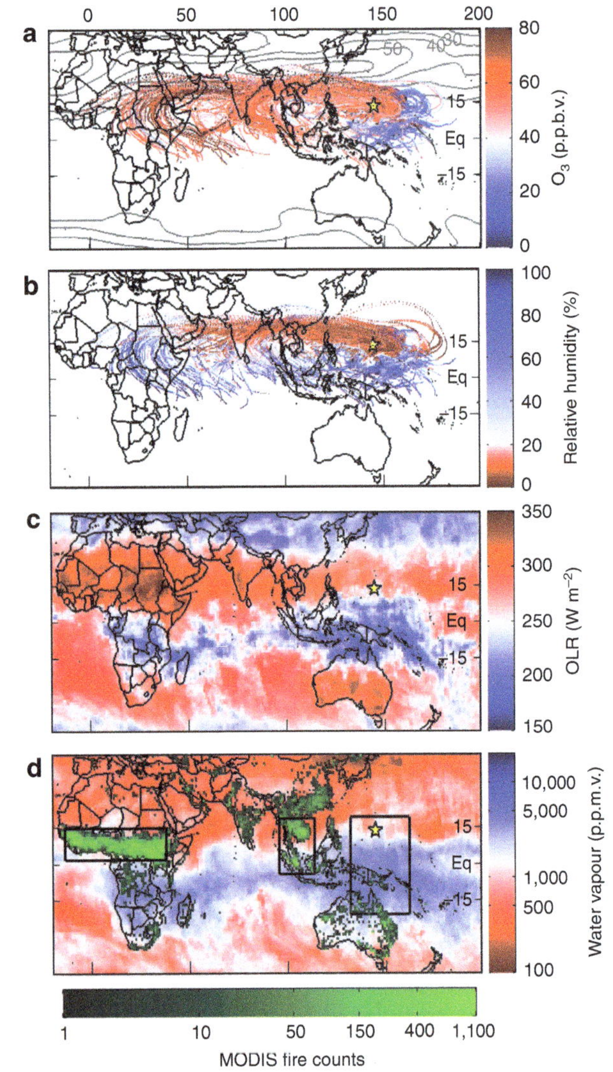

Figure 4 | Origins of air in the TWP. (**a**) 10-day, HYSPLIT back trajectories for all CAST flights analysed here and CONTRAST RF03-05 and RF07-14 for observed pressures between 300 and 700 hPa. Trajectories are stopped when encountering convective precipitation and coloured by observed O_3. For clarity, only every third is shown. Contours are zonal winds at 200 hPa averaged over January and February 2014 in 10 m s^{-1} intervals. The yellow star shows Guam. **b** Same as **a** but coloured by HYSPLIT RH along the trajectory (see methods). (**c**) AIRS daytime OLR averaged over CAST and CONTRAST flight days. **d** Same as **c** but for AIRS H_2O at 500 hPa. MODIS fire counts are the total for January and February 2014. Black rectangles represent the African and Southeast Asian tropical biomass burning regions (determined subjectively by the high fire counts) and the CAST/CONTRAST study region.

Pacific (Fig. 6). Back trajectory analysis of the two polluted regimes connects the high CO, NO and O_3 parcels to biomass burning regions in central Africa or Southeast Asia. The high CO/low NO/low O_3 parcels have various geographic origins in Southeast Asia, which could be reflective of emissions of high CO and low NO_x from two-stroke engines, dominant in this region of the world.

Analysis of other chemical tracers suggests that the observed O_3 in the HOLW structures is likely produced photochemically in biomass burning plumes. Both HCN and CH_3CN are emitted almost exclusively by biomass burning[36]. Elevated mixing ratios of either species in an air parcel therefore suggests a biomass burning origin. Figure 7 shows the profiles of these two species as well as O_3 and the industrial tracer tetrachloroethylene (C_2Cl_4) for the same flights as Fig. 2. Panels b–d show very tight correlations of O_3 with HCN and CH_3CN, while, in panel a, HCN and CH_3CN are elevated between 400 and 700 hPa. Over the entire campaign, air parcels with back trajectories that trace back to Africa and Southeast Asia have a high correlation between O_3 and HCN ($r^2 = 0.80$), demonstrating significant biomass burning influence. The enhancement ratio of CH_3CN to CO (4.02 p.p.t.v./p.p.b.v.) for the HOLW structures is consistent with CH_3CN emissions from tropical forest burning (Supplementary Fig. 3)[37] and is significantly higher than the enhancement ratio for CH_3CN emissions from fossil fuel combustion (<0.1 p.p.t.v./p.p.b.v.)[38]. The biomass burning tracers benzene (C_6H_6) and ethyne (C_2H_2) also show strong correlation with CO ($r^2 > 0.6$) as seen in Fig. 8, further confirming the origin of these air parcels.

Photochemical aging is consistent with the back trajectory analysis. Both C_6H_6 and C_2H_2 have lifetimes much shorter than that of CO (ref. 34). On the basis of modelled OH from CONTRAST, C_6H_6, C_2H_2 and CO in the TWP have lifetimes of ~6, 12 and 35 days, respectively. All species are primarily removed by reaction with OH, so the change in CO with respect to either C_6H_6 or C_2H_2 relative to the initial emissions ratio allows determination of a photochemical age[39]. The photochemical age for C_6H_6 with respect to CO as determined by the Total Organic Gas Analyzer (TOGA) and Advanced Whole Air Sampler (AWAS) measurements were 13 and 8 days, respectively (see Methods section). The photochemical age for C_2H_2 was 11 days. These photochemical ages are broadly consistent with the elapsed time between detrainment of biomass burning plumes over Africa and Southeast Asia and transit to the TWP indicated by the back trajectory analysis (5 ± 4 and 7 ± 2 for Southeast Asia and Africa, respectively). The difference between the age indicated by back trajectory analysis and that of photochemical aging is likely due to dilution of biomass burning plumes with ambient air. Dilution would tend to artificially inflate the photochemical age since only CO is present in appreciable amounts in the background TWP. These relatively short values for photochemical age show the composition of air in the HOLW structures is of recent origin.

Tagged tracers for biomass burning CO in the CAM-Chem model[40], run using assimilated meteorology, support our interpretation. A strong African biomass burning influence is frequently seen in the upper troposphere over much of the TWP and extending as far east as Hawaii, at times accounting for 17% of total CO (Supplementary Fig. 4). Deep convection can loft emissions from fires into the upper troposphere, where strong westerlies transport pollutants long distances[41]. Southeast Asian emissions are prominent throughout the tropospheric column (Supplementary Fig. 4).

Photochemical ozone production. Analysis of ozonesonde observations, CAM-Chem O_3, and photochemical box model output quantitatively show the high O_3 likely originates within the tropical troposphere. Regions of tropical biomass burning have elevated O_3 as compared with the rest of the tropics, with an O_3 maximum over Africa and the Atlantic basin and a minimum over the TWP[42]. Median O_3 in central Africa and Southeast Asia from CAM-Chem was ~50 p.p.b.v. (Fig. 9a) and ~40 p.p.b.v. (Supplementary Fig. 5a), respectively, a factor of 2 greater than background O_3 in the TWP. We compare the CAM-Chem output to ozonesonde[43] measurements over Nairobi (Fig. 9b) and Hanoi (Supplementary Fig. 5b), both strongly influenced by biomass burning, to evaluate CAM-Chem model performance. Mean O_3 from the ozonesondes generally lie within 1σ of the mean

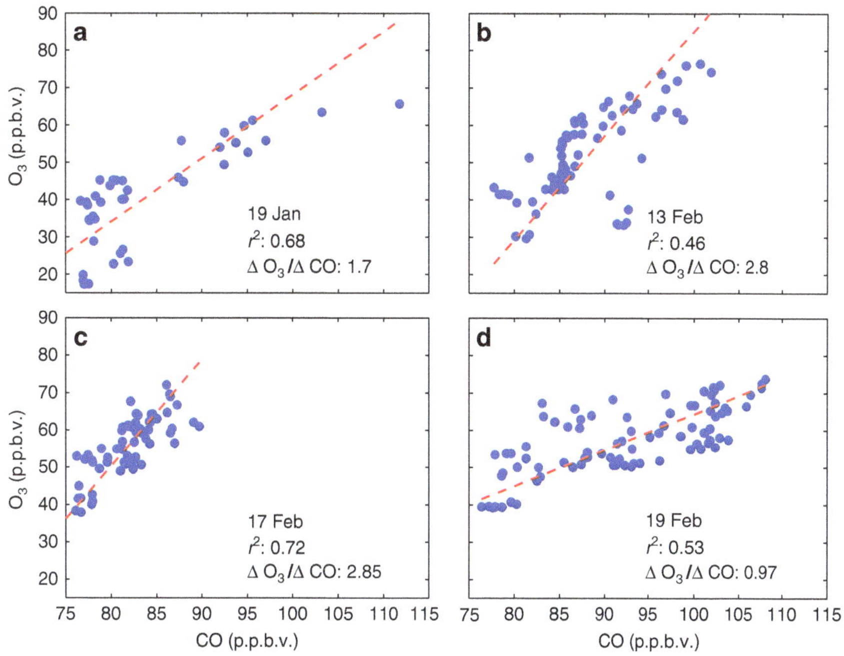

Figure 5 | Enhancement ratios of CO and ozone from CAST and CONTRAST. Regression of CO against O_3 for the data shown in Fig. 2 for pressures between 300 and 700 hPa (**a-d**). The $\Delta O_3/\Delta CO$ ratio for all profiles suggests significantly aged air, consistent with the back trajectory and photochemical aging analyses. The dashed red line is the best fit via orthogonal linear regression. Flight dates, slope and r^2 values are shown for all panels.

CAM-Chem value, suggesting the model accurately captures the O_3 profile in these locations. Means are used because only four or five ozonesonde profiles are available for the study period from each site. Median ozone profiles from CAM-Chem and SHADOZ show comparable agreement in the middle and upper troposphere and substantial differences near the surface (Supplementary Fig. 6). Transport of O_3 from these biomass burning regions cannot explain all of the observed O_3 in the HOLW structures, however, as values frequently peaked at ~75 p.p.b.v., implying there must be photochemical production as the air parcels travel from the biomass burning region.

To estimate the net O_3 production in the HOLW structures, a box model constrained by CONTRAST observations (see Methods section) was run for the profiles shown in Fig. 2. Net O_3 production in the HOLW structures was on the order of ~2 p.p.b.v. per day (Fig. 9c and Supplementary Fig. 7). The value calculated here is a lower bound to net O_3 production in the plume. It is likely that photochemical production along the flight track is significantly lower than in fresh biomass burning plumes, which would be enriched in both NO_x and volatile organic compounds (VOCs) as compared with the more aged air observed in CONTRAST. Lightning, a significant contributor to upper-tropospheric NO_x (ref. 44), would likely further enrich fresh plumes that were lofted into the upper troposphere through deep convection. These fresh plumes can have O_3 production rates of 7 p.p.b.v. per day or higher[42]. As the air ages, however, the abundance of HNO_3 in the plume increases[34], making NO_x unavailable for O_3 production. OH will also oxidize VOCs to less reactive species, and dilution with background air will tend to counteract photochemical O_3 production. Nevertheless, a net O_3 production rate of ~2 p.p.b.v. per day over 5–10 days, combined with initial O_3 of 40–50 p.p.b.v. in the outflow of biomass burning, is consistent with the observed O_3 in the HOLW structures.

Low water vapour origin. We now turn to the origin of low H_2O. As with O_3, H_2O observed during CAST and CONTRAST has two distinct modes, suggesting that H_2O in the TWP is controlled both locally and by large-scale processes from outside the study region. For air parcels with HCN <150 p.p.t.v., the H_2O profile follows the saturation vapour pressure (Fig. 3b, blue boxes), suggesting that the H_2O mixing ratio is controlled by local thermodynamics associated with deep convection. CONTRAST flights designed to measure fresh convective outflow were the only tropical flights where HOLW structures were not observed[15]. Parcels with HCN >150 p.p.t.v. had H_2O mixing ratios an order of magnitude lower than the local saturation vapour pressure. This is consistent with transport of dry air from outside of the TWP. Potential mechanisms to produce the dry air include horizontal advection from the mid-latitudes as well as large-scale descent in the tropics. The association of low H_2O with high HCN indicates these air parcels originate from biomass burning regions.

AIRS measurements[45] and trajectory analysis strongly support our supposition that the observed departures from background H_2O result from large-scale descent in the tropics. The ascending and descending branches of the Hadley Cell, as determined by AIRS OLR, are shown in Fig. 4c as regions with OLR <250 W m^{-2} (blue) and OLR >250 W m^{-2} (red), respectively[46]. AIRS H_2O averaged over the ascending branch agrees well with *in situ* H_2O in the low HCN air parcels (Fig. 3b). Regression of *in situ* H_2O observations from CONTRAST to co-located AIRS retrievals of H_2O (Supplementary Fig. 8) shows these data sets are in excellent agreement ($r^2 = 0.98$), allowing for direct comparison of the satellite-retrieved H_2O to the *in situ* observations. These profiles are characteristic of deep convection, leading to RH >70%. Back trajectories show the RH <20% air parcels frequently originate in the ascending branch of the Hadley cell, flow anti-cyclonically towards the descending branch, and then reach the TWP along the prevailing westerlies (Fig. 4). Air parcels originating over Africa descend 202 ± 64 hPa, on average, leading to significant decline in RH during transit to the TWP (Fig. 4b). Figure 3c shows the AIRS H_2O profile over Africa (orange lines and squares). The final H_2O profile over the TWP

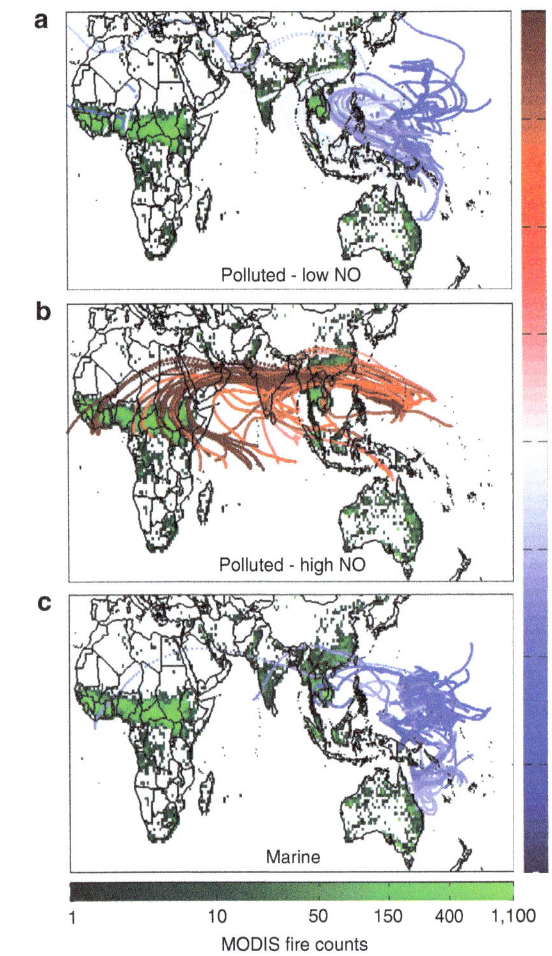

Figure 6 | Back trajectories by chemical regime. 10-day, HYSPLIT back trajectories for all CAST flights analysed here and CONTRAST RF03-05 and RF07-14 for observed pressures between 300 and 700 hPa. Trajectories are stopped when encountering convective precipitation and coloured by observed O_3. Trajectories are separated by the CO regimes illustrated in Supplementary Fig. 2: (**a**) Polluted—low NO (CO >95 p.p.b.v. and NO <40 p.p.t.v.), (**b**) Polluted—high NO (CO >95 p.p.b.v. and NO >50 p.p.t.v.), and (**c**) Marine (CO <80 p.p.b.v. and NO <30 p.p.t.v.). The sum of January and February 2014 MODIS fire counts is shown in green.

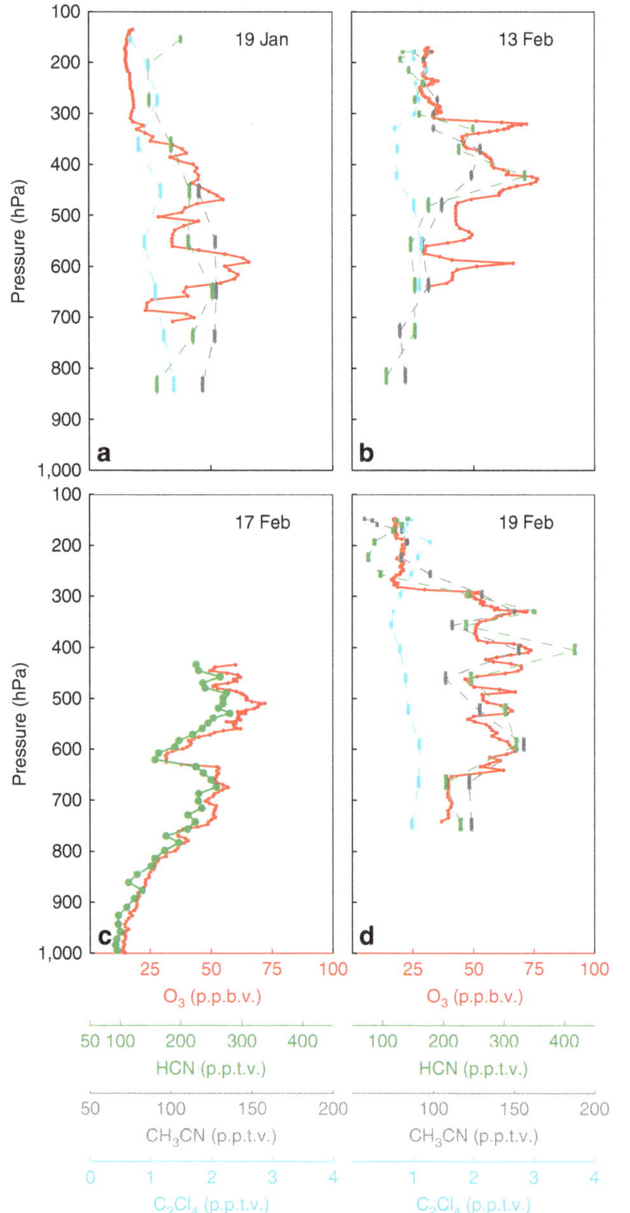

Figure 7 | Anthropogenic tracers in the HOLW structures. Sample profiles from CONTRAST (**a**,**b**,**d**) and CAST (**c**) from the four flights shown in Fig. 2. O_3 (red), CH_3CN (grey), HCN (green) and C_2Cl_4 (light blue) are shown. O_3 data are 10 s averages. CH_3CN and C_2Cl_4 were not measured during CAST. CONTRAST HCN, CH_3CN and C_2Cl_4 were sampled for 35 s at 2 min intervals. CAST HCN was sampled for 30 s. Vertical bars show the pressure range traversed during sampling.

after subsidence (circles), assuming conservation of the H_2O mixing ratio in transit, quantitatively agrees with *in situ* H_2O for the enhanced HCN mode (grey symbols, Fig. 3b). Similar agreement is found for AIRS H_2O profiles over Southeast Asia (Fig. 3d). This analysis demonstrates large-scale descent of tropical biomass burning plumes, on their transit to the TWP, can produce the dry element of the HOLW structures.

Drying of the air parcels by mixing with mid-latitude air or transport to higher latitudes is not supported by the back trajectory analysis. To determine whether air parcels were moistened or dried during transit, the difference between the trajectory starting point (that is, the trajectory initialization point along the flight track) and the H_2O mixing ratios at the trajectory end point (that is, 10 days before observation or the point of last precipitating convection) was calculated for all HOLW structures. Water vapour mixing ratios were calculated from the RH and temperature output by the HYSPLIT model along each trajectory. Supplementary Fig. 9 shows a histogram of these changes in H_2O. For the majority of air parcels (>80%), the H_2O mixing ratio either increased or did not change in transit to the TWP. This moistening indicates that the majority of the HOLW structures encountered by the CAST and CONTRAST aircraft did not require additional condensation after detrainment to account for the low RH. In fact, these air parcels moistened (experienced a modest increase in H_2O mixing ratio) in transit, likely due to mixing with background air. The low RH is due to large-scale descent in the tropics and does not require condensation in the mlUT.

Potential origins outside the tropical troposphere. Mixing with mlUT air is inconsistent with the air mass history and observed composition of the HOLW structures. Parcel trajectories with

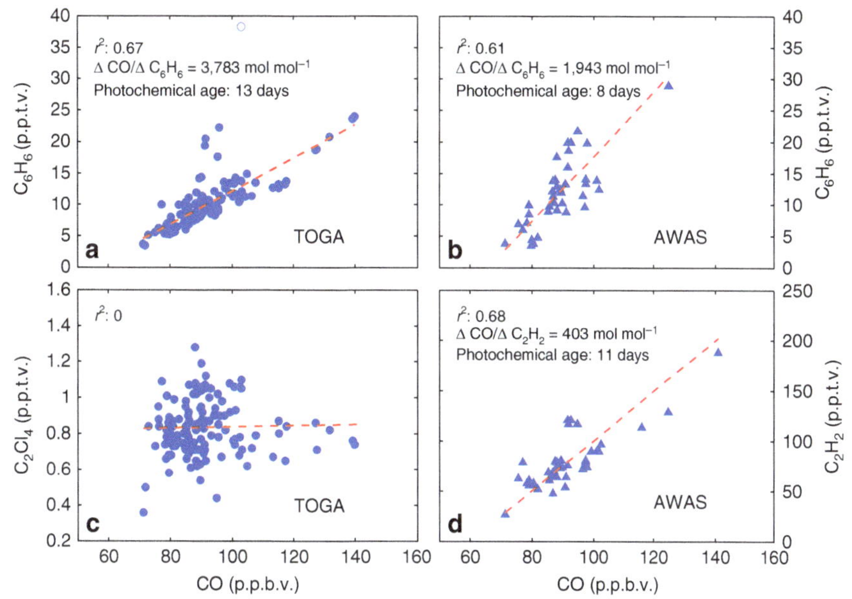

Figure 8 | Photochemical age of observed air parcels. Regression of CO against: TOGA C_6H_6 (**a**); AWAS C_6H_6 (**b**); TOGA C_2Cl_4 (**c**); and AWAS C_2H_2 (**d**). Data are only for the HOLW structures (O_3 >40 p.p.b.v. and RH <20%). The dashed red line is the best fit via orthogonal linear regression. A single data outlier (open circle, **a**) has been excluded from the analysis. Values of r^2 are shown for all panels; enhancement ratios and photochemical ages are shown for panels **a**, **b** and **d**. Excluding the data point for which C_6H_6 >25 p.p.t.v. for panel **b** results in an r^2 of 0.47, a CO to C_6H_6 enhancement ratio of 1,823 mol mol^{-1}, and a photochemical age of 7 days.

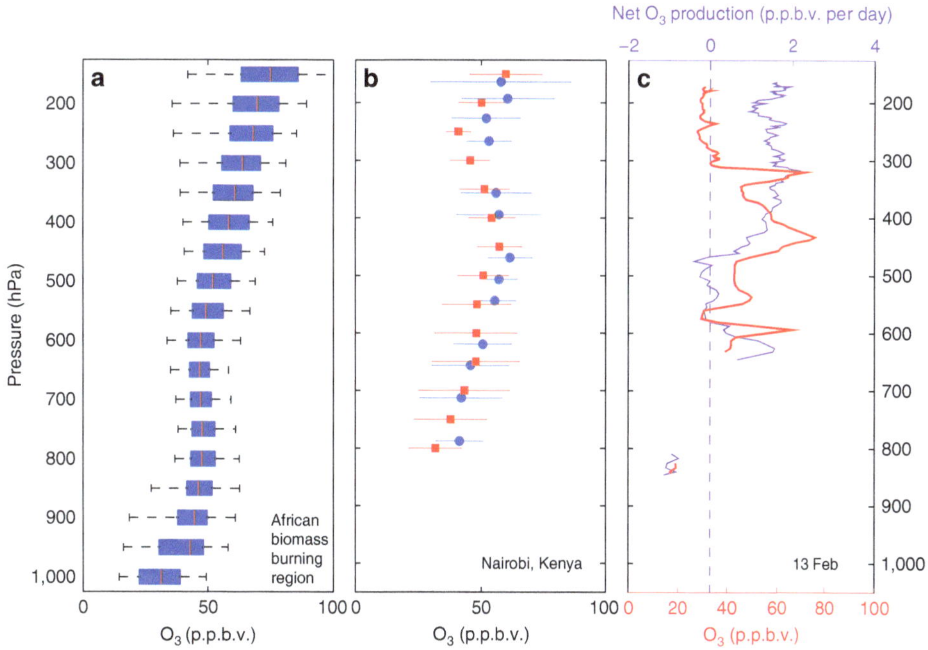

Figure 9 | Ozone over Africa and *in situ* potochemical ozone production in the TWP. (**a**) Vertical distribution of CAM-Chem O_3 in the African biomass burning region (that is, black box Fig. 4); 5th, 25th, median, 75th and 95th percentiles are shown. (**b**) Mean ±1σ of SHADOZ ozonesonde observations over Nairobi, Kenya for January and February 2014 (red). Mean ±1σ CAM-Chem O_3 modelled over Nairobi sampled on the same days as the ozonesondes (blue). (**c**) O_3 profile from Fig. 2b (red) and net O_3 production in the profile (purple, top axis), calculated using the DSMACC photochemical box model (see Methods section).

high O_3 began in the tropics and remained south of the jet core (Fig. 4a), indicating minimal contact with mid-latitude air. None of the back trajectories connect the HOLW air parcels to the mid-latitudes. This is consistent with the chemical composition of the filaments. Tetrachloroethylene (C_2Cl_4), a tracer of industrial pollution, has an atmospheric lifetime on the order of 4 months and is primarily emitted in the Northern Hemisphere mid-latitudes, creating a strong latitudinal gradient. Air masses in the tropics influenced by mid-latitude emissions from China frequently have C_2Cl_4 mixing ratios >2 p.p.t.v. versus a tropical background of 1 p.p.t.v. or less[47]. Tetrachloroethylene mixing ratios in the HOLW structures are a factor of 1.7 times smaller

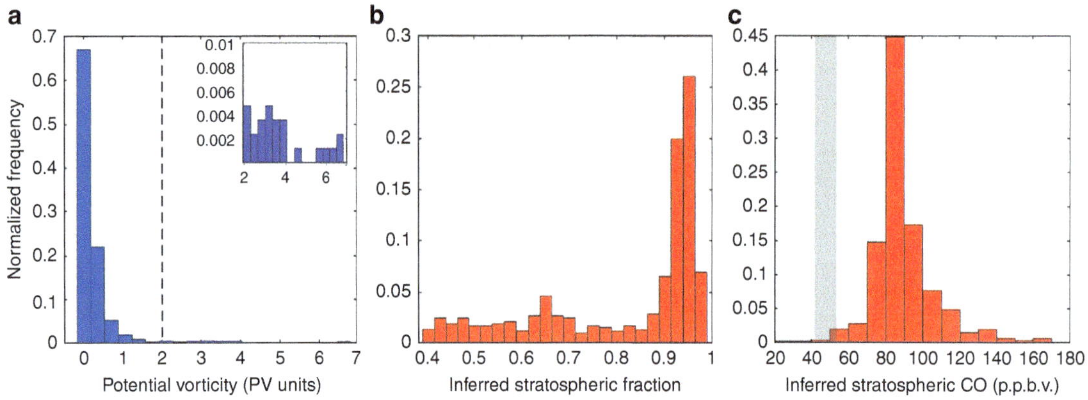

Figure 10 | Negligible stratospheric influence on the composition of the TWP troposphere. (**a**) Distribution of the maximum absolute value of PV along the back trajectories for all air parcels observed between 300 and 700 hPa that have travelled over the southeast Asian and/or African biomass burning regions. (**b**) Distribution of the inferred stratospheric fraction of air (f_{STRAT}) needed to explain the low H_2O in the HOLW filaments, if the depression in H_2O were solely due to stratospheric intrusion. (**c**) Distribution of inferred stratospheric mixing ratio of CO ($CO_{STRAT\ INFERRED}$) assuming the fraction of stratospheric air shown in **b**. Grey area represents the mean $\pm 2\sigma$ of CO observed in the stratosphere during CONTRAST RF15 ($O_3 > 200$ p.p.b.v.). This mixing line analysis suggests negligible stratospheric influence on the composition of the TWP mid-troposphere during CAST and CONTRAST.

than that observed in the mid-latitude free troposphere during CONTRAST (0.89 and 1.52 p.p.t.v., respectively), suggesting little influence from the mid-latitudes. This is further confirmed by the lack of correlation between C_2Cl_4 and either high O_3 (Fig. 7) or CO (Fig. 8c).

Stratospheric origin is also inconsistent with the observed composition of the HOLW structures. Elevated O_3 in the remote TWP can have plausible sources from the polluted troposphere or the O_3 rich stratosphere, resulting in enhancements of a similar magnitude. Mixing line analysis using measurements of at least two other tracers with distinctly different abundances in the troposphere and stratosphere can be used to assess the relative contribution of each potential source[48]. Here we use CO and H_2O as tracers of the polluted troposphere and the stratosphere, respectively (see Methods section). The mixing line analysis shows that to reproduce the observed H_2O profile, mixing of stratospheric and background air would require >90% stratospheric air for the majority of observed parcels (Fig. 10b) and, in turn, a stratospheric CO mixing ratio of >75 p.p.b.v. for the vast majority of the air parcels (red bars, Fig. 10c). Observed stratospheric CO was 47.6 ± 6 p.p.b.v. (2σ; grey bar, Fig. 10c). Only ~1% of parcels had an inferred stratospheric CO within the observed range of stratospheric CO, demonstrating a negligible role for stratospheric influence on the HOLW structures using mixing line analysis. Stratospheric influence was also estimated by interpolating PV to the calculated back trajectories. Only 4% of observed air parcels encountered stratospheric air, defined as intersecting a PV surface >2 PVU (ref. 49) along the trajectory (Fig. 10a). Relaxing the tropopause definition to 1.5 PVU does not significantly alter this percentage. The combination of low PV air along the trajectories and the mixing line analysis indicates negligible stratospheric influence on the composition of the TWP mid-troposphere during CAST and CONTRAST.

Frequent deep convection in the eastern Pacific likely prevents stratospheric air from reaching the study region. The westerly duct near 180–200° E, a region of preferential transport from the mlUT to the tropics, could potentially supply stratospheric HOLW air to the study area[50]. However, trajectories originating from the SH and eastern Pacific encountered precipitating convection ~2.2 days before observation, that is, these trajectories do not extend back to the westerly duct. Convection promotes mixing with MBL air, removing any extra-tropical signature, as evidenced by the low O_3 mixing ratios for air originating from the eastern Pacific as discussed above.

Discussion

We have shown that the high O_3 in the HOLW structures sampled in the TWP during winter 2014 is quantitatively consistent with a tropical, biomass burning source and that the low H_2O mixing ratio is consistent with large-scale descent in the tropics. In a sense, low RH acts as a tropospheric age of air indicator in the tropics. Photochemical O_3 production driven by emissions from biomass burning regions, in combination with large-scale descent of tropical, tropospheric air parcels that do not experience active precipitation, leads to a strong anti-correlation of H_2O and O_3. Prior analyses of HOLW structures, which suggested an extra-tropical tropospheric origin[12,18–20,25], lacked the chemical sophistication of CAST and CONTRAST, relying primarily on ozonesondes or a limited number of chemical tracers in their analyses. Dynamical features suggested as possible mechanisms to bring dry air into the tropics—intrusions of high PV[27], Rossby wave breaking[26], and drying through mixing with the subtropical jet[28,29]—are inconsistent with the results presented here. These prior studies tend to focus on the subtropics rather than the tropics, use trajectories calculated without consideration of convective precipitation and at times in isentropic coordinates, or provide an interpretation for H_2O that is qualitative rather than quantitative. Our aircraft and satellite data indicate, in agreement with Dessler and Minschwaner[30], that large-scale descent in the tropics associated with the Hadley circulation exerts primary control on the H_2O composition of the TWP troposphere for air parcels that have not experienced recent convection.

The attribution of the high O_3 in these HOLW structures suggests a potentially larger role for biomass burning in the radiative forcing of climate in the remote TWP than is commonly appreciated. Tropical tropospheric O_3 is a greenhouse gas, exerting a strong radiative forcing on global climate[2,3]. However, present efforts to limit emissions of O_3 precursors are primarily focused on industrial activities and fossil fuel combustion that occur outside the tropics[51]. If the high O_3 in these structures is primarily of tropical origin, then present legislation to limit the emission of O_3 precursors in the extra-tropics may have little, if any, positive impact for the radiative forcing of climate due to tropospheric O_3. While it is beyond the scope of this paper to estimate the radiative effects of these HOLW structures, biomass burning[52] and HOLW[18] structures are common features of the tropics throughout the year, implying that these structures could have a substantial impact on both global and regional radiative forcing of climate.

Methods

Field campaigns. The CONTRAST campaign consisted of 13 research and 4 transit flights conducted using the National Center for Atmospheric Research (NCAR) Gulfstream V aircraft during January and February 2014. Objectives of the campaign included determining the budget and speciation of very short-lived halogen compounds in the TWP and investigating the transport pathways of these and other chemicals from the MBL to the tropical tropopause layer in a strongly convective environment. Research flights (RF) were either based out of Guam (13.5° N, 144.8° E) or conducted in transit between Broomfield, CO (39.95° N, 105.1° W), the home-base of the Gulfstream V, and Guam. Flights from Guam spanned latitudes from the northern coast of Australia to Japan and altitudes from ~0.5 to 15.5 km. Transit flights (RF01, RF02, RF16 and RF17) as well as flights that sampled primarily mid-latitude air (RF06 and RF15) have been excluded from our analysis, unless otherwise indicated. Tracks of the 11 CONTRAST flights considered here are shown in Fig. 1; these flights include all that sampled exclusively in the tropics. In all, 63 vertical profiles were conducted during these flights, offering unprecedented sampling in the middle and upper troposphere of the TWP. RF15 sampled the lower, mid-latitude stratosphere near Japan, as evidenced by O_3 mixing ratios that approached 1 p.p.m.v. The mixing ratio of stratospheric species quoted here are taken from this flight segment, where stratospheric air is defined as having O_3 >200 p.p.b.v..

The NCAR Gulfstream V aircraft was outfitted to measure various trace gases, meteorological parameters and radiative flux. O_3, NO and NO_2 were measured by chemiluminescence at 1 Hz (ref. 53). The 1σ precisions of O_3, NO and NO_2 at 1 Hz sampling frequency, in the troposphere, are below 0.5 p.p.b.v., 10 p.p.t.v. and 20 p.p.t.v., respectively. Total uncertainties are 5% for NO and O_3 and is 20% for NO_2. O_3 has been corrected for quenching owing to ambient H_2O (ref. 54). CO was also measured at 1 Hz, with an Aero-Laser 5002 vacuum ultraviolet fluorescence instrument[55] with a 2σ uncertainty of 3 p.p.b.v. ± 3%. Water vapour was measured by an open path, laser hygrometer at two wavelengths (1,853.37 and 1,854.03 nm), allowing for the sampling of H_2O mixing ratios spanning 5 orders of magnitude[32]. Data were reported at 1 Hz with a 2σ precision of <3%. RH was calculated from observed H_2O and temperature. Reported RH is with respect to water for temperatures above 0 °C and with respect to ice for temperatures below 0 °C. Formaldehyde (HCHO), necessary for modelling OH and HO_2, was measured at high frequency using laser induced by the *in situ* airborne formaldehyde (ISAF) instrument[56]. HCHO was reported by ISAF at 1 Hz with a 2σ uncertainty of ± 20 p.p.t.v..

TOGA measured a suite of trace gases via gas chromatography/quadrupole mass spectrometry with a sampling time of 35 s and 2 min between sampling periods[57]. Measured species relevant to this study are hydrogen cyanide (HCN), acetonitrile (CH_3CN), tetrachloroethylene (C_2Cl_4), acetone (CH_3COCH_3), acetaldehyde (CH_3CHO), propane (C_3H_8) and HCHO. AWAS also measured a suite of trace gases, including ethyne (C_2H_2) and benzene (C_6H_6). AWAS acquires up to 60 samples of ambient air per flight in electropolished stainless-steel canisters. Sampling time is pressure dependent. Canisters were analysed post-flight using gas chromatography mass spectrometry. All CONTRAST data used in this study have been averaged over the TOGA observation time, unless otherwise indicated. The sampling resolution for vertical flight segments for data averaged over the TOGA sampling period is ~210 m.

Photolysis frequencies were calculated from up and downwelling, spectrally resolved actinic flux density by the HIAPER Airborne Radiation Package (HARP). The system uses independent, 2π steradian optical collectors connected via ultraviolet enhanced fiber optics to charged-coupled device detectors. Spectra were collected every 6 s at ~0.8 nm resolution between 280 and 600 nm with a full-width at half maximum of 1.7 and 2.4 nm in the ultraviolet and visible, respectively. Total photolysis frequencies were calculated from the actinic flux as well as laboratory determinations of molecular cross sections and quantum yields[58].

The CAST campaign was conducted simultaneously with CONTRAST. Whereas observations during CONTRAST were concentrated in the mid- to upper troposphere, the goal of CAST was to observe O_3, CO, NO, very short-lived organic halogen species and radicals in the MBL and lower troposphere. Together, the two campaigns provide coverage of the TWP from just above the ocean surface to the base of the tropical tropopause layer. CAST flights were based out of Guam, Chuuk (7.4° N, 151.8° E), and Palau (7.5° N, 134.5° E) and covered altitudes from near the surface (30 m above mean sea level) to 8 km. The research portion of this campaign consisted of 23 flights during January and February 2014; 17 of these flights, as shown in Figure 1, provided observations between pressures of 300–700 hPa and are analysed here. CONTRAST and CAST flights were jointly coordinated at a shared operations center in Guam.

CAST was conducted using the Facility for Airborne Atmospheric Measurements BAe-146 aircraft. NO was measured with an air quality design, 2-channel chemiluminescence instrument through reaction with O_3 with a 2σ uncertainty of 15%. O_3 was measured at 0.1 Hz with a thermo environmental 49c O_3 analyser by ultraviolet absorption with a 2σ uncertainty of ~0.8 p.p.b.v. CO was measured with an Aero-Laser 5002 instrument at 1 Hz with a 2σ uncertainty of ~1.4 p.p.b.v.. Water vapour mixing ratios were calculated from the observed dew point, measured with a General Eastern dew point hygrometer. HCN was measured by chemical ionization mass spectrometry[59].

No wingtip-to-wingtip comparisons of observations for the CAST and CONTRAST campaign instruments were acquired, due to air traffic concerns in the remote TWP. To compare HCN observations, data from TOGA and the chemical ionization mass spectrometry instrument were selected for background conditions (RH >70% and O_3 <25 p.p.b.v.) and sorted into 0.5 km bins. The mean ± 1σ values of HCN for both campaigns strongly overlap, with CAST HCN slightly lower. The mean ratio of CAST to CONTRAST HCN for all altitude bins was 0.90 ± 0.21. A similar process was used to compare O_3 observations, selecting for measurements with RH >70%. The mean CAST to CONTRAST ratio of O_3 values was 0.98 ± 0.26. Since flights were conducted in different air masses and often on different days, this agreement indicates the measurements of O_3 and HCN obtained during CAST and CONTRAST are directly comparable.

Sondes. Profiles of ozone were measured with electrochemical concentration cell ozonesondes from SHADOZ[43]. Data used here are from observations over Hanoi, Vietnam (21.03° N, 105.85° E) and Nairobi, Kenya (1.28° S, 36.82° E) and were downloaded from the SHADOZ archive (http://croc.gsfc.nasa.gov/shadoz). Nairobi data were acquired over 5 days in January and February 2014 using a 0.5% half buffer KI solution with launch times near 8 UTC. Hanoi data were taken over 4 days in January and February 2014, using a 0.5% unbuffered KI solution and launch times near 6 UTC.

Satellite data. Fire count data are from MODIS onboard the Terra satellite, with a local overpass time of ~10:30 (ref. 52). The version 1 monthly product from collection 5, available at a 1 × 1° (latitude, longitude) spatial resolution, is used. MODIS data were downloaded from ftp://neespi.gsfc.nasa.gov. The Level 3, Version 6 daily AIRS product for OLR and water vapour, at 1 × 1° horizontal resolution, is also used[60,61]. AIRS is onboard the Aqua satellite with a local overpass time of ~13:30. AIRS data were downloaded from ftp://acdisc.sci.gsfc.nasa.gov/ftp/data/s4pa/Aqua_AIRS_Level3. Water vapour data cited here are pressure layer averages.

Back trajectories. Ten day kinematic back trajectories along the flight track were calculated using the NOAA HYSPLIT model[33]. RH, temperature and pressure were output along the trajectory, and H_2O mixing ratios were calculated using the Clausius–Clapeyron relation. RH, as output by HYSPLIT, is with respect to ice for temperatures below −20 °C and a linear blend of RH with respect to ice and water for temperatures between 0 and −20 °C. HYSPLIT RH was post-processed to convert all points with temperatures between 0 and −20 °C to RH with respect to ice, to render HYSPLIT RH directly comparable to *in situ* RH.

The trajectories allowed for vertical displacement, using estimates of the vertical wind from assimilated meteorological fields. Trajectories from CONTRAST were computed along the flight track at 2 min intervals for pressures between 300 and 700 hPa, corresponding with the time between TOGA observations. CAST data were averaged over 35 s (TOGA integration time) and trajectories were calculated at 2 min intervals (TOGA sampling interval) along the flight track, to make our analysis of CAST data using trajectories analogous to the CONTRAST analysis. *Global data assimilation system* meteorological fields at 1 × 1° resolution drove the HYSPLIT model. PV, at 6 h resolution from the National Center for Environmental Prediction final analysis, was interpolated to the back trajectory through a bilinear interpolation in the horizontal and a linear interpolation in the vertical and time coordinates.

Trajectories were stopped at the point of last precipitating convection. In the TWP, convection promotes mixing with MBL air, altering air parcel composition. Precipitation rates from the tropical rainfall measuring mission satellite were combined with cloud top heights calculated from geostationary satellite infrared measurements. This technique provides coverage of the entire tropics. The convective precipitation product is available at 0.25 × 0.25° (latitude, longitude) resolution with a time step of 3 h. Intersection of a trajectory with precipitating convection was defined as a point on the trajectory being within 25 km of convection in the horizontal and being at or below the cloud top height. The 25 km radius allows for the uncertainty in the back trajectory calculation.

CAM-Chem. The Community Atmosphere Model version 4.0 (CAM4) is the atmospheric component of the global chemistry-climate model Community Earth System Model[40]. When run with active chemistry, it is known as CAM-Chem. Here the model was run offline, with meteorological fields specified by the NASA GEOS5 model, with a horizontal resolution of 0.94° latitude × 1.25° longitude and 56 vertical levels. The model chemistry scheme includes a detailed representation of tropospheric and stratospheric chemistry (~180 species; ~500 chemical reactions), including very short-lived halogens. Fernandez *et al.*[62] provide details on surface emissions, wet and dry deposition, heterogeneous reactions and photochemical processes of halogens used within CAM-Chem.

Anthropogenic emissions are from the RCP 8.5 scenario and biomass burning emissions are from the Fire INventory for NCAR (FINN)[63]. FINN combines observations of biomass burning and vegetation/land cover type from MODIS and emissions factors from multiple data sets to produce a gridded global product with a 1 × 1 km resolution. To determine the relative contributions of biomass burning from individual regions, CO emitted from fires in Africa as well as CO emitted from fires in Southeast Asia were treated as separate variables (referred to as tagged CO in the main paper).

Box model. Net photochemical production of O_3 was calculated using equation (1) (ref. 31) for the CONTRAST profiles shown in Fig. 2, where brackets indicate concentration and k_i represents the rate constant for a given reaction:

$$dO_3/dt = k_1[NO][HO_2] + k_2[NO][CH_3O_2] - k_3[O_3][OH] - k_4[O_3][HO_2] - k_5[H_2O][O^1D] \quad (1)$$

CH_3O_2 comprised >95% of RO_2 for the majority of modelled points, so O_3 production from other RO_2 species has been ignored. HO_2, CH_3O_2 and O^1D were not measured; to determine net production of O_3, these species were calculated using the Dynamically Simple Model of Atmospheric Chemical Complexity (DSMACC) box model[64]. Model runs were only conducted for CONTRAST flights because of lack of necessary VOC data for CAST. The model uses a subset of the Master Chemical Mechanism v3.3 (ref. 65) and was initialized with observations of methyl vinyl ketone, methacrolein, acetone, isoprene, methanol and acetaldehyde. NO_2 was estimated using observations of $j(NO_2)$, O_3, NO and modelled values of HO_2 and CH_3O_2 and then used to initialize the model. The box model was constrained by Gulfstream V observations of meteorological parameters, $j(NO_2)$, $j(O^1D)$, O_3, CO, NO, HCHO, H_2O, C_3H_8 and CH_4. The abundance of NO and j-values was allowed to vary with time of day. Photolysis frequencies for all reactions were calculated with the Tropospheric Ultraviolet and Visible Radiation model version 4.2 (ref. 66) and then scaled using observations of $j(O^1D)$ and $j(NO_2)$. Data were averaged over 10 s. Data from the TOGA instrument were linearly interpolated in time to create a 10 s data set. All output has been integrated over the diel cycle to produce 24 h mean photochemical production of O_3, because of the diurnal variation of radical species. Model results were compared with output from another box model, the University of Washington Chemical Model (UWCM)[67], which was constrained and initialized with the same set of input parameters. Between 300 and 700 hPa, the mean ratio of net O_3 production found using DSMACC and UWCM was 1.01.

Mixing line analysis. Elevated O_3 in the remote TWP can have plausible sources from the polluted troposphere or the O_3 rich stratosphere, imposing enhancements of a similar magnitude. Mixing line analysis using measurements of at least two other tracers with distinctly different abundances in the troposphere and stratosphere can be used to assess the relative contribution of each potential source[48]. Here we use CO and H_2O as tracers of the polluted troposphere and the stratosphere, respectively. The fraction of stratospheric air, f_{STRAT} was calculated from equation (2), where H_2O_{OBS} is the observed H_2O mixing ratio, H_2O_{STRAT} is the stratospheric H_2O mixing ratio, and H_2O_{TROP} is the altitude dependent background tropospheric H_2O mixing ratio:

$$H_2O_{OBS} = f_{STRAT}H_2O_{STRAT} + (1 - f_{STRAT})H_2O_{TROP}(z) \quad (2)$$

A constant stratospheric H_2O mixing ratio of 3 p.p.m.v. was assumed based on the mean observed H_2O (3.1 ± 0.1 p.p.m.v.) from RF15, which probed the lower stratosphere. Background tropospheric H_2O was calculated by filtering 10 s averaged data for air parcels with RH >70% and O_3 <20 p.p.b.v., leading to a profile similar to that shown by the blue bars in Fig. 3b. These data were then averaged into 1 km altitude bins.

An inferred stratospheric CO mixing ratio, denoted $CO_{STRAT\ INFERRED}$ was calculated based on f_{STRAT} found using equation (2). The relation for $CO_{STRAT\ INFERRED}$ is:

$$CO_{STRAT\ INFERRED} = [CO_{OBS} - (1 - f_{STRAT})CO_{TROP}]/f_{STRAT} \quad (3)$$

where variables have analogous definitions to those in equation (2). A value for CO_{TROP} of 85 p.p.b.v. was used, based on the median mixing ratio of CO for air parcels with RH >70%. Since CO_{TROP} showed little variation with altitude, a constant value was used.

Photochemical aging. Both benzene (C_6H_6) and ethyne (C_2H_2) are tracers of biomass burning pollution and have lifetimes much shorter than that of CO (ref. 34). All species are primarily removed by reaction with OH (see reactions 1–5), so the change in CO with respect to either C_6H_6 or C_2H_2, relative to the initial emissions ratio, allows photochemical age to be determined[39]. Rate constants for reactions R3 and R4 are from IUPAC[68] and for reactions R1, R2, and R5 from NASA JPL 2011 (ref. 69).

$$OH + CO \rightarrow CO_2 + H \quad (R1)$$

$$OH + CO \rightarrow HOCO \quad (R2)$$

$$OH + C_6H_6 \rightarrow H_2O + C_6H_5 \quad (R3)$$

$$OH + C_6H_6 \rightarrow HOC_6H_6 \quad (R4)$$

$$OH + C_2H_2 \rightarrow HOCHCH \quad (R5)$$

All reactions have first-order kinetics, so the ratio of the CO concentration at time t, $[CO(t)]$, to that of benzene, $[C_6H_6(t)]$, is described by equation (4) (ref. 39), where k_{CO} is the sum of the rate constants for reactions R1 and R2, k_{C6H6} is the sum of the rate constants for reactions R3 and R4, ER(t) is the enhancement ratio of CO to C_6H_6 at time t, and ER_0 is the ratio of [CO] to $[C_6H_6]$ at the time of emission. The expression for C_2H_2 is analogous. If all the other variables are known, the expression can be solved for t, the photochemical age.

$$ER(t) = [CO(t)]/[C_6H_6(t)] = ER_0 e^{-[OH]t(k_{CO} - k_{C_6H_6})} \quad (4)$$

The enhancement ratios of CO to C_6H_6 and CO to C_2H_2 over the entire campaign were determined by using an orthogonal linear regression for parcels where O_3 >40 p.p.b.v. and RH <20% (Fig. 8). The slope of these lines is the enhancement ratio. This was done over a campaign-wide basis and not for each profile because of the limited sampling frequency of C_6H_6 and C_2H_2. Separate regressions were done for the TOGA and AWAS C_6H_6 measurements. Values of r^2 for all regressions were >0.61. The value of ER_0 for the two relations was assumed to be 724 mol CO per mol C_6H_6 and 241 mol CO per mol C_2H_2, which are characteristic of a tropical forest[70]. The photochemical age varied by <0.5 days when emission ratios for other vegetation types[70] were considered. Constant values of temperature (247 K), pressure (350 hPa) and [OH] ($1.7 \times 10^6\ cm^{-3}$) were assumed when calculating the photochemical age. Temperature and pressure values were the average along the back trajectory for the HOLW filaments, and $1.7 \times 10^6\ cm^{-3}$ is the 24-h mean OH concentration for parcels between 300 and 700 hPa from our box model runs. Equation (4) also assumes that any mixing with ambient air dilutes both species equally.

Code availability. The CAM-Chem code is available for download at www2.cesm.ucar.edu. An online version of the HYSPLIT model is available at http://ready.arl.noaa.gov/HYSPLIT.php. The DSMACC photochemical box model can be downloaded at http://wiki.seas.harvard.edu/geos-chem/index.php/DSMACC_chemical_box_model and the UWCM box model can be downloaded at https://sites.google.com/site/wolfegm/models.

References

1. IPCC. *Climate Change 2013: The Physical Science Basis. Contribution of Working Group I to the Fifth Assessment Report of the Intergovernmental Panel on Climate Change* (Cambridge University Press, 2013).
2. Shindell, D. & Faluvegi, G. Climate response to regional radiative forcing during the twentieth century. *Nat. Geosci.* **2,** 294–300 (2009).
3. Stevenson, D. S. *et al.* Tropospheric ozone changes, radiative forcing and attribution to emissions in the Atmospheric Chemistry and Climate Model Intercomparison Project (ACCMIP). *Atmos. Chem. Phys.* **13,** 3063–3085 (2013).
4. Bell, M. L., Peng, R. D. & Dominici, F. The exposure-response curve for ozone and risk of mortality and the adequacy of current ozone regulations. *Environ. Health Perspect.* **114,** 532–536 (2006).
5. Avnery, S., Mauzerall, D. L., Liu, J. & Horowitz, L. W. Global crop yield reductions due to surface ozone exposure: 1. Year 2000 crop production losses and economic damage. *Atmos. Environ.* **45,** 2284–2296 (2011).
6. Parrish, D. D. Critical evaluation of US on-road vehicle emission inventories. *Atmos. Environ.* **40,** 2288–2300 (2006).
7. Vestreng, V. *et al.* Evolution of NO_x emissions in Europe with focus on road transport control measures. *Atmos. Chem. Phys.* **9,** 1503–1520 (2009).
8. Cooper, O. R., Gao, R.-S., Tarasick, D., Leblanc, T. & Sweeney, C. Long-term ozone trends at rural ozone monitoring sites across the United States, 1990-2010. *J. Geophys. Res.* **117** doi:10.1029/2012jd018261 (2012).
9. Oltmans, S. J. *et al.* Long-term changes in tropospheric ozone. *Atmos. Environ.* **40,** 3156–3173 (2006).
10. Read, K. A. *et al.* Extensive halogen-mediated ozone destruction over the tropical Atlantic Ocean. *Nature* **453,** 1232–1235 (2008).
11. Dickerson, R. R. *et al.* Ozone in the remote marine boundary layer: a possible role for halogens. *J. Geophys. Res.* **104,** 21385–321395 (1999).
12. Kley, D. *et al.* Tropospheric water-vapour and ozone cross-sections in a zonal plane over the central equatorial Pacific Ocean. *Q. J. R. Meteorol. Soc.* **123,** 2009–2040 (1997).
13. Rex, M. *et al.* A tropical West Pacific OH minimum and implications for stratospheric composition. *Atmos. Chem. Phys.* **14,** 4827–4841 (2014).

14. Oltmans, S. J. *et al.* Ozone in the Pacific tropical troposphere from ozonesonde observations. *J. Geophys. Res.* **106,** 32503–32525 (2001).
15. Pan, L. L. *et al.* Bimodal distribution of tropical free tropospheric ozone over the Western Pacific revealed by airborne observations. *Geophys. Res. Lett.* **42,** 7844–7851 (2015).
16. Mapes, B. E. & Zuidema, P. Radiative-dynamical consequences of dry tongues in the tropical troposphere. *J. Atmos. Sci.* **53,** 620–638 (1995).
17. Parsons, D. B., Yoneyama, K. & Redelsperger, J. L. The evolution of the tropical western Pacific atmosphere-ocean system following the arrival of a dry intrusion. *Q. J. R. Meteorol. Soc.* **126,** 517–548 (2000).
18. Newell, R. E. *et al.* Ubiquity of quasi-horizontal layers in the troposphere. *Nature* **398,** 316–319 (1999).
19. Stoller, P. *et al.* Measurements of atmospheric layers from the NASA DC-8 and P-3B aircraft during PEM-Tropics A. *J. Geophy. Res.* **104,** 5745–5764 (1999).
20. Hayashi, H., Kita, K. & Taguchi, S. Ozone-enhanced layers in the troposphere over the equatorial Pacific Ocean and the influence of transport of midlatitude UT/LS air. *Atmos. Chem. Phys.* **8,** 2609–2621 (2008).
21. Blake, N. J. *et al.* Influence of southern hemispheric biomass burning on midtropospheric distributions of nonmethane hydrocarbons and selected halocarbons over the remote South Pacific. *J. Geophys. Res.* **104,** 16213–16232 (1999).
22. Singh, H. B. *et al.* Biomass burning influences on the composition of the remote South Pacific troposphere: analysis based on observations from PEM-Tropics-A. *Atmos. Environ.* **34,** 635–644 (2000).
23. Kondo, Y. *et al.* Impacts of biomass burning in Southeast Asia on ozone and reactive nitrogen over the western Pacific in spring. *J. Geophys. Res.* **109,** D15S12 (2004).
24. Folkins, I., Chatfield, R., Baumgardner, D. & Proffitt, M. Biomass burning and deep convection in southeastern Asia: results from ASHOE/MAESA. *J. Geophys. Res. Atmos.* **102,** 13291–13299 (1997).
25. Ridder, T. *et al.* Ship-borne FTIR measurements of CO and O_3 in the Western Pacific from 43° N to 35° S: an evaluation of the sources. *Atmos. Chem. Phys.* **12,** 815–828 (2012).
26. Yoneyama, K. & Parsons, D. B. A proposed mechanism for the intrusion of dry air into the Tropical Western Pacific region. *J. Atmos. Sci.* **56,** 1524–1546 (1999).
27. Waugh, D. W. Impact of potential vorticity intrusions on subtropical upper tropospheric humidity. *J. Geophys. Res. Atmos.* **110,** D11305 (2005).
28. Cau, P., Methven, J. & Hoskins, B. Origins of dry air in the tropics and subtropics. *J. Clim.* **20,** 2745–2759 (2007).
29. Galewsky, J., Sobel, A. & Held, I. Diagnosis of subtropical humidity dynamics using tracers of last saturation. *J. Atmos. Sci.* **62,** 3353–3367 (2005).
30. Dessler, A. E. & Minschwaner, K. An analysis of the regulation of tropical tropospheric water vapor. *J. Geophys. Res. Atmos.* **112,** D10120 (2007).
31. Jacob, D. J. *et al.* Origin of ozone and NO_x in the tropical troposphere: a photochemical analysis of aircraft observations over the South Atlantic basin. *J. Geophys. Res. Atmos.* **101,** 24235–24250 (1996).
32. Zondlo, M. A., Paige, M. E., Massick, S. M. & Silver, J. A. Vertical cavity laser hygrometer for the National Science Foundation Gulfstream-V aircraft. *J. Geophys. Res. Atmos.* **115,** D20309 (2010).
33. Stein, A. F. *et al.* NOAA's HYSPLIT atmospheric transport and dispersion modeling system. *Bull. Am. Meteorol. Sci.* doi: http://dx.doi.org/10.1175/BAMS-D-14-00110.1 (2015).
34. Mauzerall, D. L. *et al.* Photochemistry in biomass burning plumes and implications for tropospheric ozone over the tropical South Atlantic. *J. Geophys. Res. Atmos.* **103,** 8401–8423 (1998).
35. Parrington, M. *et al.* Ozone photochemistry in boreal biomass burning plumes. *Atmos. Chem. Phys.* **13,** 7321–7341 (2013).
36. Holzinger, R. *et al.* Biomass burning as a source of formaldehyde, acetaldehyde, methanol, acetone, acetonitrile, and hydrogen cyanide. *Geophys. Res. Lett.* **26,** 1161–1164 (1999).
37. Akagi, S. K. *et al.* Emission factors for open and domestic biomass burning for use in atmospheric models. *Atmos. Chem. Phys.* **11,** 4039–4072 (2011).
38. de Gouw, J. A. *et al.* Emission sources and ocean uptake of acetonitrile (CH_3CN) in the atmosphere. *J. Geophys. Res.* **108,** 4329 (2003).
39. Parrish, D. D. *et al.* Effects of mixing on evolution of hydrocarbon ratios in the troposphere. *J. Geophys. Res. Atmos.* **112,** D10S34 (2007).
40. Lamarque, J. F. *et al.* CAM-chem: description and evaluation of interactive atmospheric chemistry in the Community Earth System Model. *Geosci. Model Dev.* **5,** 369–411 (2012).
41. Pickering, K. E. *et al.* Convective transport of biomass burning emissions over Brazil during TRACE A. *J. Geophys. Res. Atmos.* **101,** 23993–24012 (1996).
42. Thompson, A. M. *et al.* Where did tropospheric ozone over southern Africa and the tropical Atlantic come from in October 1992? Insights from TOMS, GTE TRACE A, and SAFARI 1992. *J. Geophys. Res. Atmos.* **101,** 24251–24278 (1996).
43. Thompson, A. M. *et al.* Southern Hemisphere Additional Ozonesondes (SHADOZ) ozone climatology (2005–2009): Tropospheric and tropical tropopause layer (TTL) profiles with comparisons to OMI-based ozone products. *J. Geophys. Res. Atmos.* **117,** D23301 (2012).
44. Staudt, A. C. *et al.* Global chemical model analysis of biomass burning and lightning influences over the South Pacific in austral spring. *J. Geophys. Res. Atmos.* **107,** 4200 (2002).
45. Gettelman, A. *et al.* Validation of Aqua satellite data in the upper troposphere and lower stratosphere with in situ aircraft instruments. *Geophys. Res. Lett.* **31,** L22107 (2004).
46. Johanson, C. M. & Fu, Q. Hadley cell widening: model simulations versus observations. *J. Clim.* **22,** 2713–2725 (2009).
47. Ashfold, M. J. *et al.* Rapid transport of East Asian pollution to the deep tropics. *Atmos. Chem. Phys.* **15,** 3565–3573 (2015).
48. Pan, L. L. *et al.* A set of diagnostics for evaluating chemistry-climate models in the extratropical tropopause region. *J. Geophys. Res.* **112,** D09316 (2007).
49. Holton, J. R. *et al.* Stratosphere-troposphere exchange. *Rev. Geophys.* **33,** 403–439 (1995).
50. Waugh, D. W. & Polvani, L. M. Climatology of intrusions into the tropical upper troposphere. *Geophys. Res. Lett.* **27,** 3857–3860 (2000).
51. Granier, C. *et al.* Evolution of anthropogenic and biomass burning emissions of air pollutants at global and regional scales during the 1980-2010 period. *Clim. Change.* **109,** 163–190 (2011).
52. Giglio, L., Csiszar, I. & Justice, C. O. Global distribution and seasonality of active fires as observed with the Terra and Aqua Moderate Resolution Imaging Spectroradiometer (MODIS) sensors. *J. Geophys. Res. Biogeosci.* **111,** G02016 (2006).
53. Ridley, B. A. & Grahek, F. E. A Small, low-flow, high-sensitivity reaction vessel for NO chemiluminescence detectors. *J. Atmos. Ocean. Technol.* **7,** 307–311 (1990).
54. Boylan, P., Helmig, D. & Park, J. H. Characterization and mitigation of water vapor effects in the measurement of ozone by chemiluminescence with nitric oxide. *Atmos. Meas. Tech.* **7,** 1231–1244 (2014).
55. Gerbig, C. *et al.* An improved fast-response vacuum-UV resonance fluorescence CO instrument. *J. Geophys. Res. Atmos.* **104,** 1699–1704 (1999).
56. Cazorla, M. *et al.* A new airborne laser-induced fluorescence instrument for in situ detection of formaldehyde throughout the troposphere and lower stratosphere. *Atmos. Meas. Tech.* **8,** 541–552 (2015).
57. Apel, E. C. A fast-GC/MS system to measure C_2 to C_4 carbonyls and methanol aboard aircraft. *J. Geophys. Res.* **108,** 8794 (2003).
58. Shetter, R. E. & Muller, M. Photolysis frequency measurements using actinic flux spectroradiometry during the PEM-Tropics mission: Instrumentation description and some results. *J. Geophys. Res. Atmos.* **104,** 5647–5661 (1999).
59. Le Breton, M. *et al.* Airborne hydrogen cyanide measurements using a chemical ionisation mass spectrometer for the plume identification of biomass burning forest fires. *Atmos. Chem. Phys.* **13,** 9217–9232 (2013).
60. Tian, B. *et al.* Evaluating CMIP5 models using AIRS tropospheric air temperature and specific humidity climatology. *J. Geophys. Res. Atmos.* **118,** 114–134 (2013).
61. Susskind, J., Molnar, G., Iredell, L. & Loeb, N. G. Interannual variability of outgoing longwave radiation as observed by AIRS and CERES. *J. Geophys. Res. Atmos.* **117,** D23107 (2012).
62. Fernandez, R. P., Salawitch, R. J., Kinnison, D. E., Lamarque, J. F. & Saiz-Lopez, A. Bromine partitioning in the tropical tropopause layer: implications for stratospheric injection. *Atmos. Chem. Phys.* **14,** 13391–13410 (2014).
63. Wiedinmyer, C. *et al.* The Fire INventory from NCAR (FINN): a high resolution global model to estimate the emissions from open burning. *Geosci. Model Dev.* **4,** 625–641 (2011).
64. Emmerson, K. M. & Evans, M. J. Comparison of tropospheric gas-phase chemistry schemes for use within global models. *Atmos. Chem. Phys.* **9,** 1831–1845 (2009).
65. Jenkin, M. E., Young, J. C. & Rickard, A. R. The MCM v3.3.1 degradation scheme for isoprene. *Atmos. Chem. Phys.* **15,** 11433–11459 (2015).
66. Madronich, S. Implications of recent total atmospheric ozone measurements for biologically-active ultraviolet-radiation reaching the earth's surface. *Geophys. Res. Lett.* **19,** 37–40 (1992).
67. Wolfe, G. M. & Thornton, J. A. The chemistry of atmosphere-forest exchange (CAFE) model—part 1: model description and characterization. *Atmos. Chem. Phys.* **11,** 77–101 (2011).
68. Atkinson, R. *et al.* Evaluated kinetic and photochemical data for atmospheric chemistry: volume II—gas phase reactions of organic species. *Atmos. Chem. Phys.* **6,** 3625–4055 (2006).
69. Sander, S. P. *et al.* *Chemical Kinetics and Photochemical Data for Use in Atmospheric Studies, Evaluation No. 17* (Jet Propulsion Laboratory, 2011).
70. Andreae, M. O. & Merlet, P. Emission of trace gases and aerosols from biomass burning. *Global Biogeochem. Cycles* **15,** 955–966 (2001).

Acknowledgements

We thank L. Pan for coordinating the CONTRAST flights and her constructive criticism of an early version of the manuscript; S. Schauffler, V. Donets and R. Lueb for collecting and analysing AWAS samples; T. Robinson and O. Shieh for providing meteorology forecasts in the field; and the pilots and crews of the CAST BAe-146 and CONTRAST Gulfstream V aircrafts for their dedication and professionalism. CAST was funded by the Natural Environment Research Council; CONTRAST was funded by the National Science Foundation. Research at the Jet Propulsion Laboratory, California Institute of Technology, is performed under contract with the National Aeronautics and Space Administration (NASA). A number of the US-based investigators also benefitted from the support of NASA as well as the National Oceanic and Atmospheric Administration. The views, opinions, and findings contained in this report are those of the author(s) and should not be construed as an official National Oceanic and Atmospheric Administration or US Government position, policy or decision. We would like to acknowledge high-performance computing support from Yellowstone (ark:/85065/d7wd3xhc) provided by NCAR's Computational and Information Systems Laboratory. NCAR is sponsored by the National Science Foundation.

Author contributions

D.C.A., R.J.S. and J.M.N. wrote the manuscript. All authors discussed the results, commented on the manuscript and many suggested new ways to examine the data. N.R.P.H., J.D.L. and L.J.C. were CAST co-PIs; E.A. and R.J.S. were CONTRAST co-PIs. J.F.B. provided in-field meteorological forecasting and guidance on meteorological portions of the manuscript. J-.F.L., A.S-.L., D.E.K., R.P.F. and R.B.P. provided chemistry transport modelling used for flight planning in the field as well as data analysis. M.E. assisted in CAST flight planning and provided the box model used in the analysis. R.R.D. and T.P.C. assisted in the data analysis and numerous revisions of the manuscript. S.B. measured CO and O_3; M.L.B., T.B. and C.P. measured HCN; J.D.L. and A.V. measured NO, all for CAST. A.J.W. measured O_3, NO and NO_2; E.A., E.C.A., R.S.H., N.J.B. and D.D.R. measured organic trace gases; T.L.C. measured CO; T.F.H., G.M.W. and D.C.A. measured HCHO; S.R.H. and K.U. measured actinic flux and provided photolysis frequencies, all for CONTRAST. M.D.C. and B.J.B.S. provided the HYSPLIT model and guidance in its use. L.P. created the convective precipitation product. B.H.K. provided AIRS data and guidance. A.M.T. provided the ozonesonde profiles. J.M.N. and D.C.A. conducted the box modelling analysis.

Additional information

Competing financial interests: The authors declare no competing financial interests.

18

The absence of an Atlantic imprint on the multidecadal variability of wintertime European temperature

Ayako Yamamoto[1] & Jaime B. Palter[1,2]

Northern Hemisphere climate responds sensitively to multidecadal variability in North Atlantic sea surface temperature (SST). It is therefore surprising that an imprint of such variability is conspicuously absent in wintertime western European temperature, despite that Europe's climate is strongly influenced by its neighbouring ocean, where multidecadal variability in basin-average SST persists in all seasons. Here we trace the cause of this missing imprint to a dynamic anomaly of the atmospheric circulation that masks its thermodynamic response to SST anomalies. Specifically, differences in the pathways Lagrangian particles take to Europe during anomalous SST winters suppress the expected fluctuations in air-sea heat exchange accumulated along those trajectories. Because decadal variability in North Atlantic-average SST may be driven partly by the Atlantic Meridional Overturning Circulation (AMOC), the atmosphere's dynamical adjustment to this mode of variability may have important implications for the European wintertime temperature response to a projected twenty-first century AMOC decline.

[1] Department of Atmospheric and Oceanic Sciences, McGill University, 805 Sherbrooke Street West, Montreal, Quebec, Canada H3A 2K6. [2] Graduate School of Oceanography, University of Rhode Island, Narragansett Bay Campus, Narragansett, Rhode Island 02882, USA. Correspondence and requests for materials should be addressed to A.Y. (email: ayako.yamamoto@mail.mcgill.ca).

Large-scale, multidecadal variability in North Atlantic sea surface temperature (SST), frequently referred to as the Atlantic Multidecadal Oscillation (AMO), is a prominent feature of Northern Hemisphere climate[1,2]: Sahel drought[3], Atlantic hurricanes[4], large-scale atmospheric circulation[2,5–7] and summertime European temperature and precipitation[8,9] all respond sensitively to this low-frequency variability in North Atlantic SST. A number of studies suggest that the cause of this SST oscillation is internal variation in ocean heat transport, possibly related to the Atlantic Meridional Overturning Circulation (AMOC) variability[10–12], with the role of external and/or atmospheric stochastic forcing provoking recent controversy[13–16]. Evidence in support of the AMO variability being driven internally comes in the form of proxy evidence of a persistent oscillation throughout the past 8,000 years[17] and the reconstruction of this mode of variability in a number of modelling studies, even in the absence of external forcing[18–20].

It is well known that the North Atlantic strongly influences western European climate, with the most obvious manifestation being the anomalous wintertime warmth of the region relative to the zonal mean at equivalent latitudes[21]. Moreover, a recent study showed that temporal variability in western European wintertime temperature is set largely by the size of the air-sea turbulent fluxes along the trajectories of Lagrangian air parcels en route to Europe[22]. Coupled with evidence that variability in air-sea heat fluxes over the Atlantic is controlled by the ocean on decadal and longer time scales[23], it is natural to expect that decadal, basin-scale SST fluctuations should translate to variability in European temperature. Indeed, in all seasons besides winter, the imprint of the AMO is evident in European temperature[8,9]. The SST anomaly associated with the AMO persists throughout the year[5,10], making the absence of a wintertime AMO signal in western Europe all the more puzzling (Fig. 1a).

We propose here that the AMO is closely associated with variability in the position and strength of the storm track, which suppresses the influence of the anomalous SST on the heat fluxes seen by Lagrangian parcels transiting to Europe. To evaluate this possibility, we examine the National Centers for Environmental Prediction/National Center for Atmospheric Research 20th-Century Reanalysis (20CR)[24] from complementary Eulerian and Lagrangian perspectives. The 20CR is one of the longest reanalysis products currently available, and is faithful with independent observations of northeast Atlantic storminess from 1940 onward[25]. Thus, we analyse storm track and Lagrangian pathway variability in the period from 1940 to 2011, which encompasses one full AMO cycle.

Results

Eulerian perspective. The AMO index, computed by taking an area-weighted mean of the linearly detrended SST field over the North Atlantic[9,26], is generally positive from 1940 to 1963 and 1996 to 2011 and negative from 1966 to 1994 (Fig. 1a). There are various approaches to defining an AMO index[27,28], yet the main features remain almost identical to those identified here regardless of the method chosen[18,20,27]. In particular, both modelling[6,10,19] and observational studies[2,6,9,11], utilizing different methodologies to isolate only the internal mode of variability, show spatial patterns of SST anomalies similar to that shown in Fig. 2a, and the approximate timing of transitions between AMO phases is not sensitive to its definition[28].

During each multidecadal period characterized by a given phase of the AMO index, short-term variability gives rise to months in which the basin-averaged SST anomaly is near zero or of the opposite sign relative to the decade in which it is embedded (Fig. 1a). To expose the association of the atmospheric anomaly pattern with SST anomalies, we made composite periods using only the January months with the most extreme AMO index. The extreme AMO months are chosen such that they meet the criteria, $|AMOindex| > 0.15$, which is nearly one standard deviation beyond zero (Fig. 1a). All of these extreme months fall within a longer period where the 10-year running mean AMO index has the appropriate sign. In this manner, 17 positive and 18 negative AMO January months are selected (Fig. 1a). We repeated the analysis using all years within the corresponding AMO phase and present the results in the Supplementary Materials, where it is apparent that the key results and interpretation are essentially unchanged, although their statistical power is slightly weaker.

The large-scale atmospheric flow varies with the AMO index (Fig. 2b). The difference in the 500-hPa geopotential height (Z500) field, which is analogous to streamlines, shows that the direction of winds arriving in western Europe changes between the two AMO phases: winds are more northerly during the anomalous AMO-positive years, whereas they are more zonal during the AMO-negative years (Fig. 2b). The more tightly spaced isohypses during the AMO-negative years indicate a swifter flow relative to the AMO-positive years. Accordingly, the AMO-negative years see an elongated and more zonal January storm track (Supplementary Fig. 1), which is consistent with results from a free-running climate model[7]. Composite Z500 maps constructed with more complete sampling of the longer decadal periods associated with the AMO show similar, albeit weaker, anomaly patterns (Supplementary Fig. 2a).

Lagrangian perspective. The impact of the modulation of the large-scale atmospheric flow on temperature in western Europe is best evaluated in a Lagrangian framework, where the dynamic variability of the atmosphere and the variability in the air–sea turbulent exchange can be assessed simultaneously for the atmospheric particles that influence Europe. Therefore, we launch virtual Lagrangian particles from the surface of forty-one uniformly distributed points over land in western Europe (Fig. 1b) and track them backward in time for 10 days using the atmospheric dispersion model, FLEXPART[29] (see Methods for details). Ten atmospheric particles are released twice a day in January from 1940 to 2011 from each of the forty-one release points. A climatological two-dimensional histogram of the positions of the resulting Lagrangian particles is shown in Fig. 3a, in which their trajectories are seen spreading out over the North Atlantic, many stretching back to the Labrador Sea and northern Canada. The statistical significance of these results is strengthened when separating the particle launch locations into northern and southern sub-regions of western Europe (see Supplementary Fig. 3 for details).

Our previous work showed that air–sea turbulent fluxes at the base of the atmospheric planetary boundary layer (PBL) govern variability in the potential temperature change along these particle trajectories in January almost entirely: turbulent fluxes alone explain more than 80% of the variability in the potential temperature change along 10-day back trajectories from western Europe[22]. Although the fluxes through the top of the planetary boundary layer are important for closing the heat budget of the layer, they are not crucial for understanding the low-frequency variability of the temperature tendency along Lagrangian trajectories. Therefore, we track the ocean-atmosphere turbulent fluxes along each particle's trajectory by interpolating these fluxes from 20CR to each particle's hourly position when the particle is within PBL. The turbulent fluxes are a function of the temperature and moisture gradients at the air–sea interface and the surface wind speed (see Methods for details).

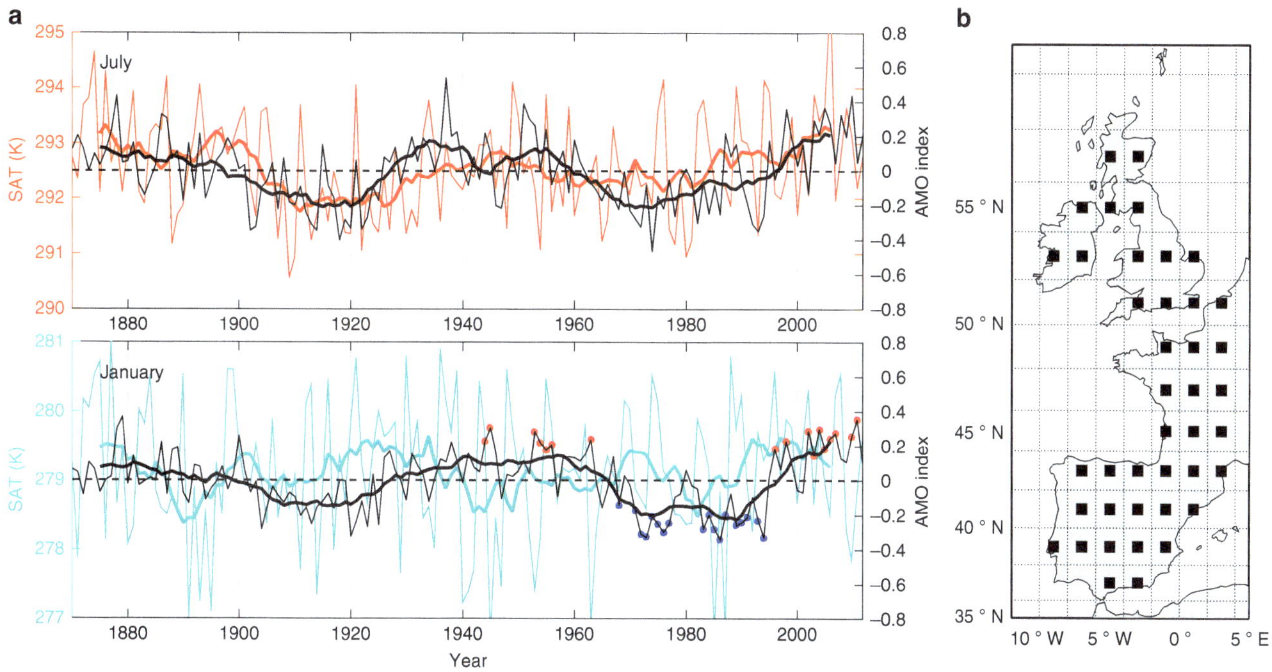

Figure 1 | Decadal variability in North Atlantic SST and western European SAT. (**a**) Time series of the linearly detrended North Atlantic SST (black lines, referred to as the AMO index) and SAT averaged over western Europe ([36N 60N] × [10W 3E]; shown in coloured lines) in July (top panel) and January (bottom panel). Bold lines show 10-year running means. The correlation coefficient between the 10-year running mean of the detrended SAT and AMO index is 0.61 in July (statistically significant at 10% confidence level even after accounting for the reduced effective degrees of freedom due to autocorrelation of the time series) and −0.02 in January; these correlations are insensitive to the averaging region chosen for western Europe. The red circles on January plot indicate the AMO-positive years chosen for the composite analysis, whereas the blue circles indicate the AMO-negative years chosen. (**b**) Study region encompassing western Europe ([36N 60N] × [10W 3E]) and locations for the backtracked Lagrangian particle release (black squares).

The turbulent fluxes are calculated in 20CR with the product's wind speeds and temperature and moisture gradients, and the size of the fluxes is influenced by correlations between the wind speed and temperatures[22].

In winter in the North Atlantic, SST is almost always warmer than the surface air temperature (SAT), so the ocean loses heat rapidly to the atmosphere over the entirety of the basin (that is, positive fluxes in our convention; Fig. 3b and Supplementary Fig. 3b). The fluxes over the warm Gulf Stream and its North Atlantic Current extension are generally a factor of five higher than found elsewhere. However, a view of the fluxes weighted by the fraction of time the particles spend in each location on their journey to western Europe (Fig. 3c and Supplementary Fig. 3c) suggests a reduced role of these strong flux regions in establishing western European wintertime temperature.

The difference in the number density of the particle positions between the composite AMO periods (Fig. 3d) shows a significant distinction in the preferred pathways, with the statistical significance increasing when results are separated by particles launched from northern and southern sub-regions of western Europe (Supplementary Fig. 3d). In the AMO-positive years, particles spend more of their 10-day trajectory recirculating locally to the southwest of Iceland. During the AMO-negative years, the pathways are anomalously long, and a greater number of trajectories originate from North America and the Arctic, before transiting over the Labrador Sea and mid-latitude North Atlantic. These differences in the atmospheric trajectories are explained mechanistically by the difference in the Z500 anomalies associated with the AMO, which shows swifter, more zonal winds during AMO-negative years (Fig. 2b). This largely barotropic anomaly pattern has been noted in a number of modelling and observational studies[5,10,30,31], and is somewhat similar to the atmospheric circulation patterns associated with the North Atlantic Oscillation (NAO)[5]. To explore whether this atmospheric anomaly pattern is linked with anomalous SST conditions of the AMO regardless of the NAO phase, we performed an additional analysis excluding the strong NAO years. This exclusion only amplifies the signal of intensified zonal flow during negative AMO years relative to positive years (c.f. Fig. 3d and Supplementary Fig. 4a). The cause of the linkage between AMO and NAO has been the subject of debate, with several papers arguing that North Atlantic SST anomalies force an atmosperic NAO response[32–34], and others arguing the reverse[12,35]. Regardless of what drives the relationship, the association between the atmospheric circulation and the AMO index is clear in the Lagrangian trajectory composites (Fig. 3d).

The difference map of turbulent fluxes along these Lagrangian trajectories points towards a leading cause of the missing AMO imprint on European wintertime temperatures. During AMO-positive years, the shorter trajectories arriving from the north and southwest (Fig. 3d and Supplementary Fig. 3d) are accompanied by high fluxes (Fig. 3e and Supplementary Fig. 3e). However, the long and zonal trajectories associated with the AMO-negative years are accompanied by even stronger turbulent fluxes over much of the mid-latitude North Atlantic. Therefore, there are partially compensating regions of elevated and depressed flux during both phases of the AMO. The same pattern is found when the difference maps are constructed from composites of the full decadal periods associated with the AMO, but with slightly weaker statistical strength (Supplementary Fig. 2b,c).

Time series constructed by averaging along Lagrangian back trajectories (Fig. 4) further reveal the net effect of the combined

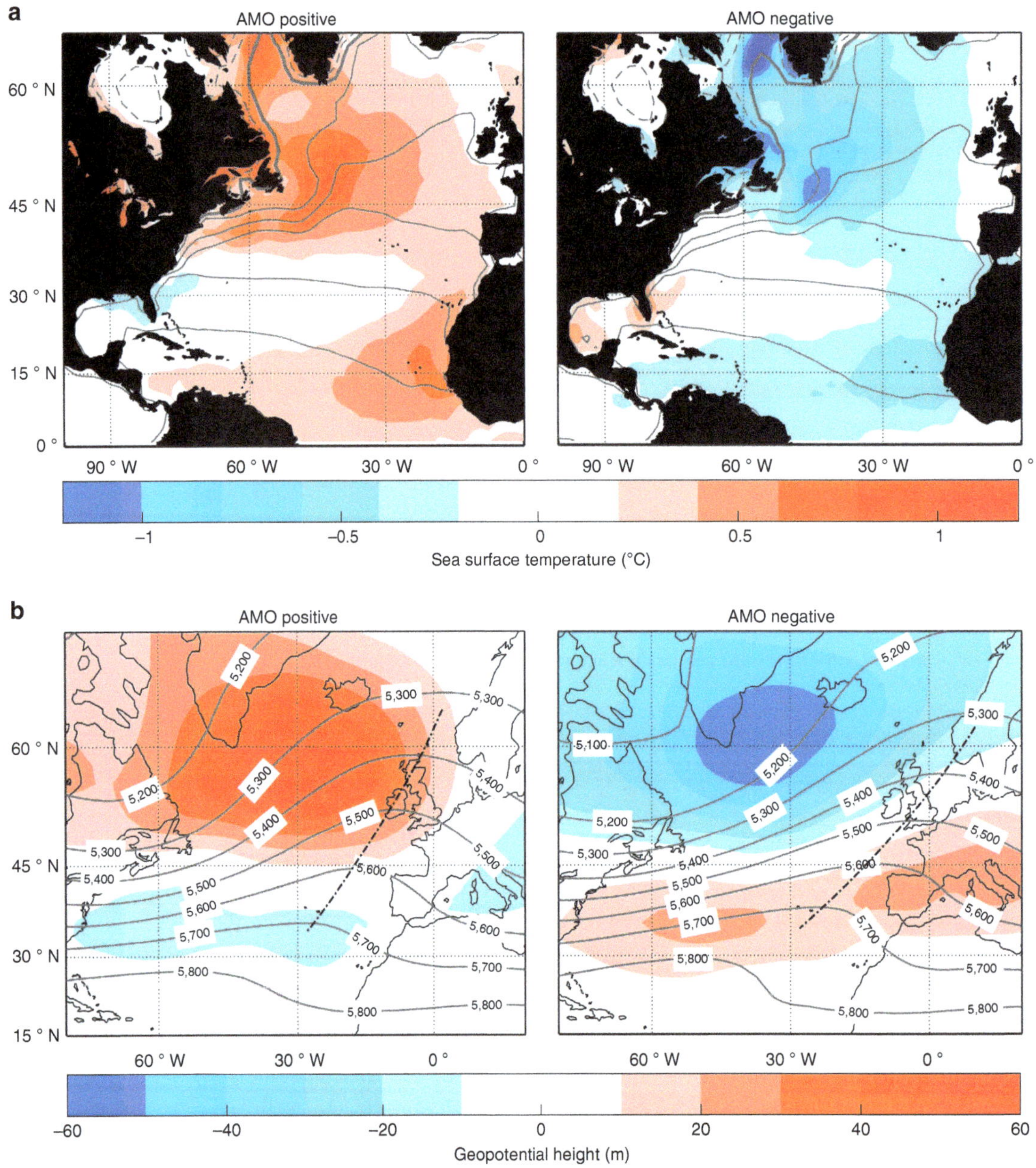

Figure 2 | The spatial pattern of the AMO index and its relationship with the atmospheric flow in January. Composite maps of (**a**) sea surface temperature (SST) field and (**b**) 500 hPa geopotential height field (Z500) for AMO anomalously positive years (left panel) and negative years (right panel). The January mean field is shown in contours, and its departure from the 72-year climatology is represented by colour shading. The thick grey contour line in **a** denotes 0 °C, whereas thin (dashed) lines denote positive (negative) SST every 5 °C. The black dashed lines in **b** are drawn through the local maxima of the geopotential height field at each latitude, which is the point where the wind changes direction from south-westerly to north-westerly.

changes to Lagrangian pathways and the properties along them. Notably, the SST sampled along the Lagrangian trajectories lacks a clear AMO signal (Fig. 4a), because decadal variability in atmospheric trajectories, which travel over a spatially variable SST field, swamps the temporal variability of the North Atlantic average SST. The SAT is highly correlated with the SST (not shown), as turbulent fluxes work to bring the surface boundary layers of the atmosphere and ocean towards equilibrium. Yet, during the negative AMO years from the mid-1970s to 1990, the air advected along Lagrangian trajectories is more anomalously cold than the SST, producing strengthened air–sea temperature gradients (Fig. 4b). Further heightened by stronger winds (Fig. 4c), the largest turbulent fluxes are achieved during these AMO-negative years (Fig. 4d). We acknowledge that at the onset of the AMO-negative period around 1968, both SST and SAT averaged along the trajectories were elevated and fluxes were approximately average. Nevertheless, the overall effect is that the 10-year running mean turbulent fluxes sampled along Lagrangian trajectories are weakly anticorrelated with the AMO index ($r = -0.39$, not significant given the few effective degrees of freedom in the smoothed time series; Fig. 4d). We conclude that, in winter, the dynamic modulation of Lagrangian pathways and the atmospheric properties transported with them oppose the influence of basin-scale SST fluctuations on turbulent air–sea fluxes, thereby concealing the temperature expression of the AMO in atmospheric particles arriving in Europe.

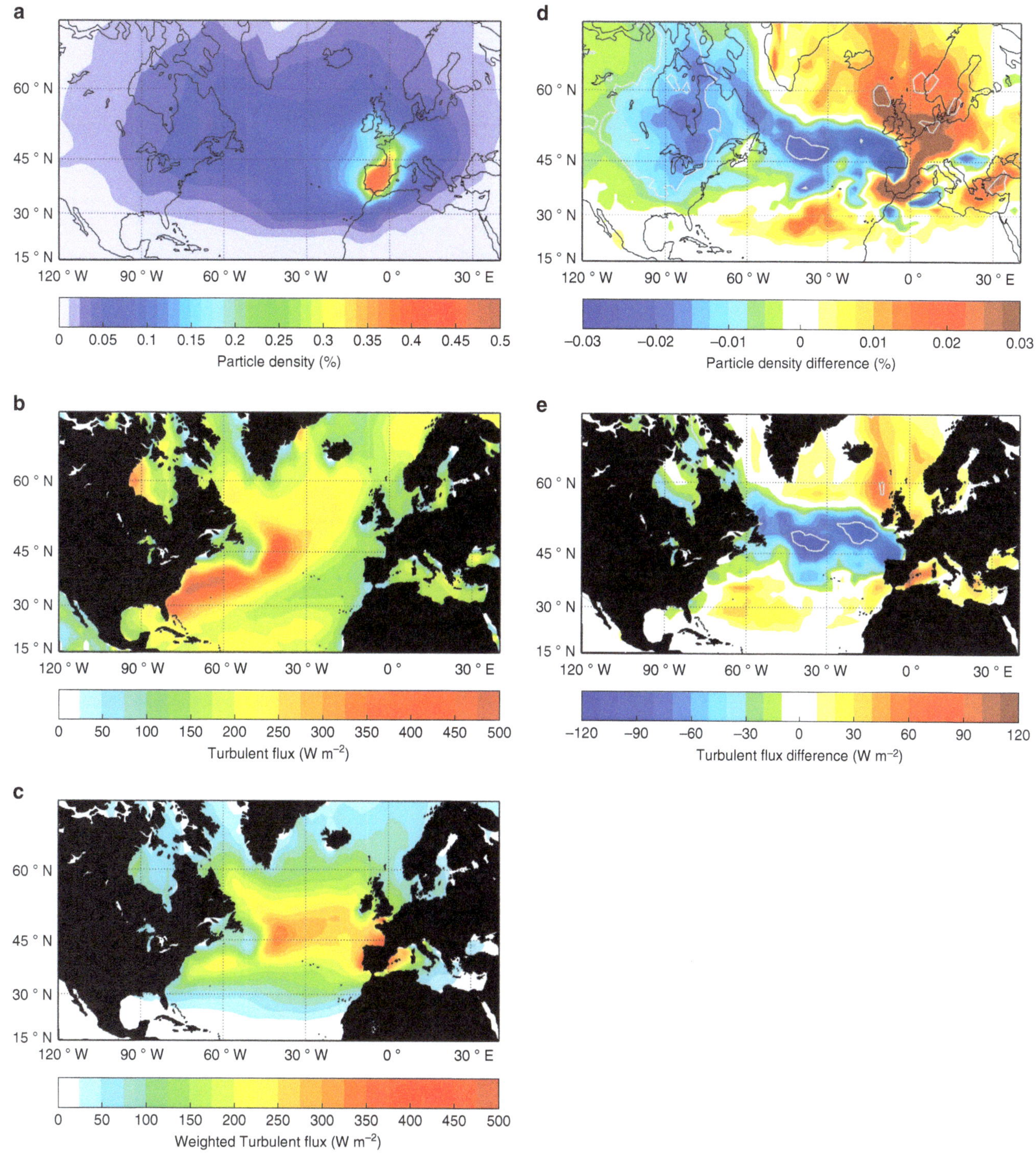

Figure 3 | Climatological mean and AMO influence on backward Lagrangian trajectories and the properties along them. (**a**) Lagrangian particle climatological number density, given as the percentage of all hourly positions that were spent in any 2° × 2° grid cell. (**b**) Climatological turbulent fluxes (sensible + latent; $W\,m^{-2}$) calculated by averaging the fluxes along the Lagrangian trajectories. (**c**) Turbulent fluxes as in **b** but weighted by the fraction of hourly particle positions spent in each grid cell, and normalized to have an equal spatial mean as the unweighted fluxes (see Methods for details; $W\,m^{-2}$). (**d**) The difference in number density for AMO-positive state minus AMO-negative state (% particle's hourly positions). (**e**) The difference in turbulent fluxes ($W\,m^{-2}$) for AMO-positive state minus AMO-negative state, weighted and normalized as in **c**. In **d**, statistical significance at 10 and 15% is shown, whereas in **e**, the 15% significance level is shown in grey contours. All the significance levels were obtained using a bootstrapping method (see Methods for details).

To further assess the degree to which the dynamic modulation of the trajectories is responsible for suppressing the AMO imprint on the fluxes, we re-ran our Lagrangian simulations with randomly selected, unvarying trajectories (see Methods for details). The time series of the 10-year running mean conditions along these random trajectories is plotted in blue in Fig. 4. The SST (Fig. 4a) averaged along these randomly selected trajectories vary in phase with the AMO index, and the anticorrelation of the

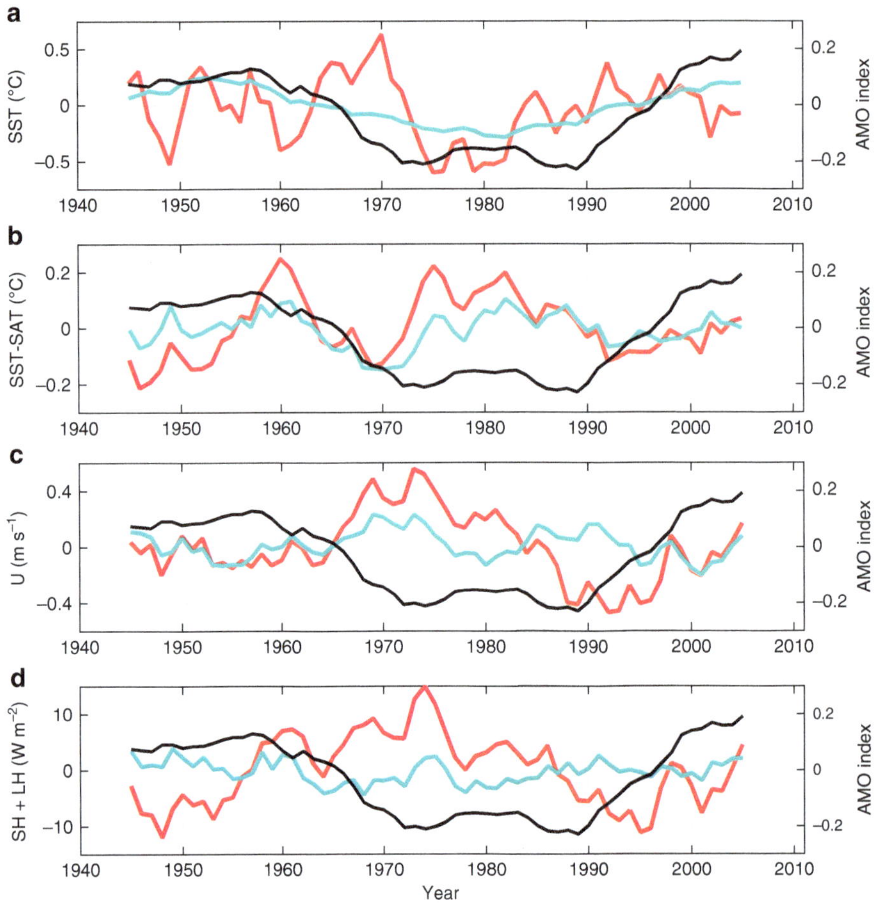

Figure 4 | Time series of properties and heat fluxes averaged along Lagrangian trajectories. 10-Year running mean of each linearly detrended variable along true trajectories (red), and along 10 sets of 10 random trajectories (blue), averaged across all 41 release locations. The January 10-year running mean AMO index is overlaid in black in each panel. (**a**) SST (°C), (**b**) SST − SAT (°C), (**c**) wind speed (m s^{-1}) and (**d**) turbulent fluxes (W m^{-2}).

fluxes with the AMO is eliminated (Fig. 4d). Finally, we confirm that, in summer, when the European temperature reflects AMO variability, the turbulent fluxes vary in phase with the AMO. In July, there is a minimal difference between the preferred pathways by AMO phase (Supplementary Fig. 5). Hence, the dominant signature of the AMO-positive phase appears to be due to higher basin-scale SST, which allows for a broad region of enhanced fluxes, consistent with an extended analysis of air–sea fluxes over the past century[23].

Discussion

The strengthening and lengthening of the storm track in sync with anomalously cooler North Atlantic SSTs has important implications for future climate. Given that decadal variability in North Atlantic SSTs may be driven partly by fluctuations in the strength of the AMOC[10–12], our result suggests the possibility of a stabilizing feedback for ocean circulation: Cooler SSTs associated with a sluggish AMOC is linked with an atmospheric adjustment that enhances turbulent heat fluxes over oceanic convective regions in winter. These larger fluxes could possibly reinvigorate convection, deep water formation and the AMOC. Moreover, the observed link of the atmospheric circulation with the cool SST anomalies of the late 1970s to early 1990s is much like the predicted change of the storm track in response to a decline of the AMOC under global warming[36]. A weakened AMOC has long been thought to cause anomalous cooling in western Europe via a decline in oceanic heat transport and associated atmospheric feedbacks[21]. However, the changes we describe here in atmospheric Lagrangian trajectories and the heat fluxes along them could provide a mechanism that reduces the magnitude of European wintertime cooling on decadal time scales, even as they might stabilize the oceanic circulation.

Methods

FLEXPART. We adapted the Lagrangian atmospheric dispersion model FLEXPART version 9.02 (ref. 29) for use with the Twentieth Century Atmospheric Reanalysis product (20CR)[24] in order to simulate the atmospheric particles released from 41 equally spaced western European locations (Fig. 1b). These release points are chosen from an evenly spaced 2° × 2° grid over the study region [36N 60N] × [10W 3E], when these points fall on land. Every January from 1940 to 2011, ten particles are released from the surface of each of these points twice daily at 0 coordinated universal time (UTC) and 12 UTC and advected backward in time for the duration of 10 days, following the three-dimensional wind field. There are three components to this wind field: (i) resolved wind, (ii) turbulent wind fluctuations and (iii) mesoscale wind fluctuations. FLEXPART accounts for the turbulent wind fluctuations by adding a perturbation to the velocity field for atmospheric particles in the PBL, where these random motions are calculated by solving Langevin equations for Gaussian turbulence. Mesoscale velocity, whose spectral interval falls between the resolved flow and the turbulent flow, is included by solving an independent Langevin equation. The PBL height is diagnosed at each particle's hourly position. The duration of 10 days for the back trajectories was chosen based on the fact that the Lagrangian decorrelation time scale is ~3 days[22]; thus, the choice of 10 days is long enough for the memory of each particle's initial temperature to be erased under the effect of diabatic processes along the trajectory.

20CR is one of the longest reanalysis products currently available. It has 6-h temporal resolution and 2° × 2° spatial resolution. The product assimilates only observations of surface pressure, monthly SST and sea-ice distributions, and we only use the ensemble mean fields. To assess the reliability of our FLEXPART results with use of 20CR, the trajectories computed using 20CR was compared with those with a default input for FLEXPART, Climate Forecast System Reanalysis

(CFSR) forecast and reanalysis data sets from National Centers for Environmental Prediction[37], which has hourly temporal and 0.5° × 0.5° spatial resolution, for the period of 1981–2009 under the same set up as Yamamoto *et al.*[22] We found that the trajectory paths computed with 20CR are generally very close to those computed with CFSR especially over the ocean, with particles in 20CR taking slightly more northern paths relative to CFSR (Supplementary Fig. 6). We note that 20CR assimilates monthly mean SST data, whereas CFSR assimilates SST every 6 h. The agreement of the amplitude and variability of the turbulent fluxes along Lagrangian trajectories constructed from the two reanalysis products (Supplementary Fig. 7) suggests that the missing sub-monthly SST variability in 20CR has a minimal impact on these fluxes on interannual and longer time scales.

Bootstrapping. Bootstrapping was used in order to gauge the statistical significance of the difference in the spatial patterns of Lagrangian trajectory pathways and the fluxes along the trajectories for the two AMO phases (Fig. 3d,e). We sample the Lagrangian particle trajectories or the fluxes along them from randomly selected January months ('pseudo-periods') for the same number of years as each AMO phase (17 years for AMO positive and 18 years for AMO negative, respectively). We then take a difference between the composite pseudo-periods. This operation was repeated 500 times. We consider differences of the Lagrangian particle density and the fluxes along them from the true AMO composites significant at the 10% (15%) level when this true difference exceeds the 90th (85th) percentile of the pseudo-period differences. We repeated the same procedure to make the composites with entire AMO periods (40 AMO positive years and 29 AMO negative years) and show these results in the Supplementary Materials.

Surface fluxes along the trajectories. Along the particle trajectories simulated using FLEXPART, the surface turbulent fluxes are interpolated from 20CR sensible heat (SH) and latent heat flux (LH) fields, whenever the particle's hourly position falls within PBL, under the assumption that the turbulent fluxes influence the entire air mass within the PBL. The turbulent fluxes in 20CR are computed using bulk formulae with a typical formulation[38]:

$$SH = \rho_a c_p C_h U(T_s - T_a) \qquad (1)$$

$$LH = \rho_a L_e C_e U(q_s - q_a) \qquad (2)$$

where ρ_a is the atmospheric density, c_p is the atmospheric heat capacity, C_h and C_e are transfer coefficients, U is the mean value of wind speed relative to the surface ocean current, T_s is sea surface temperature, T_a is the atmospheric potential temperature at a reference height, L_e is latent heat of evaporation, q_s is interfacial value of water vapour mixing ratio, and q_a is the atmospheric water vapour mixing ratio at a reference height.

Time series of the mean accumulated surface heat fluxes along the trajectories using 20CR are highly correlated with those using CFSR (Supplementary Fig. 7), with the average correlation coefficient being $r = 0.92$.

Weighting of composite fluxes. In the composite figures of weighted surface fluxes along the trajectories (Fig. 3c,e), weights are proportional to the fraction of all particle positions that pass over a each 2° × 2° grid cell (that is, the number density, given as a percentage in Fig. 3a). These weights are scaled so that the mean value of the climatological map (Fig. 3c) is equal to the mean of the unweighted climatology (Fig. 3b). The spatial mean of both mapped fields used in this scaling includes only those grid cells visited by at least 0.01% of the climatological particle positions; these collectively contain 90.1% of all hourly particle positions.

Randomly selected trajectories. The unvarying particle trajectory pathways used to produce Fig. 4 (blue lines) were chosen by randomly selecting ten particles from the set of all possible pathways generated from the full 72-year Lagrangian simulation. The truly varying surface fluxes are interpolated along these random trajectories. We then repeat this process ten times, each time by picking a different random set of ten trajectories from each release point, thereby creating a spread of particle positions. In total, 100 particles (10 particles × 10 realizations) are selected for each release location.

References

1. Schlesinger, M. E. & Ramankutty, N. An oscillation in the global climate system of period 65-70 years. *Nature* **367**, 723–726 (1994).
2. Kushnir, Y. Interdecadal variations in North Atlantic sea surface temperature and associated atmospheric conditions. *J. Clim.* **7**, 141–157 (1994).
3. Rowell, B. D. P., Folland, C. K., Maskell, K. & Ward, M. N. Variability of summer rainfall over tropical north Africa (1906-92): Observations and modelling. *Quat. J. R. Meteorol. Soc.* **121**, 669–704 (1995).
4. Goldenberg, S. B., Landsea, C. W., Mestas-Nunez, A. M. & Gray, W. M. The recent increase in Atlantic hurricane activity: causes and implications. *Science (New York, N.Y.)* **293**, 474–479 (2001).
5. Sutton, R. T. & Hodson, D. L. R. Influence of the ocean on North Atlantic climate variability 1871-1999. *J. Clim.* **16**, 3296–3313 (2003).
6. Gastineau, G., D'Andrea, F. & Frankignoul, C. Atmospheric response to the North Atlantic Ocean variability on seasonal to decadal time scales. *Clim. Dyn.* **40**, 2311–2330 (2012).
7. Zhang, R. & Delworth, T. L. Impact of the Atlantic Multidecadal Oscillation on North Pacific climate variability. *Geophys. Res. Lett.* **34**, 2–7 (2007).
8. Arguez, A., O'Brien, J. J. & Smith, S. R. Air temperature impacts over Eastern North America and Europe associated with low-frequency North Atlantic SST variability. *Int. J. Climatol.* **10**, 1–10 (2009).
9. Sutton, R. T. & Dong, B. Atlantic Ocean influence on a shift in European climate in the 1990s. *Nat. Geosci.* **5**, 788–792 (2012).
10. Delworth, T. L. & Mann, M. E. Observed and simulated multidecadal variability in the Northern Hemisphere. *Clim. Dyn.* **16**, 661–676 (2000).
11. Latif, M., Roeckner, E., Botzet, M. & Esch, M. Reconstructing, monitoring, and predicting multidecadal-scale changes in the North Atlantic thermohaline circulation with sea surface temperature. *J. Clim.* **17**, 1605–1614 (2004).
12. McCarthy, G. D., Haigh, I. D., Hirschi, J. J.-M., Grist, J. P. & Smeed, D. A. Ocean impact on decadal Atlantic climate variability revealed by sea-level observations. *Nature* **521**, 508–510 (2015).
13. Otterå, O. H., Bentsen, M., Drange, H. & Suo, L. External forcing as a metronome for Atlantic multidecadal variability. *Nat. Geosci.* **3**, 688–694 (2010).
14. Booth, B. B. B., Dunstone, N. J., Halloran, P. R., Andrews, T. & Bellouin, N. Aerosols implicated as a prime driver of twentieth-century North Atlantic climate variability. *Nature* **484**, 228–232 (2012).
15. Zhang, R. *et al.* Have Aerosols Caused the Observed Atlantic Multidecadal Variability? *J. Atmos. Sci.* **70**, 1135–1144 (2013).
16. Clement, A. *et al.* The Atlantic Multidecadal Oscillation without a role for ocean circulation. *Science* **350**, 320–324 (2015).
17. Knudsen, M. F., Seidenkrantz, M.-S., Jacobsen, B. H. & Kuijpers, A. Tracking the Atlantic Multidecadal Oscillation through the last 8,000 years. *Nat. Commun.* **2**, 178 (2011).
18. Ting, M., Kushnir, Y., Seager, R. & Li, C. Forced and Internal Twentieth-Century SST Trends in the North Atlantic. *J. Clim.* **22**, 1469–1481 (2009).
19. Ting, M., Kushnir, Y., Seager, R. & Li, C. Robust features of Atlantic multi-decadal variability and its climate impacts. *Geophys. Res. Lett.* **38**, L17705 (2011).
20. DelSole, T., Tippett, M. K. & Shukla, J. A Significant Component of Unforced Multidecadal Variability in the Recent Acceleration of Global Warming. *J. Clim.* **24**, 909–926 (2011).
21. Palter, J. B. The Role of the Gulf Stream in European Climate. *Annu. Rev. Marine Sci.* **7**, 113–137 (2015).
22. Yamamoto, A., Palter, J. B., Lozier, M. S., Bourqui, M. S. & Leadbetter, S. J. Ocean versus atmosphere control on western European wintertime temperature variability. *Clim. Dyn.* **45**, 3593–3607 (2015).
23. Gulev, S. K., Latif, M., Keenlyside, N., Park, W. & Koltermann, K. P. North Atlantic Ocean control on surface heat flux on multidecadal timescales. *Nature* **499**, 464–467 (2013).
24. Compo, G. P. *et al.* The Twentieth Century Reanalysis Project. *Quat. J. R. Meteorol. Soc.* **137**, 1–28 (2011).
25. Krueger, O., Schenk, F., Feser, F. & Weisse, R. Inconsistencies between Long-Term Trends in Storminess Derived from the 20CR Reanalysis and Observations. *J. Clim.* **26**, 868–874 (2013).
26. Enfield, D. B., Mestas-Nunez, A. M. & Trimble, P. J. The Atlantic multidecadal oscillation and its relation to rainfall and river flows in the continental U.S. *Geophys. Res. Lett.* **28**, 2077–2080 (2001).
27. Trenberth, K. E. & Shea, D. J. Atlantic hurricanes and natural variability in 2005. *Geophys. Res. Lett.* **33**, L12704 (2006).
28. Nigam, S., Guan, B. & Ruiz-Barradas, A. Key role of the Atlantic Multidecadal Oscillation in 20th century drought and wet periods over the Great Plains. *Geophys. Res. Lett.* **38**, L16713 (2011).
29. Stohl, A., Forster, C., Frank, A., Seibert, P. & Wotawa, G. Technical note: The Lagrangian particle dispersion model FLEXPART version 6.2. *Atmos. Chem. Phys.* **5**, 2461–2474 (2005).
30. Ting, M., Kushnir, Y. & Li, C. North Atlantic Multidecadal SST Oscillation: External forcing versus internal variability. *J. Marine Sys.* **133**, 27–38 (2014).
31. Gastineau, G. & Frankignoul, C. Influence of the North Atlantic SST on the atmospheric circulation during the twentieth century. *J. Clim.* **28**, 1396–1416 (2015).
32. Frankignoul, C., Chouaib, N. & Liu, Z. Estimating the observed atmospheric response to SST anomalies: maximum covariance analysis, generalized equilibrium feedback assessment, and maximum response estimation. *J. Clim.* **24**, 2523–2539 (2011).
33. Peings, Y. & Magnusdottir, G. Forcing of the wintertime atmospheric circulation by the multidecadal fluctuations of the North Atlantic ocean. *Environ. Res. Lett.* **9**, 034018 (2014).
34. Omrani, N.-E., Keenlyside, N. S., Bader, J. & Manzini, E. Stratosphere key for wintertime atmospheric response to warm Atlantic decadal conditions. *Clim. Dyn.* **42**, 649–663 (2014).

35. Mecking, J. V., Keenlyside, N. S. & Greatbatch, R. J. Stochastically-forced multidecadal variability in the North Atlantic: a model study. *Clim. Dyn.* **43,** 271–288 (2014).
36. Woollings, T., Gregory, J. M., Pinto, J. G., Reyers, M. & Brayshaw, D. J. Response of the North Atlantic storm track to climate change shaped by ocean—atmosphere coupling. *Nat. Geosci.* **5,** 313–317 (2012).
37. Saha, S. *et al.* The NCEP climate forecast system reanalysis. *Bull. Amer. Meteor. Soc.* **91,** 1015–1057 (2010).
38. Fairall, C. W., Bradley, E. F., Rogers, D. P., Edson, J. B. & Young, G. S. Bulk parameterization of air-sea fluxes for Tropical Ocean-Global Atmosphere Coupled-Ocean Atmosphere Response Experiment difference relative analysis. *J. Geophys. Res.* **101,** 3747–3764 (1996).

Acknowledgements

This work was supported by a funding from the McGill University, the NSERC Discovery Program, FQRNT's Programme Etablissement de Nouveaux Chercheurs Universitaires and Quebec-Ocean. We thank three anonymous reviewers and E. Galbraith and M. Gervais for their constructive comments that greatly improved earlier versions of this manuscript.

Author contributions

A.Y. and J.B.P. designed the experiments, analysed the results and wrote the manuscript together. A.Y. adapted FLEXPART for 20CR input and performed the experiments.

Additional information

Competing financial interests: The authors declare no competing financial interests.

19

The puzzling Venusian polar atmospheric structure reproduced by a general circulation model

Hiroki Ando[1], Norihiko Sugimoto[2], Masahiro Takagi[3], Hiroki Kashimura[4], Takeshi Imamura[1] & Yoshihisa Matsuda[5]

Unlike the polar vortices observed in the Earth, Mars and Titan atmospheres, the observed Venus polar vortex is warmer than the midlatitudes at cloud-top levels ($\sim$65 km). This warm polar vortex is zonally surrounded by a cold latitude band located at $\sim$60° latitude, which is a unique feature called 'cold collar' in the Venus atmosphere. Although these structures have been observed in numerous previous observations, the formation mechanism is still unknown. Here we perform numerical simulations of the Venus atmospheric circulation using a general circulation model, and succeed in reproducing these puzzling features in close agreement with the observations. The cold collar and warm polar region are attributed to the residual mean meridional circulation enhanced by the thermal tide. The present results strongly suggest that the thermal tide is crucial for the structure of the Venus upper polar atmosphere at and above cloud levels.

[1] Institute of Space and Astronautical Science, Japan Aerospace Exploration Agency, Sagamihara, Kanagawa 252-0222, Japan. [2] Research and Education Center for Natural Sciences, Department of Physics, Keio University, Yokohama, Kanagawa 223-8521, Japan. [3] Faculty of Science, Kyoto Sangyo University, Kita-ku, Kyoto 603-8555, Japan. [4] Japan Agency for Marine-Earth Science and Technology, Yokohama, Kanagawa 236-0001, Japan. [5] Department of Astronomy and Earth Sciences, Tokyo Gakugei University, Koganei, Tokyo 184-8501, Japan. Correspondence and requests for materials should be addressed to H.A. (email: hando@ac.jaxa.jp).

A lot of infrared and radio occultation measurements in the Venus exploration missions show that the polar vortex is warmer than the midlatitudes at cloud-top levels and surrounded by a zonal cold latitude band called the cold collar located around ~60° latitude at ~65 km (refs 1–4). The morphology seen in the polar vortex changes temporally and sometimes shows a hot dipole or S-shaped structure[3,5–7]. Furthermore, these unique structures have been observed in both north and south polar regions.

While a lot of observational studies about the atmospheric structure in the Venus polar region have been performed, there were no numerical studies that succeeded in reproducing these unique structures in realistic model settings. Lee *et al.*[8] have reported that the cold collar and the enhanced warm polar region (hot polar spot) are reproduced in their general circulation model (GCM) results and indicated that these structures are related to the reversal of the latitudinal temperature gradient, which is caused by the zonal jet produced by mean meridional circulation (MMC). However, the zonal-mean wind reproduced in their model is not consistent with recent measurements[9]. In addition, the MMC is driven by Newtonian cooling with constant relaxation time of 25 Earth days, in which diurnal variation of the solar heating is excluded. Yamamoto and Takahashi[10,11] have also reproduced the polar vortex in their GCM with an artificial forcing of zonal wavenumber one at the bottom boundary to excite Kelvin and/or Rossby waves. Although they indicated that the polar diurnal tide enhances the cold collar and warm polar region, their results are not consistent with optical and radio occultation measurements for the following reasons: (1) the horizontal temperature distribution has large non-uniformity; (2) the location of the temperature minimum does not move in the longitudinal direction but is fixed at a specific geographical longitude; and (3) the temperature minimum corresponding to the cold collar is not seen in the latitude–height distribution of temperature in their calculation.

The Venus polar region also has interesting features in terms of material transport. Recent optical measurements showed that a significant infrared absorption signature by CO, which is mainly created photochemically above the cloud layer, exists over Venus' south pole, coinciding with bright dipole features seen in the polar vortex and the CO abundance increases with latitude above Venus' nightside cloud top[12,13]. Furthermore, the clouds in the polar region are thicker than those in the cold collar[14]. By using a two-dimensional model including a simple cloud microphysics scheme, Imamura and Hashimoto[15] suggested that the clouds might become thick with latitude by the cloud material transport associated with the MMC. However, the structure of the MMC is simply assumed in their model. It is necessary to consider how its structure is related to the dynamical process.

A thermal tide is the planetary scale wave excited by the solar heating. In the Venus atmosphere, it is strongly excited at the cloud levels because a large part of the solar flux is absorbed there[16–18]. The horizontal wind velocity associated with the thermal tide is estimated to be $>40\,\mathrm{m\,s^{-1}}$ at the cloud top. It is also indicated that superrotation at the cloud levels is affected significantly by the thermal tide[19,20]. However, its effect on the atmospheric structure in the polar region has not yet been examined.

In this study, we investigate the structure of the Venus upper polar atmosphere by using a GCM named Atmospheric GCM for the Earth Simulator (AFES) for Venus (see Methods for details). To examine the dynamical effects of the thermal tide, we perform two numerical experiments with observation-based distributions of the solar heating: one with the diurnal components (Case A) and one without them (Case B). The thermal tide is excited only in Case A. Our study shows that the cold collar and warm polar region are due to the residual mean meridional circulation (RMMC) intensified by the thermal tide, implying that the atmospheric structure at and above cloud levels in the Venus polar region is strongly attributed to the thermal tide. Furthermore, our results also indicate that the observed meridional distributions of cloud thickness and CO abundance above the cloud top are associated with the RMMC enhanced by the thermal tide.

Results

Background zonal wind and temperature distributions. The model atmosphere reached quasi-steady states within approximately four Earth years, and the quasi-steady states were maintained for more than ten Earth years[21]. In the following, we focus on the atmospheric structure in the northern hemisphere, because almost the same structure was observed in the southern hemisphere.

Figure 1 shows meridional cross-sections of the zonally and temporally averaged zonal wind and temperature obtained in Case A. The fast mean zonal wind is maintained at 60–80 km. Above ~80 km, the mean zonal wind decreases very sharply in the vertical direction; the velocity is $<10\,\mathrm{m\,s^{-1}}$ at 90 km. In the temperature field, the meridional gradient changes its sign near the cloud-top level (~70 km). That is, the zonal wind and temperature in our results are consistent with the gradient wind balance. As shown later, the warm polar region is related to the positive gradient of the temperature above that level. Significant negative shear of the mean zonal wind is attributed to the thermal tide, since it is theoretically predicted that the thermal tide accelerates the mean zonal wind at 50–70 km and decelerates at 80–100 km (ref. 20). Another notable feature is that the zonal wind velocity is almost unchanged vertically at 40–80 km levels in the polar region. These features of the zonal wind and the temperature are in close agreement with those obtained in recent observations[4,9]. Figure 1 also shows that a midlatitude jet is formed around the cloud-top level (~70 km) at 20°–50° latitudes. The jet axis is located more equatorwards than that obtained in previous works[8,10,11]. This difference may also be explained by the acceleration effect of the thermal tide as noted above. The meridional distribution of the zonal wind is in close agreement with that reported by Machado *et al.*[22], who retrieved the wind speed in the dayside by the ground-based Doppler velocimetry measurements. In the work of Piccialli *et al.*[9], in which the meridional-height distributions of the zonal wind are estimated from those of the temperature by assuming the gradient wind balance, the positions of the midlatitude jets are about 15° polewards. It is noted, however, the estimated wind distribution strongly depends on the assumed meridional distribution of the zonal wind at the bottom boundary altitude. Sánchez-Lavega *et al.*[23] also observed the wind distributions during the local time from 6 to 17 h, and indicated that the zonal wind speed increases from the morning to the afternoon due to the thermal tide at the upper cloud level in a latitude range of 50° S–75° S. This result is quite consistent with our present results. The zonal wind speed at middle and high latitudes is enhanced by the thermal tide by 20–30 $\mathrm{m\,s^{-1}}$ (not shown). It is also noted that, as shown in Sugimoto *et al.*[24], the strength (or velocity) and the position of the midlatitude jet vary with time in the present model, although the variation width of the latter in the latitudinal direction is not more than ~10°. These fluctuations may contribute to the difference between the present results and the observations. As shown later, the position of this jet is important to the cold collar.

Temporal variation of temperature. Figure 2 shows temporal variations of the zonal-mean temperature in a latitude range of 30°–90° at ~75 km (the pressure level of 1×10^3 Pa), ~68 km

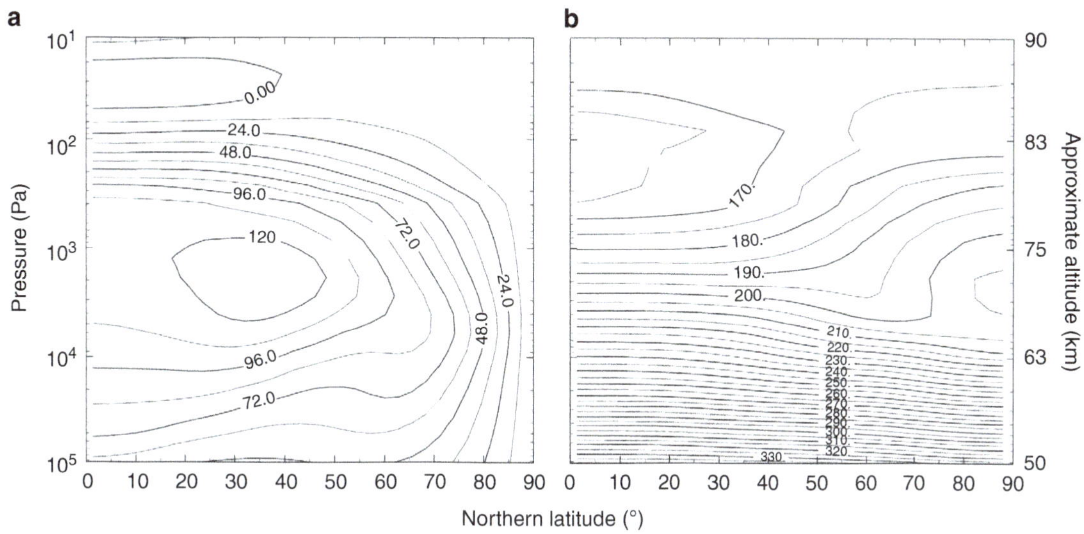

Figure 1 | Meridional cross-sections of the zonally and temporally averaged data in Case A. (**a**) Zonal wind ($m\,s^{-1}$) and (**b**) Temperature (K). Averaged period is two Venusian solar days (234 Earth days) after settling into the quasi-steady state.

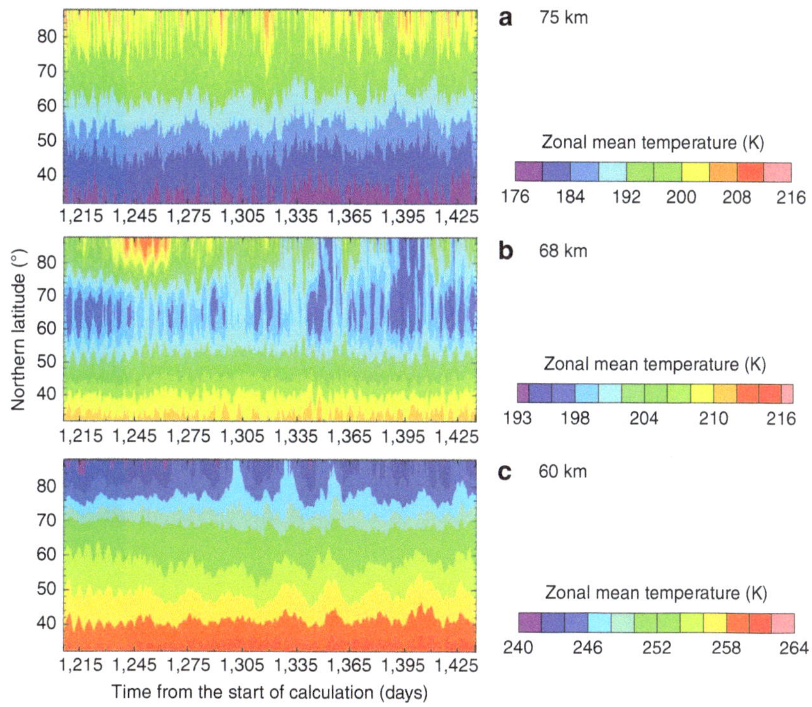

Figure 2 | Time evolutions of the zonally averaged temperature in the quasi-steady state obtained in Case A. Each panel is depicted for two Venusian solar days (234 Earth days) at the altitude of (**a**) ~75 km (the pressure level of 1×10^3 Pa), (**b**) ~68 km (4×10^3 Pa) and (**c**) ~60 km (2×10^4 Pa), respectively. The latitude range is from 30° N to 90° N.

(4×10^3 Pa) and ~60 km (2×10^4 Pa) obtained in Case A. At ~75 km, the polar region is always warmer than low latitudes with short- and long-period fluctuations (4–5 and ~30 Earth days, respectively). The meridional temperature difference is ~30 K, which is a little larger than that estimated from the radio occultation measurements[4,9]. At ~68 km, the cold collar (local minimum of the temperature) clearly appears at latitudes of 60°–70°. The temperature in the cold collar is colder than the surroundings by ~10 K, and qualitatively consistent with the radio occultation measurements[2,4,9]. The short-period fluctuation is dominant at this level, which may be caused by the baroclinic instability waves excited in the midlatitudes at the cloud levels[24]. At ~60 km, the temperature monotonically decreases with latitude. Similarly to ~75 km, fluctuations with both short and the long periods appear at this level.

Figure 3 shows time evolution of the horizontal temperature distribution at ~68 km obtained in Case A. The cold collar surrounds the warmer polar region at ~60° N. The maximum temperature difference between 60° N and the pole is ~20 K. The warm polar region rotates around the pole and varies in shape temporally. An S-shaped vortex structure also appears in Fig. 3g. It rotates around the pole, but the centre of rotation is offset from the pole. These features are in close agreement with the cold collar and the polar vortex observed by infrared measurements[1,3,5,6], although the altitude of the cold collar reproduced in the present model is slightly higher than that estimated in such observations.

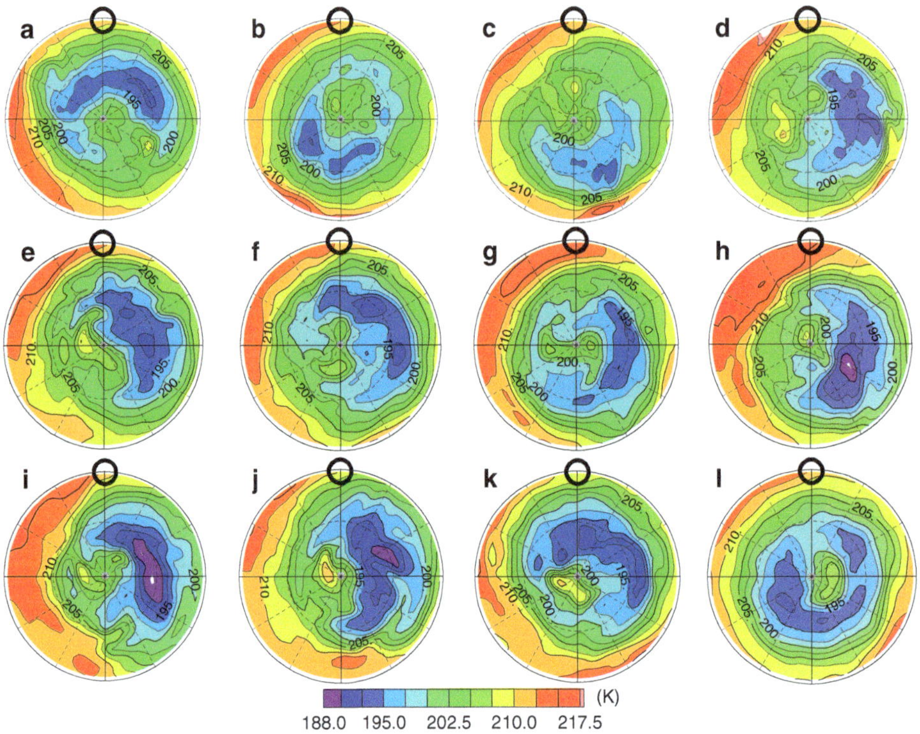

Figure 3 | Time evolution of the temperatures (K) at ~ 68 km (the pressure level of 4×10^3 Pa) in Case A in the polar plot. The latitude range is from 30°N to 90°N. The black circle represents the local solar noon. Each panel of (**a**) to (**l**) shows temperature distribution from 1414 to 1425 days from the start of the calculation. The time interval between the adjacent panels is one Earth day.

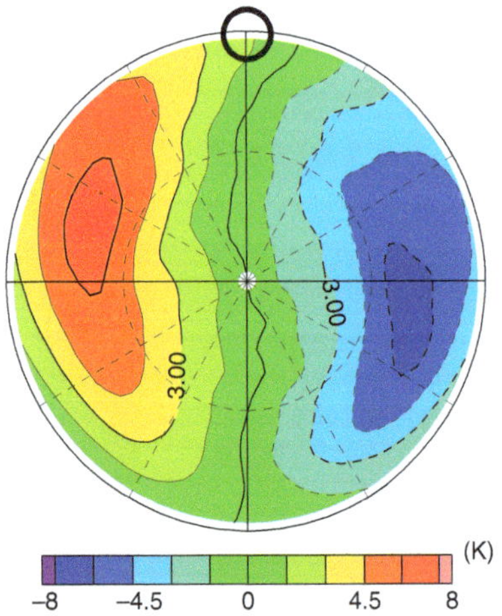

Figure 4 | Polar plot of the temperature deviation (K) associated with the thermal tide at ~ 68 km (the pressure level of 4×10^3 Pa) obtained in Case A. The latitude range is from 30° N to 90° N. The black circle represents the local solar noon. Average period of data is 12 Earth days corresponding to the time interval for **a** to **l** in Fig. 3.

Morphology in the Venus polar region. As shown in Fig. 3, zonal components with wavenumbers of zero, and one or two are predominant in the temperature distribution in the cold collar and the polar vortex. To elucidate the wave contributions, we extract the temperature components associated with the thermal tide and the short-period (transient) waves by using a frequency filter[25], which are shown in Figs 4 and 5, respectively. Figures 3 and 4 indicate that the enhanced cold part in the cold collar is caused by the cold temperature phase of the thermal tide, which is consistent with Yamamoto and Takahashi[11]. The cold part of the cold collar is nearly always expanded into the morning side, and it sometimes surrounds the polar vortex completely (for example, Fig. 3b,g,k,l). Comparing Fig. 5 with Fig. 3, we found that this expansion is induced by the cold temperature phase of the transient waves. These results show that the cold collar is composed of the superposition of the zonally averaged basic field, the thermal tide and the transient waves.

It is also suggested from Figs 3 and 5 that the small-scale fluctuations seen in the polar vortex are mainly caused by the warm temperature phase of the transient waves. The morphology seen in the polar vortex is strongly affected by the transient waves, implying that the hot dipole or S-shaped structure observed by VIRTIS, the Visible and InfraRed Thermal Imaging Spectrometer on board Venus Express, would reflect the horizontal structure of the transient waves.

Effect of thermal tide. Figure 6a,b show latitude–height distributions of the zonally and temporally averaged zonal wind and temperature over two Venusian solar days (~234 Earth days) obtained in Cases A and B. The axis of the midlatitude jet in Case A is located at a lower latitude and altitude than in Case B. This is likely due to the momentum transport by the thermal tide as mentioned above. In Case A, the temperature below 70 km altitude decreases with latitude up to 60° N in association with positive vertical shear of the mean zonal wind, whereas it increases with latitude poleward of 70° N. The remarkable cold collar is observed at 67–70 km levels at 60°–70° latitudes along with the polar warm region. In Case B, on the other hand, the temperature monotonically decreases with latitude at levels below 75 km height. Above 75 km, the temperature in the polar region is

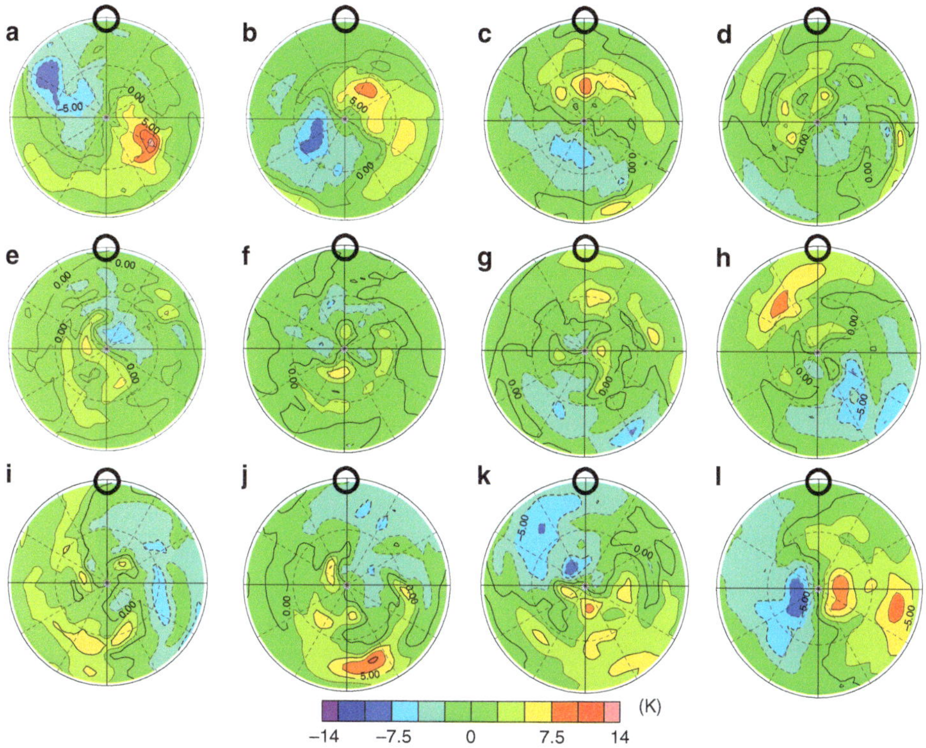

Figure 5 | Polar plot of the temperature deviations (K) associated with the transient (short period) waves. The latitude range is from 30° N to 90° N. The black circle represents the local solar noon. Each panel of (**a**) to (**l**) shows temperature distribution from 1414 to 1425 days from the start of the calculation, corresponding to the time in Fig. 3. The time interval between the adjacent panels is one Earth day.

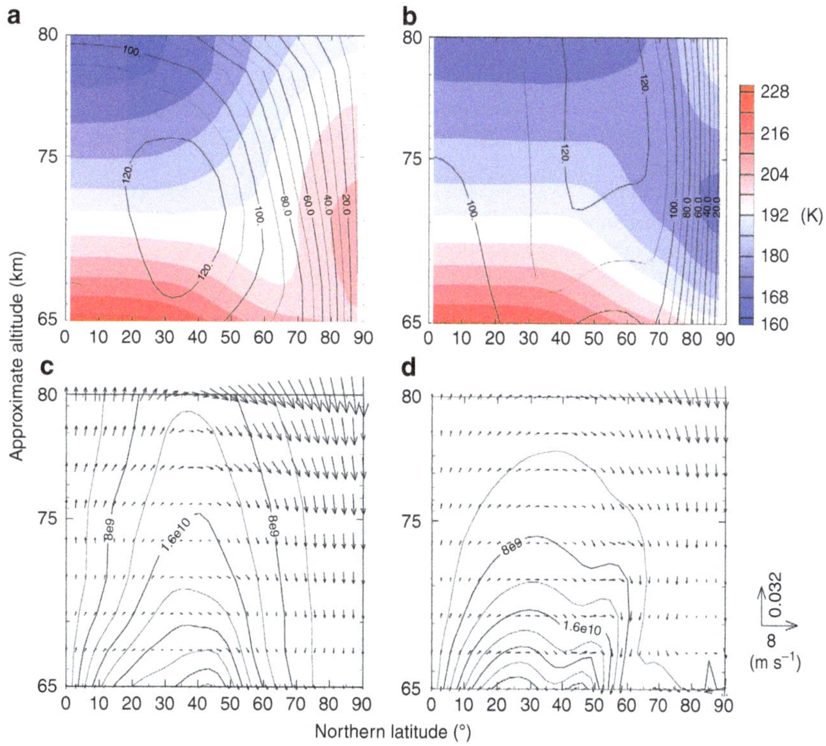

Figure 6 | Meridional cross-sections of the zonally and temporally averaged zonal wind (solid line) and temperature (colour shade) and the horizontal and vertical components of the residual mean meridional circulation (vector) and mass stream fuction (contour). (**a**) Zonal wind and temperature in Case A. (**b**) Those in Case B. (**c**) Residual mean meridonal circulation vector and mass stream function in Case A. (**d**) Those in Case B. Averaged period is two Venusian solar days (234 Earth days) after settling into the quasi-steady state.

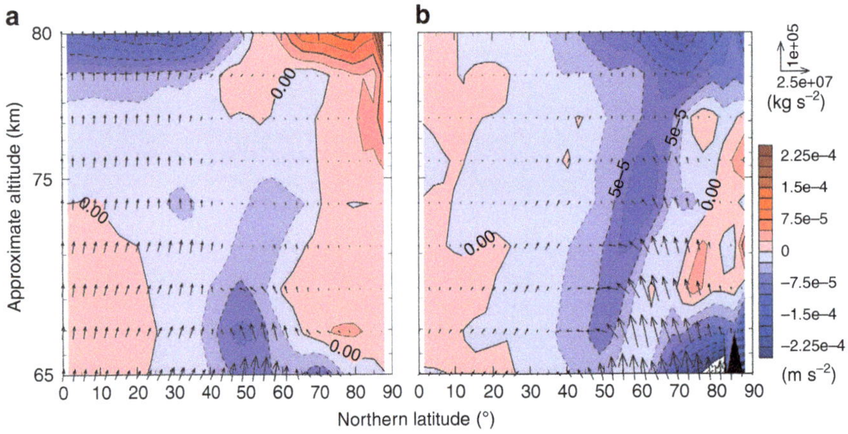

Figure 7 | Meridional cross-sections of the EP flux (vectors) and its divergence (colours). (**a**,**b**) Cases A and B, respectively. Average period of data is two Venusian solar days (234 Earth days) after settling into the quasi-steady state.

slightly warmer than in midlatitudes, but the temperature minimum corresponding to the cold collar is not reproduced, which is different from Case A.

RMMC defined by the transformed Eulerian-mean equations[26] (Supplementary Note 1) is shown in Fig. 6c,d for Cases A and B, respectively. The RMMC represents the Lagrangian mean meridional circulation approximately. In Case A, the RMMC above the cloud-top level ($\sim$70 km) reaches the polar region and remarkable downward motion exists there, which warms the atmosphere through adiabatic heating and forms the warm polar region. Also in Case B, the downward motion of the RMMC is observed in the polar region. However, it is 2–3 times slower than that in Case A (Supplementary Fig. 1). The adiabatic heating due to the downward motion in the polar region in Case B is much weaker than that in Case A.

These results imply that the thermal structure of Venus upper polar atmosphere (that is, the warm polar region and the cold collar) is attributed to the RMMC enhanced by the thermal tide. To investigate the effect of the thermal tide on the RMMC, we examined the momentum and heat balances in the transformed Eulerian-mean equations. The Eliassen–Palm (EP) flux and its divergence obtained for Cases A and B are shown in Fig. 7. In Case A, strong convergence of the EP flux (shown by blue shades) exists at latitudes equatorward of 55° N–60° N above 75 km, which is balanced mainly with upward advection of the negative vertical shear of the mean zonal wind due to the RMMC. On the other hand, the EP flux is divergent (shown by red shades) at latitudes poleward of 55° N–60° N. This divergence is balanced with the sum of poleward advection of the absolute vorticity and downward advection of the negative vertical shear of the mean zonal wind (Supplementary Fig. 2). Since these vertical motions of the RMMC induce adiabatic cooling and heating at the lower and higher latitudes, respectively, the clear structure of the cold collar has been reproduced in Case A.

In Case B, the EP flux is almost absent at lower latitudes as shown in Fig. 7b. Since the thermal tide is excluded in this case, it is inferred that the EP flux at the low latitudes obtained in Case A is induced by the thermal tide. Convergence of the EP flux is predominant at latitudes poleward of 20° N except at 68–77 km altitudes in the polar region, which is mainly balanced with poleward advection of the absolute vorticity due to the RMMC (Supplementary Fig. 2). The vertical advection can be neglected in Case B because both the vertical motions of the RMMC and the vertical shear of the mean zonal wind are much weaker than those obtained in Case A.

Furthermore, to clarify the contribution of the thermal tide in Case A, we decompose the EP flux into long- and short-period components by using a frequency filter with a period of 10 Earth days, which correspond to the thermal tide and the transient waves, respectively. Figure 8 shows the EP flux divergence due to the thermal tide and the transient waves. Compared with Fig. 7a, a large part of the total EP flux divergence can be ascribed to the thermal tide, and the transient waves may not be neglected in the polar region below 70 km. These results indicate that the different dynamical states have been established in Cases A and B due to the thermal tide. In Case A, the RMMC is enhanced by the thermal tide, which induces the adiabatic polar warming, creating the cold collar (see Supplementary Note 2 and Supplementary Figs 3 and 4 for more details).

Discussion

Our idea that the unique Venusian polar atmospheric structure is attributed to the RMMC enhanced by the thermal tide is different from Yamamoto and Takahashi[11], who suggested that warm polar region is induced by the warm temperature phase of the diurnal tide. Imamura[27] has pointed out that the RMMC above the cloud layer can be induced by upward propagating Rossby and/or gravity waves, but the thermal tide has not been taken into account in his study. Lebonnois *et al.*[28] also performed GCM simulations, which includes more realistic physical parameterization schemes of the radiative transfer and boundary layer processes. They reported that the cold collar was not reproduced in their results, and inferred that the interaction between temperature and cloud structures, which was not included in their model, might be important for the cold collar formation. However, since the realistic cold collar has been reproduced in our model with the simplified radiative process, it is suggested that the cloud-radiative feedback is not essential for the cold collar. It should be also noted that an equatorial jet appeared in their results. This is different from our results, in which the jet is located at the midlatitudes. Because the temperature distribution is closely related with the mean zonal wind through the gradient wind balance, the realistic mean zonal wind and the RMMC enhanced by the thermal tide are crucial to the unique atmospheric structure seen in the Venus polar region.

In addition, we might be able to explain the recent infrared observational results about the meridional distributions of CO abundance and cloud thickness by taking into account the fact that the meridional and vertical motions of the RMMC shown in

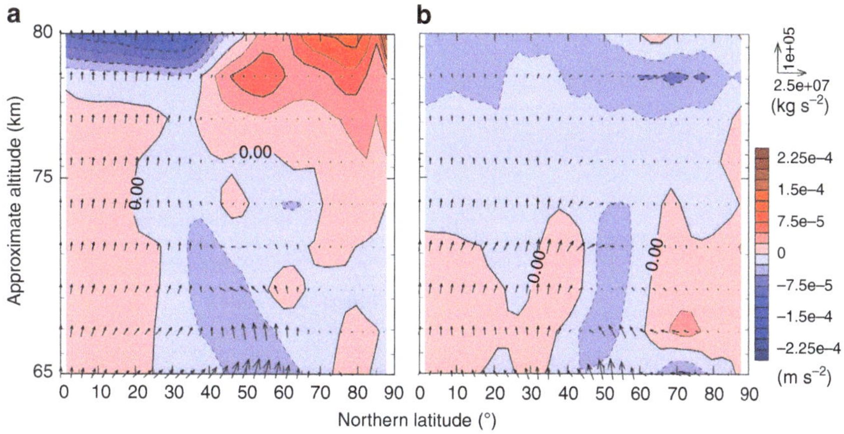

Figure 8 | Meridional cross-sections of the EP flux (vectors) and its divergence (colour) in Case A. (**a**) Thermal tide and (**b**) Transient (short period) waves. Averaged period of data is two Venusian solar days (234 Earth days) after settling into the quasi-steady state.

Fig. 6c approximately represent the material transport. It is inferred that the poleward material transport above the cloud top due to the RMMC leads to high CO abundance in the polar region, because on the way from the equator to the pole, an air parcel keeps gaining CO, which is created photochemically above the cloud layer, until it sinks in the polar region. Furthermore, H_2SO_4 liquid particles as well as CO are transported towards the polar region and grow in the same manner. The temperature at which H_2SO_4 liquid particles evaporate is almost equal to that at the cloud bottom altitude of $\sim$50 km (refs 15,29). This temperature is much higher than that observed at altitudes above 65 km, where the RMMC transports the H_2SO_4 particles towards the poles in this study. Therefore, the clouds in the polar region are expected to become thicker than those in the cold collar. This material transport expected from the present result is qualitatively consistent with Imamura and Hashimoto[15].

Our simulations elucidate the importance of the thermal tide for the Venus atmospheric circulation around the cloud-top level ($\sim$70 km). The cold collar and the warm polar region, which are related to the zonal wind distribution through gradient wind balance, are realistically reproduced in the presence of the diurnal heating and can be explained by the downward motion of the RMMC enhanced by the thermal tide. This is qualitatively analogous to a sudden stratospheric warming in the Earth atmosphere, which is related to the meridional circulation induced by upward propagating Rossby waves[30]. In addition, the observed meridional distributions of cloud thickness and CO abundance above the cloud top are expected to be associated with the RMMC enhanced by the thermal tide. This is a new insight provided by the present work and is helpful for interpreting numerous observations and features of the Venus upper atmosphere.

Although the importance of the thermal tide is revealed in this study, further details of the mechanism how the structure of the RMMC is determined should be elucidated by investigating the wave propagation in our model and analysing observational data in the near future. In this December, the Venus Climate Orbiter 'Akatsuki' will start observations[31,32]. Data provided by Akatsuki will also greatly help us to understand the real Venus atmosphere.

Methods

Set-up of AFES for Venus. AFES for Venus is a full nonlinear GCM on the sphere constructed for the Venus atmosphere[21,24]. The basic equations are primitive equations in sigma coordinates without topography. The physical parameter values applied in this study are as follows: planet radius is 6,050 km; gravity acceleration 8.87 m s^{-2}; planetary rotation period 243 Earth days; and standard surface pressure 92×10^5 Pa. The specific heat at constant pressure C_p is assumed to be constant at 1,000 J kg^{-1} K^{-1} for simplicity. It is noted that C_p varies approximately from 1.18×10^3 to 7.38×10^2 J kg^{-1} K^{-1} in the Venus atmosphere[33]. The resolution is set to T42L60 (128 × 64 horizontal grids and 60 vertical levels). The vertical domain extends from the ground to $\sim$120 km, with an almost constant grid spacing of 2 km. The model includes horizontal and vertical eddy diffusion. The former is represented by horizontal hyper-viscosity ∇^4 and has a relaxation time for the largest wavenumber of 0.1 Earth days. The vertical diffusion coefficient is assumed to be constant at 0.15 m^2 s^{-1}. Rayleigh friction is employed at the lowest level to mimic surface friction. A sponge layer is used to damp eddy components of the wind above 80 km. However, its effect becomes dominant above 100 km because the damping coefficient is very small at 80–100 km levels.

Radiative processes. Radiative transfer in the infrared region is represented simply by Newtonian cooling with the coefficients based on previous studies[20,34]. Furthermore, the temperature field is relaxed to the prescribed horizontally uniform distribution, whose vertical profile is based on the Venus International Reference Atmosphere. The period of the data analysis in this study is two Venusian solar days (234 Earth days) after settling into the quasi-steady state. The solar heating is based on previous studies[34,35]. In the present study, it is decomposed into a zonal-mean component and deviation from the zonal mean (diurnal component), which excites the mean meridional (Hadley) circulation and the thermal tide, respectively. Case A includes both the components, whereas Case B includes only the zonal mean. See Sugimoto *et al.*[21,24] for further details of the model settings.

References

1. Taylor, F. W. *et al.* Structure and meteorology of the middle atmosphere of Venus: infrared remote sensing from the Pioneer Orbiter. *J. Geophys. Res.* **85,** 7963–8006 (1980).
2. Newman, M., Schubert, G., Kliore, A. J. & Patel, I. R. Zonal winds in the middle atmosphere of Venus from Pioneer Venus radio occultation data. *J. Atmos. Sci.* **41,** 1901–1913 (1984).
3. Piccioni, G. *et al.* South-polar features on Venus similar to those near the north pole. *Nature* **450,** 637–640 (2007).
4. Tellmann, S., Pätzold, M., Häusler, B., Bird, M. K. & Tyler, G. L. Structure of the Venus neutral atmosphere as observed by the radio science experiment VeRa on Venus Express. *J. Geophys. Res.* **114,** E00B36 (2009).
5. Garate-Lopez, I. *et al.* A chaotic long-lived vortex at the southern pole of Venus. *Nat. Geosci.* **6,** 254–257 (2013).
6. Luz, D. *et al.* Venus's southern polar vortex reveals precessing circulation. *Science* **332,** 577–580 (2011).
7. Garate-Lopez, I., Garcia Munoz, A., Hueso, R. & Sanchez-Lavega, A. Instantaneous three-dimensional thermal structure of the Southern Polar Vortex of Venus. *Icarus* **245,** 16–31 (2015).
8. Lee, C., Lewis, S. R. & Read, P. L. A numerical model of the atmosphere of Venus. *Adv. Space Res.* **36,** 2142–2145 (2005).
9. Piccialli, A. *et al.* Dynamical properties of the Venus mesosphere from the radio-occultation experiment VeRa onboard Venus Express. *Icarus* **217,** 669–681 (2012).
10. Yamamoto, M. & Takahashi, M. Venusian middle-atmospheric dynamics in the presence of a strong planetary-scale 5.5-day wave. *Icarus* **217,** 702–713 (2012).

11. Yamamoto, M. & Takahashi, M. Dynamics of polar vortices at cloud top and base on Venus inferred from a general circulation model: case of a strong diurnal thermal tide. *Planet Space Sci.* **113,** 109–119 (2015).
12. Irwin, P. G. *et al.* Spatial variability of carbon monoxide in Venus' mesosphere from Venus Express/visible and infrared thermal imaging spectrometer measurements. *J. Geophys. Res.* **113,** E00B01 (2008).
13. Grassi, D. *et al.* The Venus nighttime atmosphere as observed by the VIRTIS-M instrument. Average fields from the complete infrared data set. *J. Geophys. Res.* **119,** 837–849 (2014).
14. Lee, Y. J. *et al.* Vertical structure of the Venus cloud top from the VeRa and VIRTIS observations onboard Venus Express. *Icarus* **217,** 599–609 (2012).
15. Imamura, T. & Hashimoto, G. Venus cloud information in the meridional circulation. *J. Geophys. Res.* **103,** 31349–31366 (1998).
16. Pechmann, A. P. & Ingersoll, J. B. Venus lower atmosphere heat balance. *J. Geophys. Res.* **85,** 8219–8222 (1980).
17. Takagi, M. & Matsuda, Y. Sensitivity of thermal tides in the Venus atmosphere to basic zonal flow and Newtonian cooling. *Geophys. Res. Lett.* **32,** L02203 (2005).
18. Takagi, M. & Matsuda, Y. Dynamical effect of thermal tides in the lower Venus atmosphere. *Geophys. Res. Lett.* **33,** L13102 (2006).
19. Newman, M. & Leovy, C. Maintenance of strong rotational winds in Venus' middle atmosphere by thermal tides. *Science* **257,** 647–650 (1992).
20. Takagi, M. & Matsuda, Y. Effects of thermal tides on the Venus atmospheric superrotation. *J. Geophys. Res.* **112,** D09112 (2007).
21. Sugimoto, N., Takagi, M. & Matsuda, Y. Baroclinic instability in the Venus atmosphere simulated by GCM. *J. Geophys. Res.* **119,** 1950–1968 (2014).
22. Machado, P., Luz, D., Widemann, T., Lellouch, E. & Witasse, O. Mapping zonal winds at Venus's cloud tops from ground-based Doppler velocimetry. *Icarus* **221,** 248–261 (2012).
23. Sánchez-Lavega, A. *et al.* Variable winds on Venus mapped in three dimensions. *Geophys. Res. Lett.* **35,** L13204 (2008).
24. Sugimoto, N., Takagi, M. & Matsuda, Y. Waves in a Venus general circulation model. *Geophys. Res. Lett.* **41,** 7461–7467 (2014).
25. Duchon, C. E. Lanczos filtering in one and two dimensions. *J. Atmos. Sci.* **18,** 1016–1022 (1979).
26. Andrews, D. G., Holton, J. R. & Leovy, C. B. *Middle Atmosphere Dynamics* (Academic, 1987).
27. Imamura, T. Momentum balance of the Venusian midlatitude mesosphere. *J. Geophys. Res.* **102,** 6615–6620 (1997).
28. Lebonoiss, S. *et al.* Superrotation of Venus' atmosphere analyzed with a full general circulation model. *J. Geophys. Res.* **115,** E06006 (2010).
29. Ayers, G. P., Gillett, R. W. & Gras, J. L. On the vapor pressure of sulfuric acid. *Geophys. Res. Lett.* **7,** 433–436 (1980).
30. Matsuno, T. A dynamical model of the stratospheric sudden warming. *J. Atmos. Sci.* **28,** 1479–1494 (1971).
31. Nakamura, M. *et al.* Overview of Venus orbiter, Akatsuki. *Earth Planets Space* **63,** 443–457 (2011).
32. Nakamura, M. *et al.* Return to Venus of the Japanese Climate Orbiter AKATSUKI. *Acta Astronaut.* **93,** 384–389 (2014).
33. Seiff, A. *et al.* Models of the structure of the atmosphere of Venus from the surface to 100 kilometers altitude. *Adv. Space Res.* **5,** 3–58 (1985).
34. Crisp, D. Radiative forcing of the Venus mesosphere: I. solar fluxes and heating rates. *Icarus* **67,** 484–514 (1986).
35. Tomasko, M. G., Doose, L. R., Smith, P. H. & Odell, A. P. Measurements of the flux of sunlight in the atmosphere of Venus. *J. Geophys. Res.* **85,** 8167–8186 (1980).

Acknowledgements

We are grateful to Dr Y.J. Lee, Dr S. Murakami and Dr K. Takaya for the useful suggestions. This study was conducted under the joint research project of the Earth Simulator Center entitled 'Simulations of Atmospheric General Circulations of Earth-like Planets by AFES' and partly supported by JSPS KAKENHI Grant Number 25400470.

Author contributions

H.A. and H.K. analysed the simulation data; N.S. performed the numerical simulations; M.T. developed the numerical model; all the authors contributed to the analysis and theoretical interpretations.

Additional information

Competing financial interests: The authors declare no competing financial interests.

20

Transition to a Moist Greenhouse with CO_2 and solar forcing

Max Popp[1,2,†], Hauke Schmidt[1] & Jochem Marotzke[1]

Water-rich planets such as Earth are expected to become eventually uninhabitable, because liquid water turns unstable at the surface as temperatures increase with solar luminosity. Whether a large increase of atmospheric concentrations of greenhouse gases such as CO_2 could also destroy the habitability of water-rich planets has remained unclear. Here we show with three-dimensional aqua-planet simulations that CO_2-induced forcing as readily destabilizes the climate as does solar forcing. The climate instability is caused by a positive cloud feedback and leads to a new steady state with global-mean sea-surface temperatures above 330 K. The upper atmosphere is considerably moister in this warm state than in the reference climate, implying that the planet would be subject to substantial loss of water to space. For some elevated CO_2 or solar forcings, we find both cold and warm equilibrium states, implying that the climate transition cannot be reversed by removing the additional forcing.

[1] Max Planck Institute for Meteorology, Bundesstrasse 53, Hamburg 20146, Germany. [2] Program in Atmospheric and Oceanic Sciences, Princeton University, 300 Forrestal Road, Sayre Hall, Princeton, New Jersey 08544, USA. † Present address: NOAA's Geophysical Fluid Dynamics Laboratory, Princeton, New Jersey, USA. Correspondence and requests for materials should be addressed to M.P. (email: mpopp@princeton.edu).

Water-rich planets such as Earth lose water by photo-dissociation of water vapour in the upper atmosphere and the subsequent escape of hydrogen. On present-day Earth, the loss occurs very slowly, because the mixing ratio of water vapour in the upper atmosphere is very low. But significant loss of water could occur over geological timescales if the surface temperature were around 70 K warmer than it is today[1–3]. For these high surface temperatures, the tropopause is expected to climb to high altitudes. As a consequence, the cold trapping of water vapour at the tropopause becomes ineffective, because the mixing ratio of water vapour increases with the rising tropopause. Steady states in which the mixing ratio in the upper atmosphere is sufficiently high for a water-rich planet to lose most of its water inventory in its lifetime are known as Moist-Greenhouse states[4]. A planet in this state would eventually become uninhabitable as all water is lost to space. For an Earth-like planet around a Sun-like star, a Moist Greenhouse would be attained if the mixing ratio in the upper atmosphere exceeds ~0.1 % (ref. 1). For comparison, the mixing ratio in Earth's stratosphere is presently around two orders of magnitude smaller.

Moist-Greenhouse states were found and described in several studies with one-dimensional models[1–3,5,6] and have recently been found for terrestrial planets with three-dimensional models in different setups[7–9]. However, not all three-dimensional studies found stable Moist-Greenhouse states[10–12]. Instead the climate of these models would destabilize into a Runaway Greenhouse, a self-reinforcing water-vapour feedback-loop, before the Moist Greenhouse is attained. A few studies applied large forcing but the employed models became numerically unstable before the Moist-Greenhouse regime was attained[13–15]. Therefore, it remains unclear whether planets would attain a Moist-Greenhouse state before a Runaway Greenhouse occurs, especially for planets on an Earth-like orbit, where the only two previous studies with state-of-the-art general circulation models (GCM) gave contradicting results[12,9]. Moreover, all three-dimensional studies investigating Moist-Greenhouse states only applied solar forcing without considering greenhouse-gas forcing. Several studies have applied strong greenhouse-gas forcing, but either did not run their simulations to sufficiently high temperatures[16–19] or did not investigate the emergence of a Moist Greenhouse[20]. Greenhouse-gas forcing has long been assumed to be ineffective at causing Moist-Greenhouse states, because the greenhouse effect of any additional greenhouse gas would eventually be rendered ineffective by the increasing greenhouse effect of water vapour with increasing temperatures. Furthermore, large greenhouse-gas forcing would lead to a cooling of the upper atmosphere, which would push the Moist-Greenhouse limit to much higher surface temperatures[5]. However, if clouds are considered, these arguments may not apply, because clouds themselves can contribute to the climate becoming unstable[21].

Here we compare for the first time with a state-of-the-art GCM, namely ECHAM6 (ref. 22), how effective solar and CO_2 forcing are at causing a transition to a Moist Greenhouse. We couple the atmosphere to a slab ocean and choose an aqua-planet setup (fully water-covered planet) in perpetual equinox. This idealized framework is better suited than a present-day Earth setting to understand the involved dynamics while preserving the major feedback mechanisms of the Earth[23]. It also avoids conceptual problems with the representation of land-surface processes at high temperatures. We turn off sea ice in order to investigate the possibility of solely cloud-induced multiple steady states that were recently found in a one-dimensional study[21]. We modify the model such that it can deal with surface temperatures of up to 350 K (see Methods).

Thus we show that cloud-radiative effects (CRE) destabilize a present-day Earth climate as readily with CO_2 as with solar forcing. The changes in CRE are a consequence of the weakening of the large-scale circulation with increasing global-mean surface temperature (gST). However, the resulting climate transition does not lead to a Runaway Greenhouse, but instead a new regime of warm steady state with gST above 330 K is attained. This warm regime differs substantially in its dynamics from a present-day Earth-like climate and, most importantly, the upper atmosphere exceeds the Moist-Greenhouse limit in this regime. Hence a planet in such a state would lose water at a fast rate to space. Furthermore, there is hysteresis in the warm regime and removing the imposed forcing does therefore not necessarily cause a transition back to an Earth-like climate.

Results

Simulations with increased TSI. To assess the dependence of the climate state on total solar irradiance (TSI) for fixed CO_2 levels (at 354 p.p.m. volume mixing ratio), we apply a total of five different TSI-values that range from the present-day value on Earth (S_0) to 1.15 times that value. We find two regimes of steady states that are separated by a range of gST for which stable steady states are not found (Fig. 1). The regime of steady states with gST of up to ~298 K exhibits similar features as present-day Earth climate, such as a large pole-to-equator surface-temperature contrast (Fig. 2a) and a similar meridional distribution of cloud cover (Fig. 2b). Hence this regime of steady state can be considered to be Earth-like. In contrast, the warm regime of steady states with gST above 334 K is characterized by a considerably smaller pole-to-equator surface-temperature difference and a substantially different meridional distribution of cloud cover (Fig. 2a,b). This illustrates that the dynamics in the warm regime are quite different from the present-day Earth regime. Most importantly, the mixing ratio of water vapour at the uppermost level at 0.01 hPa is considerably higher in the warm regime and exceeds the Moist-Greenhouse limit (Fig. 2c). Therefore, a planet in such a state would be losing water to space at a fast rate. The minimum TSI required to cause a climate transition from the Earth-like to the warm regime lies between 1.03 S_0 and 1.05 S_0, whereas the maximum TSI to cause a climate transition from the warm back to the Earth-like regime lies

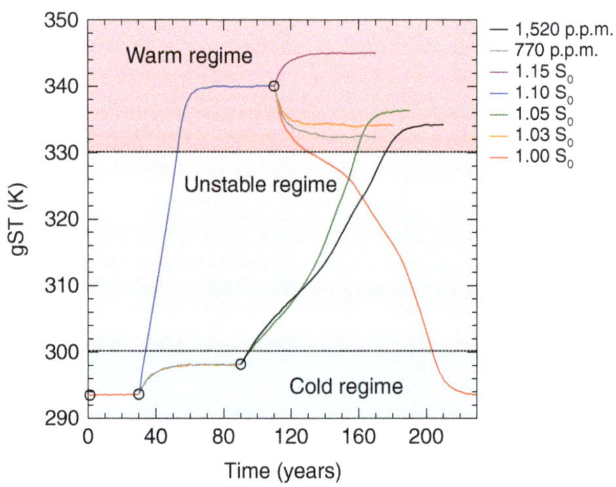

Figure 1 | Temporal evolution of gST. The circles denote the four states from which new simulations are started. For both a TSI of 1.00 S_0 and 1.03 S_0 as well as for a CO_2 concentration of 770 p.p.m., two simulations with different initial conditions are performed. The lines are interpolated from annual means.

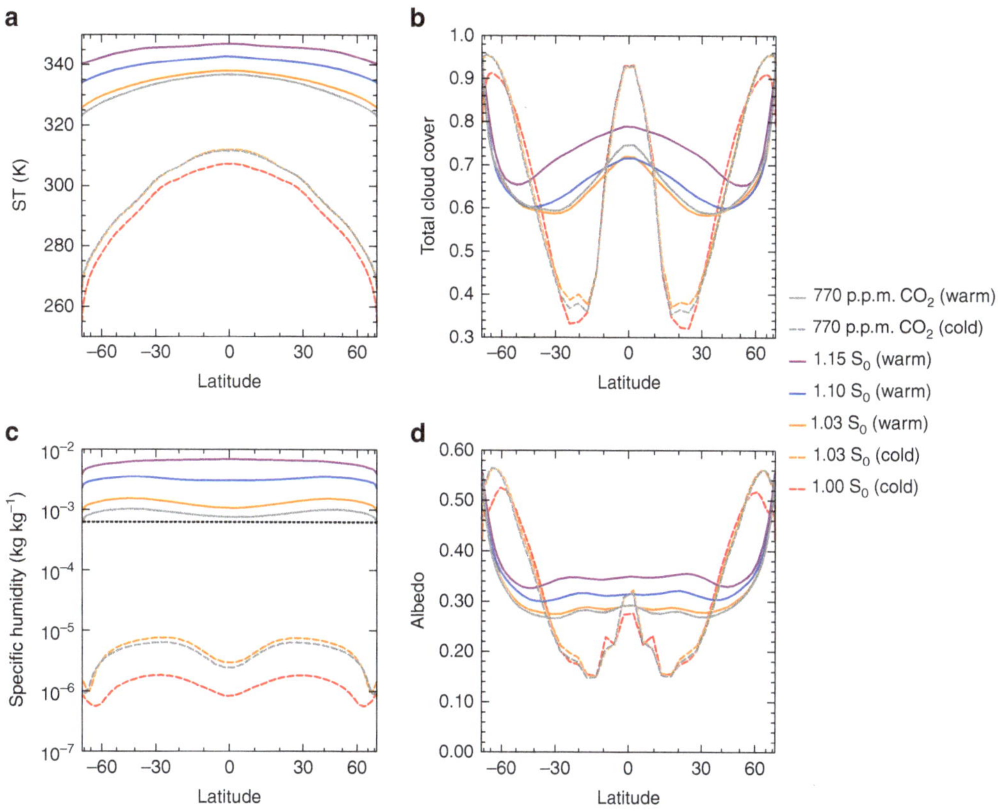

Figure 2 | Zonal means in steady state. (**a**) Surface temperature (ST); (**b**) the total cloud cover; (**c**) the specific humidity at the top level; and (**d**) the effective albedo. The effective albedo is defined as the ratio of the zonal and temporal means of the reflected solar radiation divided by the zonal and temporal means of the incoming solar radiation. The black horizontal line in **c** indicates the Moist-Greenhouse limit[1]. The temporal mean is taken over a period of 30 years. The horizontal axes are scaled with the sine of the latitude.

between 1.00 S_0 and 1.03 S_0 (Fig. 1). Consequently, there are two different stable steady states for a TSI of 1.03 S_0. Since sea ice is turned off in our model, this double steady state is entirely a consequence of atmospheric processes. In the warm regime, the cloud albedo increases at all latitudes with TSI, thus providing an efficient way to stabilize the climate against increased radiative forcing (Fig. 2d).

Energetics of the climate transition. To understand the processes governing the climate transition from the cold to the warm regime, we focus now on the transient simulation with a TSI of 1.05 S_0. The climate instability is evidenced by an increase in the total (shortwave plus longwave) net (downward minus upward component of the) global-mean top-of-the-atmosphere (TOA) radiative flux with increasing gST for gST between 300 and 330 K (Fig. 3). Note that a positive TOA radiative flux is net downward and a negative radiative flux is net upward. Therefore, the aqua-planet takes up energy for a positive TOA radiative flux and loses energy for a negative flux. The instability is caused by the cloud-radiative contribution to the total radiative flux that increases with increasing gST for gSTs below 330 K. The clear-sky contribution to the total radiative flux is decreasing with increasing gST and does thus not contribute to the climate instability. This decrease in clear-sky contribution is caused by an increase in clear-sky contribution to the outgoing longwave radiation, which is upward and thus decreases the net downward flux. At gST above 330 K, the cloud-radiative contribution decreases again with increasing gST. This allows together with the clear-sky contribution to attain a new steady state. Thus, clouds destabilize the climate at lower gST and then stabilize again at higher gST. This change in sign of the cloud feedback is also responsible for the existence of the bistability.

Dynamics of the climate transition. The changes in cloud-radiative contribution to the total net TOA radiative flux (henceforth simply referred to as CRE) are caused by the weakening of the large-scale circulation (Fig. 4a) and the increase of water vapour in the atmosphere with increasing gST. The weakening of the circulation causes tropical convection to spread more evenly around the tropics, with less convection occurring around the equator and more convection occurring in the subsidence region. As a consequence, deep convective clouds with low cloud-top temperatures become more frequent in the subsidence region of the Hadley circulation (Fig. 4b). This in turn leads to a very strong increase in longwave CRE in this region, which dominates the increase in shortwave CRE (Fig. 5a,b). However, as the gST increases further and the specific humidity in the atmosphere increases, the clouds become thicker and thus more reflective, whereas the longwave CRE does not increase as fast anymore leading to a decrease in total CRE in the tropics for gST above 320 K. This decrease in tropical CRE with increasing gST contributes to the stabilization of the climate at gST above 330 K (Fig. 5c). In general, the changes in total CRE dominate the changes in clear-sky radiative effect in the tropics (Fig. 5c,d). In the extra-tropics, the weakening of the large-scale circulation leads to a steady decrease in cloud cover everywhere except at very high latitude (Fig. 4a,b). Therefore, the shortwave CRE increases in the extra-tropics (Fig. 5a). Since the tropopause deepens with increasing surface temperatures (not shown), the difference between the temperature at the surface and at the cloud tops increases as well and leads also to an increase in longwave CRE despite the decrease in cloud cover (Fig. 5b). Whereas the changes in the clear-sky radiative effect dominate the changes in CRE in most of the extra-tropics for gST up to 315 K, the changes in CRE increasingly dominate the extra-tropical response at gST above. This supports the idea that at high gST changes in CRE are more important than changes in clear-sky radiative effect and dominate the climate response. In general, the weak large-scale circulation in the warm regime leads to a much more uniform meridional distribution of cloud condensate at all levels than in the cold regime (Fig. 6). Note that there is a decrease in global-mean convective precipitation but a slight increase in total precipitation from the cold to the warm regime (not shown). This may indicate that in the warm regime convection is overall less frequent but more intense, such that a significant fraction of condensate is not converted to convective precipitation but detrained to form large-scale precipitation. This trend continues as surface temperatures increase further in the warm regime.

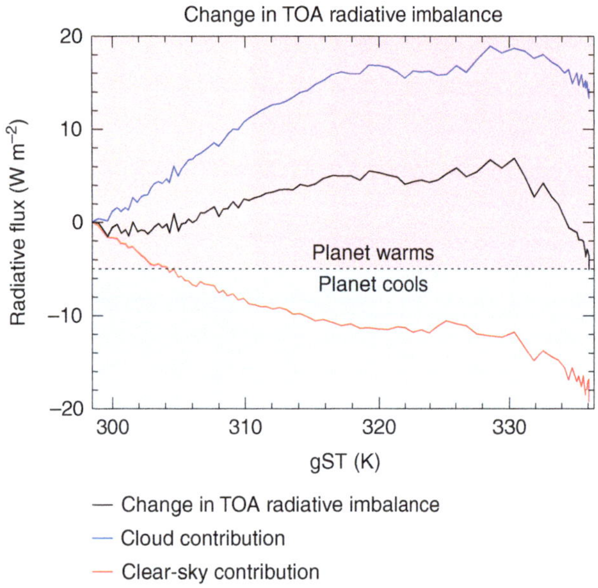

Figure 3 | Change of TOA radiative flux as a function of gST. The black line shows the change in global and annual mean of the total (shortwave plus longwave) net (downward minus upward component of the) TOA radiative flux, the blue line shows the cloud-radiative contribution and the red line the clear-sky contribution from the transient simulation with a TSI of 1.05 S_0. Note that a positive slope indicates an unstable state, because any warming/cooling would lead to an increase/decrease in energy uptake by the aqua-planet and thus to an additional warming/cooling. The cloud-radiative and clear-sky contributions sum up to the net TOA radiative flux. The changes are calculated by subtracting the global and annual means over the first year of the respective quantities. The horizontal line corresponds to the negative value of the initial total net TOA radiative flux. Hence, for a steady state to be attained, the total net TOA radiative flux must touch or intersect the horizontal line.

Simulations with increased CO_2 concentrations. We start our comparison of CO_2-induced to solar forcing by increasing CO_2 concentrations to 770 p.p.m., while keeping the TSI fixed to 1.00 S_0. This corresponds to an equivalent adjusted forcing as is caused by an increase of TSI from 1.00 S_0 to 1.03 S_0. The adjusted forcing is defined to be the temporal and global mean of the energy uptake over the first year of simulation. The results suggest that the increase of the CO_2 concentrations leads to an equivalent warming and a similar meridional distribution of surface temperatures and clouds as the increase in TSI does (Fig. 2a,d). Since the aqua-planet warms by 4.57 K for an increase in CO_2 concentrations from 354 to 770 p.p.m., the climate sensitivity of the aqua-planet for a doubling of CO_2 concentrations is 4.08 K, if a $\log_2$ scaling is assumed. Starting from the final state of the simulation with a TSI of 1.03 S_0, we then increase the CO_2 concentrations to 1,520 p.p.m. and set the TSI back to 1.00 S_0 (Fig. 1). The combined effect leads to an adjusted forcing that is equivalent to increasing the TSI to 1.05 S_0. In this case the aqua-planet undergoes a climate transition into the warm regime (Fig. 1). Thus, the aqua-planet can as readily be forced to transition from the Earth-like to the warm regime by increasing CO_2 concentrations as by increasing the TSI. When starting in the warm regime, a reduction of CO_2 concentrations to 770 p.p.m. does not cause the planet to fall back into the Earth-like regime, but the aqua-planet remains in the warm regime. Therefore, the aqua-planet also exhibits a bistability of the climate for a TSI of 1.00 S_0 and a CO_2 concentration of 770 p.p.m.

Overall, the results suggest that the aqua-planet behaves similarly for solar forcing and CO_2-induced forcing (Figs 1,2 and 6). The most notable difference is that the steady-state gST in the warm regime is ~2 K lower for CO_2-induced forcing. The

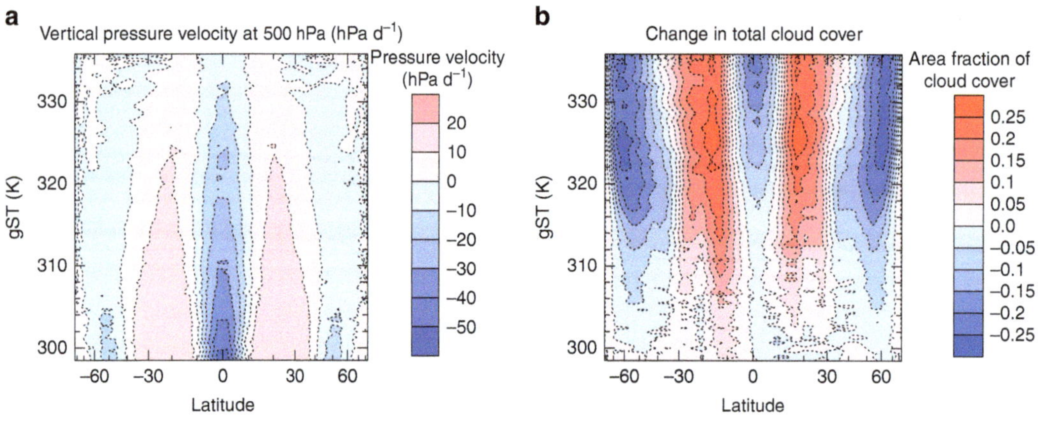

Figure 4 | Large-scale circulation and cloud cover during the climate transition. (**a**) Zonal and annual mean of the vertical pressure velocity as a function of latitude and gST for the transient period of the simulation with a TSI of 1.05 S_0. (**b**) Same as in **a** but for the change in zonal and annual mean of cloud cover from the first year of simulation.

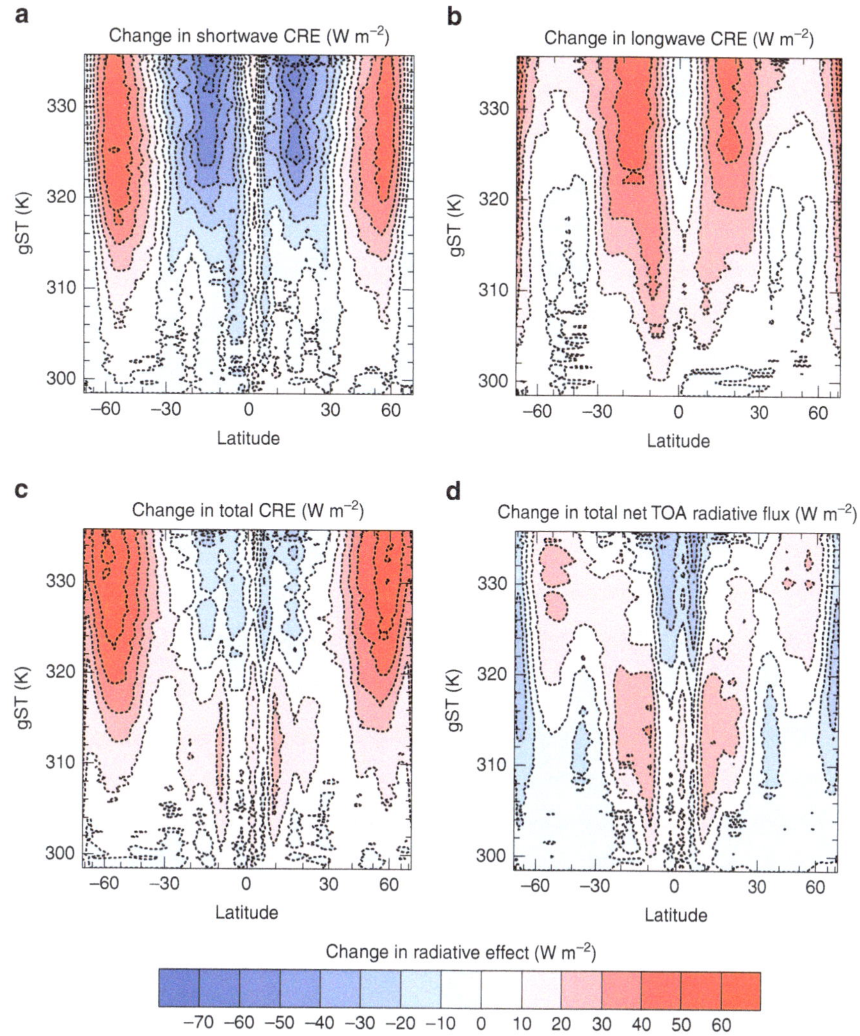

Figure 5 | Zonal means of changes in radiative effects during the climate transition. (**a**) Change of zonal and annual mean of the shortwave CRE; (**b**) change of the longwave CRE; (**c**) change of the total CRE; and (**d**) the change of the total net TOA radiative flux. All panels use the same colour bar. The horizontal axes are scaled with the sine of the latitude.

likely reason for this is that the thermal absorption by water vapour overlaps with the thermal absorption by CO_2 in the warm moist atmosphere, which renders the greenhouse effect of CO_2 less effective. However, since the climate instability in our simulations is caused by CRE at a gST at which the atmosphere is not yet sufficiently opaque to cancel the greenhouse effect of CO_2, CO_2-induced forcing can as easily cause a climate transition to the Moist Greenhouse as solar forcing does.

Sensitivity experiments. Two of the assumptions made in this study could potentially have a large influence on the results and may contribute to the differences in the results between our study and previous ones[17,12,9,20]. These assumptions concern the treatment of ozone and oceanic heat transport of the model. For these two cases, we show with sensitivity experiments that the qualitative nature of the results is not changed by the assumptions (Supplementary Figs 1–3). The neglect of sea ice should have no influence on the qualitative results, because there is no sea ice in the warm regime and because the sea-ice albedo feedback is positive and would thus favour a climate instability in the cold regime. However, the absence of sea ice may explain why our control simulation is warmer than present-day Earth.

Discussion

A recent study using the same model but in a different version found that the Earth's climate remains stable for CO_2 concentrations of at least 4,480 p.p.m. (ref. 17), whereas our study suggests that such concentrations would lead to a climate transition. Studies of Earth with other GCMs also found the climate to remain stable for higher CO_2 concentrations than we do[16,18]. However, the initial climate of our aqua-planet is ~6 K warmer than the one of present-day Earth. Such a warming would be attained by a quadrupling of CO_2 in the different version of our model used in ref. 17. By a simple estimate, this other study would thus have explored CO_2 concentrations of up to a fourth of 4,480 p.p.m.; hence, 1,120 p.p.m., if the simulations were started from a climate similar to ours. Therefore, if we account for the difference in the initial climates, the results of the two studies are not in contradiction. Indeed, the climate of the model version used in ref. 17 was recently shown to become unstable when the CO_2 concentrations were increased from 4,480 to 8,960 p.p.m. (eventually leading to numerical failure of their model)[19]. Nonetheless, the forcing required to cause a climate transition would certainly be higher on present-day Earth than on our aqua-planet, even with our version of the model. Several other studies of Earth have found lower climate sensitivities to

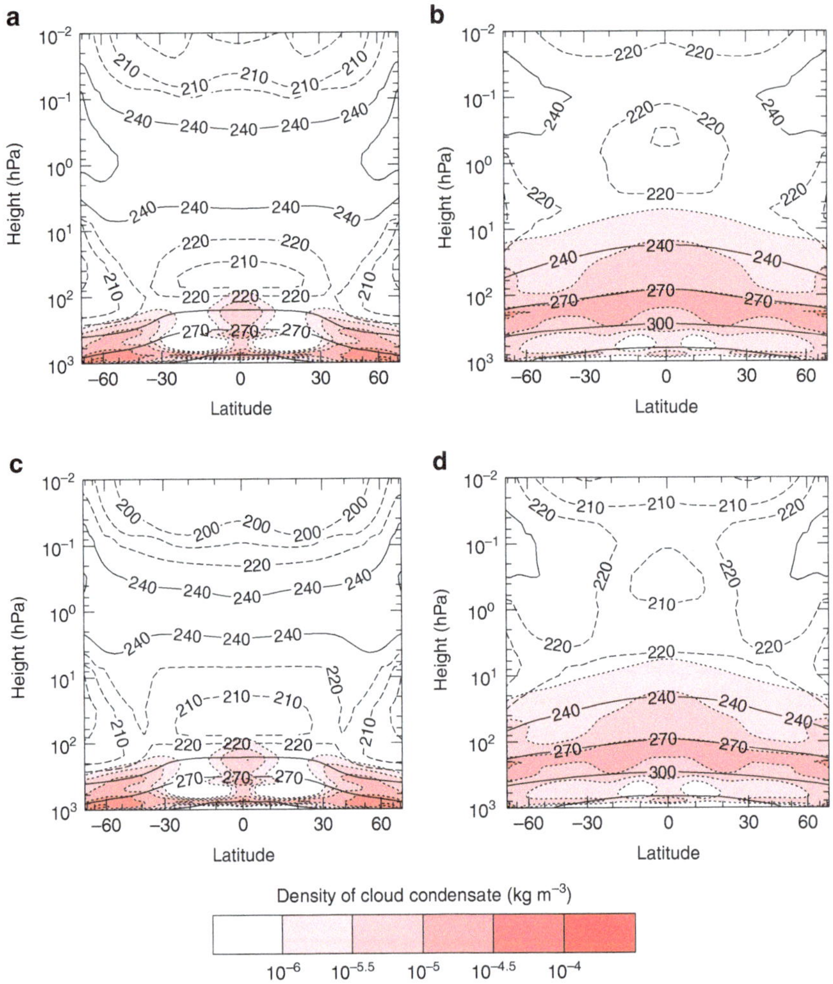

Figure 6 | Zonal means of cloud condensate in steady state. (**a**) Temporal mean over the last 30 years of simulation of the cloud condensate for a TSI of 1.03 S_0 in the cold regime and (**b**) same quantity for the same TSI but in the warm regime. (**c**) Same quantity obtained with a TSI of 1.00 S_0 but with atmospheric CO_2 concentrations of 770 p.p.m. in the cold regime (**d**) with the same CO_2 concentration and TSI in the warm regime. The contours denote temperatures in Kelvin, with the solid lines denoting the contours for 240, 270, 300 and 330 K and the dashed lines for the contours of 200, 210 and 220 K. The vertical axes are the height in terms of pressure of dry air and the horizontal axes are the latitudes scaled with their sines. The pressure of dry air is defined to be the pressure the atmosphere would have at a given level if no water vapour was present. Since the total mass of dry air does not change between simulations, the mean of the dry surface pressure is constant across simulations.

relatively large CO_2 forcing than we do which supports this notion[16,18,20].

Two studies recently investigated climates at gST above 330 K with state-of-the-art GCMs for Earth-like planets[12,9]. Wolf and Toon[9] used a modified version of the Community Atmosphere Model version 4 (CAM4) and Leconte *et al.*[12] a modified version of the Laboratoire de Météorologie Dynamique Generic (LMDG) climate model to investigate the climate response to strong solar forcing and both studies also found a region of increased climate sensitivity[12,9]. Therefore, a region of gST with increased climate sensitivity surrounded by regions of lower climate sensitivity appears to be a robust result, despite some differences in the magnitude of the region (Fig. 7c,e). We calculated the climate sensitivity here following[9], which yields a small climate sensitivity in the cold and in the warm regimes because the method uses the instantaneous forcing. If we use the adjusted forcing, the climate sensitivity is considerably larger, because the fast atmospheric adjustments reduce the initial TOA radiative imbalance quickly. Our model encounters the region of high climate sensitivity for smaller values of TSI (Fig. 7a), because our control climate is the warmest and because the climate instability is encountered at ~300 K, whereas the region of increased climate sensitivity starts at ~310–315 K in the two other studies. Our steady-state albedo is a monotonically increasing function of both TSI and gST, in contrast to the two other studies where the albedo decreases with both increasing TSI and gST until either the Moist-Greenhouse state is attained or a Runaway Greenhouse occurs (Fig. 7b,d). However, in our transition from the cold to the warm regime the albedo decreases as well with increasing gST between 300 and 320 K. Only one of the two studies found a stabilizing cloud feedback similar to ours and a moist stratosphere at high gST (ref. 9), whereas the other one found that the cloud feedback is rather destabilizing and that the stratosphere remains dry[12]. In general, our results of the simulations with increased solar forcing are qualitatively similar to the ones found in ref. 9 in that we find two regimes of steady states, in that the warm regimes have a similar temperature structure (Fig. 8a), in that the cloud albedo increases with gST and most importantly in the existence of a stable Moist-Greenhouse regime. Similarly to ref. 9, the troposphere in the warm regime is characterized in our model by a particular

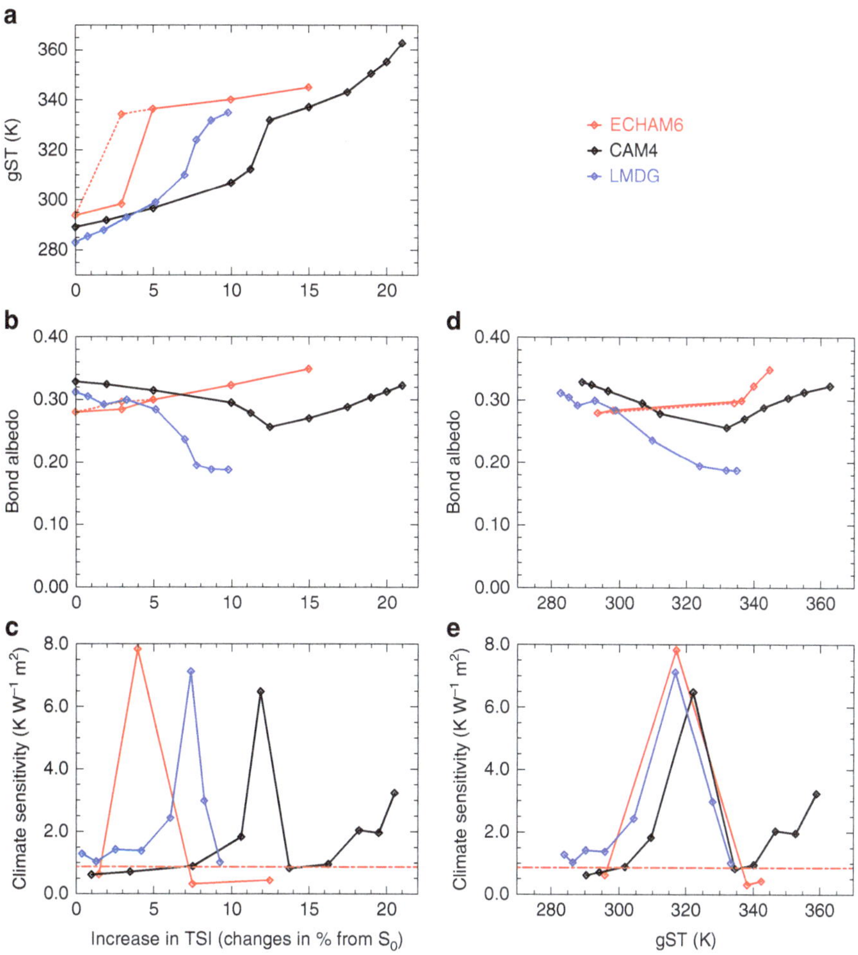

Figure 7 | Model inter-comparison. (**a**) gST; (**b**) the Bond albedo; and (**c**) the climate sensitivity as a function of the increase in TSI in per cents of S_0. The red solid lines and marks denote the results obtained with ECHAM6 when increasing TSI, whereas the dashed lines and mark denote the results obtained with ECHAM6 but when decreasing TSI. The black lines and marks denote the results obtained with CAM4 in ref. 9 and the blue lines and marks denote the results obtained with LMDG in ref. 12. (**d**) Bond albedo and (**e**) the climate sensitivity as a function of the gST. The red dashed-dotted line denotes the climate-sensitivity parameter for doubling CO_2 and is calculated from the simulation in which CO_2 concentrations are increased from 354 to 770 p.p.m. using $\log_2$ scaling. Note that the Bond albedo is equal to the global-mean of the effective albedo.

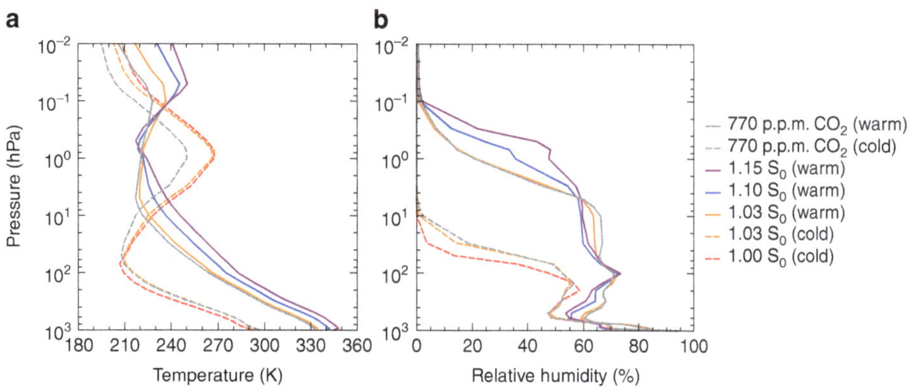

Figure 8 | Atmospheric profiles of temperature and relative humidity. (**a**) Global-mean vertical profiles of gST and (**b**) of relative humidity for the different steady states. The vertical axes are the height in terms of pressure of dry air.

radiative-convective equilibrium with a temperature inversion close to the surface, a somewhat drier region above and a more humid region up to the tropopause. A similar structure has also been found and discussed in two one-dimensional studies[5,21]. Wolf and Toon[9] argues that the change to the aforementioned radiative-convective equilibrium is crucial for the emergence of stable Moist-Greenhouse states, but our results suggest that the weakening of the large-scale circulation is equally important by allowing the radiative-convective regime to spread over the entire tropics. This spread of the convective region over a large fraction of the planet, namely the tropics, also explains some of the similarities between the three- and the one-dimensional models in the warm regime. Compared with ref. 9, the dry region is more humid in our warm regime and in general more humid than in our cold regime (Fig. 8b, Supplementary Fig. 4). Given the different dynamical cores, radiative transfer schemes, convection schemes and cloud schemes, it is, however, remarkable how many similarities the three models share.

In the two aforementioned studies of large solar forcing, ozone was entirely removed from the whole atmosphere in all simulations[12,9], whereas we do include ozone in our calculations. Since ozone has a strong warming influence in the middle atmosphere due to its strong absorption of solar radiation, the inclusion of ozone in our model may explain why our atmosphere is moister in the upper levels than theirs for similar values of gSST. To test this hypothesis, we perform a simulation with a TSI of 1.10 S_0 without atmospheric ozone. Both the upper atmospheric temperatures and specific humidity are lower but still above the Moist-Greenhouse limit in the experiments where ozone is removed (Supplementary Fig. 1). Thus, the specific humidity is still considerably higher than in our cold regime and especially than in the simulations of Wolf and Toon[9] and Leconte *et al.*[12] at similar gSST. In general, removing ozone does not appear to alter the results substantially.

One study investigated strong CO_2 forcing and gST above 330 K with the Fast Atmosphere-Ocean Model (FAOM) developed at the Goddard Institute for Space Studies and did not find any region of strongly increased climate sensitivity[20]. The climate sensitivity increases somewhat but not nearly as much as with our or with the other two state-of-the-art models used for solar forcing. As a consequence, considerably higher values of CO_2 are required in that study compared with ours to attain a gST of 330 K. Furthermore, the control climate is colder in that study, and it takes around one doubling of CO_2 to attain the gST of our control simulation. The humidity at the top level of FAOM remains roughly one order of magnitude below the Moist-Greenhouse limit in the warmest steady states found in ref. 20. This may be partly due to the CO_2 cooling of the upper atmosphere, but could also be a consequence of not running the model to sufficiently high temperatures. We perform a sensitivity experiment, where we increase CO_2 concentrations to 9,000 p.p.m. in order to assess whether increased CO_2 concentrations could cause a substantial drying of our upper atmosphere (not shown). The upper atmosphere-specific humidity stays; however, well above the Moist-Greenhouse limit also in that case. Some of the differences between FAOM and ECHAM6 may simply be caused by the use of different setups and parameterizations. But FAOM is a simplified model designed for fast computation and uses simplified cloud physics, which may be the cause for the absence of a region of strongly increased climate sensitivity. So, whereas our version of ECHAM6 is rather on the low side of CO_2 concentrations required to cause the gST to rise above 330 K, FAOM likely is on the high side of the concentrations.

To conclude, we have demonstrated with a state-of-the-art climate model that a water-rich planet might lose its habitability as readily by CO_2 forcing as by increased solar forcing through a transition to a Moist Greenhouse and the implied long-term loss of hydrogen. We confirm previous results that a region of increased climate sensitivity exists and show that the climate is unstable in our model in that region due to positive cloud feedbacks caused by a weakening of the large-scale circulation. We also demonstrate that there is hysteresis and that once a transition to the Moist-Greenhouse regime has occurred, the process may not simply be reversed by removing the additional forcing.

Methods

General setup. We employ a modified version of the GCM ECHAM6 (ref. 22) in an aqua-planet setting in which the whole surface is covered by a 50-m-deep mixed-layer ocean. We run the model with a spectral truncation of T31, which corresponds to a Gaussian grid with a grid-point spacing of 3.75°. The atmosphere is resolved vertically by 47 layers up to a pressure (of dry air) of 0.01 hPa. The oceanic heat transport is prescribed by a sinusoidal function of latitude. There is no representation of sea ice in our model and as a consequence water may be colder than the freezing temperature. The orbit of the aqua-planet is perfectly spherical with a radius of 1 AU. The obliquity of the aqua-planet is 0°. For simplicity, a year is set to be 360 days. The rotation velocity of the aqua-planet corresponds to present-day Earth.

Parameterized oceanic heat transport. To account for the meridional oceanic heat transport, we introduce an additional energy flux from the mixed-layer ocean to the atmosphere, which mimics the divergence of a prescribed meridional oceanic heat flux. This energy flux is commonly called q-flux. Imposing a q-flux is necessary to prevent the atmosphere from having to take over all the meridional energy transport, which would result in an amplified atmospheric large-scale circulation. However, since the oceanic circulation depends also on atmospheric properties, the imposed q-flux may be inaccurate in a warming climate. The weak atmospheric large-scale circulation suggests that the oceanic heat transport would be weaker in the warm regime and that thus the absolute values of the q-flux are too large. To assess whether the imposed q-flux affects the results in the warm regime, a number of simulations without q-flux are performed. Despite small differences in the large-scale circulation (Supplementary Fig. 2), the warm steady states exist and are stable even without q-flux (Supplementary Fig. 3).

Treatment of ozone. If the tropopause climbs, regions with high ozone concentrations may come to lie in the troposphere of the model, because ozone concentrations are prescribed to climatological values. High ozone concentrations could, however, not occur in the presence of tropospheric water vapour concentrations. Therefore, we limit the tropospheric ozone concentrations to a volume mixing ratio of 1.5×10^{-7}. As a consequence, ozone is taken out of the atmosphere, if the tropopause rises to levels where the climatological values would exceed this limit. This process is reversible if the tropopause descends again.

Modifications to the grid-point physics. Our version of the model incorporates several changes to the grid-point physics, such that we obtain a more accurate representation of several physical processes in warm climates[21]. The grid-point physics include representation of surface exchange, turbulence and vertical diffusion[24,25], gravity-wave drag[26,27], radiative transfer and radiative heating[28,29], convection[30,31], cloud cover[32] and cloud microphysics[33]. In summary, these changes are the inclusion of the mass of water vapour when calculating the total pressure and the omission of all approximations where small specific humidities are assumed (as for example in the calculation of density). The pressure effects of water vapour are not considered for the horizontal transport. So, the model is in sorts a hybrid model, with water vapour adding to the total pressure for local effects but not so for the large-scale transport. A detailed description of the modified model thermodynamics can be found in the appendix of ref. 34. We will give here a short overview of the modified radiative transfer scheme as well as of the convection, cloud-cover and cloud-microphysical schemes.

Radiative transfer. The radiative transfer scheme has recently been described and evaluated extensively in ref. 21, but as a courtesy to the reader we will repeat some of the major features here. It is based on the Rapid Radiative Transfer Model[28,29], but includes some small modifications. It uses the correlated-k method to solve the radiative transfer equations in the two-stream approximation. The k-coefficients are calculated from the HITRAN (1996 and 2000) database using a line-by-line radiative transfer model[28,29]. The water vapour continuum is based on CKD_v2.4. The shortwave radiation spectrum is divided into 14 bands, and the longwave radiation spectrum is divided into 16 bands. Since the lookup-tables of the molecular absorption coefficients are designed for a limited range of temperatures only, an exponential extrapolation for temperatures up to 400 K for the longwave radiation scheme is performed. The same extrapolation scheme is also applied to the lookup-tables for the absorption coefficients of the water vapour self-broadened continuum in the shortwave radiation scheme, but the original linear extrapolation scheme is kept for all the other absorption coefficients. The lookup-tables for the bandwise spectrally integrated Planck function and the derivative thereof with respect to temperature have been extended to 400 K. Furthermore the water vapour self-broadened continuum is introduced in the upper atmosphere radiation calculations, to account for the increase in water vapour with increasing gST. The effect of pressure broadening by water vapour on the molecular absorption coefficients is neglected, as is scattering by water vapour. The thus modified radiation scheme is not as accurate as line-by-line radiative transfer models or models using recalculated k-coefficients for the higher temperatures, but still sufficiently accurate for the task at hand as has recently been demonstrated[21].

Convective scheme. ECHAM6 uses a mass-flux scheme for cumulus convection[30], with modifications for penetrative convection to the original scheme[31]. The contribution of cumulus convection to the large-scale budgets of heat, moisture and momentum is represented by an ensemble of clouds consisting of updrafts and downdrafts in a steady state. Depending on moisture convergence at the surface and depth of the convection cell, the model will either run in penetrative, mid-level, or shallow convection mode. The scheme allows for the formation of precipitation, but not for radiatively active convective clouds. Instead, detrained water is passed to the cloud-microphysical scheme which creates or destroys cloud condensate in a further step.

Cloud-cover scheme. ECHAM6 uses the Sundqvist scheme for fractional cloud cover[32]. This cloud scheme has been tuned for the use with present-day Earth's climate[33,35]. However, the scheme is well suited for simulations of cloud cover in warm climates as it diagnoses cloud cover directly from relative humidity, which should be crucial to cloud formation irrespective of the temperature.

Cloud-microphysical scheme. The cloud-microphysical scheme is described in detail in ref. 33. The scheme consists of prognostic equations for the vapour, liquid, and ice phases. There are explicit microphysics for warm-phase, mixed-phase and ice clouds. The cloud-condensation-nuclei concentration follows a prescribed vertical profile which is typical for the present-day Earth's maritime conditions. Since we have no estimate of the aerosol load in a hypothetical warm climate, we assume that this profile is also a reasonable choice for warmer climates. The microphysics do not require changes for the use of the scheme in warm climates, since cloud formation in a warm-phase cloud, in which potential changes may occur, is not directly temperature-dependent (at least not in the range of temperatures we consider).

Cloud-radiative interactions. Clouds are represented in the radiative transfer calculations, assuming the so-called maximum-random-overlap assumption. Under this assumption cloud layers are assumed to be maximally overlapping if they are adjacent to one another, and randomly overlapping if they are separated by a clear layer. The absorptivity of clouds depends on their combined optical depths, the gas in which they are embedded, and the interstitial aerosol. The microphysics to determine the optical properties of the cloud particles involve the liquid water and ice paths, cloud-drop radii, as well as liquid water and ice content. Cloud scattering is represented as a single-scattering albedo by assuming Mie scattering from cloud droplets in the shortwave calculations, but is neglected in the longwave calculations. Clouds are not considered in the radiative transfer routines if the specific mass of the cloud condensate does not exceed 10^{-7} kg per kg of air.

Special settings for high-temperature simulations. To run the model at high temperatures, a few special settings are necessary. The time step is reduced from 2,400 to 600 s, and the radiation time step is reduced from 7,200 to 2,400 s, except for the simulation with a TSI of 1.15 S_0, where the time step is reduced to 360 and the radiation time step to 1,440 s. Nonetheless, we sometimes encounter problems with resolved waves propagating to the top levels of the model, where their amplitude grows and where they may be reflected. To avoid frequent model failure due to these effects, we introduce Rayleigh friction to the vorticity and the divergence as well as increased horizontal diffusion in the top six layers (above $\sim$0.75 hPa). The time constant of the Rayleigh friction is (10800 s) at the sixth layer and is increased by a factor of 3.2 per layer towards the top and is hence increased by a factor of $\sim$1,000 in the top layer. The horizontal diffusion is increased by a factor of 3.2 per layer. The values of the time constant of the Rayleigh friction and the magnitude of horizontal diffusion are determined by trial and error. Since we investigate a large range of climates, it is difficult to find suitable values for these time constants, and despite these modifications the model fails occasionally. In these cases, however, the runs can be continued by slightly changing their trajectory, which is achieved by reducing the factor of multiplication per level for the Rayleigh friction from 3.2 to 3.199 for 30 days.

References

1. Kasting, J. F. Runaway and moist greenhouse atmospheres and the evolution of earth and venus. *Icarus* **74,** 472–494 (1988).
2. Kasting, J. F., Whitmire, D. P. & Reynolds, R. T. Habitable zones around main-sequence stars. *Icarus* **101,** 108–128 (1993).
3. Kopparapu, R. K. *et al.* Habitable zones around main-sequence stars: new estimates. *Astrophys. J.* **765,** 131 (2013).
4. Kasting, J. F., Pollack, J. B. & Ackerman, T. P. Response of Earth's atmosphere to increases in solar flux and implications for loss of water from venus. *Icarus* **57,** 335–355 (1984).
5. Wordsworth, R. D. & Pierrehumbert, R. T. Water loss from terrestrial planets with CO2-rich atmospheres. *Astrophys. J.* **778,** 154 (2013).
6. Ramirez, R. M., Kopparapu, R. K., Lindner, V. & Kasting, J. F. Can increased atmospheric CO2 levels trigger a runaway greenhouse? *Astrobiology* **14,** 714–731 (2014).
7. Abe, Y., Abe-Ouchi, A., Sleep, N. H. & Zahnle, K. J. Habitable zone limits for dry planets. *Astrobiology* **11,** 443–460 (2011).
8. Yang, J., Cowan, N. B. & Abbot, D. S. Stabilizing cloud feedback dramatically expands the habitable zone of tidally locked planets. *Astrophys. J. Lett.* **771,** L45 (2013).
9. Wolf, E. & Toon, O. The evolution of habitable climates under the brightening sun. *J. Geophys. Res. Atmos.* **120,** 5775–5794 (2015).
10. Ishiwatari, M., Takehiro, S., Nakajima, K. & Hayashi, Y. Y. A numerical study on appearance of the runaway greenhouse state of a three-dimensional gray atmosphere. *J. Atmos. Sci.* **59,** 3223–3238 (2002).
11. Ishiwatari, M., Nakajima, K., Takehiro, S. & Hayashi, Y.-Y. Dependence of climate states of gray atmosphere on solar constant: From the runaway greenhouse to the snowball states. *J. Geophys. Res.* **112,** D13120 (2007).
12. Leconte, J., Forget, F., Charnay, B., Wordsworth, R. & Pottier, A. Increased insolation threshold for runaway greenhouse processes on earth-like planets. *Nature* **504,** 268–271 (2013).
13. Boer, G. J., Hamilton, K. & Zhu, W. Climate sensitivity and climate change under strong forcing. *Clim. Dyn.* **24,** 685–700 (2005).
14. Heinemann, M. Warm and sensitive paleocene-eocene climate. *Reports on Earth System Science 70/2009* (Max-Planck-Institute for Meteorology, 2009).
15. Wolf, E. T. & Toon, O. Delayed onset of runaway and moist greenhouse climates for earth. *Geophys. Res. Lett.* **41,** 167–172 (2014).
16. Hansen, J. *et al.* Efficacy of climate forcings. *J. Geophys. Res.* **110,** D18104 (2005).
17. Meraner, K., Mauritsen, T. & Voigt, A. Robust increase in equilibrium climate sensitivity under global warming. *Geophys. Res. Lett.* **40,** 5944–5948 (2013).
18. Wolf, E. T. & Toon, O. B. Hospitable archean climates simulated by a general circulation model. *Astrobiology* **13,** 656–673 (2013).
19. Bloch-Johnson, J., Pierrehumbert, R. T. & Abbot, D. S. Feedback temperature dependence determines the risk of high warming. *Geophys. Res. Lett.* **42,** 4973–4980 (2015).
20. Russell, G. L., Lacis, A. A., Rind, D. H. & Colose, C. Fast atmosphere-ocean model runs with large changes in CO2. *Geophys. Res. Lett.* **40,** 5787–5792 (2013).
21. Popp, M., Schmidt, H. & Marotzke, J. Initiation of a runaway greenhouse in a cloudy column. *J. Atmos. Sci.* **72,** 452–471 (2015).
22. Stevens, B. *et al.* The atmospheric component of the MPI-M earth system model: ECHAM6. *J. Adv. Model. Earth Syst.* **5,** 146–172 (2013).
23. Medeiros, B., Stevens, B. & Bony, S. Using aquaplanets to understand the robust responses of comprehensive climate models to forcing. *Clim. Dyn.* **44,** 1957–1977 (2015).
24. Brinkop, S. & Roeckner, E. Sensitivity of a general circulation model to parameterizations of cloudturbulence interactions in the atmospheric boundary layer. *Tellus A* **47,** 197–220 (1995).
25. Giorgetta, M. A. *et al.* The atmospheric general circulation model ECHAM6 - model description. *Technical Report 135* (Max-Planck-Institute for Meteorology, 2013).
26. Hines, C. O. Doppler-spread parameterization of gravity-wave momentum deposition in the middle atmosphere. Part 1: basic formulation. *J. Atmos. Sol.-Terr. Phys.* **59,** 371–386 (1997).
27. Hines, C. O. Doppler-spread parameterization of gravity-wave momentum deposition in the middle atmosphere. Part 2: Broad and quasi monochromatic spectra, and implementation. *J. Atmos. Sol.-Terr. Phys.* **59,** 387–400 (1997).
28. Mlawer, E. J., Taubman, S. J., Brown, P. D., Iacono, M. J. & Clough, S. A. Radiative transfer for inhomogeneous atmospheres: RRTM, a validated correlated-k model for the longwave. *J. Geophys. Res.* **102,** 16663–16682 (1997).
29. Iacono, M. J. *et al.* Radiative forcing by long-lived greenhouse gases: calculations with the AER radiative transfer models. *J. Geophys. Res.* **113,** D13103 (2008).
30. Tiedtke, M. A comprehensive mass flux scheme for cumulus parameterization in large-scale models. *Mon. Wea. Rev.* **117,** 1779–1800 (1989).
31. Nordeng, T.-E. Extended versions of the convective parametrization scheme at ECMWF and their impact on the mean and transient activity of the model in the tropics. *Technical Memorandum 206, ECMWF* (European Centre for Medium-Range Weather Forecasts, Shinfield Park, Reading, RG2 9AX, UK, 1994).
32. Sundqvist, H., Berge, E. & Kristjansson, J. E. Condensation and cloud parameterization studies with a mesoscale numerical weather prediction model. *Mon. Weather Rev.* **117,** 1641–1657 (1989).
33. Lohmann, U. & Roeckner, E. Design and performance of a new cloud microphysics scheme developed for the ECHAM general circulation model. *Clim. Dyn.* **12,** 557–572 (1996).

34. Popp, M. Climate instabilities under strong solar forcing. *Reports on Earth System Science 152/2014* (Max-Planck-Institute for Meteorology, 2014).
35. Mauritsen, T. *et al.* Tuning the climate of a global model. *J. Adv. Model. Earth Syst.* **4,** M00A01 (2012).

Acknowledgements

We thank Thorsten Mauritsen for a thorough internal review. We thank Eric Wolf for an insightful discussion and for making the data from CAM4 available to us. We thank Jérémy Leconte for making the data from LMDG available to us. We thank Dorian Abbot and two anonymous reviewers for very constructive reviews of this manuscript. We thank the Max Planck Society for the Advancement of Science for financial support.

Author contributions

J.M., H.S. and M.P. conceived the study. M.P. performed the changes to the model and conducted the modelling work. J.M., H.S. and M.P. all contributed substantially to the analysis and the discussion. M.P. wrote the manuscript with comments from J.M. and H.S.

Additional information

Competing financial interests: The authors declare no competing financial interests.

21

Top-down assessment of the Asian carbon budget since the mid 1990s

R.L. Thompson[1], P.K. Patra[2], F. Chevallier[3], S. Maksyutov[4], R.M. Law[5], T. Ziehn[5], I.T. van der Laan-Luijkx[6], W. Peters[6,7], A. Ganshin[8,9], R. Zhuravlev[8,9,10], T. Maki[11], T. Nakamura[12], T. Shirai[4], M. Ishizawa[4], T. Saeki[2], T. Machida[4], B. Poulter[13], J.G. Canadell[14] & P. Ciais[3]

Increasing atmospheric carbon dioxide (CO_2) is the principal driver of anthropogenic climate change. Asia is an important region for the global carbon budget, with 4 of the world's 10 largest national emitters of CO_2. Using an ensemble of seven atmospheric inverse systems, we estimated land biosphere fluxes (natural, land-use change and fires) based on atmospheric observations of CO_2 concentration. The Asian land biosphere was a net sink of -0.46 (-0.70-0.24) PgC per year (median and range) for 1996-2012 and was mostly located in East Asia, while in South and Southeast Asia the land biosphere was close to carbon neutral. In East Asia, the annual CO_2 sink increased between 1996-2001 and 2008-2012 by 0.56 (0.30-0.81) PgC, accounting for $\sim$35% of the increase in the global land biosphere sink. Uncertainty in the fossil fuel emissions contributes significantly (32%) to the uncertainty in land biosphere sink change.

[1] Norsk Institutt for Luftforskning (NILU), Kjeller, Norway. [2] Department of Environmental Geochemical Cycle Research, Japan Agency for Marine Earth Science and Technology (JAMSTEC), Yokohama 236-0001, Japan. [3] Laboratoire des Sciences du Climat et l'Environnement (LSCE, CEA-CNRS-UVSQ), Gif-sur-Yvette, France. [4] National Institute for Environmental Studies (NIES), Tsukuba 305-8506, Japan. [5] Oceans and Atmosphere, Commonwealth Scientific and Industrial Research Organisation (CSIRO), 3195 Aspendale, Australia. [6] Department of Meteorology and Air Quality, Environmental Sciences Group, Wageningen University (WU), 6708 PB Wageningen, The Netherlands. [7] University of Groningen, Centre for Isotope Research, 9747 AG Groningen, The Netherlands. [8] Department of Upper Atmospheric Layers Physics, Central Aerological Observatory (CAO), Moscow 141700, Russia. [9] National Research Tomsk State University (TSU), 634050 Tomsk, Russia. [10] Department of Atmospheric Physics and Microwave Diagnostics, Institute of Applied Physics of the Russian Academy of Sciences, Nizhny Novgorod 603950, Russia. [11] Atmospheric Environment and Applied Meteorology Research Department, Meteorological Research Institute (MRI), Tsukuba 305-0052, Japan. [12] Japan Meteorological Agency (JMA), Global Environment and Marine Department, Tokyo 100-8122, Japan. [13] Institute on Ecosystems and Department of Ecology, Montana State University (MSU), 59717 Bozeman, Montana, USA. [14] Global Carbon Project, Commonwealth Scientific and Industrial Research Organisation (CSIRO), 2601 Canberra, Australia. Correspondence and requests for materials should be addressed to R.L.T. (email: rona.thompson@nilu.no).

Increasing atmospheric carbon dioxide (CO_2) is the single most important driver of climate change, and CO_2 accounts for ~64% of the radiative forcing from well-mixed greenhouse gases[1]. Hence reducing CO_2 emissions is one of the principal concerns of efforts to mitigate anthropogenic climate change. The global mean atmospheric concentration of CO_2 has increased from 277 parts-per-million (p.p.m.) in 1750 (ref. 2) to 396 p.p.m. in 2013 (ref. 3), due to emissions from fossil fuel combustion, cement production and land-use change. In the 1990s, the first international agreements to limit emissions of CO_2 were ratified, the United Nations Framework Convention on Climate Change (UNFCCC), in 1992, and subsequently the Kyoto Protocol, in 1997. Very recently at the Conference of the Parties (COP21) meeting in Paris 2015, it was agreed that further efforts are needed to limit global warming to a maximum of 2 °C. To achieve this, peak greenhouse gas emissions should be reached as soon as possible and by the second half of this century anthropogenic emissions should be balanced by sinks (UNFCCC, CP21, L.9). For such agreements to be effectual, accurate means for assessing the effectiveness of regional and national emission reduction policies are critical, and require better estimates of CO_2 emissions. Also, of upmost importance are improved estimates of the land and ocean biosphere sources and sinks and a better scientific understanding of the carbon cycle[4].

Asia has undergone rapid economic growth over the past two decades, which has been associated with large increases in fossil fuel emissions. Fossil fuel emissions increased between 1990 and 2010 by ~190% in India (to 0.55 PgC per year; 1 PgC = 1×10^{15} g carbon in CO_2) and ~240% in China (to 2.26 PgC per year) (Global Carbon Atlas, http://www.globalcarbonatlas.org) and in 2006 China surpassed the USA as the largest national emitter of CO_2 (ref. 5). Asia has also seen substantial changes in land-cover over the same time period. Between 1990 and 2010, forest extent increased in South Asia, by ~3% (2.1 Mha) and in East Asia by ~22% (45.4 Mha), whereas in Southeast Asia deforestation occurred with a loss in forest extent of ~13% (33.2 Mha)[6]. Deforestation rates are particularly extensive in Indonesia, where the rate of forest clearance (including forested peatlands) reached 0.84 Mha per year in 2012, surpassing that in Brazil[7]. These changes have important consequences for the regional and global carbon budget. Deforestation in Southeast Asia is a large source of CO_2, releasing an estimated average of 0.23 PgC per year for 1990–2007 (ref. 8), while, East Asian ecosystems appear to be a substantial sink of CO_2, taking up an estimated 0.16–0.33 PgC per year due to afforestation/reforestation and regional climate change, especially in Southern China[8–10]. The uncertainty associated with land-use change fluxes, however, is very large.

Top–down approaches, specifically atmospheric inversions, can help to reduce the uncertainty on land biosphere fluxes. Atmospheric inversions use observations of CO_2 concentrations with a model of atmospheric transport to constrain CO_2 fluxes between the land/ocean and atmosphere. Because the problem is under-determined, a prior estimate of the fluxes is needed, and this estimate is updated based on the atmospheric information to give a *posterior* estimate (see Methods). Although atmospheric inversions of CO_2 have been used since the 1980s[11,12], there have been few inversion studies focusing on Asia owing to the fact that until recently few atmospheric CO_2 observations were available for this region. Recent programmes, such as CARIBIC (Civil Aircraft for the Regular Investigation of the atmosphere. On the basis of an Instrument Container)[13], CONTRAIL (Comprehensive Observation Network for TRace gases by AIrLiner)[14,15] and Asian Global Atmospheric Watch (GAW) have helped ameliorate this. CARIBIC data have been utilized in an inversion for CO_2 fluxes in South Asia[16] and CONTRAIL data in inversions focusing on Asian CO_2 fluxes[17–19]. These inversions[16–18] consistently found a net sink for South Asia of −0.11 to −0.37 PgC per year (note that a negative flux is in the direction from the atmosphere to the land/ocean, that is, a sink). In contrast, for Southeast Asia, one recent study[17] found a source of ~0.45 PgC per year, while another[18] found a sink of ~−0.28 PgC per year. Differences among inversion results can arise from differences in modelled atmospheric transport, prior information and its assumed uncertainty, as well as from the observations used. Therefore, to establish confidence in the magnitude and sign of regional CO_2 fluxes, and to determine the range of uncertainty accounting not only for random errors, it is essential to examine ensembles of inversions.

In this study, we assess the carbon budget of Asia using an ensemble of seven atmospheric CO_2 inversions focusing on the regions of East, South and Southeast Asia (see Fig. 1). The inversions use different prior information, transport models and observation data (see Supplementary Table 1) to estimate the land biosphere CO_2 fluxes (including land-use change and fires) from 1996 to 2012. (Note that only four inversions are available for 1996 to 2000). All the inverse systems solve for the global CO_2 fluxes with varying resolutions over the Asian regions. By using global inverse systems it is possible to infer the Asian fluxes consistently with the global carbon budget and the atmospheric growth rate of CO_2 as well as assess how the optimization of the Asian regional fluxes influences the fluxes in the rest of the world. The inversion performance is examined by comparing the CO_2 concentration from the optimized models against independent observations from the CONTRAIL programme between 2008 and 2010. This study focuses on the temporal evolution of the fluxes and the quantification of trends and their uncertainty. This study is part of an effort examining the Asian greenhouse gas budget initiated by RECCAP (REgional Carbon Cycle Assessment and Processes[20]) as part of the international Global Carbon Project (www.globalcarbonproject.org/reccap/).

Results

Uncertainties from fossil fuel and cement inventories. Atmospheric inversions optimize the total land/ocean to atmosphere CO_2 flux (that is, the net biosphere plus the fossil fuel and cement (FFC) flux), therefore, the FFC fluxes must be subtracted *a posteriori* to obtain the land biosphere fluxes. However, the uncertainty in the FFC fluxes is not negligible. Although globally, the FFC flux is quite consistent among all inversions and with the CDIAC (Carbon Dioxide Information Analysis Center, 2013; http://cdiac.ornl.gov) inventory (within ±2%), this is not the case in Asia (Supplementary Fig. 1). In East Asia, the modelling groups assumed different mean values and trends in their FFC fluxes leading to disparities of up to ~1 PgC per year (33%) in 2012, while in South and Southeast Asia, the absolute disparities were smaller but still significant, ~0.2 PgC per year (33–50%) in 2012. The disparities arise from the different inventories on which their estimates were based, for example, CDIAC or EDGAR (Emission Database for Global Atmospheric Research, version 4.2; http://edgar.jrc.ec.europa.eu) and whether a static or evolving spatial distribution was used (see Supplementary Note 1).

Owing to the differences in prior FFC fluxes between inverse systems, it is recommended practice to subtract a standard FFC flux estimate from the total posterior fluxes when comparing land biosphere fluxes from different CO_2 inversions[21]. However, since the total posterior flux may still be biased, and depends to a certain degree on the prior information, biases in the prior FFC fluxes may be propagated into the posterior land biosphere fluxes. In this case, the bias in the land biosphere flux will be proportional to the difference between the prior FFC flux used

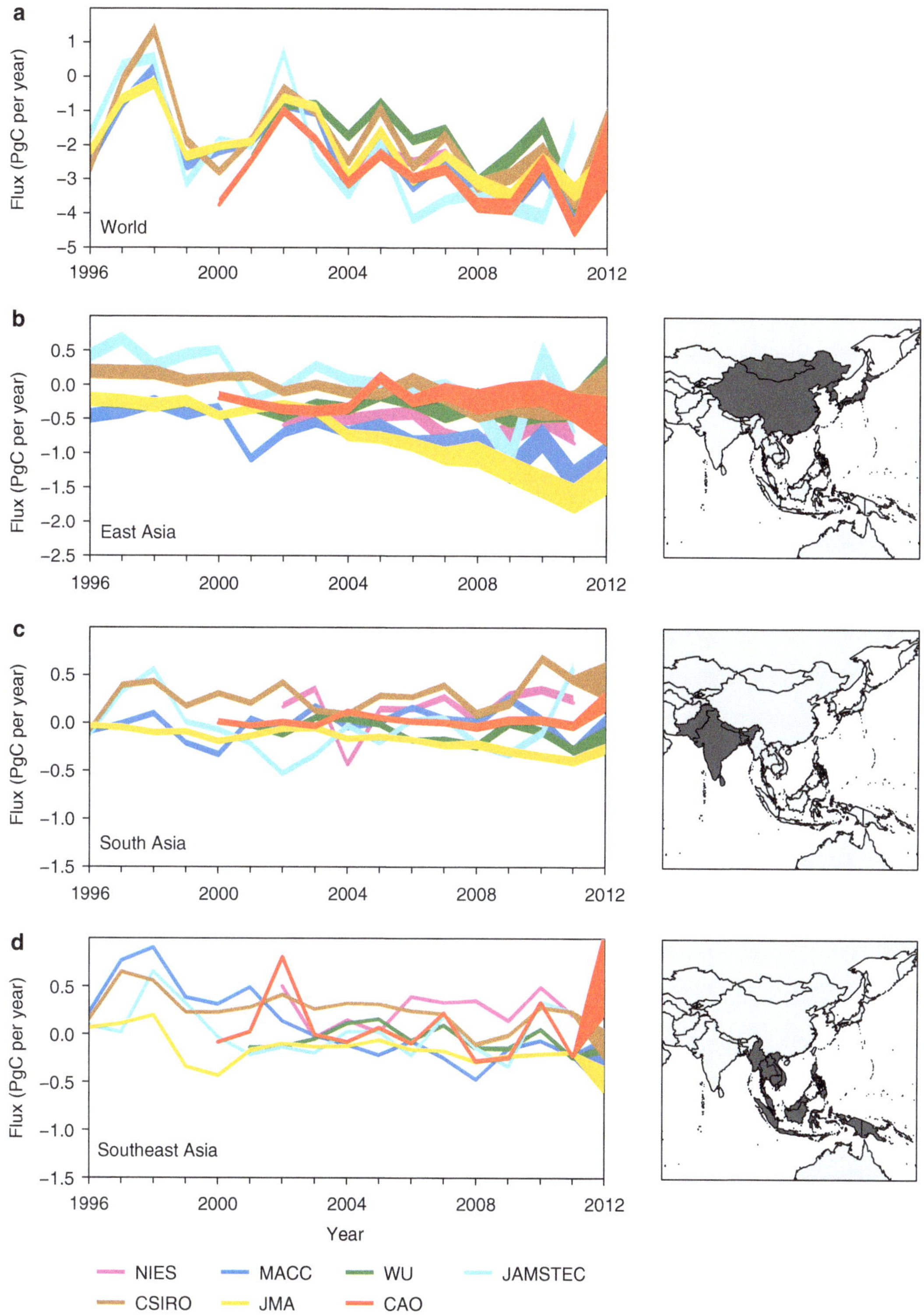

Figure 1 | Land biosphere fluxes estimated by the atmospheric CO_2 inversions. The posterior land biosphere (natural, land-use change and fire) fluxes (PgC per year) from each inversion. The width of each curve shows the range obtained by subtracting the FFC emissions estimates of the CDIAC, EDGAR and IEA inventories from the total posterior flux. Note that a negative flux indicates a carbon sink. The fluxes are shown globally (**a**) and for the regions: East Asia (**b**), South Asia (**c**) and Southeast Asia (**d**) as indicated in the adjacent maps. The inversions are named after the institute that performed them (NIES is the GELCA 64-region inversion system, MACC is the MACC-v13 inversion system used at LSCE, WU is CarbonTracker Europe, JAMSTEC is the ACTM-based inversion system, CSIRO is the CCAM-based inversion system, JMA is the JMA-CDTM-based inversion system, and CAO is the GELCA EOF-based inversion system, for details see Supplementary Table 1).

in each inversion, and the standard FFC flux (see Supplementary Fig. 2). An additional source of uncertainty comes from the choice of standard FFC flux that is subtracted; an overestimate of the standard FFC flux would lead to an overestimate of the land biosphere sink and vice versa. Commonly used FFC fluxes from inventories, i.e., CDIAC, EDGAR and IEA (International Energy Agency, 2014 edition; http://www.iea.org), differ substantially in the trend for East Asia from circa 2002 resulting in differences of ~0.5 PgC per year (~17%) in 2012, much higher than the uncertainty of the inventories of 8% (2σ) reported in ref. 22. (Note that IEA inventory does not include cement emissions so these were accounted for by using the cement emission estimates from CDIAC). Large inconsistencies have been previously reported in Chinese energy-use statistics and emission factors, leading to differences between the reported emissions of 14 to 18% in recent years[23,24], which highlights the considerable systematic uncertainty in the FFC fluxes, especially for East Asia.

Figure 1 shows the posterior (optimized using atmospheric CO_2 observations) land biosphere CO_2 fluxes from each of the inversions (see also Table 1). Note that the land biosphere fluxes include natural, land-use change, and fire fluxes. We show the range of posterior land biosphere fluxes for each inversion calculated using three standard FFC fluxes from the most commonly used inventories: CDIAC, EDGAR and IEA (see Supplementary Note 2). The uncertainty in the posterior land biosphere fluxes increases over time, especially in East Asia, owing to disparities in the FFC fluxes. For the analyses in the following sections, we use an ensemble of seven inversions with three different standard FFC fluxes.

Global land biosphere and ocean fluxes. Before discussing the Asian land biosphere fluxes, we first consider the global total as this provides a baseline for evaluating the inversion performance. The global posterior land fluxes from all inversions exhibit a high degree of consistency in the inter-annual variability and long-term trend (Fig. 1). All inversions show a reduction in the land sink (compared with the decadal mean) in 1997–1998 corresponding to a strong El Niño event (four inversions available) and in 2002–2003 corresponding to persistent El Niño conditions (7 inversions available). El Niño is associated with reduced precipitation, lower rates of net primary productivity (NPP), and increased fire disturbance in tropical and subtropical South America and Africa, Southeast Asia and Australia[25–27]. Six of the inversions also find an increased land sink in 2011, corresponding to a sustained La Niña period, which has been attributed to increased precipitation and net ecosystem uptake in semi-arid regions of South America, South Africa and Australia[28]. The 2011 land sink of -3.6 ± 0.5 (-4.6 to -1.1) PgC per year (median $\pm$ median absolute deviation, and full range of the ensemble) is the largest over 1996–2012 and is in good agreement with a recent process-based model estimate[28] with a land sink of -3.9 ± 1.3 PgC per year, which is greater than the 97.5th percentile over 1981–2012.

Over the period 1996–2007, we find a global land sink of -1.7 ± 0.3 (-2.5 to -1.2) PgC per year consistent with two other estimates, based on different methods, for the same time period. The first is from the Global Carbon Project (GCP) and is calculated as the residual between the FFC emissions and the change in atmospheric CO_2 plus the ocean flux based on ocean data[29]. GCP estimates a land sink (that is, the balance of land-use change and natural fluxes) of -1.00 ± 0.64 PgC per year. This, however, does not include the riverine 'loop' of the natural carbon cycle that existed before human perturbation and which is included in the inversion estimates. The riverine loop consists of a land biosphere uptake of CO_2 of 0.45 PgC per year, which is transported via rivers to the ocean where it is outgassed to the atmosphere[30]. Adding the land biosphere flux associated with the riverine loop to the GCP estimate gives a land sink of -1.45 ± 0.64 PgC per year, which is comparable to our estimate. The second estimate of the global land sink, based on the changes in atmospheric O_2 and CO_2 concentrations[31], is -1.1 ± 0.8 PgC per year, and again is not significantly different from our estimate owing to the large uncertainty. Furthermore, we find that the annual land biosphere sink increased by 1.7 ± 0.2 (1.4–2.0) PgC between 1996–2001 and 2008–2012 (based on the four inversions with data before 2000). In most inversions, the sink increased in tropical South America, eastern North America, Siberia and East Asia and, in three inversions, also in Europe. A trend of increasing land sink over the past circa 15 years has also been identified in previous studies[32,33]. At least part of the observed trend may be attributable to the change in occurrence frequency of El Niño and La Niña events between 1996–2004 and 2005–2012: in the second period there were less frequent and intense El Niño events and a sustained La Niña event. To ascertain whether the land sink is increasing also in response to land management changes, CO_2 and nitrogen fertilization, one would need a longer time series to filter-out climate oscillation effects, such as ENSO, and the influence of volcanic aerosols[34]. Process-based ecosystem models and forest inventories suggest that CO_2 fertilization, nitrogen deposition and forest regrowth are largely driving a long-term increase in the land biosphere sink[33].

The mean global ocean sink over 1996–2012 from all inversions is -1.6 ± 0.3 (-2.1 to -1.2) PgC per year and is not statistically different from the estimate from a recent synthesis study[35] for 1990–2009 of -1.55 ± 0.6 PgC per year (note this includes a correction of 0.45 PgC per year for the riverine transport of carbon from the land to the ocean[30], where it is eventually released to the atmosphere). We also find a small trend of increasing ocean sink (Mann Kendall test P value <0.1) with a mean annual increase of 0.01 PgC per year and is within the

Table 1 | Fossil fuel and cement emissions and land and ocean fluxes of CO_2.

Period	Air-borne fraction	Global FFC	Global Land	Global Ocean	East Asia FFC	East Asia Land	South Asia FFC	South Asia Land	Southeast Asia FFC	Southeast Asia Land
1996-2001	0.52 *0.54* (0.48-0.54)	6.69 (6.64-6.83)	−1.58 (−3.11 to −1.13)	−1.70 (−2.45 to −1.29)	1.48 (1.35-1.65)	−0.06 (−0.55 to 0.44)	0.35 (0.28-0.46)	−0.06 (−0.10 to 0.12)	0.22 (0.17-0.28)	0.26 (−0.09 to 0.53)
2002-2007	0.56 *0.57* (0.53-0.60)	7.86 (7.55-7.94)	−1.94 (−2.51 to −1.09)	−1.57 (−2.13 to −0.65)	2.03 (1.59-2.31)	−0.33 (−0.77 to 0.13)	0.45 (0.33-0.59)	−0.09 (−0.18 to 0.08)	0.27 (0.22-0.37)	0.02 (−0.12 to 0.31)
2008-2012	0.46 *0.46* (0.44-0.49)	9.12 (8.87-9.24)	−3.17 (−3.65 to −2.28)	−1.77 (−2.36 to −1.31)	2.85 (1.85-3.01)	−0.46 (−1.18 to −0.01)	0.64 (0.39-0.83)	−0.07 (−0.33 to 0.04)	0.34 (0.26-0.45)	−0.03 (−0.25 to 0.32)

FFC, fossil fuel and cement.
The median (and range) of the FFC emissions, land biosphere and ocean flux estimates from the inversions (PgC per year) are given. Note that negative numbers indicate a flux from the atmosphere to the land/ocean. The land biosphere fluxes (natural, land-use change and fire) are from the inversions and using the CDIAC, EDGAR, and IEA FFC emission estimates. For East Asia, JMA was excluded for the period 2008-2012. For South Asia, NIES, CSIRO and CAO were excluded. Also shown is the median (and range) of the global air-borne fraction of the FFC emissions from the inversions and calculated directly from atmospheric observations (in italics; data from the Global Carbon Project).

range of trends found by the synthesis study[35] for the past two decades of 0.01–0.05 PgC per year. However, the ocean sink is somewhat dependent on the prior ocean flux estimates used in the inversions, and thus the two estimates are not completely independent.

Assessment of the Asian fluxes. While the global fluxes from inversions are mass conserving, this is not the case on regional scales. Therefore, we evaluate the total posterior flux from each inversion by comparing atmospheric concentrations from forward simulations with independent observations (that is, not assimilated in the inversion) from CONTRAIL aircraft measurements[14]. The CONTRAIL measurements were made using an *in situ* analyser onboard commercial aircraft with regular flights from Japan to other Asian countries (for further details see Supplementary Note 3). We selected CONTRAIL data at seven locations where vertical profiles are available, that is, above the international airports at Bangkok, Singapore, Jakarta and Manila (used to evaluate the Southeast Asian fluxes), Seoul and Taipei (East Asian) and Delhi (South Asian) for 2008–2010. The models were sampled at the times and locations of the observations (for JMA and CSIRO only monthly mean model values were available).

To determine the importance of systematic errors in the inverse models for the regional flux estimates, we compared the annual mean posterior flux with the modelled and observed afternoon CO_2 vertical gradients and planetary boundary layer (PBL) concentrations (Fig. 2). The vertical gradient was calculated between 1 and 4 km, that is, from the PBL to the free troposphere. As well as the annual mean, the comparison was made for winter (December to February, DJF) and summer (June to August, JJA) vertical gradients and PBL concentrations (Supplementary Fig. 3). Systematic over-/underestimation of the CO_2 vertical gradient compared with that observed can indicate a bias in the posterior fluxes and may be due to too weak/strong vertical transport[36].

For South and Southeast Asia, the annual vertical CO_2 gradient and PBL concentration differences are positively correlated with the posterior fluxes, such that large modelled gradients and PBL concentrations correspond to large fluxes, and vice versa. This suggests that the model-observation differences are related to errors in the fluxes rather than to systematic biases in the vertical mixing between the PBL and free troposphere. In summer and winter, the model-observation differences for South Asia are much larger than for the other regions and may owe to the fact that the comparison is based on only one site, Delhi, which is continental and strongly influenced by local fluxes. For East Asia, there is no correlation between the annual (and winter) vertical gradients and the posterior fluxes, and the modelled gradients and PBL concentrations are close to those observed, suggesting that there is no significant bias in the annual mean (and winter) fluxes. In summer, however, the correlation is negative and the s.d. of the modelled and observed gradients is larger. This may result from the stronger influence from the southeasterly direction during this season, and hence the vertical gradients and PBL concentrations represent a combination of land and ocean influences.

Since the summer and winter model-observation differences are generally in the same direction for each model, that is, the summer and winter errors do not compensate, we use the annual mean model-observation differences to evaluate the posterior fluxes from each inversion. The annual means also contain the largest number of data points and are, therefore, more robust. For East and Southeast Asia, we do not find any strong bias in the gradients or PBL concentrations. Therefore, on this basis we do not omit any of the inversions from the ensemble for these two regions. However, for South Asia, the inversions NIES, CSIRO and CAO show large positive biases for the vertical gradient and PBL concentrations (>2 p.p.m.), therefore, we have omitted these results from the ensemble for this region.

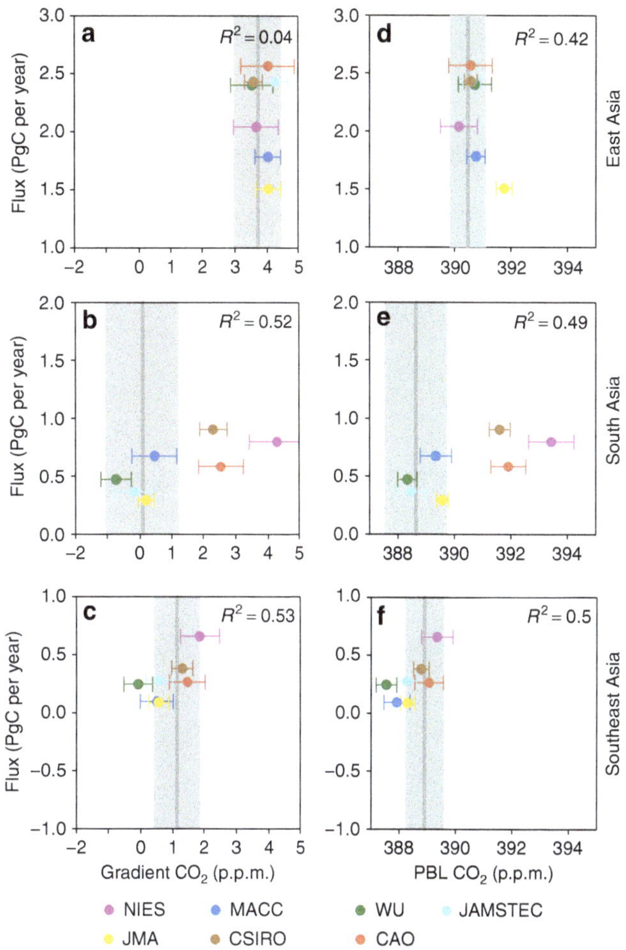

Figure 2 | Posterior total fluxes versus CO_2 vertical gradients and PBL concentrations for 2008-2010. The annual mean flux (PgC per year) versus CO_2 gradient (p.p.m.) between 1 and 4 km for each inversion is shown for East Asia (**a**), South Asia (**b**) and Southeast Asia (**c**). Also shown is the annual mean flux versus the PBL CO_2 concentration (p.p.m.) for East Asia (**d**), South Asia (**e**) and Southeast Asia (**f**). Each of the inversions was sampled at the times and locations of the CONTRAIL aircraft data and the error bars show the 1-σ s.d. of the inversion results. The grey vertical lines indicate the values of the observed gradients and PBL concentrations and the shading shows the 1-σ s.d. The correlation (R^2) between the fluxes and the gradients and PBL concentrations is given in each plot.

East Asia. We find that the East Asian land biosphere was on average a sink of −0.29 (−0.81 to 0.18) PgC per year for 1996–2012. Some of the differences among the inversion results for East Asia can be related to differences in the prior information (Supplementary Figs 1, 2 and 4). The JMA inversion used FFC emission estimates with a fixed spatial distribution based on that in 1995 and only scaled the global total. Because the growth rate in East Asia for the 2000s was higher than the global mean, this resulted in lower FFC and total fluxes in the JMA inversion for East Asia; the total posterior flux in JMA for 2008-2012 was 36% lower than the mean of the other inversions. Thus, subtracting the standard FFC flux from the JMA posterior total results in a more negative land biosphere flux compared with the other inversions. As CONTRAIL data were not available for the latter part of this period, when the bias is largest, this could not be seen in the comparison with independent atmospheric observations. In any case, to avoid potentially biasing the results, we exclude the JMA inversion from the ensemble statistics for East Asia for 2008–2012.

The three inversions with data before 2001 (and excluding JMA), all indicate an increasing land sink (that is, the flux became more negative) over time (Mann Kendall test P value <0.01). From the ensemble, we estimate that the land sink increased from -0.06 (-0.55 to 0.44) PgC per year in 1996–2001 to -0.46 (-1.18 to -0.01) PgC per year in 2008–2012; with a median difference of 0.56 (0.30–0.81) PgC per year (Fig. 3). The three inversions (MACC, JAMSTEC and CSIRO) also show a small increase in the amplitude of the seasonal cycle in the East Asian land biosphere flux in 2008–2012 relative to 1996–2001 (Supplementary Fig. 5), while there was no trend in the amplitude of the seasonal cycle in the FFC fluxes used in the inversions.

Our land sink estimate for 2001–2005, of -0.33 (-0.77 to 0.10) PgC per year, is commensurate with a study using an ecosystem model[37], which found an average sink for China of approximately -0.27 PgC per year for the same time period. Our results are also consistent with a recent study[38] based on a combination of satellite measurements of greenness, process-based ecosystem models, and atmospheric inversions, which found a sink for East Asia of -0.22 to -0.29 PgC per year for 1990–2009, compared with our estimate of -0.25 (-0.74 to 0.18) PgC per year for 1996–2009. The increase in the land sink is supported by forest inventory data, which show an increase in the forest sink in China of ~34% between 1990–1999 and 2000–2007 owing to intensive national afforestation/reforestation programmes[8]. To investigate this further, we examine the Normalized Difference Vegetation Index (NDVI; from the Advanced Very High Resolution Radiometer from NOAA satellites, see Supplementary Note 4) and an independent estimate of Net Ecosystem Exchange (NEE) based on an up-scaling model of CO_2 eddy covariance data[39]. Figure 4 shows the median and range of the East Asian land biosphere flux given by the six inversions (excluding JMA) for the three standard FFC fluxes, as well as NDVI and the eddy-covariance-based NEE anomaly over East Asia. Although NDVI showed no significant trend, NEE became more negative (that is, a larger sink) over the same period by ~0.2 PgC. Thus, a sink increase is supported by

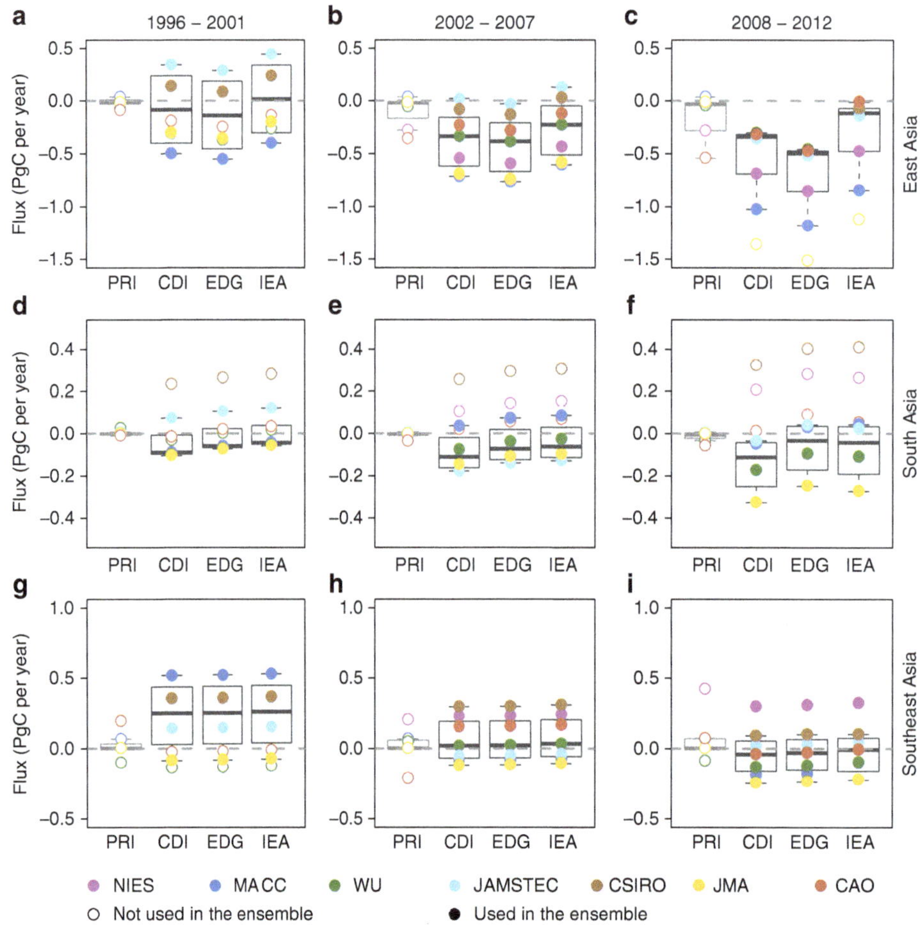

Figure 3 | Summary of the land biosphere fluxes based on the inversion ensemble. Prior and posterior land biosphere fluxes (PgC per year) for intervals 1996-2001 (**a**,**d**,**g**), 2002-2007 (**b**,**e**,**h**) and 2008-2012 (**c**,**f**,**i**). The land biosphere fluxes are calculated by subtracting CDIAC (CDI), EDGAR (EDG) and IEA FFC emission estimates, respectively, from the total posterior fluxes. The thick horizontal bar within the box shows the median, the box extent shows the upper and lower quartiles, and the whiskers extend to the furthest point within the range of the median plus/minus the interquartile range. The open circles indicate the prior (PRI) and the solid circles the posterior estimate from the individual inversions. For East Asia, JMA was excluded for 2008-2012, while for South Asia, NIES, CSIRO and CAO were excluded from the ensembles and are also shown as open circles. Inversions that only started in 2000 or 2001 were also excluded from the 1996-2001 ensembles. The grey dashed line indicates the level of a neutral land biosphere (that is, zero flux).

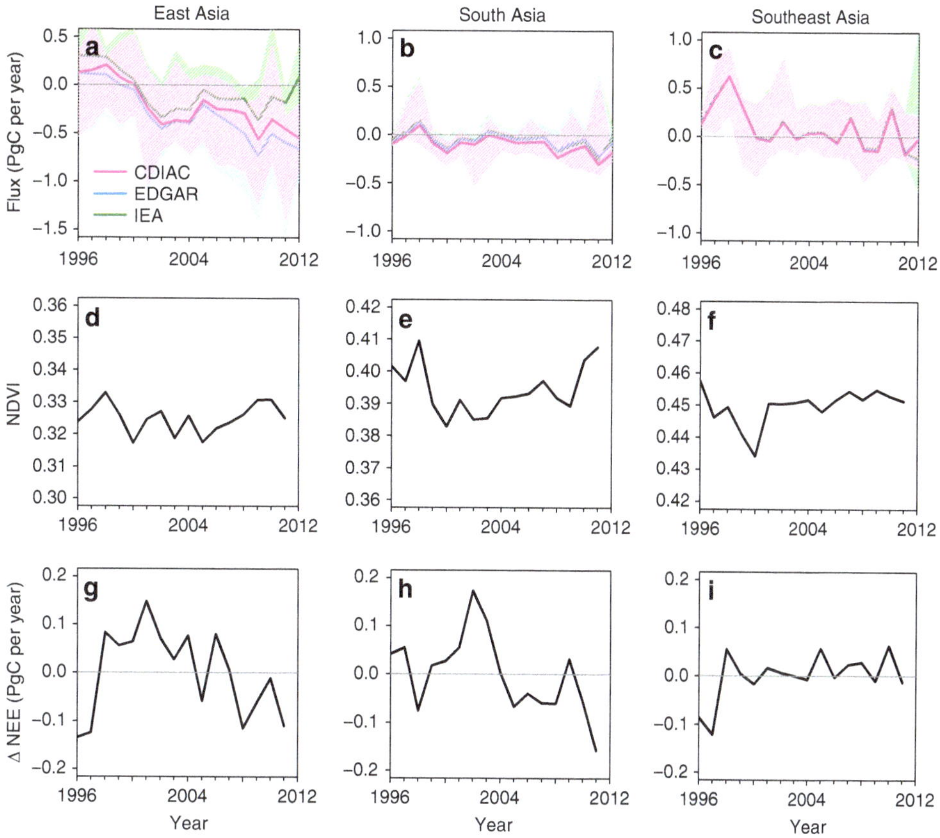

Figure 4 | Trends in the posterior land biosphere fluxes. The land biosphere fluxes (PgC per year), NDVI and eddy-covariance-based NEE anomalies (PgC per year) are shown as annual means for East Asia (**a**,**d**,**g**), South Asia (**b**,**e**,**h**) and Southeast Asia (**c**,**f**,**i**). The fluxes are calculated using FFC emission estimates from CDIAC (purple), EDGAR (blue) and IEA (green), the solid line shows the median and the shading indicates the range of the inversion ensemble. For East Asia, JMA was excluded, while for South Asia, NIES, CSIRO and CAO were excluded. (Negative NEE indicates a carbon sink.)

independent data and, in addition to afforestation/reforestation, may be driven by regional climate change; the temperature in East Asia increased by 0.4 °C between 1999 and 2009 (ref. 38).

The increase in the East Asian sink from the inversion ensemble is larger than suggested by either the eddy-covariance-based NEE or the forest inventory estimates. The larger sink increase in our estimates may be partly due to the uncertainty in the FFC fluxes. For example, using FFC fluxes with a low growth rate, such as IEA, the sink increase between 1996–2001 and 2008–2012 is 0.45 (0.30–0.59) PgC, while using a high growth rate estimate, such as EDGAR, it is 0.63 (0.56–0.81) PgC (Fig. 4). In summary, while atmospheric observations support an increase in the land biosphere sink in East Asia, the amount by which it has increased cannot be accurately determined until the uncertainty in FFC emissions is substantially reduced.

South Asia. From the inversion ensemble, we find that the land biosphere in South Asia was close to carbon neutral, − 0.05 (− 0.18 to 0.03) PgC per year for 1996–2012. Similarly to East Asia, some differences between the inversion results are apparent owing to the FFC flux estimates. The JMA inversion used an FFC flux with a low growth rate (lower than the other inversions and the three standard FFC fluxes) and correspondingly this inversion shows the largest increase in land biosphere sink. On the other hand, the CSIRO inversion used a high FFC flux and this inversion also shows the most positive land biosphere flux.

While we find that the South Asian land biosphere was nearly carbon neutral, previous inversion studies[16–18] found it to be a carbon sink of − 0.10 to − 0.37 PgC per year between 2002 and 2008. However, for the overlapping period of these studies, that is, 2007–2008, our estimate of − 0.13 (− 0.24–0.07) PgC per year is not significantly different. We also compare our results with a study based on a synthesis of 10 ecosystem models[40] for the period 2000–2009. We find a smaller sink of − 0.11 (− 0.19 to 0.06) PgC per year compared with the ecosystem model estimate of − 0.19 ± 0.19 PgC per year.

Of the inversions with data before 2001, only one (JMA) show a significant trend in the land sink for South Asia and this may be due to the low growth rate in its FFC fluxes (Supplementary Fig. 2). Furthermore, we found no change in the amplitude of the seasonal cycle in land biosphere flux over the 17-year period, consistent with the findings of ecosystem models[40]. The overall absence of a trend is in contrast to the decrease in the eddy-covariance-based NEE estimate since 2002 (Fig. 4). Aside from possible errors in the eddy-covariance-based NEE (the discussion of which is outside the scope of our study) a possible explanation for the discrepancy is the poor atmospheric observation coverage for South Asia, and thus modest uncertainty reduction, 27% (9–37%, Supplementary Fig. 6), which may mean that the trend could not be detected. Alternatively, if the trend in FFC emissions were underestimated, this would also lead to an underestimation of the increase in the land sink from the inversions.

Southeast Asia. For 1996–2001, three of the four inversions with data throughout this time period found the Southeast Asian land biosphere to be a source, and the ensemble median and range was 0.26 (− 0.09 to 0.53) PgC per year. This period included the strong El Niño event of 1997–1998, which was associated with drought and lower rates of NPP, as well as increased fire disturbance[27,41]. For 2002–2012, however, we could not determine any difference from carbon neutrality for the land biosphere, −0.02 (−0.18 to 0.27) PgC per year. Two inversions (MACC and CSIRO) show a significantly increasing land sink in Southeast Asia over 1996–2012 (Mann Kendall test P value < 0.1), which mostly results from the large positive land biosphere flux (source) in 1997–1998 related to strong El Niño conditions. Only one inversion (NIES) estimates a significant source for 2002–2012. This can be explained by the use of a positive prior land flux, the influence of which can be seen in the posterior land fluxes.

Previous inversion studies have given contradicting results for Southeast Asia, that is, some estimate a source[17] and others a sink[18]. For the average over 1996–2012, we find a weak source for the Southeast Asian land biosphere of 0.06 (−0.15 to 0.28) PgC per year, which is consistent with observed land-use changes (13% of forested area was lost in Southeast Asia between 1990 and 2010 (ref. 6)) and with estimates of above ground biomass loss of 0.09 PgC per year for the average over 1993–2012 based on satellite observations[42]. Our results, however, suggest that the Southeast Asian source reduced in 2002–2012 compared with 1996–2001, and all of the inversions with the data before 2001, show lower mean fluxes over the latter period. Furthermore, since 2001 the eddy-covariance-based NEE has been close to zero, while NDVI has been stable, supporting our finding of a near-neutral land biosphere for 2002–2012. The change to a neutral land biosphere may reflect the slowing rate of deforestation in Southeast Asia; according to FAO statistics the rate of deforestation slowed to 1.1 Mha per year between 2000 and 2010 compared with 2.7 Mha per year between 1990 and 2000 (ref. 6).

Discussion

We have used an ensemble of seven atmospheric CO_2 inversions and three global FFC flux estimates, based on the inventories of CDIAC, EDGAR and IEA, to infer the land biosphere fluxes for East, South and Southeast Asia, and to ascertain the robustness and overall uncertainty of the results. Asia as a whole emitted an average of 2.7–2.9 PgC per year over 1996–2012 from fossil fuel combustion and cement production with an average increase of ~5% per year. Of this, an average of 17% (0–26%) of the emitted carbon was sequestered by the land biosphere, with most of the sink located in East Asia (Supplementary Fig. 7). This is smaller than the mean fraction, 28% (20–38%) of global fossil fuel emissions sequestered by the global land biosphere.

The East Asian land biosphere was on average a carbon sink of −0.29 (−0.81 to 0.18) PgC per year or equivalently 14% (0–39%) of East Asia's total FFC emissions, over 1996–2012. Between 1996–2001 and 2008–2012, we find an increase in the annual sink of 0.56 (0.30–0.81) PgC, which explains 35% of the increase in the global land sink. However, the magnitude of this increase in the East Asian sink is contingent on the assumed increase in FFC emissions (the inventories differ by up to 17% for East Asia). The uncertainty in the growth rate of the FFC emissions in the inventories contributes 32% to the uncertainty in the land sink change; if China's emissions have grown commensurate with the low rate estimate of for example, IEA, then the land sink increase will be at the lower end of the range, and vice versa if they have followed the high rate estimate of for example, EDGAR. This highlights the need for better regional estimates of FFC emissions and trends for determining land biosphere fluxes from inverse systems. For South Asia, we find that on average the land biosphere was close to carbon neutral, −0.05 (−0.18 to 0.03) PgC per year over 1996–2012 and that there was no significant trend. During this time period, there was only a modest net increase, ~2 Mha, in forested area as afforestation/reforestation in India was nearly counter balanced by deforestation in Nepal and Pakistan[6]. For Southeast Asia, we find that the land biosphere was on average a carbon source between 1996 and 2001 of 0.26 (−0.09 to 0.53) PgC per year but between 2002 and 2012, we cannot determine any difference from carbon neutrality with a flux of −0.02 (−0.18 to 0.27) PgC per year.

Our results contrast with the general findings of land surface models[33], which allocate the increase in the global land sink to tropical and southern regions with little change to even a decrease in the temperate land sink over the past two decades. This disparity is likely owing to land-use changes; for instance, the regrowth of forests is not accounted for in land surface models but probably contributes substantially to the increasing land sink in East Asia[8]. While in tropical Asia the land biosphere models likely overestimate the increase in the sink, again by not accounting for land-use and land-cover changes, and because few land surface models include a coupled carbon–nitrogen cycle and as a consequence may have a too strong CO_2 fertilization response.

Methods

Inversion methodology. All of the inversion systems used in this study have been previously described[16,21,43–48] and are summarized in Supplementary Table 1. Atmospheric CO_2 inversions optimize the land/ocean to atmosphere fluxes using spatiotemporal gradients in atmospheric CO_2 concentration from observations. The relationship between the fluxes, $\mathbf{x}$ (the quantity to be optimized), and CO_2 concentration, $\mathbf{y}$, is described by an atmospheric transport model, H, and can be written as:

$$\mathbf{y} = H(\mathbf{x}) + \varepsilon \tag{1}$$

where ε is error associated with model representation (such as scale differences between the model and the observations), transport and measurements. Since there is insufficient information to solve equation (1) for the fluxes, statistical optimization methods are used, which require prior information. Using the Bayesian formalism, the problem can be expressed as the maximization of the probability density function of the fluxes given the prior information and observations and is equivalent to finding the minimum of the cost function:

$$J(\mathbf{x}) = \frac{1}{2}(\mathbf{x}-\mathbf{x}_b)^T\mathbf{B}^{-1}(\mathbf{x}-\mathbf{x}_b) + \frac{1}{2}(\mathbf{y}-H(\mathbf{x}))^T\mathbf{R}^{-1}(\mathbf{y}-H(\mathbf{x})) \tag{2}$$

where $\mathbf{B}$ and $\mathbf{R}$ are the error covariance matrices for the prior fluxes and the observations, respectively. The fluxes that minimize the cost function can be found by solving the first order derivative of equation (2), and requires the gradient of the transport operator. The inversion systems can be grouped into three basic categories according to the way they minimize the cost function: (1) inversions using a variational method and an adjoint model of transport and which resolve the fluxes at the spatial resolution of the transport model (MACC); (2) inversions using a Bayesian synthesis for a number of predefined regions (JMA, JAMSTEC and CSIRO); and (3) inversions using Kalman smoother methods (WU, NIES and CAO). In each framework, the total CO_2 flux is optimized and the posterior land biosphere fluxes are found by subtracting a standard FFC emissions estimate from the total. For access to the inversion and/or processing code, a request should be sent to the authors.

Prior information and observations. For the prior land biosphere fluxes, JAMSTEC, CSIRO and JMA used estimates based on the ecosystem model, CASA[49] while WU used the SiBCASA model[50], MACC used the ORCHIDEE model[51] and NIES and CAO used the VISIT model[52]. For the prior ocean fluxes, JAMSTEC, CSIRO, JMA and MACC used a pCO_2 observation-based estimate[53], WU used ocean fluxes estimated from an ocean-data inversion[30], while NIES and CAO used fluxes from the OTTM ocean-data inversion model[54]. The observation dataset used also differed among the inversions (Supplementary Fig. 8). NIES, CAO and WU used observations from the NOAA ESRL GMD Observation Package[55] and assimilated values at the time of sampling, or in the case on WU, as daily afternoon averages for sites below 1,000 m and daily morning averages for sites above 1,000 m. CSIRO and JAMSTEC used the GlobalView-CO_2 package[56] at monthly intervals. JMA also used monthly mean observations but used its own set of observations including 69 ground-based sites and 21 ocean-based sites as well as aircraft profiles and ship transects. MACC used 33 *in situ* sites and 106 flask sites and assimilated the data as daily afternoon averages for sites below 1,000 m and daily morning averages for sites above 1000 m.

References

1. Myhre, G. *et al.* in *Working Group I Contribution to the IPCC Fifth Assessment Report Climate Change 2013: The Physical Science Basis* (eds Jacob, D., Ravishankara, A. R. & Shine, K.) (2013).
2. Etheridge, D. M. *et al.* Natural and anthropogenic changes in atmospheric CO_2 over the last 1000 years from air in Antarctic ice and firn. *J. Geophys. Res.Atmos* **101,** 4115–4128 (1996).
3. WMO. The state of greenhouse gases in the atmosphere based on global observations through 2013. *WMO Greenhouse Gas Bull.* **10,** 8 (2014).
4. Ciais, P. *et al.* Current systematic carbon-cycle observations and the need for implementing a policy-relevant carbon observing system. *Biogeosciences* **11,** 3547–3602 (2014).
5. Gregg, J. S., Andres, R. J. & Marland, G. China: Emissions pattern of the world leader in CO_2 emissions from fossil fuel consumption and cement production. *Geophys. Res. Lett.* 35, L08806 (2008).
6. FAO. *Global Forest Resource Assessment* (Food and Agriculture Organization of the United Nations, 2010).
7. Margono, B. A., Potapov, P. V., Turubanova, S., Stolle, F. & Hansen, M. C. Primary forest cover loss in Indonesia over 2000-2012. *Nat. Clim. Change* **4,** 730–735 (2014).
8. Pan, Y. *et al.* A large and persistent carbon sink in the World's forests. *Science* **333,** 988–993 (2011).
9. Piao, S. *et al.* The carbon balance of terrestrial ecosystems in China. *Nature* **458,** 1009–1013 (2009).
10. Zhang, H. F. *et al.* Net terrestrial CO_2 exchange over China during 2001-2010 estimated with an ensemble data assimilation system for atmospheric CO_2. *J. Geophys. Res.Atmos* **119,** 3500–3515 (2014).
11. Enting, I. G. & Mansbridge, J. V. Seasonal sources and sinks of atmospheric CO_2 direct inversion of filtered data. *Tellus B* **41,** 111–126 (1989).
12. Tans, P. P., Conway, T. J. & Nakazawa, T. Latitudinal distribution of the sources and sinks of atmospheric carbon dioxide derived from surface observations and an atmospheric transport model. *J. Geophys. Res.Atmos* **94,** 5151–5172 (1989).
13. Brenninkmeijer, C. A. M. *et al.* Civil Aircraft for the regular investigation of the atmosphere based on an instrumented container: The new CARIBIC system. *Atmos. Chem. Phys.* **7,** 4953–4976 (2007).
14. Machida, T. *et al.* Worldwide measurements of atmospheric CO_2 and Other trace gas species using commercial airlines. *J. Atmos. Ocean. Technol.* **25,** 1744–1754 (2008).
15. Matsueda, H. *et al.* Evaluation of atmospheric CO_2 measurements from new flask air sampling of JAL airliner observations. *Pap. Meteor. Geophys.* **59,** 1–17 (2008).
16. Patra, P. K. *et al.* Carbon balance of South Asia constrained by passenger aircraft CO_2 measurements. *Atmos. Chem. Phys.* **11,** 4163–4175 (2011).
17. Niwa, Y. *et al.* Imposing strong constraints on tropical terrestrial CO_2 fluxes using passenger aircraft based measurements. *J. Geophys. Res. Atmos* **117,** D11303 (2012).
18. Jiang, F. *et al.* Carbon balance of China constrained by CONTRAIL aircraft CO_2 measurements. *Atmos. Chem. Phys.* **14,** 10133–10144 (2014).
19. Zhang, H. F. *et al.* Estimating Asian terrestrial carbon fluxes from CONTRAIL aircraft and surface CO_2 observations for the period 2006-2010. *Atmos. Chem. Phys.* **14,** 5807–5824 (2014).
20. Canadell, J. G. *et al.* An international effort to quantify regional carbon fluxes. *EOS Trans. Am. Geophys. Union* 92, 81–82, 2011).
21. Peylin, P. *et al.* Global atmospheric carbon budget: results from an ensemble of atmospheric CO_2 inversions. *Biogeosciences* **10,** 6699–6720 (2013).
22. Andres, R. J., Boden, T. A. & Higdon, D. A new evaluation of the uncertainty associated with CDIAC estimates of fossil fuel carbon dioxide emission. *Tellus B* **66,** 23616 (2014).
23. Guan, D., Liu, Z., Geng, Y., Lindner, S. & Hubacek, K. The gigatonne gap in China's carbon dioxide inventories. *Nat. Clim. Change* **2,** 672–675 (2012).
24. Liu, Z. *et al.* Reduced carbon emission estimates from fossil fuel combustion and cement production in China. *Nature* **524,** 335–338 (2015).
25. Myneni, R. B., Los, S. O. & Tucker, C. J. Satellite-based identification of linked vegetation index and sea surface temperature anomaly areas from 1982-1990 for Africa, Australia and South America. *Geophys. Res. Lett.* **23,** 729–732 (1996).
26. Woodward, F. I., Lomas, M. R. & Quaife, T. Global responses of terrestrial productivity to contemporary climatic oscillations. *Philos. Trans. R. Soc. Lond. B Biol. Sci.* **363,** 2779–2785 (2008).
27. Patra, P. K., Maksyutov, S. & Nakazawa, T. Analysis of atmospheric CO_2 growth rates at Mauna Loa using CO_2 fluxes derived from an inverse model. *Tellus B* **57,** 357–365 (2005).
28. Poulter, B. *et al.* Contribution of semi-arid ecosystems to interannual variability of the global carbon cycle. *Nature* **509,** 600–603 (2014).
29. Le Quéré, C. *et al.* Global carbon budget 2013. *Earth Syst. Sci. Data* **6,** 235–263 (2014).
30. Jacobson, A. R., Mikaloff Fletcher, S. E., Gruber, N., Sarmiento, J. L. & Gloor, M. A joint atmosphere-ocean inversion for surface fluxes of carbon dioxide: 1. methods and global-scale fluxes. *Global Biogeochem. Cycles* **21,** GB1019 (2007).
31. Keeling, R. F. & Manning, A. C. *Studies of Recent Changes in Atmospheric O_2 Content in Treatise on Geochemistry* 2nd Edn 385–404 (Elsevier, 2014).
32. Ballantyne, A. P., Alden, C. B., Miller, J. B., Tans, P. P. & White, J. W. C. Increase in observed net carbon dioxide uptake by land and oceans during the past 50 years. *Nature* **488,** 70–72 (2012).
33. Sitch, S. *et al.* Recent trends and drivers of regional sources and sinks of carbon dioxide. *Biogeosciences* **12,** 653–679 (2015).
34. Raupach, M. R., Canadell, J. G. & Le Quéré, C. Anthropogenic and biophysical contributions to increasing atmospheric CO_2 growth rate and airborne fraction. *Biogeosciences* **5,** 1601–1613 (2008).
35. Wanninkhof, R. *et al.* Global ocean carbon uptake: magnitude, variability and trends. *Biogeosciences* **10,** 1983–2000 (2013).
36. Stephens, B. B. *et al.* Weak northern and strong tropical land carbon uptake from vertical profiles of atmospheric CO_2. *Science* **316,** 1732–1735 (2007).
37. Tian, H. *et al.* Net exchanges of CO_2, CH_4, and N_2O between China's terrestrial ecosystems and the atmosphere and their contributions to global climate warming. *J. Geophys. Res.* **116,** G02011 (2011).
38. Piao, S. L. *et al.* The carbon budget of terrestrial ecosystems in East Asia over the last two decades. *Biogeosciences* **9,** 3571–3586 (2012).
39. Jung, M. *et al.* Global patterns of land-atmosphere fluxes of carbon dioxide, latent heat, and sensible heat derived from eddy covariance, satellite, and meteorological observations. *J. Geophys. Res. Biogeosci.* **116,** G00J07 (2011).
40. Patra, P. K. *et al.* The carbon budget of South Asia. *Biogeosciences* **10,** 513–527 (2013).
41. van der Werf, G. R. *et al.* Continental-scale partitioning of fire emissions during the 1997 to 2001 El Niño/La Niña period. *Science* **303,** 73–76 (2004).
42. Liu, Y. Y. *et al.* Recent reversal in loss of global terrestrial biomass. *Nat. Clim. Change* **5,** 470–474 (2015).
43. Chevallier, F. *et al.* Inferring CO_2 sources and sinks from satellite observations: Method and application to TOVS data. *J. Geophys. Res.* **110,** D24309 (2005).
44. Maki, T. *et al.* New technique to analyse global distributions of CO_2 concentrations and fluxes from non-processed observational data. *Tellus B* **62,** 797–809 (2010).
45. Rayner, P. J. *et al.* Interannual variability of the global carbon cycle (1992-2005) inferred by inversion of atmospheric CO_2 and $\delta^{13}CO_2$ measurements. *Global Biogeochem. Cycles* **22,** GB3008 (2008).
46. Peters, W. *et al.* Seven years of recent European net terrestrial carbon dioxide exchange constrained by atmospheric observations. *Global Change Biol.* **16,** 1317–1337 (2010).
47. Bruhwiler, L. M. P., Michalak, A. M., Peters, W., Baker, D. F. & Tans, P. An improved Kalman Smoother for atmospheric inversions. *Atmos. Chem. Phys.* **5,** 2691–2702 (2005).
48. Zhuravlev, R. V., Ganshin, A. V., Maksyutov, S. S., Oshchepkov, S. L. & Khattatov, B. V. Estimation of global CO_2 fluxes using ground-based and satellite (GOSAT) observation data with empirical orthogonal functions. *Atmos. Oceanic Opt.* **26,** 507–516 (2013).
49. Randerson, J. T., Thompson, M. V., Conway, T. J., Fung, I. Y. & Field, C. B. The contribution of terrestrial sources and sinks to trends in the seasonal cycle of atmospheric carbon dioxide. *Global Biogeochem. Cycles* **11,** 535–560 (1997).
50. van der Velde, I. R. *et al.* Terrestrial cycling of $^{13}CO_2$ by photosynthesis, respiration, and biomass burning in SiBCASA. *Biogeosciences* **11,** 6553–6571 (2014).
51. Krinner, G. *et al.* A dynamic global vegetation model for studies of the coupled atmosphere-biosphere system, Global Biogeochem. *Cycles* 19, GB1015 (2005).
52. Saito, M., Ito, A. & Maksyutov, S. Optimization of a prognostic biosphere model for terrestrial biomass and atmospheric CO_2 variability. *Geosci. Model Dev* **7,** 1829–1840 (2014).
53. Takahashi, T. *et al.* Climatological mean and decadal change in surface ocean pCO_2, and net sea-air CO_2 flux over the global oceans. *Deep Sea Res. II* **56,** 554–577 (2009).
54. Valsala, V. & Maksyutov, S. Simulation and assimilation of global ocean pCO_2 and air-sea CO_2 fluxes using ship observations of surface ocean pCO_2 in a simplified biogeochemical offline model. *Tellus B* **62,** 821–840 (2010).
55. Cooperative Global Atmospheric Data Integration Project, updated annually. Multi-laboratory compilation of atmospheric carbon dioxide data for the period 2000–2012 (obspack_co2_1_PROTOTYPE_v1.0.4_2013-11-25), NOAA Global Monitoring Division, Boulder, CO, USA.
56. Cooperative Global Atmospheric Data Integration Project. 2013, updated annually. Multi-laboratory compilation of synchronized and gap-filled atmospheric carbon dioxide records for the period 1979–2012 (obspack_co2_1_GLOBALVIEW-CO2_2013_v1.0.4_2013-12-23). Compiled by NOAA Global Monitoring Division: Boulder, CO, USA.

Acknowledgements
We thank the support of the Asia Pacific Network for Global Change for the grant 'Greenhouse gas budgets of South and Southeast Asia'. J.G.C. thanks the Australian Climate Change Science Program for their support. P.K.P., T.S., M.I. and S.M. acknowledge the support of the Environment Research and Technology Development Fund (2-1401) of the Ministry of the Environment, Japan. I.T.L.L. and W.P. received funding from the China Exchange Program of the Dutch Royal Academy of Research (KNAW) through travel grant no. 12CDP006 and a grant for computing time (SH-060-13) from the Netherlands Organization for Scientific Research (NWO). CSIRO inversions were undertaken on the NCI National Facility, supported by the Australian Commonwealth Government. We also gratefully acknowledge the following people who provided atmospheric observations that were used in the inversions: J. A. Morgui, E. Brunke, S. Hammer, M. Delmotte, M. Leuenberger, H. Meijer, H. Matsueda, E. Gloor, P. Krummel, M. van der Schoot, B. Stephens, M. Takahashi, M. Torn, K. Tsuboi, M. Ramonet, D. Worthy, S. Morimoto, A. J. Gomez-Pelaez, T. Nakazawa, L. V. Gatti, S. Aoki, K. Masarie, A. Andrews, R. Langenfelds, S. Biraud, C. Sweeney, M. Steinbacher, P. Steele, E. Dlugokencky, R. Keeling, E. Scheel, T. Aalto, H. Koide, S. Wofsy, M. Schmidt, L. Haszpra, I. Levin, P. Tans, K. Thoning, J. B. Miller, C. Labuschagne, J. Hatakka, S. Fang, M. Alemanno, M-K. Park, B. Wastine, L. Zhou, A. di Diodato, K. R. Kim, H. Mukai, J. Page, L. Lauria, H. Lee, N. Paramonova, G. Brailsford, L. Ciattaglia, Y. Fukuyama, F. Apadula, A. Lanza, M. Froehlich, E. Friedbacher, K. Uhse, C. Rafanelli, B. Noone, M. Yosuke, T. Conway, S. Piacentino, and T. Mkololo.

Author contributions
R.L.T. performed the data processing, analyses, synthesis of the results and prepared the manuscript; P.K.P assisted in the data processing, contributed inversion results and assisted in the preparing the manuscript; F.C. assisted in the data processing, contributed inversion results and assisted in the preparing the manuscript; S.M. assisted in data processing and in organizing the study; R.M.L contributed inversion results and assisted in the preparing the manuscript; T.Z. contributed inversion results and assisted in the preparing the manuscript; I.T.L.L contributed inversion results and assisted in the preparing the manuscript; W.P. contributed inversion results and assisted in the preparing the manuscript; A.G. contributed inversion results; R.Z. contributed inversion results; T.M. contributed inversion results; T.N. contributed inversion results; T.S. contributed inversion results; M.I. contributed inversion results; T.S. contributed inversion results; T.M. contributed observations for the inversion evaluation, B.P. contributed climate, remote-sensing and NEE data; J.G.C assisted in preparing the manuscript; P.C. assisted in preparing the manuscript.

Additional information

Competing financial interests: The authors declare no competing financial interests.

22

European land CO_2 sink influenced by NAO and East-Atlantic Pattern coupling

Ana Bastos[1,2], Ivan A. Janssens[3], Célia M. Gouveia[2], Ricardo M. Trigo[2], Philippe Ciais[1], Frédéric Chevallier[1], Josep Peñuelas[4,5], Christian Rödenbeck[6], Shilong Piao[7], Pierre Friedlingstein[8] & Steven W. Running[9]

Large-scale climate patterns control variability in the global carbon sink. In Europe, the North-Atlantic Oscillation (NAO) influences vegetation activity, however the East-Atlantic (EA) pattern is known to modulate NAO strength and location. Using observation-driven and modelled data sets, we show that multi-annual variability patterns of European Net Biome Productivity (NBP) are linked to anomalies in heat and water transport controlled by the NAO-EA interplay. Enhanced NBP occurs when NAO and EA are both in negative phase, associated with cool summers with wet soils which enhance photosynthesis. During anti-phase periods, NBP is reduced through distinct impacts of climate anomalies in photosynthesis and respiration. The predominance of anti-phase years in the early 2000s may explain the European-wide reduction of carbon uptake during this period, reported in previous studies. Results show that improving the capability of simulating atmospheric circulation patterns may better constrain regional carbon sink variability in coupled carbon-climate models.

[1] Laboratoire des Sciences du Climat et de l'Environnement, LSCE/IPSL, CEA-CNRS-UVSQ, Université Paris-Saclay, F-91191 Gif-sur-Yvette, France. [2] Instituto Dom Luiz, IDL, Faculdade de Ciências, Universidade de Lisboa, Lisboa 1749-016, Portugal. [3] Department of Biology, University of Antwerp, Universiteitsplein 1, 2610 Wilrijk, Belgium. [4] CREAF, Cerdanyola del Vallès, Catalonia, 08193 Barcelona, Spain. [5] CSIC, Global Ecology Unit CREAF-CSIC-UAB, Cerdanyola del Vallès, Catalonia, 08193 Barcelona, Spain. [6] Max Planck Institute for Biogeochemistry, Jena 07701, Germany. [7] Department of Ecology, College of Urban and Environmental Sciences, Peking University 5 Yiheyuan Road, Haidian District, Beijing 100871, China. [8] College of Engineering, Mathematics and Physical Sciences, University of Exeter, Exeter EX4 4QF, UK. [9] Numerical Terradynamic Simulation Group, University of Montana, Missoula, Montana 59812, USA. Correspondence and requests for materials should be addressed to A.B. (email: ana.bastos@lsce.ipsl.fr).

Fundamental patterns of atmosphere-ocean variability are explained by teleconnections at the global[1] and regional[2–4] scales. At the global scale, the El-Niño/Southern-Oscillation has been shown to influence inter-annual variability in the global land carbon sink[5,6] with an evident fingerprint in atmospheric CO_2 growth rate fluctuations[7], highlighting their impact on ecosystem functioning. Hallet *et al.*[8] proposed that teleconnections explain ecological processes even better than single climate variables, because they influence simultaneously the range of weather variables that elicit interacting, and sometimes opposing, responses by ecosystems[9–12].

European and North-American climate is particularly influenced by the North-Atlantic Oscillation (NAO), which corresponds to a meridional dipole of sea-level pressure (SLP) variability between Iceland and the Azores that controls the location and strength of storm tracks in the North-Atlantic[3] (Fig. 1a). NAO drives patterns and extremes in temperature, precipitation, snow cover and wind in Europe, especially during winter[2,4], although its effects may propagate through the following seasons[13]. NAO is usually associated with warmer and wetter (colder and drier) winters in northern (southern) Europe during positive phase and approximately the reverse pattern in negative phase[2]. Some studies have analysed impacts of NAO on European ecosystems[14,15]. In the Iberian Peninsula, Carnicer *et al.*[16] found an increase in tree defoliation in response to a period of drought (1990–1995) that coincided with the persistence of positive NAO. However, other teleconnections are known to also influence the European climate, particularly the East-Atlantic (EA) pattern that has a similar configuration as NAO, albeit displaced southwards (Fig. 1a). Although EA and NAO indices are computed as independent variables (Fig. 1b), the EA pattern has been proposed to modulate the multi-decadal variability of the location and strength of the NAO dipole[17]. Comas-Bru and McDermott[4] showed that the combined analysis of NAO and EA more appropriately describes winter climate variability in Europe.

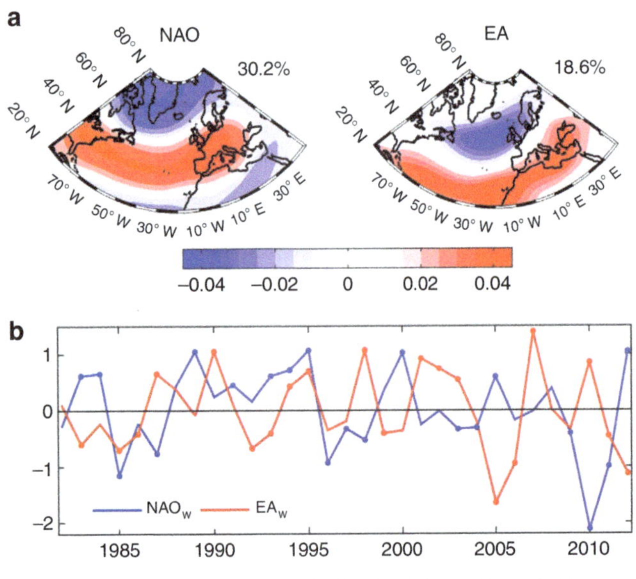

Figure 1 | Main large-scale atmospheric circulation patterns in Europe. (**a**) Spatial patterns of the two first components of 500 mb geopotential height (from NCEP/DOEII Reanalysis) variability over the North-Atlantic: NAO and EA teleconnections. (**b**) Time series of the winter composites of NAO (blue line) and EA (red line) from NOAA; circle markers indicate the positive and negative phases of each mode, defined by the upper and lower tercile thresholds (Supplementary Table 3).

Northern Hemisphere ecosystems are important carbon sinks[18,19] that partially offset anthropogenic CO_2 emissions[20]. The inter-annual variability in the terrestrial carbon sink is poorly understood[21], hindering better future land-sink strength projections[7,22] and making regional trends in sinks difficult to detect and hence controversial[23,24]. Various estimates of European Net Biome Productivity (NBP) have shown that European ecosystems have been a CO_2 sink during the past decades (1980–2005)[25,26], although signs of saturation in temperate forest carbon uptake in the early 2000s have been reported[27]. Moreover, Piao *et al.*[28] have found a weakening relationship between temperature and growing season vegetation greenness in Northern ecosystems, especially in Europe, during 1982–2011, attributable to an increase in drought conditions. Their results highlight the importance of evaluating inter-annual variability of NBP within a framework that takes into account the co-variations of the climate variables that drive ecosystem dynamics.

Here we analyse the combined impact of NAO and EA on NBP over the European area extending to the Ural Mountains (Supplementary Fig. 1), using three state-of-the-art atmospheric CO_2 inversions: two versions of the Monitoring Atmospheric Composition and Climate—Interim Implementation (MACCII) inversion[29,30] and the Jena s81 v3.6 inversion[31] (Supplementary Table 1). Atmospheric inversions are known to provide consistent estimates of inter-annual variability of the land sink on large aggregated regions[19]. The uncertainty of the annual anomalies is expected to be lower than that of the long-term mean, since these errors on annual fluxes are positively correlated from 1 year to the next, and Europe is one of the regions with most robust inter-annual variability estimates[32]. As inversions present increasing uncertainty for smaller scale regions[33] we use a complementary bottom-up estimation of NBP from eleven Dynamic Global Vegetation Models (DGVMs)[20] (Supplementary Table 2). The DGVMs were forced with climate observations and include land-use changes (LUCs), but differ on a number of characteristics, which makes them independent estimations of NBP. These data sets now report more than 30 years of CO_2 flux variability, a timely opportunity to revisit the relationship between carbon fluxes and climate signals. Our results show that the combined NAO–EA variability explains variations in the European CO_2 sink from 1982 to 2012, through their control on heat and water-vapour transport towards the continent and consequent climate anomalies. We find that the European sink is enhanced only when the two modes are in negative phase due to enhanced photosynthesis, while for both anti-phase combinations NBP is kept below average through different responses of photosynthesis and respiration.

Results

Net biome production variability. European NBP from inversions confirms that European ecosystems were on average a net CO_2 sink between 1982 and 2012 (Supplementary Fig. 2), in line with previous estimates[19,25], with strong inter-annual and multi-year variability (Fig. 2a). The late 1980s and early 1990s were characterized by generally positive NBP anomalies, that is, CO_2 uptake above average. A strong sink was observed during the years 1996–1997, followed by a decade of mostly below-average NBP, consistent with the satellite observations of Normalized Difference Vegetation Index (NDVI) data, a proxy for Gross Primary Productivity (GPP)[27]. From 2009 onwards, results from inversions suggest that European ecosystems became a stronger C sink again. The set of DGVMs suggests a weaker and less-variable European sink than the inversions (Supplementary Fig. 2), although the average NBP anomalies from these models

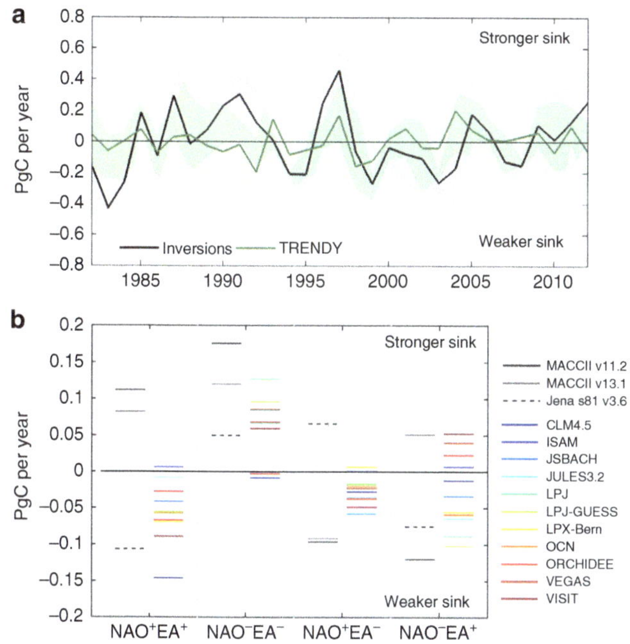

Figure 2 | Impact of NAO-EA variability on NBP. (**a**) Annual NBP anomalies integrated over Europe (bold lines) and corresponding spread, that is, the minimum and maximum NBP estimated by each set (shaded area), for the atmospheric transport model inversions (black) and the DGVMs from TRENDY project (green). (**b**) Average European NBP anomalies for the four NAO-EA composites (Supplementary Table 3), assessed by inversions (greyscale lines) and DGVMs (colour lines).

remain in most years within the uncertainty range of the inversions and also present no significant trend (Fig. 2a).

Despite the lack of long-term NBP trend at continental scale, Supplementary Fig. 3 shows distinct temporal patterns in annual NBP anomalies in the four large regions encompassing most of continental Europe. The continental integral is mainly influenced by the dynamics in central Europe (33% and 34% of variance explained on average for inversions and DGVMs, respectively) and western Russia (29%, inversions). A principal component analysis (PCA) applied on NBP fields from inversions and DGVMs confirms the importance of central Europe and western Russia dynamics, by identifying a predominant variability pattern (explaining about 30% of spatio-temporal variance) with two centres located in these regions (Supplementary Fig. 4). Particularly, anomalies in central Europe dominate the multi-year pattern observed at the continental scale, while in western Russia NBP presents an increasing trend in the first decade followed by a peak in 1996–1997 and a stalling onwards (Supplementary Fig. 3). However, it must be kept in mind that atmospheric inversions do not perform as well at the regional scale as at the continental scale due to the limited, albeit still greater than in most regions, number of observation sites in Europe and to limitations of transport models.

Impact of NAO and EA on NBP. A large spread in NBP response to climate is expected from the natural variability of the climate conditions (for example, NAO^+ years differ in NAO strength and therefore in its impacts), as well as from carryover effects from previous year's climate conditions. For example, warm and moist conditions in 1 year promote organic matter decomposition, thereby increasing nutrient availability with impact on plant growth during subsequent years[12]. Because of this large spread of the NBP response to current year's climate, we here analyse the multi-year average NBP values within the four NAO–EA phase composites (NAO^+EA^+, NAO^-EA^-, NAO^+EA^- and NAO^-EA^+), rather than individual years (Fig. 2b; Supplementary Table 3).

Our analysis reveals that the sole use of NAO does not suffice to explain NBP variability, because its impacts are strongly modulated by the sign of the EA. Inversions and DGVMs agree on significant differences in NBP composite anomalies during NAO^- years (Supplementary Tables 4 and 5), depending on the state of the EA. NBP is enhanced when EA is also negative, and is weaker than average when EA is positive (Fig. 2a; compare NAO^-EA^- with NAO^-EA^+). In fact, both sets of NBP estimates identify NAO^-EA^- as the single combination with consistent estimates of NBP enhancement (except two DGVMs), even for an extended record from 1950 to 2012 (Supplementary Fig. 5). During the other phase combinations, inversions and DGVMs report NBP anomalies of similar magnitude for the two anti-phase states (NAO^+EA^- and NAO^-EA^+), which are significantly lower than for NAO^-EA^-. For NAO^+EA^+, MACCII inversions disagree with DGVMs and Jena inversion on the sign of the anomalies. However, the limited number of years (only four) in this composite highlights the need for caution in the interpretation of NBP anomalies for this combination. It is worth pointing that inversions present different sensitivities to NAO–EA variability: while the sign of NBP anomaly from the two MACCII versions tends to depend on the phase/anti-phase dichotomy, NBP anomalies from the Jena inversion appear to respond more strongly to EA phases.

Mechanism. Having shown the need to consider both NAO and EA when assessing European CO_2 uptake variability, we now propose the main components of the physical mechanisms associated, that is, linking variations in atmospheric circulation and climate variables to the anomalies observed in NBP for each NAO–EA phase combination.

Variability in the land sink is linked to that in the teleconnections through the control of the latter on one or more of the meteorological variables influencing NBP. The four NAO–EA phases are characterized by distinct winter SLP anomaly patterns (December–April (Dec–Apr); Fig. 3a, black contours) and 500 mb geopotential height anomalies (Fig. 3b, black contours) during late winter/early spring. Although NAO clearly controls the sign of the North/South SLP anomaly dipole, with NAO^+ (NAO^-) imposing negative (positive) anomalies in the North and positive (negative) in the south, the EA modulates their strength and configuration, consistent with (ref. 4). During in-phase combinations, a sharp North/South pressure gradient is observed at the surface and in altitude, while during anti-phase combinations anomalies tend to be less intense and present a meandering configuration.

By influencing the strength and shape of the North-Atlantic jet, the interplay between NAO and EA controls the atmospheric transport of heat and humidity over the region (Fig. 3a,b; colours). The stronger pressure gradients during in-phase combinations create a channel of enhanced (weaker) eastward transport of heat and water-vapour spanning from the UK to eastern Europe and including Scandinavia, and reduced (higher) heat and moisture advection over the Mediterranean during NAO^+ (NAO^-). In contrast, during anti-phase periods, the weaker pressure gradients at the surface and in altitude leads to a northward displacement and attenuation of eastward heat flux, and water-vapour transport is close to average values in most of the continental Europe. The impact of these fluxes on cloud cover was evaluated as a means to estimate the combined variability of

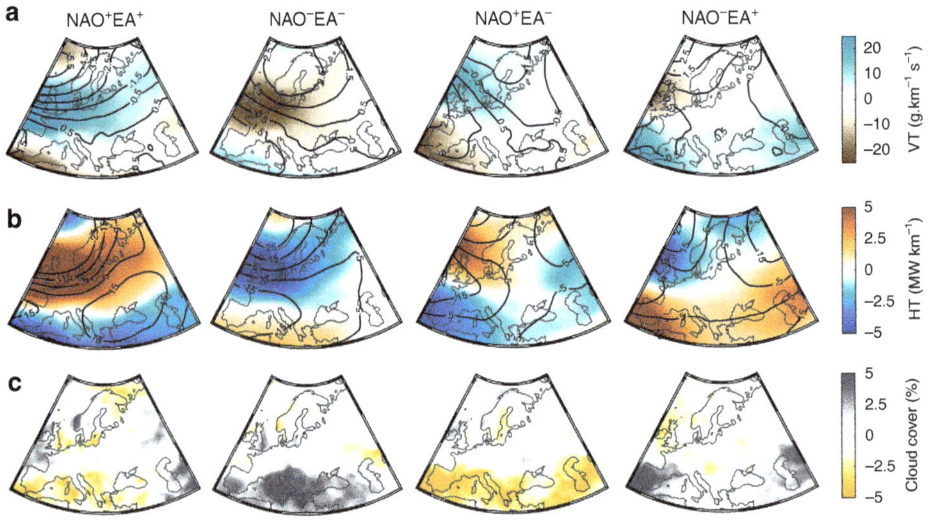

Figure 3 | Impact of NAO-EA variability on atmospheric circulation. Anomaly fields during extended winter (Dec-Apr) for each of the four NAO-EA phases of (**a**) sea-level pressure (black contours, in mb) and eastward water-vapour transport (colours); (**b**) 500-mb geopotential height (black contours, in g.p.m.) and heat transport (colours); (**c**) Cloud cover fraction.

precipitation and sunlight, both of which are relevant (with opposing effects) for vegetation growth during these months[9]. The patterns of winter cloud cover anomalies roughly match the ones of heat and vapour transport, with increased cloudiness in the regions affected by stronger heat and water flux anomalies. It must be noted that although precipitation anomalies are expected to be related to cloud cover, the dependency is not direct, since precipitation also depends on vertical instability of the atmosphere.

The differences in regional advection of heat and humidity transport associated with a given phase impose distinct and spatially heterogeneous climate conditions during late winter/early spring, which may propagate until summer (May–September) due to the memory effects of snow cover[13] and soil moisture[34] (Fig. 4). During NAO^{-}EA^{-}, the inhibition of heat and water-vapour transport in winter establishes very cold conditions over most of Europe, which lead to more precipitation falling as snow rather than rain, explaining the negative soil moisture anomalies and the generally higher-than-average snow depth registered in most areas. In the Mediterranean region, soil moisture is enhanced during December–April due to the stronger advection of water vapour and prevalence of cloudiness (Fig. 3). In contrast to the lower soil moisture during winter/spring, soil water was enhanced during spring/summer in the NAO^{-}EA^{-} composite in most of central Europe (Fig. 4; Supplementary Table 6). This shift in water availability may be related to later snow melt and lower evaporative losses during the colder winter/spring periods. Higher soil moisture levels support stronger latent heat exchange and thus evaporative cooling in most regions, except the Iberian Peninsula, Scandinavia and the southern section of western Russia, in which dry conditions promote higher summer temperatures. For NAO^{+}EA^{+} these patterns are approximately the opposite.

The meandering SLP pattern and the corresponding energy transport during anti-phase periods impose a north-east/south-west gradient (rather than north/south as during in-phase combinations) in temperature anomalies with warmer-than-average (colder) winter/spring conditions over Scandinavia and western Russia and cooler (warmer) temperatures in the western-Mediterranean sector during NAO^{+}EA^{-} (NAO^{-}EA^{+}). The patterns of soil moisture for NAO^{+}EA^{-} (NAO^{-}EA^{+}) reflect the northward displacement and attenuation of water-vapour transport anomalies: most regions register close to average winter soil moisture, with drier (wetter) winter conditions being registered only in western Russia and Iberian Peninsula. Most of central Europe registers lower (higher) levels of snow cover, followed by dry (wet) soil conditions in summer. As for NAO^{-}EA^{-}, the snow-depth winter patterns generally match the ones observed in summer soil moisture and temperature anomalies, although the moisture–temperature coupling is most evident for NAO^{-}EA^{+}. The Iberian peninsula is also affected by very low winter cloud cover during NAO^{+}EA^{-}, which promotes higher evaporation rates, explaining the strong water deficits in winter.

The anomalies in NBP are related to two largely offsetting fluxes, photosynthesis and respiration, which respond differently to climate anomalies[12]. It is worth comparing the continental NBP response to NAO–EA phases with observations of vegetation greenness anomalies assessed by annually integrated NDVI fields (Fig. 5a) from the Global Inventory Modeling and Mapping Studies (GIMMS)[35], as well as the corresponding seasonal evolution (Fig. 5b, black lines). Since DGVMs allow partitioning the NBP anomalies into GPP and respiration responses, the seasonal dynamics of these two components of NBP is also evaluated (Fig. 5b, coloured lines). It must be noted that since the climate anomalies are spatially heterogeneous, and Europe is composed of diverse biomes, these responses must also be assessed at the regional scale (Supplementary Figs 6 and 7, Supplementary Table 4). Regional NBP anomalies from inversions and DGVMs are compared, although inversions present higher uncertainty for smaller regions and in most cases do not present significant results. At the regional scale, DGVMs are expected to better capture ecological variability patterns in response to climate anomalies (see Methods).

Inversions and most DGVMs consistently estimate enhanced NBP during NAO^{-}EA^{-} years in all regions, except in the Iberian Peninsula, where inversions and DGVMs diverge. The clear positive NBP anomaly during NAO^{-}EA^{-} cannot be explained solely by positive NDVI anomalies, since large areas in Europe present NDVI anomalies of small magnitude, and in many regions even slight browning. On the seasonal scale, the cold and dry winter and spring conditions affecting most of

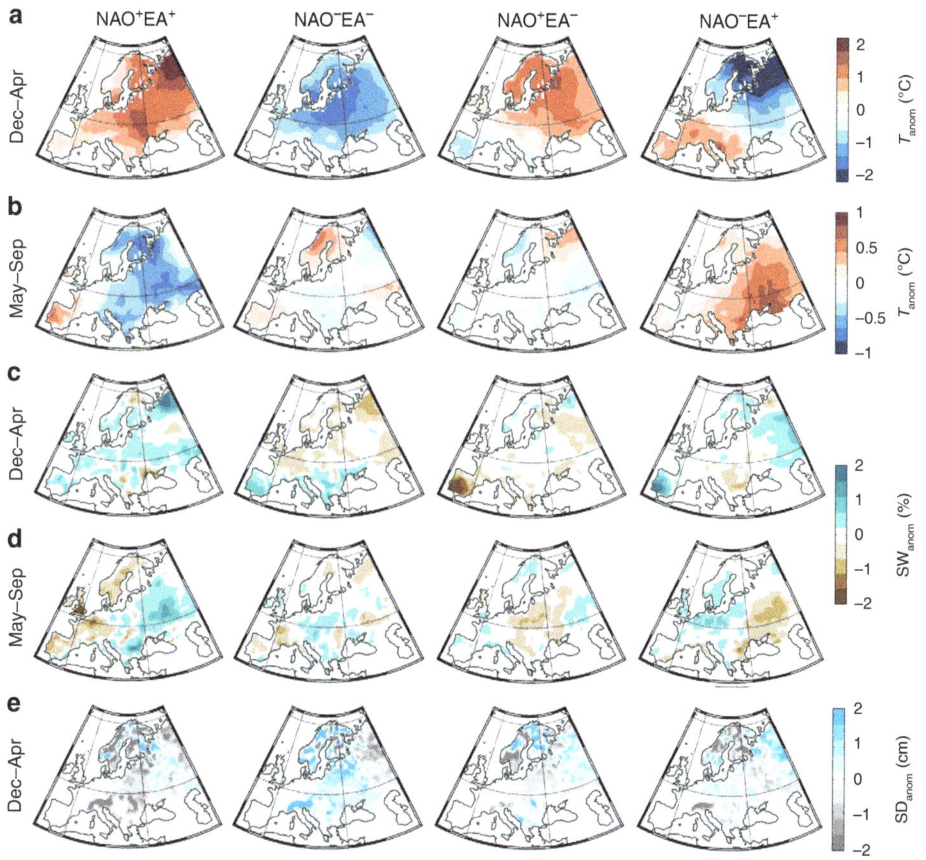

Figure 4 | Impact of NAO-EA variability on climate variables. Average anomalies in late winter/early spring (Dec-Apr) and summer (May-Sep) of temperature (*T*, (**a**,**b**)), soil water content (SW, (**c**,**d**)) and snow depth (SD, only Dec–Apr, (**e**)) for the four NAO-EA composites.

Europe impose negative NDVI anomalies and lower-than-average GPP (especially in western Russia; Fig. 5b; Supplementary Fig. 6). High water availability due to retarded snow melt in spring and reduced evaporative soil-water loss during the colder spring sustains increased GPP from late spring to early autumn, while respiration is kept close to average values due to the mild summer temperatures, especially in parts of central Europe and western Russia. Indeed, most DGVMs estimate a strong positive dependence of summer GPP in soil water in all regions except Scandinavia (Supplementary Table 7; Supplementary Fig. 8), while only four are able to capture the known dependence of GPP in winter water availability in the Iberian Peninsula[36]. The only two DGVMs that estimate negative NBP anomalies at continental scale for NAO^-EA^- (ISAM and VEGAS) present very weak control of water availability on GPP. The enhanced CO_2 uptake during the NAO^-EA^- phase combination is thus mostly due to enhanced GPP during the growing season, not to reduced respiration.

The highest positive integrated NDVI anomalies are observed for NAO^+EA^+ in most of Europe (except Iberia). However, the higher annual NDVI does not coincide with increased GPP, mainly because positive NDVI anomalies occur when radiation is sub-optimal (that is, late winter and early spring) and hence photosynthesis is weakly stimulated. DVGMs report higher GPP during spring, which offsets the increase in respiration during these warmer conditions. In summer, NDVI is about average, however, DGVMs estimate strong negative GPP and respiration anomalies, the former being stronger than the latter, and leading the negative NBP response (Fig. 2b). This is mainly due to the response observed in central Europe, where very dry conditions lead to very low GPP, despite the low temperatures inducing lower-than-average respiration rates (Supplementary Figs 6 and 7).

Despite both anti-phase composites resulting in negative NBP anomalies, they diverge in the corresponding climate patterns (Fig. 4). For NAO^+EA^-, DGVMs (inversions are consistent, but not significant) report a sink reduction in all regions except western Russia, which matches the marked vegetation 'browning' and decrease in GPP during the year (Fig. 5) imposed by the cold and dry winter/spring in central Europe and the Mediterranean, especially in the Iberian Peninsula, and the drier summers in central Europe. This decrease in photosynthesis is accompanied by a smaller reduction in respiration, explaining the reduced NBP during this phase combination (Fig. 2; Supplementary Fig. 7). On the contrary, NAO^-EA^+ is characterized by overall high annual NDVI fields, associated with increased photosynthesis during spring (Fig. 5), promoted by warm and wet conditions over most of Europe, except in western Russia, where very cold late winter and spring inhibits vegetation activity (Supplementary Fig. 7; Supplementary Table 6). However, the GPP enhancement during spring is offset by greater increase in respiration during summer due to widespread positive temperature anomalies that reinforce, and are amplified by, soil dryness. This is especially evident in western Russia, where below average summer GPP is accompanied by enhanced respiration.

Discussion

Changes in NBP are the result of variation in land cover, in land management and/or in ecosystem activity. Changes in land cover

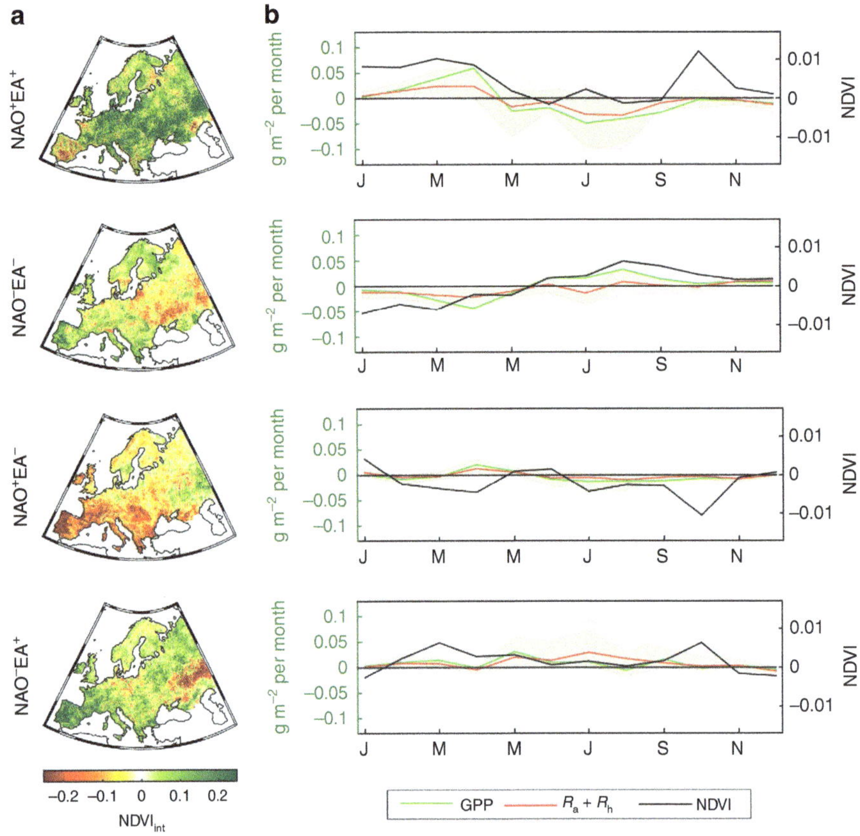

Figure 5 | Ecological response to NAO-EA phases. (**a**) Anomaly fields of annually integrated NDVI from GIMMS for each of the four NAO-EA composites; (**b**) seasonal anomalies of NDVI (black line, right *yy*-axis), GPP (green, left *yy*-axis) and aggregated autotrophic and heterotrophic respiration (red, left *yy*-axis). For GPP and respiration, bold line corresponds to the average of the eleven DGVMs and the shaded area indicates the model spread.

and land management are slow and not very intense in Europe, being more likely to induce long-term trends rather than short-term variation in NBP. On the inter-annual scale, NBP responds to a number of physical drivers which present co-variability patterns that are ultimately controlled by large-scale atmospheric circulation. Here, a mechanism driving part of the inter-annual and multi-year variability in the European sink is proposed, based on the interplay between the two main patterns of atmospheric circulation in the North-Atlantic sector (NAO and EA) and their control on heat and water-vapour transport across Europe.

The NBP enhancement during NAO^-EA^- conditions suggests that the highest NBP (largest uptake) should occur during periods with predominantly negative phases of both teleconnections. In fact, the largest NBP peak observed in 1996–1997 in the inversions corresponds to two consecutive years of NAO^-EA^-. The strong increase in NBP during 1985 also occurred during this combination of NAO and EA. In contrast, the years with lowest NBP are associated with years of opposing phases of NAO and EA: the first 3 years of the period (1982–1984) and 1998–1999, when a strong decrease in NBP is observed. Anti-phase combinations dominated the late 1990s and the 2000s, with some exceptions, with a period of 6 consecutive years of anti-phase lasting from 1998 to 2003, coinciding with a weaker sink over the European continent[11] (Fig. 2).

When computing future trends in NBP, the presence of multi-annual variability and trends in the most prominent modes of atmospheric circulation should be taken into account. In this context, anthropogenic forcing has been shown to lead to more frequent positive phases of NAO[37], associated with lower NBP, especially when EA is negative. However CMIP3 and CMIP5 models predict opposite trends in the NAO sign over the 21*st* century under similar scenarios[38]. Shepherd[39] pointed out that atmospheric dynamics still constitutes an important source of uncertainty in the state-of-the-art CMIP5 models, and showed that the climate-change response of the wintertime North-Atlantic jet largely differs among models, both in magnitude and sign. Other works have shown the influence of other teleconnections such as the Scandinavian Pattern[17] or the Pacific Decadal Oscillation[40] on North-Atlantic sea-level pressure, adding further complexity to the study of long-term NBP variability.

This study shows that the non-stationary relationship between teleconnections may strongly affect the relationships found between climate and ecosystem activity. This highlights the importance of correctly reproducing the mid-latitude atmospheric circulation patterns in Earth System Models to account for inter-annual and decadal variability in the land carbon sink. More long-term high-quality observation-driven data sets are needed to further improve our understanding of the multi-year variability of the land sink and its relation to climate variability.

Methods

Inversions. Atmospheric CO_2 inversions estimate spatial distribution of surface CO_2 fluxes using a top-down approach, in which surface fluxes are estimated from atmospheric CO_2 concentration measurements across the globe. An atmospheric transport model that represents the global atmospheric circulation is required, as well as prior-information about the fluxes. Flux estimates over land include the carbon exchange in ecosystems, LUC related fluxes and forest fire CO_2 emissions as well as fossil-fuel emissions. To estimate NBP, the latter are subtracted from the flux estimates.

We use monthly CO_2 fluxes of two versions of the atmospheric inversion of the MACCII from the LSCE[29,30]. Version 11.2 covers the period from 1979 until 2011 and is used here from January 1982 onwards. It has a 2.5×3.75 (latitude, longitude) spatial resolution. Version 13.1 has a slightly higher latitudinal resolution (1.9) and covers year 2012 as well (Supplementary Table 1). Apart from the archive length, the main difference between v11.2 and v13.1 is the number of vertical layers in the underlying atmospheric transport model, the LMD General Circulation Model (LMDZ)[29], that was doubled in the latter (39 versus 19), thus allowing a better representation of the variability of CO_2 in the planetary boundary layer, where the assimilated measurements are located.

Since MACC inversions are based on a large but variable number of stations during the study period (134 in total), with fewer observations in the beginning of the time series, part of the variability observed on the estimated fluxes may be affected by the changes in the number of sites[41]. Thus, we use the Jena s81 (v3.6) inversion that is based on fewer stations and is provided at 4×5 spatial resolution (Supplementary Table 1), but uses a consistent set of sites for the 1982–2012 period. Jena inversion computes NBP on each grid cell using the atmospheric transport model TM3, as described in ref. 31.

To subtract fossil fuel emissions, MACCII inversions use the EDGAR3.2 FastTrack 2000 emission database[42] with scaling factor to account for trends: MACC uses the annual global totals of the Carbon Dioxide Information Analysis Center and prescribes an increase of 1.4% per year after 2001. Jena inversion uses EDGAR 4.2, with FastTrack 2010 for 2009 and 2010, extrapolation based on British Petrol global totals for 2011 and 2012, and 2% year increase afterwards.

Uncertainties of annual NBP at the continental scale are given, for each model, in Supplementary Table 1. Uncertainties of the annual fluxes account for errors in the prior fluxes, network configuration (number of sites, Supplementary Fig. 1) and in the transport model. The uncertainty of the annual anomalies is expected to be lower than that of the absolute fluxes, since these errors are positively correlated from 1 year to the next[32]. We use two inversions computed with the same model but variable number of sites, and a second with a fixed number of observations, and use the spread between inversions as an indicator of the uncertainty in the anomalies.

In order to test the available signals from the smaller number of sites in Europe used by the Jena s81 inversion, we performed three synthetic-data runs (1982–2010) by transporting net ecosystem exchange (NEE) fields as simulated by the Biome-BGC DGVM in the atmosphere using the same transport model as in the Jena inversion (v3.7), though on coarser 10×8 spatial resolution. To create pseudo data, modelled atmospheric CO_2 is sampled at the same locations and times as the real data for three different station sets: as in Jena s81 (14 sites) and s90 (25 sites) as well as mostly the same sites used for MACCII inversion (Supplementary Table 2). The CO_2 concentrations sampled in this way were inverted back using the Jena scheme (setting ocean and fossil fuel priors to zero). As the same transport model is used for synthetic data and inversion, results are not contaminated by model errors. By comparing with the original fluxes (Supplementary Fig. 9), the sensitivity of the inversion skill to the site network can then be evaluated. Despite the presence of discrepancies, the run with fewer stations (Jena s81) is able to capture the variability patterns of the CO_2 fluxes and differences in the magnitude of the fluxes are on average 0.04–0.05 PgC per year for each of the three runs.

Dynamic global vegetation models. DGVMs simulate the main processes of vegetation dynamics and decomposition associated with energy, water and carbon balances at the ecosystem scale and provide a bottom-up approach to evaluate NBP variability, as well as the corresponding GPP and respiration components.

The TRENDY project[43] compiles outputs from a group of DGVMs to evaluate trends and drivers in land–atmosphere carbon exchange[20]. We use 11 DGVMs (Supplementary Table 2) from simulation S3, in which all models are forced by the same values of changing atmospheric CO_2 concentrations from ice core data and observations, historical climate observations from the CRU-NCEP v4 data set and LUC from the History Database of the Global Environment (HYDE) database of human-induced land-cover changes[44].

Supplementary Table 2 provides a summary of the characteristics of the 11 DGVMs used in simulation S3, which is performed over the period 1860–2012. All models include deforestation, afforestation and regrowth, but differ in the way they represent disturbances (fire), nutrient limitation and do not realistically simulate land-management practices or crop seasonality, which strongly interplays with climate in determining NBP of European ecosystems. Nevertheless, DGVMs provide an independent data set to evaluate the variability of the land carbon sink. NBP from DGVMs was selected for two periods: the common period with the inversions and satellite data (1982–2012) for the main analysis, and the common period with the NAO and EA indices (1950–2012), except for JSBACH whose record ends in 2005. To partition NBP, continental and regional GPP and respiration (computed as the sum of autotrophic and heterotrophic respiration components) anomalies from the DGVMs were also evaluated between 1982 and 2012.

NDVI. Biweekly NDVI fields over 1982–2012 from the GIMMS NDVI data set, provided at 8 km spatial resolution[35]. NDVI values were integrated over each year on a pixel basis from the biweekly fields. NDVI anomalies were then computed as the departure from the average annual integral values, and used as a proxy for GPP anomalies[28].

Monthly NBP values from inversions and DGVMs were first deseasonalized (mean seasonal-cycle removed) on a pixel basis to compute monthly anomaly and integrated over the year for annual NBP anomaly fields, for the European continent and for the regional boxes (Supplementary Fig. 3). A positive (negative) sign of NBP anomalies correspond to either an enhanced (weaker) sink or a smaller (increased) source. To evaluate the main pattern of NBP variability, a PCA was performed on NBP fields for inversions and DGVMs. Since results from PCA depend on the resolution of the data set, all fields were resampled to a common 1° spatial resolution. The first principal component of NBP variability (PC_1), the corresponding empirical orthogonal function (EOF_1) and explained variance were selected for each inversion and model. EOF_1 presented in all cases dipolar pattern. The consistency of these patterns was compared by locating the corresponding centres of action for each data set (Supplementary Fig. 4).

All climate variables used for this analysis were extracted from ERA-Interim reanalysis[45]: vertical integral of eastward transport water vapour (VT, in $g\,km^{-1}\,s^{-1}$) and heat (HT, in $MW\,km^{-1}$), fraction of cloud cover (%), SLP (in mb), 500-mb geopotential height (z500, in geopotential metres, g.p.m.), average air temperature at 2 m (T, in °C), volumetric soil-water content (SW, in % of volume) in the top layer[46] and average snow depth (SD, in cm). The data comprise monthly average fields from 1982 to 2012, at 0.75 spatial resolution.

As complementary information about soil-water conditions, monthly values of the self-calibrated Palmer Drought Severity Index (PDSI)[47], provided by the NOAA/OAR/ESRL PSD, Boulder, CO, USA at 2.5 spatial resolution was used. Since PDSI presented regional dependence (negative for Iberian Peninsula and central Europe and positive for western Russia and Scandinavia), we estimated drought conditions as the PDSI departure from the regional average (Supplementary Table 6).

Atmospheric circulation. We focus our analysis on the impacts of winter state Dec-Feb (DJF) of two teleconnections, using the monthly indices from NOAA's Climate Prediction Centre, which are calculated by rotated PCA of monthly means of 500-mb geopotential height anomalies from NCEP/DOE II (ref. 48) as described in (ref. 49). This procedure allows the calculation of orthogonal (that is, non-correlated) indices for each month. The first mode corresponds to the NAO pattern, the leading mode of SLP variability in the North Atlantic and the main climate teleconnection affecting European weather. The second mode corresponds to EA circulation pattern. These modes affect more significantly European weather during winter[2,50]. For visualization of the spatial patterns associated with the two teleconnections used in this work, we performed a PCA on standardized monthly means of 500-mb geopotential height from NCEP/DOEII Reanalysis over the North-Atlantic region between 20 to 80 N and 90 W to 50 E for the winter months (DJF) as in ref. 49.

The phases of an index are defined as those exceeding the lower (negative) or upper (positive) terciles, and identified in superscript (for example, NAO^+ for a positive phase of winter NAO). To evaluate the combined impacts of both teleconnections, we identify four subsets of years corresponding to the four possible NAO–EA phase combinations, together with a neutral composite (Supplementary Table 3).

Regional fluxes. Since data sets are provided at very different spatial resolutions (Supplementary Tables 1 and 2), and because inversions perform better on larger scales, we consider regionally integrated NBP from inversions and DGVMs for the European continent defined in the TransCom inter-comparison project[51] (Supplementary Fig. 1). We additionally define four large regions encompassing most of Europe: Iberian Peninsula extending from −11° to 3.5° E and 34° to 44° N; continental central Europe encompassing the region between −5° to 25° E and 44° to 53° N, but excluding Great Britain; western Russia and eastern Europe extending from 29° to 60° E and 46° to 62° N; Scandinavia covering the region between 4.5° to 29° E and 56° to 71° N (Supplementary Fig. 1).

To evaluate the skill of inversions in capturing regional fluxes, we used the same synthetic runs from Jena inversion to evaluate: (i) the ability of the inversion to reproduce fluxes in large regions over Europe; (ii) the influence of the observational network. Despite the very coarse resolution, results (Supplementary Fig. 10) indicate that inversions do have moderate skill in distinguishing sub-regions within Europe. As expected from the better observational constrain, the skill increases with increasing number of stations. Nevertheless, even the s81 run (with few sites) is able to capture part of the regional differences, especially in eastern, central and western Europe. In south-western Europe (corresponding to the Iberian Peninsula), the inversion is not able to reproduce the fluxes, as expected from the very low observational constraint (only two sites in the complete MACCII set, located in the Pyrenees).

NBP response to NAO–EA. All data fields were selected for the study region (Supplementary Fig. 1). Continental and regional NBP anomalies were calculated for each inversion and DGVM and evaluated for each of the four NAO–EA combinations (Fig. 2). In the case of MACC v11.2 inversion and JSBACH, the data sets do not encompass the whole period, therefore, the composites were calculated with fewer years. To partition annual NBP anomalies, continental and regional GPP and total respiration (autotrophic plus heterotrophic respiration components) were calculated at monthly scale, and compared with the corresponding NDVI anomalies (Supplementary Fig. 7).

For each composite, a one-sided analysis of variance (ANOVA) was performed to test the significance of the average anomaly value on the continental and regional scales (Supplementary Table 4), separately for inversions and DGVMs. Absolute NBP anomalies for each composite are easier to associate with NBP variations from year to year. However, it is worth assessing whether the inversions (or DGVMs) agree on the relative response, that is, the difference in NBP anomalies between two NAO–EA combinations. Therefore, a two-sided ANOVA was carried out to test the difference between NBP anomaly estimates for pairs of the four combinations (Supplementary Table 5), on the continental scale. Supplementary Tables 4 and 5 also include results of the ANOVA analyses for the neutral composite. To assess whether relationships hold for longer periods, the same analysis was extended for the period 1950–2012 using DGVMs (Supplementary Fig. 5).

Since the synthetic runs of Jena inversion present some dependence of absolute fluxes in the observational network, it is worth assessing how the number of sites may influence the results of NBP anomalies in response to NAO–EA variations. As the observational network has increased with time, Jena inversion is provided for different periods using an increasing number (always constant for each data set) of sites. Here we compare the other two inversions from Jena v3.6 that still encompass at least 20 years and use the same model but more sites: s85 (1985–2012, 19 sites), s90 (1990–2012, 25 sites). We also performed a run with the newer version of Jena inversion (v3.7) using the MACCII sites, at the same resolution as Jena s81 v3.6. The corresponding anomalies for each NAO–EA combination are presented in Supplementary Fig. 11 and show that, despite results depending on the size of the observational network, the response of continental NBP to the phases of NAO and EA for the data sets with more observations is consistent with the anomalies estimated by the two MACCII inversions and Jena s81 v3.6. In most cases, the inversion run using the MACCII sites is closer to the anomalies estimated by MACCII, highlighting the dependence of the anomalies on the observational constrains, but also the importance of other sources of uncertainty, for example, differences in the transport models.

References

1. McPhaden, M. J., Zebiak, S. E. & Glantz, M. H. ENSO as an integrating concept in earth science. *Science* **314,** 1740–1745 (2006).
2. Trigo, R. M., Osborn, T. J. & Corte-Real, J. M. The North Atlantic Oscillation influence on Europe: climate impacts and associated physical mechanisms. *Clim. Res* **20,** 9–17 (2002).
3. Scaife, A. a., Folland, C. K., Alexander, L. V., Moberg, A. & Knight, J. R. European climate extremes and the North Atlantic Oscillation. *J. Clim.* **21,** 72–83 (2008).
4. Comas-Bru, L. & McDermott, F. Impacts of the EA and SCA patterns on the European twentieth century NAO-winter climate relationship. *Quart. J. R. Meteorol. Soc.* **140,** 354–363 (2013).
5. Behrenfeld, M. J. *et al.* Biospheric primary production during an ENSO transition. *Science* **291,** 2594–2597 (2001).
6. Bastos, A., Running, S. W., Gouveia, C. & Trigo, R. M. The global NPP dependence on ENSO: La Niña and the extraordinary year of 2011. *J. Geophys. Res. Biogeosci.* **118,** 1247–1255 (2013).
7. Friedlingstein, P. & Prentice, I. Carbon-climate feedbacks: a review of model and observation based estimates. *Curr. Opin. Environ. Sustainability* **2,** 251–257 (2010).
8. Hallett, T. B. *et al.* Why large-scale climate indices seem to predict ecological processes better than local weather. *Nature* **430,** 71–75 (2004).
9. Nemani, R. R. *et al.* Climate-driven increases in global terrestrial net primary production from 1982 to 1999. *Science* **300,** 1560–1563 (2003).
10. Angert, A. *et al.* Drier summers cancel out the CO_2 uptake enhancement induced by warmer springs. *Proc. Natl Acad. Sci. USA* **102,** 10823–10827 (2005).
11. Piao, S. *et al.* Net carbon dioxide losses of northern ecosystems in response to autumn warming. *Nature* **451,** 49–52 (2008).
12. Xia, J. *et al.* Terrestrial carbon cycle affected by non-uniform climate warming. *Nat. Geosci.* **7,** 173–180 (2014).
13. Barriopedro, D., García-Herrera, R. & Hernández, E. The role of snow cover in the Northern Hemisphere winter to summer transition. *Geophys. Res. Lett.* **33,** L14708 (2006).
14. Gouveia, C., Trigo, R. M., DaCamara, C. C., Libonati, R. & Pereira, J. M. C. The North Atlantic Oscillation and European vegetation dynamics. *Int. J. Clim.* **28,** 1835–1847 (2008).
15. Peters, W. *et al.* Seven years of recent European net terrestrial carbon dioxide exchange constrained by atmospheric observations. *Global Change Biol.* **16,** 1317–1337 (2010).
16. Carnicer, J. *et al.* Widespread crown condition decline, food web disruption, and amplified tree mortality with increased climate change-type drought. *Proc. Natl Acad. Sci. USA* **108,** 1474–1478 (2011).
17. Moore, G. W. K., Renfrew, I. A. & Pickart, R. S. Multidecadal mobility of the North Atlantic Oscillation. *J. Clim.* **26,** 2453–2466 (2013).
18. Pan, Y. *et al.* A large and persistent carbon sink in the world's forests. *Science* **333,** 988–993 (2011).
19. Peylin, P. *et al.* Global atmospheric carbon budget: results from an ensemble of atmospheric CO_2 inversions. *Biogeosciences* **10,** 6699–6720 (2013).
20. Le Quéré, C. *et al.* The global carbon budget 1959-2011. *Earth Syst. Sci. Data* **5,** 165–185 (2013).
21. Heimann, M. & Reichstein, M. Terrestrial ecosystem carbon dynamics and climate feedbacks. *Nature* **451,** 289–292 (2008).
22. Cox, P. M. *et al.* Sensitivity of tropical carbon to climate change constrained by carbon dioxide variability. *Nature* **494,** 341–344 (2013).
23. Zhao, M. & Running, S. W. Drought-induced reduction in global terrestrial net primary production from 2000 through 2009. *Science* **329,** 940–943 (2010).
24. Ahlström, A., Miller, P. A. & Smith, B. Too early to infer a global NPP decline since 2000. *Geophys. Res. Lett.* **39,** L15403 (2012).
25. Janssens, I. A. *et al.* Europe's terrestrial biosphere absorbs 7 to 12% of European anthropogenic CO_2 emissions. *Science* **300,** 1538–1542 (2003).
26. Schulze, E. D. *et al.* Importance of methane and nitrous oxide for Europe's terrestrial greenhouse-gas balance. *Nat. Geosci.* **2,** 842–850 (2009).
27. Piao, S. *et al.* Changes in satellite-derived vegetation growth trend in temperate and boreal Eurasia from 1982 to 2006. *Global Change Biol.* **17,** 3228–3239 (2011).
28. Piao, J. *et al.* Evidence for a weakening relationship between interannual temperature variability and northern vegetation activity. *Nat. Commun.* **5,** 5018 (2014).
29. Chevallier, F. *et al.* CO_2 surface fluxes at grid point scale estimated from a global 21 year reanalysis of atmospheric measurements. *J. Geophys. Res.* **115,** D21307 (2010).
30. Chevallier, F. *et al.* Toward robust and consistent regional CO_2 flux estimates from in situ and spaceborne measurements of atmospheric CO_2. *Geophys. Res. Lett.* **41,** L058772 (2014).
31. Rödenbeck, C. *Estimating CO_2 sources and sinks from atmospheric mixing ratio measurements using a global inversion of atmospheric transport.* Technical Report No. 6 (Max Planck Institute for Biogeochemistry, Jena, 2005) Available at http://www.bgc-jena.mpg.de/mpg/websiteBiogeochemie/Publikationen/Technical_Reports/tech_report6.pdf.
32. Baker, D. F. *et al.* TransCom 3 inversion intercomparison: impact of transport model errors on the interannual variability of regional CO_2 fluxes, 1988-2003. *Global Biogeochem. Cycles* **20,** GB1002 (2006).
33. Kaminski, T. & Heimann, M. Inverse modeling of atmospheric carbon dioxide fluxes. *Science* **294,** 259–259 (2001).
34. Quesada, B., Vautard, R., Yiou, P., Hirschi, M. & Seneviratne, S. I. Asymmetric European summer heat predictability from wet and dry southern winters and springs. *Nat. Clim. Change* **2,** 736–741 (2012).
35. Tucker, C. *et al.* An extended AVHRR 8km NDVI dataset compatible with MODIS and SPOT vegetation NDVI data. *Int. J. Remote Sens.* **26,** 4485–4498 (2005).
36. Peñuelas, J. *et al.* Complex spatiotemporal phenological shifts as a response to rainfall changes. *New Phytol.* **161,** 837–846 (2004).
37. Gillett, N. P., Zwiers, F. W., Weaver, A. J. & Stott, P. A. Detection of human influence on sea-level pressure. *Nature* **422,** 292–294 (2003).
38. Cattiaux, J., Douville, H. & Peings, Y. European temperatures in CMIP5: origins of present-day biases and future uncertainties. *Clim. Dyn.* **41,** 2889–2907 (2013).
39. Shepherd, T. G. Atmospheric circulation as a source of uncertainty in climate change projections. *Nat. Geosci* **7,** 703–708 (2014).
40. Trenberth, K. E., Fasullo, J. T., Branstator, G. & Phillips, A. S. Seasonal aspects of the recent pause in surface warming. *Nat. Clim. Change* **4,** 911–916 (2014).
41. Rödenbeck, C., Houweling, S., Gloor, M. & Heimann, M. CO_2 flux history 1982-2001 inferred from atmospheric data using a global inversion of atmospheric transport. *Atmos. Chem. Phys.* **3,** 1919–1964 (2003).
42. Olivier, J. G. J. & Berdowski, J. J. M. *The Climate System* 33–78 (Swets & Zeitlinger Publishers, 2001).
43. Sitch, S. *et al.* Trends and drivers of regional sources and sinks of carbon dioxide over the past two decades. *Biogeosci. Discuss.* **10,** 20113–20177 (2013).
44. Klein Goldewijk, K., Beusen, A., van Drecht, G. & de Vos, M. The HYDE 3.1 spatially explicit database of human-induced global land-use change over the past 12.000 years. *Global Ecol. Biogeogr.* **20,** 73–86 (2011).
45. Dee, D. P. *et al.* The ERA-Interim reanalysis: configuration and performance of the data assimilation system. *Quart. J. R. Meteorol. Soc.* **137,** 553–597 (2009).
46. Balsamo, G. *et al.* A revised hydrology for the ECMWF model: verification from field site to terrestrial water storage and impact in the integrated forecast system. *J. Hydrometeorol.* **10,** 623–643 (2009).
47. Dai, A. Characteristics and trends in various forms of the Palmer Drought Severity Index during 1900-2008. *J. Geophys. Res. Atmospheres* **116,** D12115 (2011).
48. Kanamitsu, M. *et al.* NCEP/DOE AMIP-II Reanalysis (R-2). *Bull. Am. Meteorol. Soc.* **83,** 1631–1643 (2002).
49. Barnston, A. G. & Livezey, R. E. Classification, seasonality and persistence of low-frequency atmospheric circulation patterns. *Mon. Weather Rev.* **115,** 1083–1126 (1987).

50. Hurrell, J. W. Decadal Trends in the North Atlantic Oscillation: regional temperatures and precipitation. *Science* **269,** 676–679 (1995).
51. Gurney, K. R. *et al.* Towards robust regional estimates of CO_2 sources and sinks using atmospheric transport models. *Nature* **415,** 626–630 (2002).

Acknowledgements

Ana Bastos was partially funded by the Portuguese Foundation for Science and Technology (SFRH/BD/78068/2011). Célia M. Gouveia and Ricardo M. Trigo were supported by QSECA (PTDC/AAGGLO/4155/2012). This research was also supported by the European Research Council Synergy grant ERC-2013-SyG 610028 IMBALANCE-P.

Author contributions

A.B. conceived the study, conducted the analysis and wrote the paper; I.A.J., R.M.T. and C.M.G. conceived the study and supervised the project; S.P., F.C. and C.R. developed the data sets and provided technical and conceptual advice; C.R. also performed the sensitivity tests for Jena inversion. P.C, J.P., P.F. and S.W.R. provided expert advice on the different topics. All authors contributed to the revision of the paper.

Additional information

Competing financial interests: The authors declare no competing financial interests.

23

Local biomass burning is a dominant cause of the observed precipitation reduction in southern Africa

Øivind Hodnebrog[1], Gunnar Myhre[1], Piers M. Forster[2], Jana Sillmann[1] & Bjørn H. Samset[1]

Observations indicate a precipitation decline over large parts of southern Africa since the 1950s. Concurrently, atmospheric concentrations of greenhouse gases and aerosols have increased due to anthropogenic activities. Here we show that local black carbon and organic carbon aerosol emissions from biomass burning activities are a main cause of the observed decline in southern African dry season precipitation over the last century. Near the main biomass burning regions, global and regional modelling indicates precipitation decreases of 20–30%, with large spatial variability. Increasing global CO_2 concentrations further contribute to precipitation reductions, somewhat less in magnitude but covering a larger area. Whereas precipitation changes from increased CO_2 are driven by large-scale circulation changes, the increase in biomass burning aerosols causes local drying of the atmosphere. This study illustrates that reducing local biomass burning aerosol emissions may be a useful way to mitigate reduced rainfall in the region.

[1]Center for International Climate and Environmental Research-Oslo (CICERO), P.O. Box 1129 Blindern, N-0318 Oslo, Norway. [2]School of Earth and Environment, University of Leeds, Leeds LS2 9JT, UK. Correspondence and requests for materials should be addressed to Ø.H. (email: oivind.hodnebrog@cicero.oslo.no).

Black carbon (BC) aerosols, in the atmosphere, absorb solar radiation and may affect the hydrological cycle through changes in the atmospheric and surface energy budgets[1-5]. In contrast to most greenhouse gases, BC is removed from the atmosphere only after a few days[6,7]. Absorbing aerosols such as biomass burning BC modify cloud distribution, mainly by suppressing convective activity, albeit with some debate existing in the literature[8-12]. On a global scale, the altitude of added absorbing aerosol concentration is shown to influence the magnitude and sign of global average precipitation changes, with a net decrease in precipitation expected unless BC is added close to the surface[3,13-17]. Organic carbon (OC) aerosols are emitted together with BC and have approximately the same lifetime, but OC scatters sunlight to a much larger extent than BC, and therefore cools the atmosphere-surface system[6]. Biomass burning is a major source of BC and OC aerosols and normally takes place during the dry season, which is therefore the focus of this study.

Observational data reveal changes in precipitation patterns since the middle of the past century, with some regions in the world getting increased precipitation while other regions are experiencing less precipitation[18,19]. In general, increases in anthropogenic emissions have led to more precipitation on a global scale due to surface temperature warming[20]. However, the detailed effects of changes in greenhouse gases and aerosols on the hydrological cycle are complex[2,21]. Precipitation responses depend on the climate forcing mechanism, and also differ with timescale[15,16,22]. Responses in precipitation often show strong regional differences. This is mainly due to large-scale changes in circulation patterns and land-ocean contrasts, but could also arise from strictly local interactions, which are often caused by constituents that have short residence times in the atmosphere, such as BC.

Southern Africa is a region where observations indicate reduced precipitation and increased dryness since the 1950s (refs 18,19). The region is highly vulnerable to climate change (see ref. 23 and references therein), and has strong connections between climate and water, energy and food security[24]. Global climate models predict a further decrease in precipitation and a strong increase in the number of consecutive dry days near the end of the century for a business-as-usual scenario, due to widening of the Hadley Circulation[25]. While previous work on observed precipitation has focused mostly on annual trends, the focus in this study is on the dry season during which emissions from biomass burning are particularly strong. Although long-term observational data are sparse in the region, previous studies have shown evidence of declining trends in annual precipitation for several stations. In the latter half of the 20th century, significant declining precipitation trends have been shown in Zambia and southwest Tanzania, in addition to a more widespread increase in the annual maximum of consecutive dry days[26]. Annual precipitation has also decreased in tropical rainforest regions, such as in Congo, the Democratic Republic of the Congo and Gabon[27,28].

The present study focuses on explaining causes of observed precipitation decline during the dry season in southern Africa. We use a global precipitation observation data set (Global Precipitation Climatology Centre (GPCC)[29,30]) and a global (CESM1.0.4/CAM4 (refs 31,32)) and regional climate model (Weather Research and Forecasting (WRF)[33]). Our results show that the combination of BC/OC from local anthropogenic biomass burning activities, and increases in the global CO_2 concentration, can explain the precipitation decline in the region. These findings are further supported by analysis of the regional energy budget, which shows a local drying of the atmosphere from added biomass burning aerosols.

Results

Spatial distributions of precipitation changes. Some of the largest BC and OC emissions in the world take place in southern Africa due to extensive biomass burning during the dry season[34,35], approximately from June to September each year. Due to human activities, BC biomass burning emissions in the region almost doubled during the 20th century, with the majority of the increase in the latter half[35] (Supplementary Fig. 1), leading to a 60% increase in the BC burden[36] (Table 1). The increase in OC burden over the period is about 70% (Table 1).

In the present study, we show the modelled precipitation impact of increases in local BC and OC emissions due to biomass burning over the past one-and-a-half centuries, and compare them with the possible influence of increases in atmospheric CO_2 concentrations over the same period. We quantify and interpret the results from global and regional model results in terms of the regional energy budget[37] to uncover the mechanisms behind modelled changes in precipitation.

Throughout the manuscript, we show differences between model experiments (see Table 1) and denote 'BB' as the impact of changes in biomass burning emissions of BC and OC from 1850 to 2000 (experiments BASE – BB1850), 'CO2' as the impact of changes in CO_2 volume mixing ratios over the same time period (BASE – CO21850) and 'CO2BB' as the combined impact of BC/OC biomass burning and CO_2 concentration changes (BASE – CO2BB1850). Due to large variability in the precipitation changes, five-member ensemble runs were performed for each scenario.

Observations over southern Africa in the past century indicate a precipitation decline during the dry season from June to September (Fig. 1a,b), with a stronger decline in the latter part of the period, coinciding with the period of strongly increasing biomass burning emissions (Supplementary Fig. 1b).

A strong impact of anthropogenic biomass burning particles on precipitation during the dry season can be seen in the regional and the global climate model (Fig. 1c,d; Supplementary Fig. 2c,d), with precipitation decreases in the order of 10–40% in large parts of the region. In both models the strongest decline in absolute precipitation occurs in the northwest, which also is the region with the largest increase in atmospheric biomass burning aerosol

Table 1 | Overview of model experiments.

Experiment*	Short description	Global CO_2 vmr (p.p.m.)	BC burden† (mg m^{-2})	OC burden† (mg m^{-2})
BASE	Year 2000 conditions	367	1.85	11.8
BB1850	Biomass burning emissions of BC/OC for 1850	367	1.15	6.79
CO21850	CO_2 concentrations for 1850	285	1.85	11.8
CO2BB1850	Both CO_2 and BB changed to 1850	285	1.15	6.79

*BB = BASE – BB1850; CO_2 = BASE – CO21850; CO2BB = BASE – CO2BB1850.
†Burdens are averaged over southern Africa (subregion shown by dashed rectangle in Fig. 1) and over the June to September season.

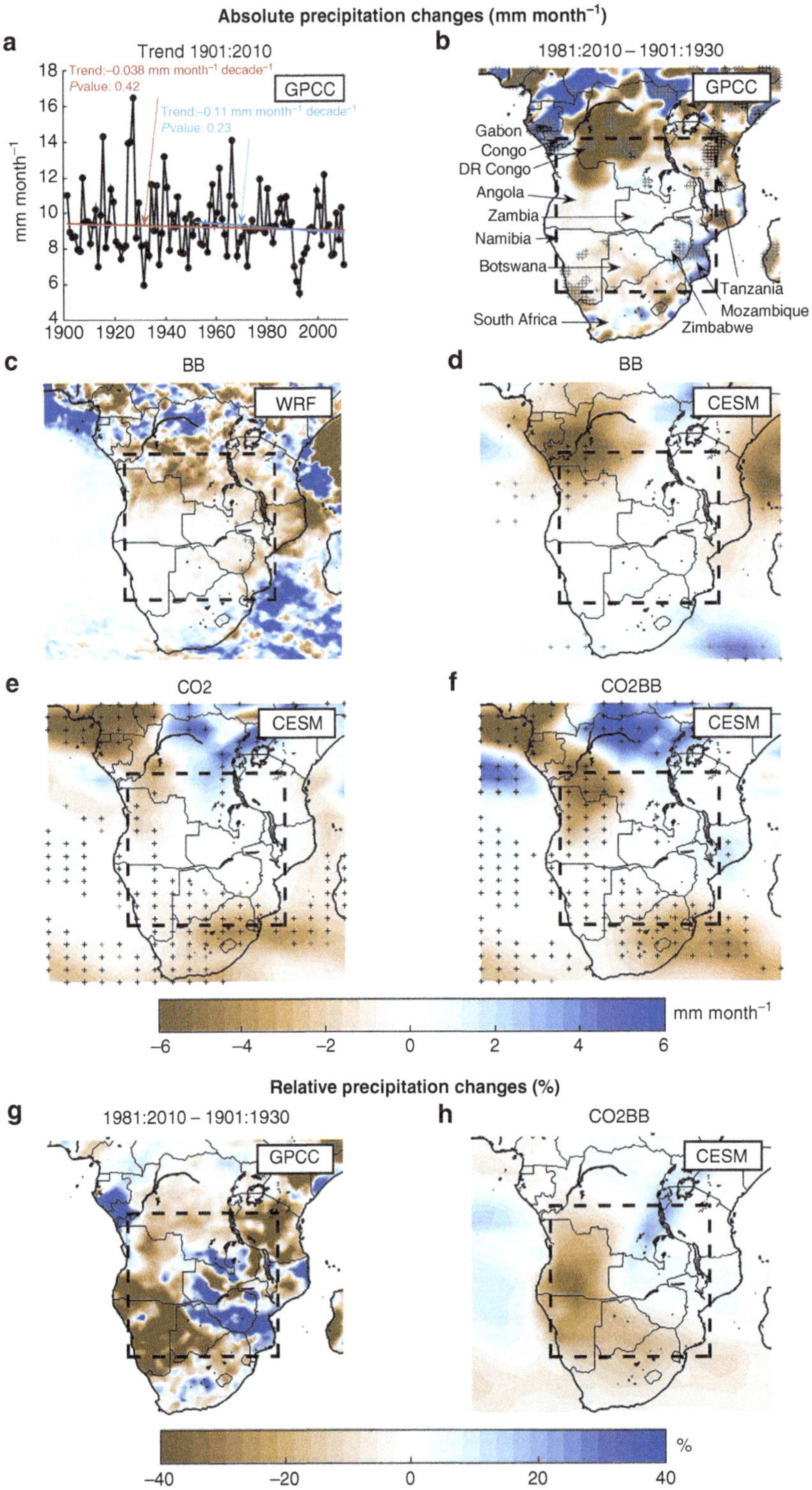

Figure 1 | Distributions of observed and modelled precipitation changes. Trend in observed precipitation (mm month^{-1}) over the June to September season (**a**), absolute precipitation changes (mm month^{-1}) for the same season from observations and various model experiments (**b-f**), and relative precipitation changes (%) from observations and the CO2BB model experiment (**g-h**). Values in (**a**) are averaged over the subregion shown by the dashed rectangle in (**b**). Panels **a**,**b** and **g** are from observations (GPCC) whereas the five other panels are from models results, one with the regional WRF model (**c**) and four from the global CESM model (**d-f**,**h**). Model results are shown as differences from the BASE simulation and show differences between 2000 and 1850 conditions in each case; see text and Table 1 for notation of the experiments. Observed precipitation trends in (**a**) are shown for the whole GPCC time period from 1901 to 2010 (red) and from 1951 to 2010 (blue). Observations in (**b**) are shown as the difference 1981:2010 – 1901:1930 because 30-year averages are needed to reduce noise (that is, the influence of natural variability). Note that in most parts of the region, the number of stations per grid cell is low in the GPCC data. In plots (**b-f**), symbols denote grid boxes where changes are significant (P value <0.05) according to a two-tailed Student's t-test (grey '+' symbols), and when also accounting for multiple statistical testing according to ref. 54 (black '+' symbols).

burden (Supplementary Fig. 1c). In this region, the global (Community Earth System Model (CESM)) model BB and CO2BB simulations show statistically significant reductions in precipitation. The BB simulation in the regional (WRF) model supports this finding, showing precipitation reduction in the same area, although not statistically significant due to a smaller sample of years. An overall decrease in precipitation can be seen in southern Africa, but with some smaller regions experiencing increased precipitation due to biomass burning. Such regions of increased precipitation downwind of regions where BB leads to precipitation decreases can be expected and have been seen before[11]. The regional model, with higher spatial resolution than the global model, shows a more inhomogeneous distribution with more local scale features.

The modelled impact of industrial-era changes in CO_2 concentrations on precipitation is an increase in the northeast of southern Africa and a decrease of up to 30% in most of the remaining part (Fig. 1e; Supplementary Fig. 2e). Thus, our results indicate that the known changes in both BB and CO_2 concentrations may have led to a precipitation decrease since 1850, but with slightly different patterns. This decrease is enhanced in the simulation with combined changes in BB and CO_2 concentrations (Fig. 1f,h). The precipitation pattern in the subregion in CO2BB resembles that of the sum of BB and CO2 (not shown), indicating that the two effects are to a large extent additive. Distributions of changes in the number of dry days (that is, days with precipitation <1 mm) and maximum number of consecutive dry days show similar tendencies of a drying due to BB and increased CO_2 concentrations (Supplementary Figs 3 and 4).

Modelled precipitation changes compare relatively well with what has been observed over the previous century (Fig. 1; Supplementary Fig. 2). Notably, observed precipitation decreases close to regions where biomass burning emissions have increased most are well reproduced by both models' BB experiment, particularly in Angola and south of the Democratic Republic of the Congo. Correlation coefficients for the subregion confirm that observed spatial patterns of absolute precipitation change correlate well with modelled precipitation changes induced by increased biomass burning (correlation coefficients >0.5 at 2.5° resolution, and significant at $P<0.05$ level), while there is no correlation with modelled precipitation change due to increased CO_2 (correlation coefficient ~ 0.0) (Supplementary Table 1). Uncertainties usually increase when going from larger to smaller scales in studies of climate change impacts, and a small displacement in simulated patterns may therefore lead to poor correlation. We note here that correlation coefficients calculated after averaging to coarser resolution are much higher (for example, 0.7–0.8 for BB at 5° resolution). It is also encouraging that the models are able to reproduce the spatial patterns of present-day precipitation climatology, with significant (P value<0.05) correlation coefficients of 0.97 and 0.78 for WRF and CESM, respectively (Supplementary Fig. 5; Supplementary Table 1).

The domination of BB aerosols in driving precipitation changes is unique for the dry season in the southern Africa region. Anthropogenic biomass burning aerosols are not found to dominate precipitation changes in other regions or seasons. In central Africa, biomass burning activities mainly take place from December to February, but the increase in biomass burning aerosol emissions since pre-industrial times has not led to a clear impact on precipitation (Supplementary Fig. 6). The third region of the world, in addition to southern and central Africa, with strong increases in biomass burning activities is the central part of South America. Here observations indicate a precipitation decline during its dry season from June to September, and global model results show that this is caused by changes in CO_2 rather than changes in BB (Supplementary Fig. 7).

Changes in mean precipitation and number of dry days. In the following we focus on the southern African subregion indicated by the dashed rectangle in Fig. 1. This region has had a strong increase in biomass burning emissions, and is a region with minimal precipitation during the June to September period, implying that a climatic change in precipitation could have important societal effects.

The reduction in precipitation in the subregion due to BB and anthropogenic CO_2 emissions is evident from Fig. 2, with a decrease in monthly mean precipitation and an increase in the number of dry days. This precipitation reduction is also observed, and although the absolute precipitation change of around 1 mm month^{-1} seems small, the relative change is as much as 5–10% (Supplementary Fig. 8). It is also an average over a relatively large area, locally observed changes are much larger (Fig. 1b,g). With the exception of one ensemble member for the BB case and one for the CO2BB case, the decrease in mean precipitation and increase in the number of dry days are consistent for all cases, between all ensemble members, and across the two models (Fig. 2; Supplementary Fig. 8). In the BB and CO2BB cases, the results are statistically significant (P value<0.05) for all ensemble members aggregated (Fig. 2). The WRF model result is based on fewer years and therefore not statistically significant at the same level, but is still in general agreement with the mean of the CESM ensembles.

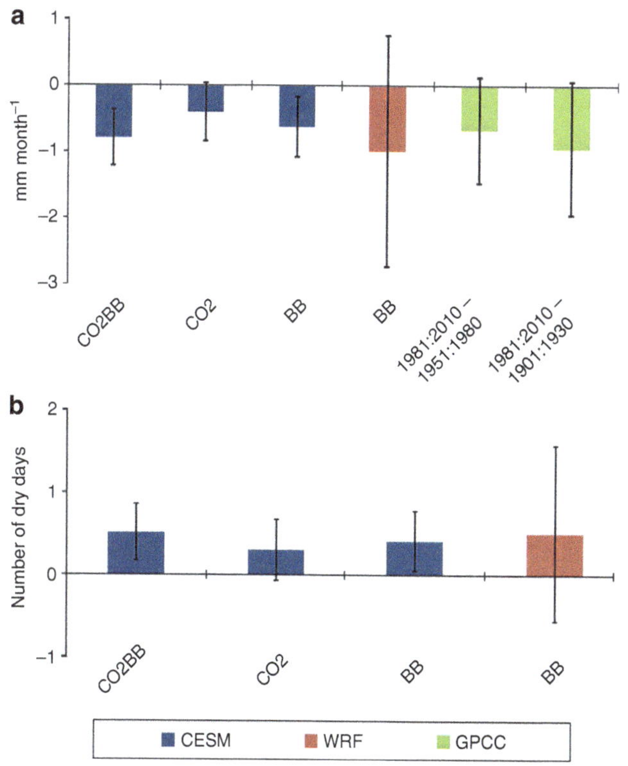

Figure 2 | Southern Africa average change in precipitation. Change in mean precipitation (mm month^{-1}) (**a**) and number of dry days (**b**) over the June to September season within the subregion (region indicated by dashed rectangle in Fig. 1) for various model experiments. Model results are shown as differences from the BASE simulation, where CESM denotes the global climate model and WRF the regional model. GPCC observations are both shown as the difference 1981:2010 – 1901:1930 and 1981:2010 – 1951:1980. Error bars show the 95% confidence interval derived from a two-tailed Student's t-test.

Precipitation and the regional atmospheric energy budget. Analysis of the energy budget, including changes in horizontal energy transport, have proved useful in understanding global and regional precipitation changes[21,37]. The local atmospheric energy budget representing the difference between two climate situations can be approximated by

$$L\delta P = \delta Q - \delta H_{\rm dry} = \delta LH + \delta H_{\rm moist}, \quad (1)$$

where L is the latent heat of condensation, δP is the change in surface precipitation flux, δQ is the change in diabatic cooling of the atmosphere, $\delta H_{\rm dry}$ and $\delta H_{\rm moist}$ denote the change in net inward horizontal dry and moist static energy transport, respectively, and δLH is the net upward latent heat flux from the surface (see Methods section and refs 21,37 for details). The diabatic cooling term (δQ) can be written $\delta Q = -(\delta SW + \delta LW + \delta SH)$, where δSW and δLW is the net atmospheric shortwave and longwave radiative heating, respectively, and δSH is the upward sensible heat flux from the surface. Equation 1 shows that precipitation changes are constrained by the atmospheric energy balance[21].

Regional energy budgets for each model experiment are shown in Fig. 3 and Supplementary Fig. 9. Net atmospheric absorption of solar radiation from added BC in southern Africa is strong and exceeds 2 W m^{-2} (Fig. 3a,b). The level of agreement between the two models and between the various ensemble members is high relative to the other terms that experience much larger natural year-to-year variability.

The atmospheric radiative heating due to solar absorption by BC, or in other words, the reduced diabatic cooling term (δQ) caused by BC, leads to a decrease in surface precipitation flux (assuming L is constant). However, this is partly balanced by reduced sensible heat flux from the surface to the atmosphere caused by reduced solar radiation reaching the surface. Scattering by OC particles also contributes to the reduced sunlight at the surface in the BB case. Absorbing aerosols have previously been shown to cause circulation changes, due to the induced temperature gradients, and this depends principally on the vertical location and magnitude of the aerosols[38]. Here, in the global model, increased net horizontal transport of dry static energy outward from the atmospheric column, presumably resulting from circulation changes, draws in the same direction as the sensible heat term (Fig. 3b). In the regional model, the horizontal transport terms are negligible (Fig. 3a), and this difference seems to be the main reason for the larger precipitation reduction in the regional model, compared with the global model, seen in the BB case in Fig. 2a.

The local energy budget for increased CO_2 concentrations is fundamentally different from that for the BB case (Fig. 3c). Increased CO_2 leads to increasing global sea surface temperatures, thus feedback effects dominate the atmospheric energy budget. The main change in the local atmospheric energy budget is from the LW radiative cooling term, caused by increased surface temperature, partly offset by atmospheric absorption[39], and contributes to increased precipitation. However, this precipitation increase is counterbalanced by increases in atmospheric SW heating caused mainly by increased solar absorption by water vapour[37], sensible heat flux from the surface, and net inward horizontal energy flux due to large scale circulation changes. Increase in CO_2 leads to a precipitation decline; despite a small increase in latent heat flux from the surface to the atmosphere, a net loss of moist static energy in the horizontal direction leads to an overall decrease. Although the effect of increased CO_2 concentrations is a precipitation increase on a global scale, this local decrease in precipitation is not unexpected as the spatial pattern of precipitation changes due to anthropogenic climate change is to a large extent affected by circulation changes[37]. In fact, analysis of modelled vertical velocity shows a widening of the Hadley cell in the increased CO_2 case, consistent with what has been observed[40]. This widening leads to more subsidence, and presumably less precipitation, south of ~18°S in the subregion, and slightly less subsidence north of this latitude (not shown), largely consistent with modelled precipitation changes (Fig. 1e; Supplementary Fig. 2e).

When both adding biomass burning aerosols and increasing CO_2, the decrease in precipitation is driven by shortwave atmospheric absorption and horizontal transport of both dry and moist static energy (Fig. 3d). Overall the regional energy budget explains the precipitation changes illustrated in Figs 1 and 2, with dissimilar impact on the energy budget from CO_2 and BB. Comparison of an additional transient simulation that includes changes in BC/OC from biomass burning and increased CO_2 concentrations, and a transient simulation including all anthropogenic climate forcers, indicate that increases in CO_2 and biomass burning BC/OC are the main drivers of changes to the atmospheric energy budget since pre-industrial time (Supplementary Fig. 10).

Discussion

It should be noted that long-term measurements of precipitation in most of southern Africa are sparse[23], and different observed precipitation products show relatively large discrepancies[41]. There may also be other drivers additional to BB and CO_2, such as non-CO_2 greenhouse gases and other aerosols, which may have contributed to the observed precipitation changes. However, reflective aerosols such as sulphate are not likely to contribute substantially because emissions of its main precursor, SO_2, are relatively small in and around the surrounding region[35]. Furthermore, remote influences from the mid-latitudes cause only small effects in southern Africa[42].

Previous studies have also found a clear impact of biomass burning aerosols on the hydrological cycle over southern and central Africa[5,11,43,44]. Although different experimental setups and focuses make it difficult to compare, there are some contradictory results in previous studies; some studies indicate more precipitation[5,43] from biomass burning aerosols and others indicate less precipitation[11,44], as is found here. Impacts of BC on clouds and precipitation are known to depend strongly on the altitude of added BC[3,14], and this could be one reason for the differing results. The BC and OC aerosol concentrations in this study are from a model that includes a plume rise assumption for biomass burning emissions (OsloCTM2), and this leads to a portion of the aerosols being emitted above the boundary layer. Consequently, the main atmospheric heating from biomass burning BC aerosols occurs at an altitude high enough to inhibit vertical instability under certain conditions, and biomass burning BC may therefore suppress convective activity in our simulations. While we acknowledge that uncertainties in BC vertical distribution are large, it is well established that biomass burning emissions are rapidly lifted to high altitudes[45]. OsloCTM2 vertical profiles of extinction coefficients compare very well with the CALIOP satellite instrument over the southern Africa region (see Fig. 4 in ref. 46).

Although the findings in this study are supported by a regional and a global model, observations, and analysis of the regional energy budget, uncertainties exist when studying changes in precipitation on a regional scale. A multi-model study would be required to provide further robustness to the findings.

The global climate model calculations in the present study were performed using a slab ocean setup, to reduce the required computational resources. We do not expect that our results would differ largely if a full ocean rather than a slab ocean model had

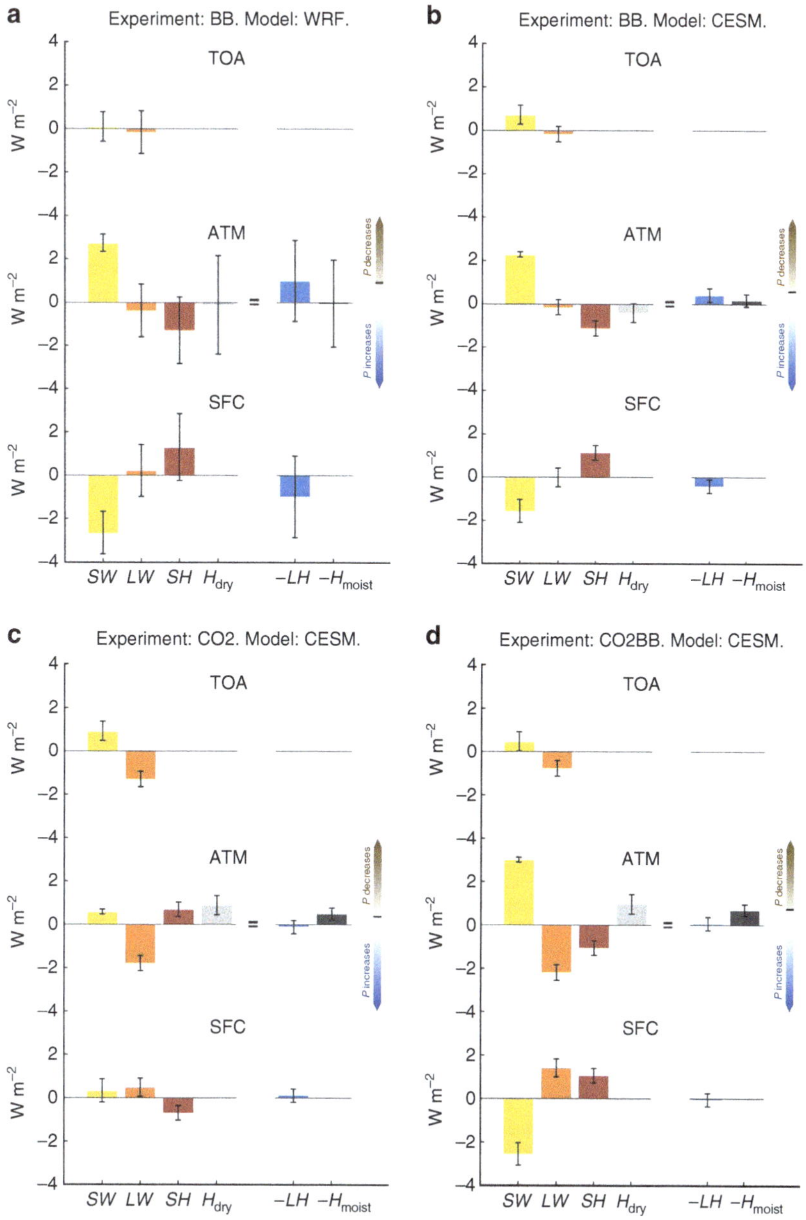

Figure 3 | Regional atmospheric energy budget for southern Africa. Atmospheric energy budget over the June to September season and averaged over the subregion (shown by the dashed rectangle in Fig. 1) for different models and experiments. Results are shown as differences from the BASE simulation, where CESM denotes the global climate model and WRF the regional model. At top-of-atmosphere (TOA) and surface (SFC), net downward energy fluxes have positive values. The energy budget of the atmosphere (ATM) is defined as the difference between the energy balance at TOA and SFC, except for the horizontal transport terms (H_{dry} and H_{moist}), which are positive when net energy transport is into the atmospheric column from the outside. Direction of brown and blue arrows on the right-hand-side of each plot denotes contribution towards decreased or increased surface precipitation according to equation 1 ($-L\delta P = \delta SW + \delta LW + \delta SH + \delta H_{dry} = -\delta LH - \delta H_{moist}$), and the total change in precipitation in each experiment is indicated by a black line embedded on these arrows. Error bars show the 95% confidence interval derived from a two-tailed Student's *t*-test.

been used. In a study using an earlier version of the CESM model[47], a fully coupled ocean model simulation led to a slightly higher climate sensitivity. Hence, a slightly stronger response in the hydrological cycle could possibly have been expected in the increased CO_2 case if a full ocean model would have been used. The impact of BC on the hydrological cycle is however dominated by rapid adjustments[15] and is therefore not likely to be strongly affected by the choice of slab ocean versus full ocean.

Our results show that local BC and OC aerosol emissions from anthropogenic biomass burning activities and increasing global CO_2 concentrations are the main causes of the observed decline in southern African dry season precipitation over the last century. This implies that, in addition to improving air quality, reducing biomass burning emissions of BC and OC may be a useful way to mitigate reduced rainfall in the southern Africa region. This conclusion is strengthened by the suggestion that these aerosols may be part of a positive feedback loop where increased drought could increase the risk of igniting more fires[11]. Precipitation in the region is expected to decrease further in the future if greenhouse gas emissions continue unabated[25]. Hence, initiation of stringent air quality legislation, involving BC mitigation, may help to counteract this precipitation decline in the short term.

Methods

Observation data. We have used long-term observation data from the GPCC[29,30], which provides monthly values of precipitation from 1901 to 2010 from 67,200 stations world-wide, gridded to a horizontal resolution of 0.5° × 0.5°. This observation dataset has been chosen because it gives a better coverage than other long-term gridded precipitation observation datasets, such as CRU and GHCN - see TFE.1, Fig. 2 in ref. 48. In regions where GPCC coverage overlaps with CRU and GHCN, they give similar estimates of annual trends. The trend analysis in Fig. 1a has been calculated using the Mann-Kendall method[49], and *P* values in the observed trends are calculated using a two-tailed *z*-test.

Global climate model. Global climate model simulations have been carried out using the National Center for Atmospheric Research (NCAR) CESM1.0.4 (ref. 31). The atmospheric component, Community Atmosphere Model (CAM4) (ref. 32), was in our simulations run with 1.9° × 2.5° horizontal resolution and 26 vertical layers, and coupled with the Community Land Model (CLM4) (ref. 50). Five different ensemble members have been run for 60 years each, whereof the first 10 years were considered as spin-up and are not part of the analysis. The ensemble members were started from different initial conditions but the same climatological state, and we have used a configuration where the atmospheric component is coupled with a slab ocean model[51]. Results are shown as averages of all ensemble members. Atmospheric distributions of aerosol concentrations have been prescribed for black and organic carbon, and sulphate, using monthly mean 3-D concentrations from the chemistry-transport model OsloCTM2 from a previous study[36]. This model has previously shown good agreement with observed pattern and magnitude of aerosol optical depth in southern Africa during the biomass burning season[34], and with vertical aerosol extinction coefficient profiles over the same region[46]. In the CESM model experiments, both the CO_2 and BC/OC concentrations have been changed globally. To test the effect of global versus local perturbation of biomass burning BC/OC, the first two ensemble members of the BB1850 simulation were re-run, but with biomass burning BC/OC concentrations changed to 1850 levels only during June to September and only in and around southern Africa. The results in terms of southern Africa dry season precipitation change were within 25% of the results obtained when biomass burning BC/OC were changed globally and annually.

Regional climate model. Additional climate model simulations have been conducted using the regional WRF model[33] version 3.5. The model has been run in 50 × 50 km horizontal resolution over Africa, and with 50 vertical layers from the surface and up to 10 hPa. The extent of the horizontal domain is the same as defined in CORDEX-Africa[52]. WRF has been run for the 30 year period, 1984–2013, in 1 year time slices, each re-initialized in December of the previous year to allow one month of spin-up. Meteorological initial and boundary conditions, as well as sea-surface temperatures, have been taken from the 6-hourly 1° × 1° resolution final reanalysis data of the Global Forecast System (GFS) model of the National Centers for Environmental Prediction (NCEP) from year 2000 onward, and NCEP/NCAR 2.5° × 2.5° resolution reanalysis for the preceding years[53]. The same radiation scheme (CAM) as in the CESM has been used in WRF. In the standard version of WRF, however, the CAM scheme only allows one fixed column value of each aerosol species to represent the entire domain for the whole year. Here, the WRF code was modified to enable prescribed monthly mean 3-D concentration fields of BC, organic carbon and sulphate from OsloCTM2, for consistency with what is used in the CESM.

Energy budget. In the calculation of δH_{dry}, we have used the same approach as in ref. 37 and calculated it as the residual of equation 1 ($\delta H_{dry} = L\delta P - \delta Q$), assuming $L = 29\,W\,m^{-2}\,mm^{-1}$ day, rather than a direct calculation of the total dry static energy flux. This approximation has been shown to introduce only small errors[37]. In Fig. 3, we have represented differences in horizontal transport of energy in and out of the atmospheric region both as dry static energy flux (δH_{dry}) and as moist static energy flux (δH_{moist}). The latter term is again calculated as a residual, $\delta H_{moist} = \delta LH\text{-}L\delta P$, and both *LH* and *P* are available from the model output.

References

1. Ramanathan, V., Crutzen, P. J., Kiehl, J. T. & Rosenfeld, D. Atmosphere - Aerosols, climate, and the hydrological cycle. *Science* **294,** 2119–2124 (2001).
2. Allen, M. R. & Ingram, W. J. Constraints on future changes in climate and the hydrologic cycle. *Nature* **419,** 224–232 (2002).
3. Ban-Weiss, G. A., Cao, L., Bala, G. & Caldeira, K. Dependence of climate forcing and response on the altitude of black carbon aerosols. *Clim. Dyn.* **38,** 897–911 (2012).
4. Kawase, H. *et al.* Physical mechanism of long-term drying trend over tropical North Africa. *Geophys. Res. Lett.* **37,** L09706 (2010).
5. Randles, C. A. & Ramaswamy, V. Direct and semi-direct impacts of absorbing biomass burning aerosol on the climate of southern Africa: a Geophysical Fluid Dynamics Laboratory GCM sensitivity study. *Atmos. Chem. Phys.* **10,** 9819–9831 (2010).
6. Boucher, O. *et al.* in *Climate Change 2013: The Physical Science Basis. Contribution of Working Group I to the Fifth Assessment Report of the Intergovernmental Panel on Climate Change* (eds Stocker, T. F. *et al.*) Ch. 7 571–658 (Cambridge University Press, 2013).
7. Samset, B. H. *et al.* Modelled black carbon radiative forcing and atmospheric lifetime in AeroCom Phase II constrained by aircraft observations. *Atmos. Chem. Phys.* **14,** 12465–12477 (2014).
8. Koren, I., Kaufman, Y. J., Remer, L. A. & Martins, J. V. Measurement of the effect of Amazon smoke on inhibition of cloud formation. *Science* **303,** 1342–1345 (2004).
9. Feingold, G., Jiang, H. L. & Harrington, J. Y. On smoke suppression of clouds in Amazonia. *Geophys. Res. Lett.* **32,** L02804 (2005).
10. Rosenfeld, D. *et al.* Flood or drought: How do aerosols affect precipitation? *Science* **321,** 1309–1313 (2008).
11. Tosca, M. G., Diner, D. J., Garay, M. J. & Kalashnikova, O. V. Human-caused fires limit convection in tropical Africa: first temporal observations and attribution. *Geophys. Res. Lett.* **42,** 6492–6501 (2015).
12. Koch, D. & Del Genio, A. D. Black carbon semi-direct effects on cloud cover: review and synthesis. *Atmos. Chem. Phys.* **10,** 7685–7696 (2010).
13. Ming, Y., Ramaswamy, V. & Persad, G. Two opposing effects of absorbing aerosols on global-mean precipitation. *Geophys. Res. Lett.* **37,** L13701 (2010).
14. Samset, B. H. & Myhre, G. Climate response to externally mixed black carbon as a function of altitude. *J. Geophys. Res.-Atmos.* **120,** 2913–2927 (2015).
15. Kvalevåg, M. M., Samset, B. H. & Myhre, G. Hydrological sensitivity to greenhouse gases and aerosols in a global climate model. *Geophys. Res. Lett.* **40,** 1432–1438 (2013).
16. Andrews, T., Forster, P. M., Boucher, O., Bellouin, N. & Jones, A. Precipitation, radiative forcing and global temperature change. *Geophys. Res. Lett.* **37,** L14701 (2010).
17. Hodnebrog, Ø., Myhre, G. & Samset, B. H. How shorter black carbon lifetime alters its climate effect. *Nat. Commun.* **5,** 5065 (2014).
18. Hartmann, D. L. *et al.* in *Climate Change 2013: The Physical Science Basis. Contribution of Working Group I to the Fifth Assessment Report of the Intergovernmental Panel on Climate Change* (eds Stocker, T. F. *et al.*) Ch. 2 159–254 (Cambridge University Press, 2013).
19. Greve, P. *et al.* Global assessment of trends in wetting and drying over land. *Nat. Geosci.* **7,** 716–721 (2014).
20. Knutti, R. & Sedlacek, J. Robustness and uncertainties in the new CMIP5 climate model projections. *Nat. Clim. Change* **3,** 369–373 (2013).
21. O'Gorman, P. A., Allan, R. P., Byrne, M. P. & Previdi, M. Energetic constraints on precipitation under climate change. *Surv. Geophys.* **33,** 585–608 (2012).
22. Bala, G., Caldeira, K. & Nemani, R. Fast versus slow response in climate change: implications for the global hydrological cycle. *Clim. Dyn.* **35,** 423–434 (2010).
23. Niang, I. *et al.* in *Climate Change 2014: Impacts, Adaptation, and Vulnerability. Contribution of Working Group II to the Fifth Assessment Report of the Intergovernmental Panel on Climate Change* (eds Barros, V. R. *et al.*) Ch. 22 1199–1265 (Cambridge University Press, 2014).
24. Conway, D. *et al.* Climate and southern Africa's water-energy-food nexus. *Nat. Clim. Change* **5,** 837–846 (2015).
25. Collins, M. *et al.* in *Climate Change 2013: The Physical Science Basis. Contribution of Working Group I to the Fifth Assessment Report of the Intergovernmental Panel on Climate Change* (eds Stocker, T. F. *et al.*) Ch. 12 1029–1136 (Cambridge University Press, 2013).
26. New, M. *et al.* Evidence of trends in daily climate extremes over southern and west Africa. *J. Geophys. Res.-Atmos.* **111,** D14102 (2006).
27. Aguilar, E. *et al.* Changes in temperature and precipitation extremes in western central Africa, Guinea Conakry and Zimbabwe, 1955-2006. *J. Geophys. Res.-Atmos.* **114,** D02115 (2009).
28. Malhi, Y. & Wright, J. Spatial patterns and recent trends in the climate of tropical rainforest regions. *Philos. Trans. R. Soc. Lond. Ser. B-Biol. Sci* **359,** 311–329 (2004).
29. Schneider, U. *et al.*GPCC Full Data Reanalysis Version 6.0 at 0.5°: Monthly Land-Surface Precipitation from Rain-Gauges built on GTS-based and Historic Data (2011).
30. Becker, A. *et al.* A description of the global land-surface precipitation data products of the global precipitation climatology centre with sample applications including centennial (trend) analysis from 1901–present. *Earth Syst. Sci. Data* **5,** 71–99 (2013).
31. Gent, P. R. *et al.* The community climate system model version 4. *J. Clim.* **24,** 4973–4991 (2011).
32. Neale, R. B. *et al.*Description of the NCAR Community Atmosphere Model (CAM 4.0). (NCAR Technical Report, NCAR/TN-485 + STR (2010).
33. Skamarock, W. C. & Klemp, J. B. A time-split nonhydrostatic atmospheric model for weather research and forecasting applications. *J. Comput. Phys.* **227,** 3465–3485 (2008).

34. Myhre, G. *et al.* Modeling the solar radiative impact of aerosols from biomass burning during the Southern African Regional Science Initiative (SAFARI-2000) experiment. *J. Geophys. Res.-Atmos.* **108,** 8501 (2003).
35. Lamarque, J. F. *et al.* Historical (1850–2000) gridded anthropogenic and biomass burning emissions of reactive gases and aerosols: methodology and application. *Atmos. Chem. Phys.* **10,** 7017–7039 (2010).
36. Skeie, R. B. *et al.* Anthropogenic radiative forcing time series from pre-industrial times until 2010. *Atmos. Chem. Phys.* **11,** 11827–11857 (2011).
37. Muller, C. J. & O'Gorman, P. A. An energetic perspective on the regional response of precipitation to climate change. *Nat. Clim. Change* **1,** 266–271 (2011).
38. Lee, S. S. *et al.* Effect of gradients in biomass burning aerosol on shallow cumulus convective circulations. *J. Geophys. Res.-Atmos.* **119,** 9948–9964 (2014).
39. Pendergrass, A. G. & Hartmann, D. L. The atmospheric energy constraint on global-mean precipitation change. *J. Clim.* **27,** 757–768 (2014).
40. Seidel, D. J., Fu, Q., Randel, W. J. & Reichler, T. J. Widening of the tropical belt in a changing climate. *Nat. Geosci.* **1,** 21–24 (2008).
41. Nikulin, G. *et al.* Precipitation climatology in an ensemble of CORDEX-Africa regional climate simulations. *J. Clim.* **25,** 6057–6078 (2012).
42. Shindell, D. T., Voulgarakis, A., Faluvegi, G. & Milly, G. Precipitation response to regional radiative forcing. *Atmos. Chem. Phys.* **12,** 6969–6982 (2012).
43. Tummon, F., Solmon, F., Liousse, C. & Tadross, M. Simulation of the direct and semidirect aerosol effects on the southern Africa regional climate during the biomass burning season. *J. Geophys. Res.-Atmos.* **115,** D19206 (2010).
44. Tosca, M. G., Randerson, J. T. & Zender, C. S. Global impact of smoke aerosols from landscape fires on climate and the Hadley circulation. *Atmos. Chem. Phys.* **13,** 5227–5241 (2013).
45. Freitas, S. R. *et al.* Including the sub-grid scale plume rise of vegetation fires in low resolution atmospheric transport models. *Atmos. Chem. Phys.* **7,** 3385–3398 (2007).
46. Koffi, B. *et al.* Application of the CALIOP layer product to evaluate the vertical distribution of aerosols estimated by global models: AeroCom phase I results. *J. Geophys. Res.-Atmos.* **117,** D10201 (2012).
47. Danabasoglu, G. & Gent, P. R. Equilibrium climate sensitivity: is it accurate to use a slab ocean model? *J. Clim.* **22,** 2494–2499 (2009).
48. Stocker, T. F. *et al.* in *Climate Change 2013: The Physical Science Basis. Contribution of Working Group I to the Fifth Assessment Report of the Intergovernmental Panel on Climate Change* (eds Stocker, T. F. *et al.*) Ch. TS 33–115 (Cambridge University Press, 2013).
49. Sen, P. K. Estimates of the regression coefficient based on Kendall's Tau. *J. Am. Stat. Assoc.* **63,** 1379–1389 (1968).
50. Lawrence, D. M. *et al.* Parameterization improvements and functional and structural advances in version 4 of the Community land model. *J. Adv. Model. Earth Sy.* **3,** M03001 (2011).
51. Kiehl, J. T., Shields, C. A., Hack, J. J. & Collins, W. D. The climate sensitivity of the community climate system model version 3 (CCSM3). *J. Clim.* **19,** 2584–2596 (2006).
52. Kim, J. *et al.* Evaluation of the CORDEX-Africa multi-RCM hindcast: systematic model errors. *Clim. Dyn.* **42,** 1189–1202 (2014).
53. Kalnay, E. *et al.* The NCEP/NCAR 40-year reanalysis project. *Bull. Amer. Meteorol. Soc.* **77,** 437–471 (1996)077<0437:tnyrp>2.0.co;2 (1996).
54. Ventura, V., Paciorek, C. J. & Risbey, J. S. Controlling the proportion of falsely rejected hypotheses when conducting multiple tests with climatological data. *J. Clim.* **17,** 4343–4356 (2004).

Acknowledgements

GPCC precipitation data are provided by the NOAA/OAR/ESRL PSD, Boulder, Colorado, USA, from their web site at http://www.esrl.noaa.gov/psd/. Funding from the Research Council of Norway through the NAPEX (no. 229778) and HYPRE (no. 243942) projects is acknowledged. P.M.F. acknowledges IAGP grant EP/I014721/ and a Royal Society Wolfson Merit Award.

Author contributions

Ø.H. wrote the paper, but with contributions from all the authors. G.M. developed the idea. G.M. and Ø.H. designed the GCM and RCM model experiments and Ø.H. carried out the experiments. All authors participated in analysis of the model results.

Additional information

Competing financial interests: The authors declare no competing financial interests.

24

Tornado outbreak variability follows Taylor's power law of fluctuation scaling and increases dramatically with severity

Michael K. Tippett[1,2] & Joel E. Cohen[3,4]

Tornadoes cause loss of life and damage to property each year in the United States and around the world. The largest impacts come from 'outbreaks' consisting of multiple tornadoes closely spaced in time. Here we find an upward trend in the annual mean number of tornadoes per US tornado outbreak for the period 1954-2014. Moreover, the variance of this quantity is increasing more than four times as fast as the mean. The mean and variance of the number of tornadoes per outbreak vary according to Taylor's power law of fluctuation scaling (TL), with parameters that are consistent with multiplicative growth. Tornado-related atmospheric proxies show similar power-law scaling and multiplicative growth. Path-length-integrated tornado outbreak intensity also follows TL, but with parameters consistent with sampling variability. The observed TL power-law scaling of outbreak severity means that extreme outbreaks are more frequent than would be expected if mean and variance were independent or linearly related.

[1] Department of Applied Physics and Applied Mathematics, Columbia University, New York, New York 10027, USA. [2] Center of Excellence for Climate Change Research, Department of Meteorology, King Abdulaziz University, Jeddah 21589, Saudi Arabia. [3] Laboratory of Populations, Rockefeller University, New York, New York 10065, USA. [4] The Earth Institute, Columbia University, New York, New York 10027, USA. Correspondence and requests for materials should be addressed to M.K.T. (email: mkt14@columbia.edu) or to J.E.C. (email: cohen@rockefeller.edu).

Hazardous convective weather (tornadoes, hail and damaging wind) associated with severe thunderstorms affects large portions of the United States. Tornadoes cause particularly intense damage. Over a recent 10-year period (2005–2014), tornadoes in the United States resulted in an average of 110 deaths per year and annual losses ranging from $500 million to $9.6 billion[1]. The largest societal impacts from tornadoes are from 'outbreaks' in which multiple tornadoes occur in a single weather event. Tornado outbreaks across the eastern two-thirds of the United States were associated with 79% of all tornado fatalities over the period 1972–2010 (ref. 2) and are routinely responsible for billion-dollar loss events[3].

Whether US tornado activity will change in the future remains uncertain. Climate change projections indicate that environments favourable to severe thunderstorms will be more frequent in a warmer climate[4], and high-resolution convection-permitting numerical modelling indicates increased activity and year-to-year variability of March–May US tornado occurrence as measured by severe weather proxies derived from explicitly depicted storms[5]. However, to date, there is no upward trend in the number of reliably reported US tornadoes per year[6]. Interpretation of the US tornado report data requires some caution. For instance, the total number of US tornadoes reported each year has increased dramatically over the last half century, but most of that increase is due to more reports of weak tornadoes and is believed to reflect changing reporting practices and other non-meteorological factors rather than increased tornado occurrence[7]. The variability of reported tornado occurrence has increased over the last few decades with more tornadoes being reported on days when tornadoes are observed[8,9]. In addition, greater year-to-year variability in the number of tornadoes reported per year has been associated with consistent changes in the monthly averaged atmospheric environments favourable to tornado occurrence[10]. Likewise, US large-event severe thunderstorm losses and the frequency of the most extreme environments have increased[11]. Regional changes in the seasonality of tornado occurrence have also been reported[12,13].

Despite their importance, relatively little is known about the current or projected statistics of tornado outbreak severity beyond their climatology[2]. Most studies have considered only the statistics of tornadoes occurring during a single day[14] and have not considered outbreaks over multiple dates. Here we show that the annual mean number of tornadoes per outbreak increased during 1954–2014, and the annual variance increased more than four times faster than the mean. We show that the mean and variance of the number of tornadoes per outbreak are related by Taylor's power law of fluctuation scaling (TL)[15,16] with parameters that are consistent with multiplicative growth[17]. TL scaling in tornado outbreak statistics was not previously known. Although power-law scaling is present in the probability distributions of various tornado characteristics[18,19], the presence of power-law scaling in such probability distributions is neither necessary nor sufficient for TL scaling of the variance in relation to the mean (see also Supplementary Discussion for examples of TL scaling without power-law probability distributions). We also find that a tornado-related atmospheric proxy shows a similar power-law scaling and multiplicative growth. Path-length-integrated tornado outbreak intensity follows TL as well, but with parameters predicted by sampling variability[20]. The findings are similar when we restrict the analysis to the more recent period 1977–2014 and to more intense tornadoes. The observed TL scaling of outbreak severity means that extreme outbreaks are more frequent than would otherwise be expected if mean and variance were independent or linearly related.

Results

Data and sensitivity analyses. We use data from 1954 to 2014, which are generally considered reliable. Because of concerns regarding the data before 1977 ('Methods' section, Supplementary Fig. 1), we repeat some analysis using the more recent period 1977–2014 to test the robustness of the results (Supplementary Figs 2–5). We exclude the weakest tornadoes from our analysis and denote the remaining tornadoes as F1 + tornadoes ('Methods' section). We repeat some of the analysis restricted to more intense tornadoes (F2 +; Supplementary Figs 7–10).

Number of tornadoes per outbreak. The annual number of F1 + tornadoes shows no significant trend over the period 1954–2014 (Fig. 1a). Generally, trends have not been found in the number of severe tornadoes when severity is defined using the Fujita scale, but upward trends have been found when severity is defined using path length[18]. The percentage of F1 + tornadoes that occur in outbreaks ('Methods' section) is increasing by 0.34 percentage points ±0.13 percentage points per year (Fig. 1b), consistent with upward trends in the proportion of tornadoes occurring on days with many tornadoes[8,9]. Here and in all results, ± intervals are 95% confidence intervals. The fraction of F1 + tornadoes that occur in outbreaks is less than one because not all F1 + tornadoes occur in outbreaks. During the period 1977–2014, the number of F1 + tornadoes also shows no significant trend, and the percentage of F1 + tornadoes occurring in outbreaks is also increasing, at a larger estimated rate (Supplementary Fig. 2). The US tornado reports show no statistically significant trend in the frequency of tornado outbreaks (Fig. 2a). Since the number of

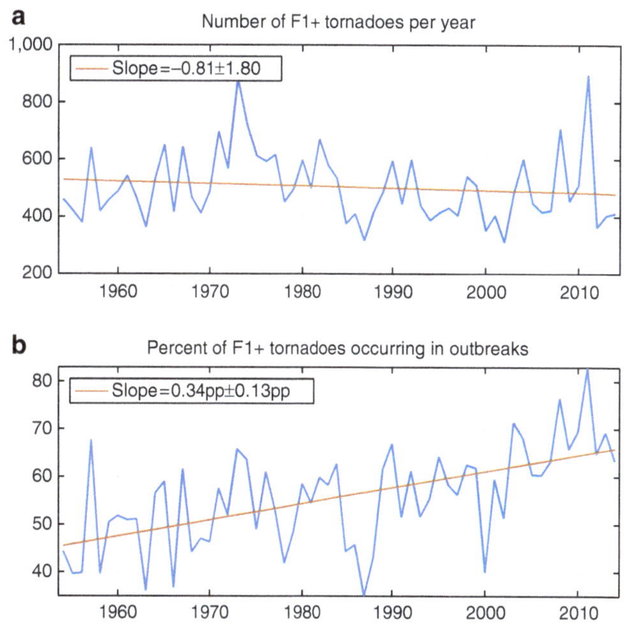

Figure 1 | Time series of counts and clustering of F1+ tornadoes 1954–2014 in the contiguous US. (**a**) Number of F1+ tornadoes per year. The slope of the least-squares regression indicates that the number of F1+ tornadoes per year declined by 0.81 per year on average from 1954 to 2014 inclusive. This rate of decline is not statistically significantly different from 0 (no change). (**b**) Annual percentage of F1+ tornadoes occurring in outbreaks. The slope of the least-squares regression indicates that the percentage of F1+ tornadoes per year that occurred as part of outbreaks increased by 0.34 percentage points (pp) per year on average from 1954 to 2014 inclusive. This increase is statistically significantly greater than 0. In both **a** and **b**, ± intervals are 95% confidence intervals.

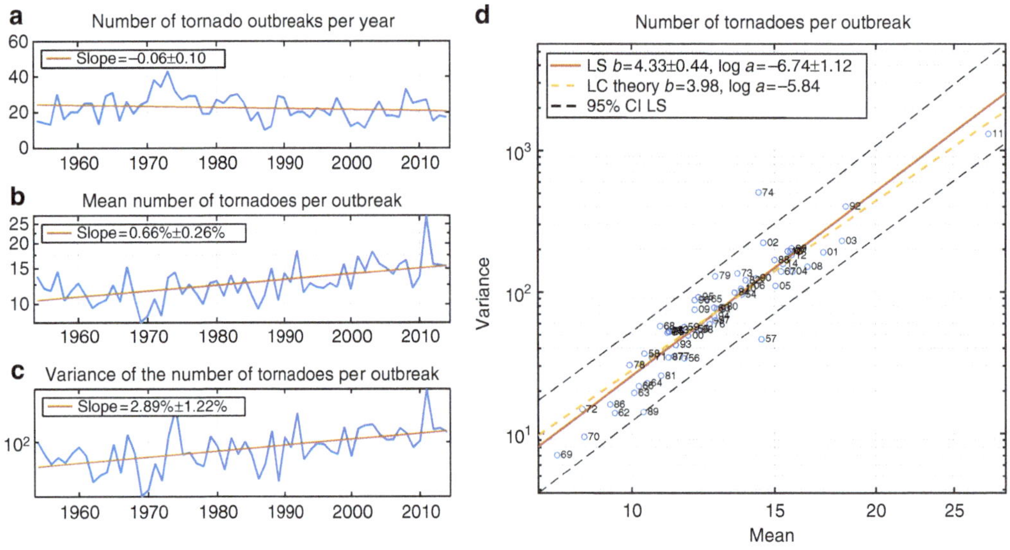

Figure 2 | Numbers of F1+ tornadoes per outbreak 1954–2014. (**a**) Number of tornado outbreaks per year. The rate of decline is not statistically significantly different from 0 (no change). (**b**) Annual mean number of tornadoes per outbreak. Vertical axis is on logarithmic scale, so the rate of increase in the annual mean is expressed as a percentage per year. This rate of increase is statistically significantly greater than 0. (**c**) Annual variance of the number of tornadoes per outbreak. Vertical axis is on logarithmic scale, so the rate of increase in the annual mean is expressed as a percentage per year. This rate of increase is statistically significantly greater than 0. (**d**) Scatter plot of the annual mean number of tornadoes per outbreak versus the annual variance of the number of tornadoes per outbreak. Both axes are on logarithmic scale. The solid red line is the least-squares (LS) regression line (Taylor's power law of fluctuation scaling) and the dashed yellow line has the slope and intercept predicted by LC theory[17]. The two-digit number following the plotting symbol 'o' gives the calendar year in the second half of the twentieth century or first half of the twenty-first century. In all the panels, ± intervals are 95% confidence intervals.

F1+ tornadoes and the number of outbreaks are not changing (on average, over time), the increasing percentage of F1+ tornadoes occurring in outbreaks means that the number of F1+ tornadoes per outbreak must be increasing, and indeed, the annual mean number of F1+ tornadoes per outbreak shows a significant upward trend (Fig. 2b). The annual mean number of tornadoes per outbreak is increasing by 0.66% ±0.26% per year, and the variance is increasing more than four times as fast, 2.89% ±1.22% per year (Fig. 2b,c) over the period 1954–2014. The growth rates are greater over the recent period 1977–2014, with similar ratio between the growth rates of mean and variance (Supplementary Fig. 3b,c).

The fact that the variance is increasing several times faster than the mean is especially noteworthy: it indicates a changing distribution in which the likelihood of extreme outbreaks is increasing faster than what the trend in mean alone would suggest. The coefficient of dispersion of a probability distribution with a positive mean is the ratio of its variance to its mean. Values greater than one (over-dispersion) indicate more clustering than a Poisson variable. For instance, European windstorms exhibit over-dispersion and serial clustering that increases with intensity[21] with implications for the return intervals of rare events[22]. Taylor's law (TL) relates the mean and variance of a probability distribution by

$$\text{variance} = a(\text{mean})^b, a > 0, \tag{1}$$

where a and b are constants[15,16]. A value of $b > 1$ indicates that the coefficient of dispersion increases with the mean. The annual mean and annual variance of the number of tornadoes per outbreak approximately satisfy TL with $b = 4.3 \pm 0.44$ and $\log a = -6.74 \pm 1.12$ (Fig. 2d); consistent values are seen over the period 1977–2014 (Supplementary Fig. 3d). (Throughout log is the natural logarithm.) The value of b here is remarkable since in most ecological applications, the TL exponent seldom exceeds 2. The TL exponent can be greater than 2 for lognormal distributions with changing parameters (Supplementary Discussion and Supplementary Fig. 11). The TL scaling of tornado outbreak severity reveals a remarkably regular relation between annual mean and annual variance that extends over the full range of the data, even for years like 2011 which are extreme in mean and variance. The data from 1974 deviate most from TL scaling, with the excessive variance reflecting the 3–4 April 'Super Outbreak.'

The upward trend in the number of tornadoes per outbreak provides an interpretation for the observed TL scaling since TL scaling arises in models of stochastic multiplicative growth[17]. In such models, the quantity $N(t+1)$ at time $t+1$ is related to its previous value $N(t)$ by

$$N(t+1) = A(t)N(t), \tag{2}$$

where $A(t)$ is the random multiplicative factor by which $N(t)$ grows or declines from one time to the next. Here $N(t)$ is the annual average number of tornadoes per outbreak, and each integer value of t represents one calendar year. The Lewontin–Cohen (LC) model for stochastic multiplicative growth assumes that the $A(t)$ are independently and identically distributed for all $t \geq 0$ with finite mean $M > 0$ and finite variance V. If $M \neq 1$, $N(t)$ follows TL asymptotically with[17]

$$a = \frac{E[(N(0))^2]}{[E(N(0))]^b} \text{ and } b = \frac{\log[V + M^2]}{\log M}. \tag{3}$$

Here we estimate ('Methods' section) $M = 1.03$ and $V = 0.068$, which leads to TL parameters $b = 3.98$ and $\log(a) = -5.84$. Both values are consistent with the least-squares (LS) estimates of the corresponding parameters of TL (Fig. 2d). The LS estimates are also consistent with the values from LC theory during 1977–2014 (Supplementary Fig. 3d). 95% confidence intervals for M and V show that the hypothesis of no growth ($M = 1$) under which equation (3) is not valid cannot be rejected (Supplementary Table 1). The Supplementary Discussion provides additional

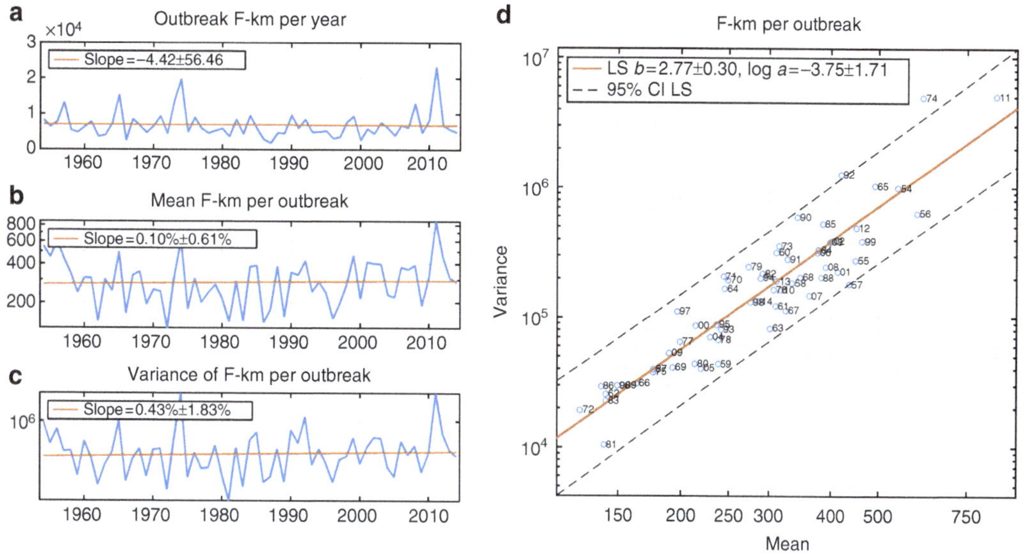

Figure 3 | F-km per outbreak 1954–2014. (**a**) Total outbreak F-km per year. The rate of decline is not statistically significantly different from 0 (no change) among F1+ tornadoes. (**b**) Annual mean F-km per outbreak. Vertical axis is on logarithmic scale, so the rate of increase in the annual mean is expressed as a percentage per year. This rate of increase is not statistically significantly greater than 0. (**c**) Annual variance of F-km per outbreak. Vertical axis is on logarithmic scale, so the rate of increase in the annual mean is expressed as a percentage per year. This rate of increase is not statistically significantly greater than 0. (**d**) Scatter plot of the annual mean of F-km per outbreak versus the annual variance of F-km per outbreak. Both axes are on logarithmic scale. The solid red line is the least-squares (LS) regression line (Taylor's power law of fluctuation scaling). The two-digit number following the plotting symbol 'o' gives the calendar year in the second half of the twentieth century or first half of the twenty-first century. In all the panels, ± intervals are 95% confidence intervals.

description of how the LC model leads to TL scaling with exponent approximately 4.

Fujita-kilometers per outbreak. Another measure of outbreak severity is Fujita-kilometers (ref. 2; F-km) which is the sum (over all tornadoes in an outbreak) of each tornado's path length in kilometers multiplied by its Fujita or Enhanced Fujita rating ('Methods' section). Annual totals of outbreak F-km, mean number of F-km per outbreak and the variance of F-km per outbreak do not show significant trends over the period 1954–2014 (Fig. 3a–c). The mean number of F-km per outbreak and the variance of F-km per outbreak show marginally significant trends over the recent period 1977–2014 (Supplementary Fig. 4a–c). The TL parameters relating the mean and variance of F-km per outbreak are $b = 2.77 \pm 0.30$ and $\log a = -3.75 \pm 1.71$ (Fig. 3d). The lack of robust trends means that LC theory is not appropriate to explain the TL scaling of F-km. However, TL scaling also arises from the sampling of stationary skewed distributions[20]. For a distribution with mean m, variance v, skewness γ_1 and coefficient of variation CV, theory[20] predicts

$$a = \log v - \frac{\gamma_1}{CV}\log m \text{ and } b = \frac{\gamma_1}{CV}. \quad (4)$$

Here, excluding two outlier outbreaks from the calculation of the distribution parameters ('Methods' section, Supplementary Fig. 5), equation (4) gives $b = 2.71$ and $\log a = -3.05$, both of which are consistent with the LS estimates of the TL parameters for F-km per outbreak. (We use 'outlier' to indicate values far from other observations, not to suggest that the unusual values are the result of measurement error.) Therefore TL scaling of F-km per outbreak could be explained by sampling variability.

A tornado environment proxy. A reasonable concern is that the findings here represent properties of the tornado report database that are not meteorological in origin, especially since other prominent features of the tornado report database are not meteorological in origin[7]. Environmental proxies for tornado occurrence and number of tornadoes per occurrence provide an independent, albeit imperfect, measure of tornado activity for the period 1979–2013 ('Methods' section). At a minimum, the environmental proxies provide information about the frequency and severity of environments favourable to tornado occurrence. The correlation between the annual average number of tornadoes per outbreak and the proxy for number of tornadoes per occurrence is 0.56 (Supplementary Fig. 6a). This correlation falls to 0.34, still significant at the 95% level, when the data from 2011 are excluded. Applying a 5-year moving average to the data highlights their common trends and increases the correlation to 0.88 (Supplementary Fig. 6b). The annual mean of the occurrence proxy, a surrogate for number of tornadoes per year, shows a marginally significant upward trend (Fig. 4a). The annual mean and annual variance of the proxy for number of tornadoes per occurrence show upward trends of $0.63 \pm 0.30\%$ and $2.43 \pm 1.12\%$, respectively (Fig. 4b,c), values strikingly similar to those for number of tornadoes per outbreak (Fig. 2b,c). Moreover, the TL parameters of the proxy for number of tornadoes per occurrence are 3.54 ± 0.42 and $\log a = -5.87 \pm 1.43$, which are consistent with the LC multiplicative growth theory estimates for the proxy (Fig. 4d) and quite similar to those for the number of tornadoes per outbreak. Extreme environments associated with tornado occurrence display the TL scaling and multiplicative growth similar to those of the number of tornadoes per outbreak. This similarity plausibly suggests that the changes in the number of tornadoes per outbreak reflect changes in the physical environment.

Sensitivity to outbreak definition. Another concern is that the results are sensitive to the details of the outbreak definition. We

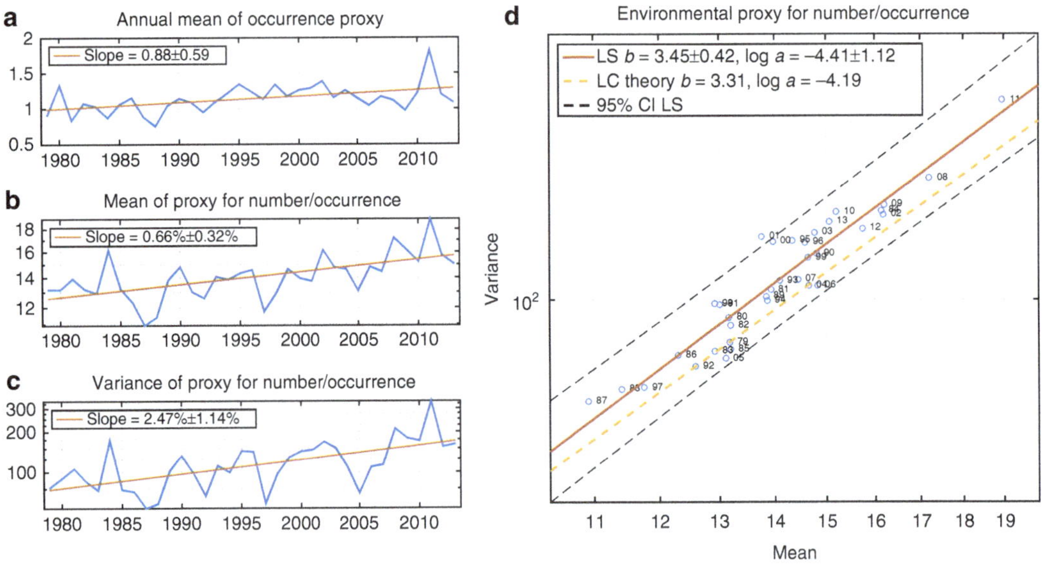

Figure 4 | Environmental proxies 1979–2013. (**a**) Annual mean of occurrence proxy in per cent. The rate of increase is statistically significantly greater than 0. (**b**) Annual mean of the environmental proxy for number of tornadoes per occurrence. Vertical axis is on logarithmic scale, so the rate of increase in the annual mean is expressed as a percentage per year. This rate of increase is statistically significantly greater than 0. (**c**) Annual variance of the environmental proxy for number of tornadoes per occurrence. Vertical axis is on logarithmic scale, so the rate of increase in the annual mean is expressed as a percentage per year. This rate of increase is statistically significantly greater than 0. (**d**) Scatter plot of the annual mean of proxy for number of tornadoes per occurrence versus the annual variance of proxy for number of tornadoes per occurrence. Both axes are on logarithmic scale. The solid red line is the least-squares (LS) regression line and the dashed yellow line has slope and intercept predicted by LC theory[17]. The two-digit number following the plotting symbol 'o' gives the calendar year in the second half of the twentieth century or first half of the twenty-first century. In all the panels, ± intervals are 95% confidence intervals.

assess the robustness of the results to the E/F1 threshold by repeating the analysis with tornadoes rated E/F2 and higher, denoted F2+ (Supplementary Figs 7–10). We use the period 1977–2014 because the annual number of F2+ tornadoes display a substantial decrease (not shown) around the 1970s that is likely related to the introduction of the F-scale. Overall the F2+ results are remarkably similar to the F1+ ones. The annual number of F2+ tornadoes has an insignificantly negative trend during 1977–2014 (Supplementary Fig. 7a), and the percentage of F2+ tornadoes occurring in F2+ outbreaks has a significant positive trend (Supplementary Fig. 7b). Although the number of F2+ outbreaks shows no significant trend, the mean number of tornadoes per F2+ outbreak and its variance both have significant upward trends (Supplementary Fig. 8a–c). The TL exponent for number of tornadoes per F2+ outbreak is 3.65 and is consistent with LC theory (Supplementary Fig. 8d). Annual totals of F2+ outbreak F-km have no significant trend (Supplementary Fig. 9a). Mean F-km per F2+ outbreak does have a significant upward trend, but variance does not (Supplementary Fig. 9b,c). The TL scaling of F2+ outbreak F-km (Supplementary Fig. 9d) is consistent with sampling variability when 2011 is excluded (Supplementary Fig. 10).

Discussion

These findings have important implications for tornado risk in the United States and perhaps elsewhere, though we have examined only the US data. First, the number of tornadoes per outbreak is increasing. However, there is less evidence that F-km per outbreak are increasing. Both the number of tornadoes per outbreak and F-km per outbreak follow TL, which relates mean and variance. We find that TL scaling for the number of tornadoes per outbreak is compatible with multiplicative growth, and that TL scaling for F-km per outbreak could be due to sampling variability. Finally, the key implication of TL scaling is that both number of tornadoes per outbreak and F-km per outbreak exhibit extreme over-dispersion which increases with mean. When the average tornado outbreak severity gets worse, the high extreme of severity rises even faster and the low extreme falls even faster, by either measure of severity.

Methods

Outbreak data. Tornado reports come from the NOAA Storm Prediction Center (http://www.spc.noaa.gov/wcm/#data). There are no reports from either Alaska or Hawaii; one report from Puerto Rico is excluded. Tornadoes are rated by estimated or reported damage using the Fujita (F) scale, introduced in the mid-1970s, and since March 2007, the enhanced Fujita (EF) scale, with 0 being the weakest and 5 the strongest. Only reports of tornadoes rated F/EF1 or greater are used and are denoted as F1+. To compute outbreak statistics, tornado reports are first sorted in chronological order taking into account the time zone. Outbreaks are sequences of 6 or more F1+ tornadoes (regardless of location in the contiguous United States) whose successive start times are separated by no more than 6 h (ref. 2). A total 1,361 outbreaks are found over the period 1954–2014. Outbreaks spanning more than 1 day are possible with this definition. The median outbreak duration is about 8 h, and 95% of the outbreaks last less than 24 h. Additional climatological features can be found in ref. 2 The number of tornadoes and F-km are computed for each outbreak. Then the annual mean and variance of two outbreak severity measures, number of tornadoes per outbreak and F-km per outbreak, are calculated. The distribution of outbreak tornadoes by F/EF-scale shows considerable variation before 1977 (Supplementary Fig. 1) and may well reflect lower reliability in the earlier period[23] before damage surveys were a routine part of the rating procedure. Therefore, we repeat some of our analysis using the recent period 1977–2014 (Supplementary Figs 2,3).

The same outbreak calculation procedure is repeated but considering only reports of tornadoes rated F/EF2 or greater, denoted as F2+. These outbreak events are referred to as F2+ outbreaks (Supplementary Figs 7–10). Only F2+ tornadoes are used to calculate tornado numbers and F-km of F2+ outbreaks.

Trends. All trends and 95% confidence intervals are assessed using linear regression and ordinary least squares, assuming approximately normal distributions of residuals. The growth rates of the annual mean and variance of the outbreak severity measures in Fig. 2b,c, Fig. 3b,c and Supplementary Figs 3b,c; 4b,c; 8b,c and 9b,c are computed by assuming exponential growth and fitting a linear trend to the logarithms of the data. All the other trends are fitted using untransformed data.

TL parameters. The TL parameters in Figs 2d, 3d and 4d and Supplementary Figs 3d, 4d, 8d and 9d and 95% confidence intervals are estimated using ordinary least-squares regression with the logarithms of the mean and variance.

TL parameters implied by LC theory. The growth factor in the LC model is computed from $A(t) = N(t+1)/N(t)$, where $N(t)$ is the average number of tornadoes per outbreak in calendar year t. The 95% confidence intervals for the mean and variance of $A(t)$ are computed from 10,000 bootstrap samples and reported in Supplementary Table 1. The mean and variance of $A(t)$ are used to compute a prediction of the slope b using equation (3). The TL parameter a is estimated from equation (1) evaluated at the initial year, either 1954 or 1977.

TL parameters implied by sampling variability. There are 1,361 cases of outbreak F-km, and their distribution is highly right-skewed (Supplementary Figs 5a and 10a). The F-km values for the 1974 Super Outbreak and the 25–28 April 2011 tornado outbreak are more than 26 standard deviations above the mean of the data on an arithmetic scale and more than six standard deviations above the mean of the log-transformed data, when means and standard deviations are calculated after withholding the two extreme values. These outliers (values that are far from other observations) have a substantial impact on the estimates of the mean, variance and skewness of the F-km distribution. Despite their rarity, about 86% $(1-(1-2/1{,}361)^{1361})$ of the bootstrap samples will contain one or both of these two events. The presence of the outliers results in bimodal distributions of the TL slope and intercept estimates (Supplementary Figs 5b,c and 10b,c) computed from equation (4), depending on whether or not the outlier values are in the particular bootstrap sample. Removal of the outliers results in unimodal distributions, whose ranges are consistent with the least-squares estimates of the TL slope and intercept for F-km (Supplementary Figs 5d,e and 10d,e).

Environmental proxies. We use two environmental proxies: one for tornado occurrence and one for the number of tornadoes. Tornadoes are often associated with elevated values of convective available potential energy (CAPE; $\mathrm{J\,kg^{-1}}$) and a measure of vertical wind shear, called storm relative helicity (SRH; $\mathrm{m^2\,s^{-2}}$). Here 0–180 hPa CAPE and 0–3,000 m SRH data are taken from the North American Regional Reanalysis[24]. The data are interpolated to a 1×1 degree latitude–longitude grid over the continental United States and daily averages are computed. The environmental proxy for tornado occurrence is defined using the energy-helicity index[25] (EHI), which is the product of CAPE and SRH divided by $160{,}000\,\mathrm{J\,kg^{-1}\,m^2\,s^{-2}}$. Values of EHI greater than one indicate the potential for supercell thunderstorms and tornadoes[26], and accordingly we take our proxy for tornado occurrence to be the condition that EHI is greater than 1. Selection of an environmental proxy for the number of tornadoes is more challenging. The only example of a proxy calibrated to predict the number of tornadoes from the surrounding environment uses monthly averaged environment and monthly number of tornadoes[27,28], and therefore is not suited for outbreaks, which last much less than a month. However, previous research does provide some indications for the functional form of such a proxy. For instance, on subdaily timescales the likelihood of significant severe weather is nearly twice as sensitive to vertical wind shear as to CAPE[29,30]. This sensitivity would argue for a proxy based on the product of CAPE and the square of vertical wind shear or equivalently the product of the square root of CAPE and vertical wind shear[31]. Likewise, proxies for the monthly number of tornadoes contain SRH with an exponent ranging from 1.89 to 4.36 depending on region[27,28]. The supercell composite parameter[32] is used in weather forecasting and is the product of CAPE, SRH and vertical shear, again indicating a scaling of severe weather with the square of vertical wind shear measures. On the basis of this evidence, we take as the proxy for number of tornadoes

$$\frac{\mathrm{CAPE} \times \mathrm{SRH}^2}{3{,}600{,}000\,\mathrm{m^4\,s^{-4}}}, \tag{5}$$

which differs from the EHI in that the square of SRH appears. The normalizing factor is chosen to match overall annual outbreak numbers. The proxy for the number of tornadoes per occurrence is therefore the mean of equation (5) conditional on the occurrence proxy of EHI being greater than one.

References

1. Data from Summary of U.S. Natural Hazard Statistics 2005-2014 compiled by the National Weather Service Office of Climate, Water and Weather Services and the National Climatic Data Center. http://www.nws.noaa.gov/om/hazstats.shtml.
2. Fuhrmann, C. M. *et al.* Ranking of tornado outbreaks across the United States and their climatological characteristics. *Wea. Forecasting* **29,** 684–701 (2014).
3. Smith, A. & Matthews, J. Quantifying uncertainty and variable sensitivity within the US billion-dollar weather and climate disaster cost estimates. *Nat. Hazards* **77,** 1829–1851 (2015).
4. Diffenbaugh, N. S., Scherer, M. & Trapp, R. J. Robust increases in severe thunderstorm environments in response to greenhouse forcing. *Proc. Natl Acad. Sci. USA* **110,** 16361–16366 (2013).
5. Gensini, V. & Mote, T. Downscaled estimates of late 21st century severe weather from CCSM3. *Clim. Change* **129,** 307–321 (2015).
6. Tippett, M. K., Allen, J. T., Gensini, V. A. & Brooks, H. E. Climate and hazardous convective weather. *Curr. Clim. Change Rep.* **1,** 60–73 (2015).
7. Verbout, S. M., Brooks, H. E., Leslie, L. M. & Schultz, D. M. Evolution of the U.S. tornado database: 1954-2003. *Wea. Forecasting* **21,** 86–93 (2006).
8. Brooks, H. E., Carbin, G. W. & Marsh, P. T. Increased variability of tornado occurrence in the United States. *Science* **346,** 349–352 (2014).
9. Elsner, J. B., Elsner, S. C. & Jagger, T. H. The increasing efficiency of tornado days in the United States. *Clim. Dyn.* **45,** 651–659 (2015).
10. Tippett, M. K. Changing volatility of U.S. annual tornado reports. *Geophys. Res. Lett.* **41,** 6956–6961 (2014).
11. Sander, J., Eichner, J. F., Faust, E. & Steuer, M. Rising variability in thunderstorm-related U.S. losses as a reflection of changes in large-scale thunderstorm forcing. *Wea. Climate Soc.* **5,** 317–331 (2013).
12. Long, J. A. & Stoy, P. C. Peak tornado activity is occurring earlier in the heart of 'Tornado Alley'. *Geophys. Res. Lett.* **41,** 6259–6264 (2014).
13. Lu, M., Tippett, M. & Lall, U. Changes in the seasonality of tornado and favorable genesis conditions in the central United States. *Geophys. Res. Lett.* **42,** 4224–4231 (2015).
14. Doswell, III C. A., Edwards, R., Thompson, R. L., Hart, J. A. & Crosbie, K. C. A simple and flexible method for ranking severe weather events. *Wea. Forecasting* **21,** 939–951 (2006).
15. Taylor, L. R. Aggregation, variance and the mean. *Nature* **189,** 732–735 (1961).
16. Eisler, Z., Bartos, I. & Kertész, J. Fluctuation scaling in complex systems: Taylor's law and beyond. *Adv. Phys.* **57,** 89–142 (2008).
17. Cohen, J. E., Xu, M. & Schuster, W. S. F. Stochastic multiplicative population growth predicts and interprets Taylor's power law of fluctuation scaling. *Proc. R. Soc. B* **280,** 20122955 (2013).
18. Malamud, B. D. & Turcotte, D. L. Statistics of severe tornadoes and severe tornado outbreaks. *Atmos. Chem. Phys.* **12,** 8459–8473 (2012).
19. Elsner, J. B., Jagger, T. H., Widen, H. M. & Chavas, D. R. Daily tornado frequency distributions in the United States. *Environ. Res. Lett.* **9,** 024018 (2014).
20. Cohen, J. E. & Xu, M. Random sampling of skewed distributions implies Taylor's power law of fluctuation scaling. *Proc. Natl Acad. Sci. USA* **112,** 7749–7754 (2015).
21. Vitolo, R., Stephenson, D. B., Cook, I. M. & Mitchell-Wallace, K. Serial clustering of intense European storms. *Meteor. Z.* **18,** 411–424 (2009).
22. Karremann, M. K., Pinto, J. G., Reyers, M. & Klawa, M. Return periods of losses associated with European windstorm series in a changing climate. *Environ. Res. Lett.* **9,** 124016 (2014).
23. Agee, E. & Childs, S. Adjustments in tornado counts, F-scale intensity, and path width for assessing significant tornado destruction. *J. Appl. Meteor. Climatol.* **53,** 1494–1505 (2014).
24. Mesinger, F. *et al.* North American regional reanalysis. *Bull. Am. Meteorol. Soc.* **87,** 343–360 (2006).
25. Davies, J. M. in *17th Conference on Severe Local Storms*, 107–111 (Amer. Meteor. Soc., St. Louis, MO, 1993).
26. Rasmussen, E. N. & Blanchard, D. O. A baseline climatology of sounding-derived supercell and tornado forecast parameters. *Wea. Forecasting* **13,** 1148–1164 (1998).
27. Tippett, M. K., Sobel, A. H. & Camargo, S. J. Association of U.S. tornado occurrence with monthly environmental parameters. *Geophys. Res. Lett.* **39,** L02801 (2012).
28. Tippett, M. K., Sobel, A. S., Camargo, S. J. & Allen, J. T. An empirical relation between U.S. tornado activity and monthly environmental parameters. *J. Clim.* **27,** 2983–2999 (2014).
29. Brooks, H. E., Lee, J. W. & Craven, J. P. The spatial distribution of severe thunderstorm and tornado environments from global reanalysis data. *Atmos. Res.* **67–68,** 73–94 (2003).
30. Brooks, H. E. Severe thunderstorms and climate change. *Atmos. Res.* **123,** 129–138 (2013).
31. Gilleland, E., Brown, B. G. & Ammann, C. M. Spatial extreme value analysis to project extremes of large-scale indicators for severe weather. *Environmetrics* **24,** 418–432 (2013).
32. Thompson, R. L., Mead, C. M. & Edwards, R. Effective storm-relative helicity and bulk shear in supercell thunderstorm environments. *Wea. Forecasting* **22,** 102–115 (2007).

Acknowledgements

M.K.T. was partially supported by a Columbia University Research Initiatives for Science and Engineering (RISE) award, the Office of Naval Research award N00014-12-1-091, NOAA's Climate Program Office's Modeling, Analysis, Predictions and Projections program award NA14OAR4310185, and the Willis Research Network. J.E.C. was partially supported by U.S. National Science Foundation grant DMS-1225529. The views expressed herein are those of the authors and do not necessarily reflect the views of any of the sponsoring agencies.

Author contributions

All the authors were involved in conducting the research and the scientific discussion equally throughout the study. The data were prepared by M.K.T. The manuscript was written, edited and reviewed by both the authors.

Additional information

Competing financial interests: The authors declare no competing financial interests.

Arctic cut-off high drives the poleward shift of a new Greenland melting record

M. Tedesco[1,2], T. Mote[3], X. Fettweis[4], E. Hanna[5], J. Jeyaratnam[6], J.F. Booth[6], R. Datta[1,6,7] & K. Briggs[8]

Large-scale atmospheric circulation controls the mass and energy balance of the Greenland ice sheet through its impact on radiative budget, runoff and accumulation. Here, using reanalysis data and the outputs of a regional climate model, we show that the persistence of an exceptional atmospheric ridge, centred over the Arctic Ocean, was responsible for a poleward shift of runoff, albedo and surface temperature records over the Greenland during the summer of 2015. New records of monthly mean zonal winds at 500 hPa and of the maximum latitude of ridge peaks of the 5,700 ± 50 m isohypse over the Arctic were associated with the formation and persistency of a cutoff high. The unprecedented (1948–2015) and sustained atmospheric conditions promoted enhanced runoff, increased the surface temperatures and decreased the albedo in northern Greenland, while inhibiting melting in the south, where new melting records were set over the past decade.

[1] Lamont-Doherty Earth Observatory, Columbia University, New York, New York 10964, USA. [2] NASA Goddard Institute of Space Studies, New York, New York 10025, USA. [3] University of Georgia, Athens, Georgia 30602-2502, USA. [4] University of Liege, Liege 4000, Belgium. [5] University of Sheffield, Sheffield S10 2TN, UK. [6] The City College of New York, New York, New York 10031, USA. [7] The Graduate Center of the City University of New York, New York, New York 10016, USA. [8] University of Leeds, Leeds LS2 9JT, UK. Correspondence and requests for materials should be addressed to M.T. (email: mtedesco@ldeo.columbia.edu or cryocity@gmail.com).

Atmospheric circulation affects the energy and mass budgets of the Greenland ice sheet[1–5] by controlling cloud coverage and optical depth[6], and by driving the spatial distribution, and amount of surface melting and accumulation[7,8]. Improving our understanding of the impact of atmospheric circulation on the Greenland's surface mass balance is, therefore, crucial for the refinement of climate and ice-sheet models, and will ultimately enable improved estimates of current and future contributions to sea level, by the largest ice body in the Northern Hemisphere.

Here we show that a poleward shift of melting record over the Greenland ice sheet in 2015 was driven by the exceptional atmospheric conditions characterized by new records in mean zonal winds and jet stream wave amplitude associated with the formation and evolution of a Arctic cutoff high.

Results

Atmospheric conditions and indicators. Our analysis of the geopotential height at 500 hPa (Methods) shows that during July 2015 a persistent atmospheric ridge was centred over the Arctic Ocean (Lincoln Sea, north of Greenland), with geopotential height anomalies being up to 3.7 s.d.'s (σ, ~150 m) above the 1981–2010 long-term mean (Fig. 1a). The North Atlantic Oscillation (NAO; Methods) and Greenland Blocking Index (GBI, defined as the 500 hPa geopotential height area averaged between 60–80° N and 20–80° W (ref. 9); Methods) have been associated with extreme melting events over the Greenland[9–12]. The summer average (June-July-August) value for NAO in 2015 of −1.61 was close to the summer value in 2012 of −1.59. Differently from 2015, however, the atmospheric ridge in 2012 was centred over the Greenland ice sheet[2,10] (Supplementary Figs 1 and 2). The July monthly averaged NAO value set a new record low of −1.23 (since 1899), being 3.2σ below the 1981–2010 mean (Fig. 1c). Concurrently, the GBI also set a new record for the month of July (Fig. 1c; Supplementary Fig. 3b), being 2.8σ above the 1981–2010 mean. The June and August conditions in 2015 were not as exceptional in 2012, with mean June and August NAO values in 2015 being higher than the same quantities in 2012 (Supplementary Fig. 3a).

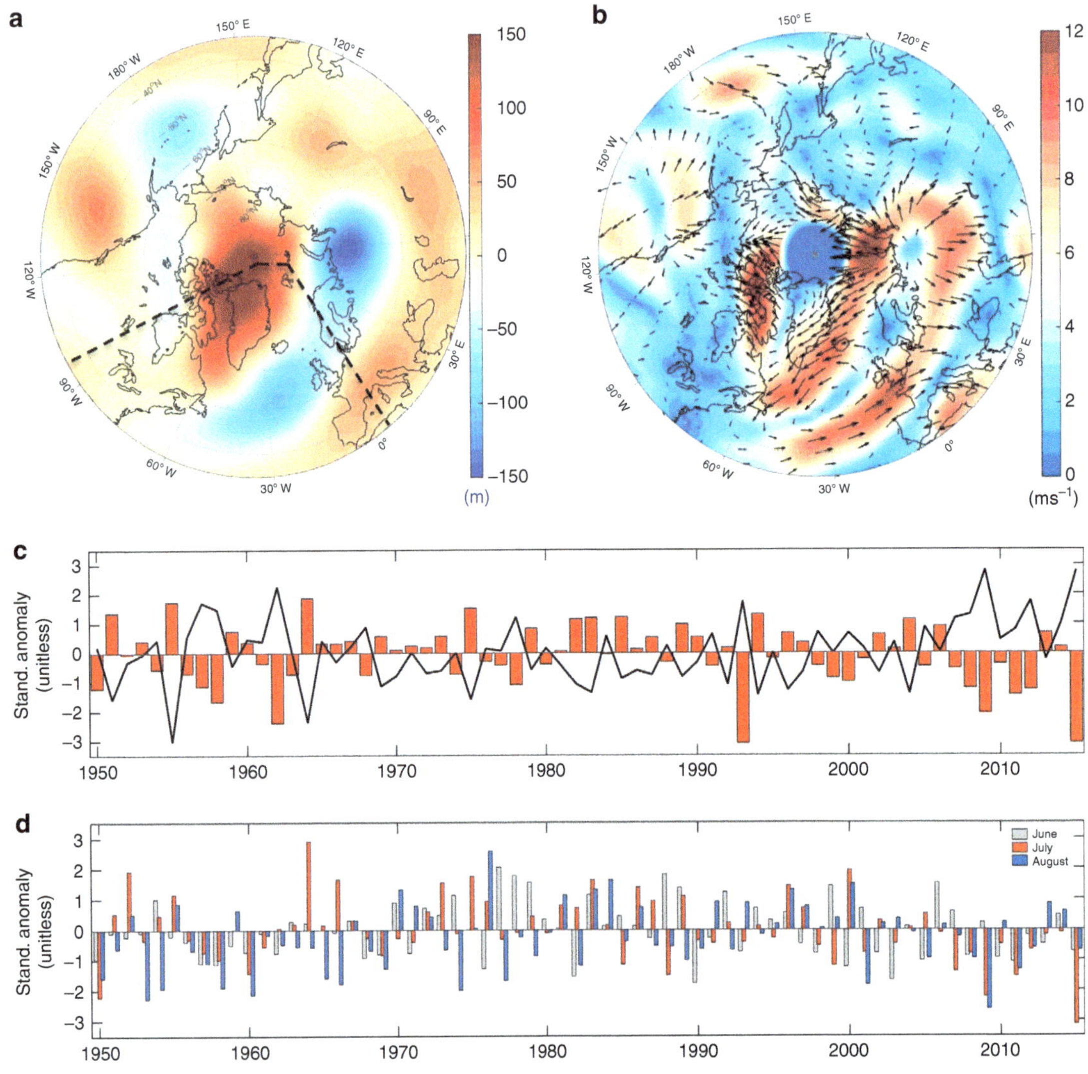

Figure 1 | Atmospheric conditions and selected quantities over the Arctic and over the Greenland ice sheet. (**a**) 500 hPa geopotential height composite anomaly (m) for the month of July 2015, with respect to the 1981-2010 baseline period (using NCEP-NCARv1 reanalysis); (**b**) same as **a**, but for the vector winds (ms^{-1}). (**c**) Time series of monthly averaged July NAO (red bars) and GBI (black line) indices (unitless) for the period 1950-2015. (**d**) Time series of standardized anomalies for the zonal winds at 500 hPa (unitless) averaged over the months of June (light gray), July (red) and August (pale blue) over the region bounded between 45-85° N and 100° W-0° E (included in the area marked by the dashed lines in **a**).

Arctic cutoff high and new surface Greenland records. The July 2015 high-pressure ridge over the Greenland evolved from a cutoff high that formed along the eastern coast of Greenland at the end of June (white circle in Fig. 2b). Over the same period, the jet stream, here characterized through the 5,700 ± 500 m 500 hPa isoheights[13], broke into three positive heights around the Northern Hemisphere (Fig. 2c). These conditions reinforced the atmospheric ridge over the Greenland ice sheet, which moved westward and persisted until mid-July (Fig. 2b–e). This promoted new records for meltwater production, runoff, albedo and surface temperature over northwest Greenland (Fig. 3; Supplementary Fig. 4), as simulated by the Modèle Atmosphérique Régionale[1,3,7] (MAR; Methods). The monthly averaged record setting values for simulated albedo and surface temperature in northwest Greenland for July 2015 were, respectively, $\sim 2.5\sigma$ below and above the 1981–2010 mean, while runoff was up to $\sim 3\sigma$ above the mean (Fig. 3). The spatial distribution of the 2015 surface albedo anomaly simulated by MAR (Supplementary Fig. 5a–c) indicates that the July 2015 negative anomaly was driven by an albedo decrease at relatively high elevations, promoted by the reduced summer snowfall (associated with anticyclonic conditions) and by increased surface melting and runoff. The same atmospheric conditions that promoted these new records over the northern Greenland also inhibited melting in the south, where enhanced melting and new records have been occurring over the past recent years[1,10]. This had implications for the surface mass balance of Greenland at both regional and ice-sheet scales (Supplementary Fig. 6). The summer exposure of bare ice and the presence of surface impurities have been suggested to be driving the enhanced melting observed over the past ~20 years[1]. However, the 2015 summer atmospheric conditions promoted the flow of cold air from the Arctic Ocean (Fig. 1b), favouring the accumulation of fresh new snow with a high albedo along western Greenland, hence offsetting the effects of bare ice exposure (Supplementary Figs 5–7).

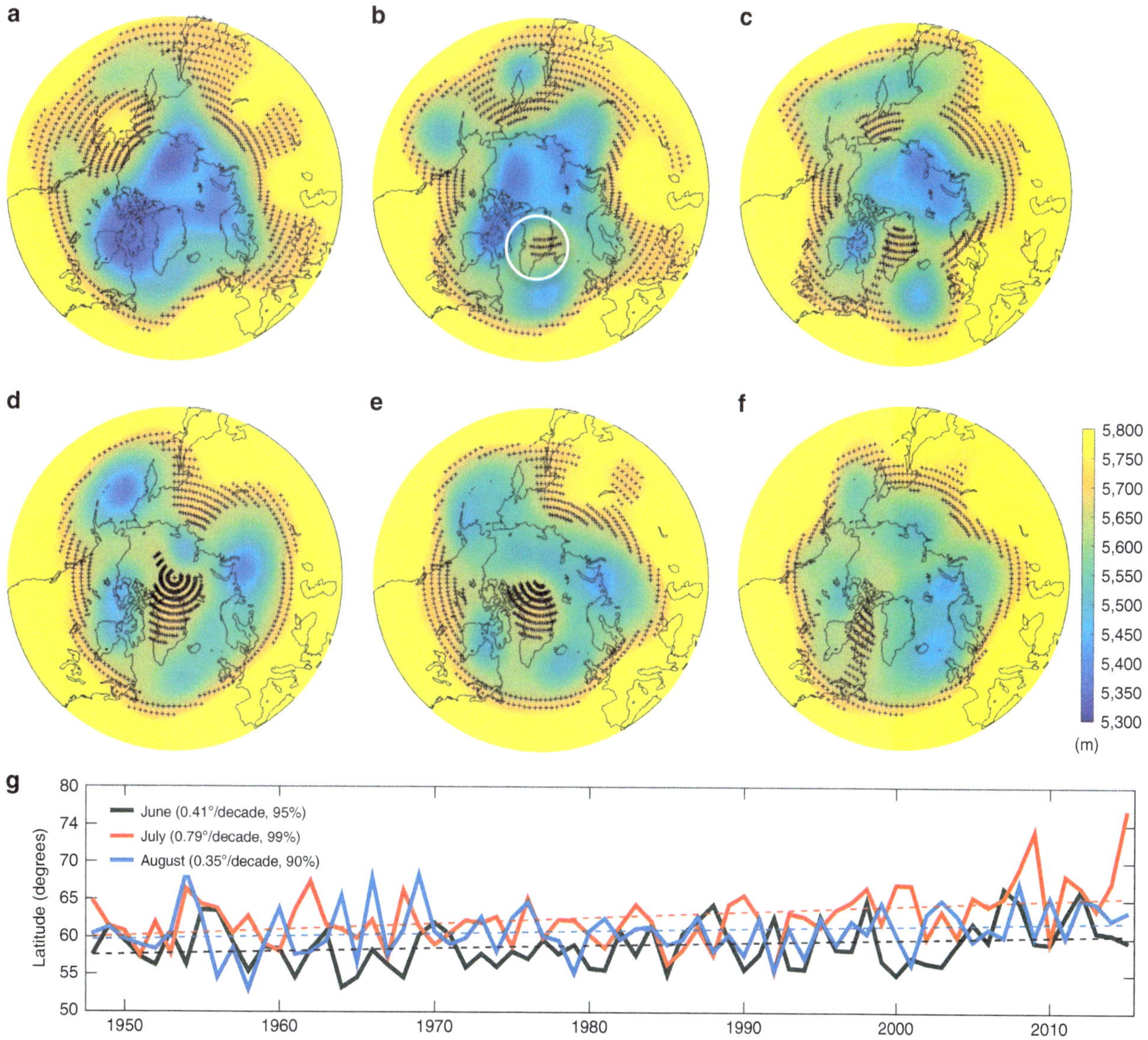

Figure 2 | Spatial and temporal evolution of the jet stream conditions and wave amplitude. Jet stream is here characterized through the 5,700 ± 50 m 500 hPa isoheights[13]. **(a–f)** Five day average geopotential height (m) at 500 hPa for the period 18th June to 22nd July 2015. For each day, the daily averaged values for the 2 days before and after were averaged with the daily average value of that day. Crosses show the locations where geopotential height values are 5,700 ± 50 m. **(g)** Maximum latitude of ridge peaks computed from the 500 hPa 5,700 ± 50 m isoheight for the period 1948-2015 over the region bounded between 45-85° N and 100° W-0° E (included in the area marked by the dashed lines in Fig. 1b), averaged over the months of June (black), July (red) and August (blue). Linear trends and significance levels are reported in the figure.

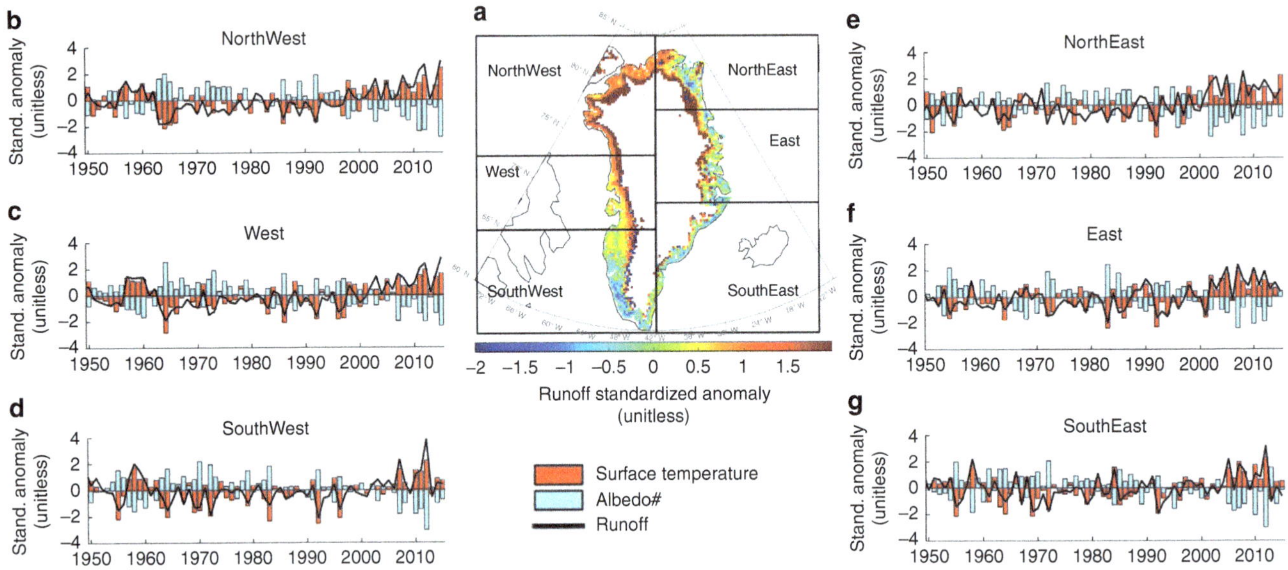

Figure 3 | Surface mass balance and energy balance quantities over the Greenland ice sheet. (**a**) Spatial distribution of the July 2015 MAR-simulated runoff anomaly (unitless) over the Greenland ice sheet (1981-2010). The boxes in the map display the boundaries of the different drainage regions. (**b-g**) Time series of standardized anomalies (unitless) for mean July runoff (black line), surface temperature (red bars) and surface broadband albedo (cyan bars) for the period 1950-2015 over the different drainage basin regions identified in **a**, as simulated by the MAR model.

Mean zonal winds and jet stream wave amplitude records. The westward shift of the cutoff high between the end of June and the beginning of July 2015 is associated with new records for both the mean zonal winds at 500 hPa (Fig. 1d) and the maximum latitude of ridging (Fig. 2g) over the region bounded between 45–85° N and 100° W–0° E. The July-averaged value of the 500 hPa zonal winds speed over the Greenland for latitudes between 60° and 80° N was $-1.5\,m\,s^{-1}$ (here, the negative sign indicates easterly flow), $\sim 3\sigma$ below the mean. The two other occurrences of easterly flow for the same quantity during the 1948–2015 period happened in 1950 and 2009. However, the mean wind speed values were only $\sim -0.15\,m\,s^{-1}$ in both cases, hence much smaller in magnitude than the 2015 record value. The monthly mean maximum latitude of ridge peaks of the 5,700 ± 50 m isohypse over the Atlantic sector (45–85° N and 100° W–0° E) also set the new record of 76.61° N (3.4σ) in July 2015, exceeding the previous record of 74.92° N (2.9σ) set in 2009 (Fig. 2g). The trend for the maximum latitude of the ridge peaks for the period 1948–2015 for the month of July is 0.79 ± 0.09° per decade. The same trends for the months of June and August are, respectively, 0.41 ± 0.14° per decade and 0.35 ± 0.19° per decade. Such trends are even larger when considering only the satellite era (1979–2015), being up to ~2.4° per decade for the month of July and are confirmed by the analysis of different global atmospheric reanalysis data sets (Supplementary Fig. 8).

Discussion

The mechanisms that created and maintained the 2015 observed ridge may be linked with forcing from very strong extratropical cyclones[14], to forcings from southern regions[15] or to latent heat release[16]. Another possibility is the local forcing related to Arctic amplification[13,17]. Although recent melt records over the Greenland have been linked to exceptional mid-tropospheric conditions, with episodes of atmospheric blocking ridges being associated with Greenland's melting extremes[9,12], little or no attention has been given to the impact of the anticipated effects of Arctic amplification on the surface mass balance of the Greenland ice sheet. In this regard, the 2015 records for both the 500 hPa zonal winds and the maximum ridging latitude are consistent with the proposed effects on upper level atmosphere characteristics associated with Arctic amplification[13,17].

The 2015 poleward shift of the surface melting record in 2015, clearly indicates that improving our understanding of the impact of exceptional atmospheric conditions on the spatial distribution of extreme melting is crucial. Besides modulating the contribution of Greenland to sea level through the volume of meltwater production, the location of enhanced melting can influence ocean/ice interaction processes and ocean circulation[18], and bio-productivity, by altering salinity and temperature profiles of the surrounding ocean. Furthermore, the evolution of surface melting strongly impacts the Greenland's hydrological system, with implications for the englacial and subglacial systems, as well as ice discharge and dynamics[19]. Currently, several general circulation global climate models and Earth System models do not properly capture summer Arctic atmospheric forcing[8], limiting our capability to properly project the evolution of the surface mass balance and melting under future warming scenarios. Our work presented here demonstrates a strong need to identify the mechanisms that create and maintain strong cutoff highs. The new atmospheric records, and the trends of mean zonal winds and wave amplitude of the jet stream are consistent with the suggested effects of Arctic amplification[13,17]. Recent studies provide theoretical arguments that slowing zonal winds might be associated with larger planetary wave amplitudes[20] and that Arctic amplification and/or sea-ice loss do intensify existing ridges, thereby contributing to their persistence[21,22]. In the event studied here, however, the exceptional melting followed the ridging, rather than preceding it in alignment with other studies, indicating that observations and models results do not support the above mentioned expected effects of Arctic amplification[23–27]. Be that as it may, understanding the impact of cutoff highs on the Greenland's surface mass balance, and studying the mechanisms driving the trends and extremes of the anticipated effects of Arctic amplification are crucial tasks in view of the potential regional and global impacts long-time effects of enhanced melting over Greenland.

Methods

NAO data set. For the NAO index, we used the Hurrell[28] NAO values distributed by the Climate Research Unit of the University of Anglia (http://www.cru.uea.ac.uk/cru/data/nao/). The Hurrell NAO index is computed from the difference of normalized sea-level pressure between Lisbon, Portugal and Stykkisholmur/Reykjavik, and Iceland. More detail on the Hurrell NAO values can be found at https://climatedataguide.ucar.edu/climate-data/hurrell-north-atlantic-oscillation-nao-index-pc-based (ref. 28).

Reanalysis data. Geopotential heights and zonal winds at 500 hPa are obtained from the National Centers for Environmental Prediction (NCEP)/National Center for Atmospheric Research (NCEP/NCAR) reanalysis data set. The NCEP/NCAR data set consists of globally, gridded data sets, incorporating observations and outputs from a numerical weather prediction model from 1948 to present[29]. Values of the GBI are computed from the daily and monthly values of the NCEP/NCAR 500 hPa geopotential height values averaged over the area between 60–80° N and 280–340° E (ref. 9). ERA-Interim reanalysis data is used to complement the results obtained with NCEP/NCAR and was downloaded from http://apps.ecmwf.int/datasets/data/interim-full-daily/levtype=sfc/.

The MAR regional climate model. Simulations of surface quantities over the Greenland ice sheet are performed using the MAR[1,7,8] MAR is a modular, hydrostatic, and compressible atmospheric model that uses the sigma-vertical coordinate to better simulate airflow over complex terrain and the Soil Ice Snow Vegetation Atmosphere Transfer scheme surface model. The snow model in MAR is the CROCUS model[30]. The MAR model configuration used here has 25 terrain-following sigma layers between the Earth's surface and the top. The horizontal resolution of the outputs used here is 20 km. The lateral boundary conditions are prescribed every 6 h from NCEP/NCAR meteorological fields. Here we use daily outputs obtained from the average of 120 s outputs produced by the model. The temporal configuration for the runs is from 1948 to the present. The sea surface temperature and the sea-ice cover are also prescribed every 6 h in the model, using NCEP/NCAR reanalysis data[29]. No nudging or interactive nesting is used in any of the experiments, with the atmospheric fields over the Greenland ice sheet computed by the atmospheric module of MAR. The atmospheric model, in turn, interacts with the CROCUS model, which provides the state of the snowpack, and associated surface mass balance and energy balance quantities (for example, albedo and runoff).

Code availability. MAR is an open-source code available to the scientific community. The source code for the MAR version used in this study is available at ftp://tedesco-dell.ldeo.columbia.edu/cryoftp/MARv3.5.2src_2015-03-18.tgz. The codes used for analysing the reanalysis data are available upon request from the authors.

References

1. Tedesco, M. *et al.* The role of albedo and accumulation in the 2010 melting record in Greenland. *Environm. Res. Lett.* **6,** 014005 (2011).
2. Tedesco, M. *et al.* Evidence and analysis of 2012 Greenland records from spaceborne observations, a regional climate model and reanalysis data. *The Cryosphere* **7**, 615–630 (2013).
3. Fettweis, X. *et al.* Brief communication 'Important role of the mid-tropospheric atmospheric circulation in the recent surface melt increase over the Greenland ice sheet'. *The Cryosphere* **7**, 241–248 (2013).
4. Hanna, E. *et al.* The influence of North Atlantic atmospheric and oceanic forcing effects on 1900–2010 Greenland summer climate and ice melt/runoff. *Int. J. Climatol.* **33,** 862–880 (2013).
5. Hanna, E. *et al.* Atmospheric and oceanic climate forcing of the exceptional Greenland ice sheet surface melt in summer 2012. *Int. J. Climatol.* **34,** 1022–1037 (2014).
6. Van Tricht *et al.* Clouds enhance Greenland ice sheet meltwater runoff. *Nat. Commun.* **7,** 10266 (2016).
7. Fettweis, X., Tedesco, M., Broeke, M. & Ettema, J. Melting trends over the Greenland ice sheet (1958–2009) from spaceborne microwave data and regional climate models. *The Cryosphere* **5,** 359–375 (2011).
8. Tedesco, M. & Fettweis, X. 21st century projections of surface mass balance changes for major drainage systems of the Greenland ice sheet. *Environ. Res. Lett.* **7,** 045405 (2012).
9. Hanna, E., Cropper, T., Hall, R. & Cappelen, J. Greenland Blocking Index 1851-2015: a regional climate change signal. *Int. J. Climatol.* doi:10.1002/joc.4673 (2016).
10. Tedesco, M. *et al.* Greenland, in states of the climate in 2013. *Bull. Amer. Meteor. Soc.* **95,** S1–S279 (2014).
11. Nghiem, S. V. *et al.* The extreme melt across the Greenland ice sheet in 2012. *Geophys. Res. Lett.* **39,** L20502 (2012).
12. McLeod, J. T. & Mote, T. L. Linking interannual variability in extreme Greenland blocking episodes to the recent increase in summer melting across the Greenland ice sheet. *Int. J. Climatol* **36,** 1484–1499 (2016).
13. Francis, J. A. & Vavrus, S. J. Evidence linking Arctic amplification to extreme weather in mid-latitudes: Arctic links to mid-latitude weather. *Geophys. Res. Lett.* **39,** L06801 (2012).
14. Archambault, H. M., Bosart, L. F., Keyser, D & Cordeira, J. M. A climatological analysis of the extratropical flow response to recurving western north pacific tropical cyclones. *Mon. Wea. Rev.* **141,** 2325–2346 (2013).
15. Krishnamurti, T. N. *et al.* A Pathway connecting the monsoonal heating to the rapid Arctic ice Melt. *J. Atmos. Sci.* **72,** 5–34 (2015).
16. Pfahl, S., Schwierz, C., Croci-Maspoli, M., Grams, C. M. & Wernli, H. Importance of latent heat release in ascending air streams for atmospheric blocking. *Nat. Geosci.* **8,** 610–614 (2015).
17. Francis, J. & Vavrus, S. J. Evidence for a wavier jet stream in response to rapid Arctic warming. *Environ. Res. Lett.* **10,** 014005 (2015).
18. Rahmstorf, S. *et al.* Exceptional twentieth-century slowdown in Atlantic Ocean overturning circulation. *Nat. Clim. Change* **5,** 475–480 (2015).
19. Tedesco, M. *et al.* Ice dynamic response to two modes of surface lake drainage on the Greenland ice sheet. *Environ. Res. Lett.* **8,** 034007 (2013).
20. Huang, C. S. & Nakamura, N. Local finite-amplitude wave activity as a diagnostic of anomalous weather events. *J. Atmos. Sci.* **73,** 211–229 doi:http://10.1175/JAS-D-15-0194.1 (2016).
21. Lee, M.-Y., Hong, C.-C. & Hsu, H.-H. Compounding effects of warm sea surface temperature and reduced sea ice on the extreme circulation over the extratropical North Pacific and North America during the 2013–2014 boreal winter. *Geophys. Res. Lett.* **42,** 1612–1618 (2015).
22. Kug, J. S. *et al.* Two distinct influences of Arctic warming on cold winters over North America and East Asia. *Nat. Geosci.* **8,** 759–762 (2015).
23. Hoskins, B. & Woollings, T. in *Current Climate Change Reports.* (eds Sobel, A. & Camargo, S. J) **1,** 115–124 (2015).
24. Barnes, E. A. & Screen, J. A. The impact of Arctic warming on the midlatitude jet-stream: Can it? Has it? Will it?: Impact of Arctic warming on the midlatitude jet-stream. *Wiley Interdiscip. Rev. Clim. Change* **6,** 277–286 (2015).
25. Barnes, E. A. Revisiting the evidence linking Arctic amplification to extreme weather in midlatitudes: Arctic Amplification and weather. *Geophys. Res. Lett.* **40,** 4734–4739 (2013).
26. Screen, J. & Simmonds., I. The central role of diminishing sea ice in recent Arctic temperature amplification. *Nature* **464,** 1334–1337 (2010).
27. Hassanzadeh, P. & Kuang, Z. Blocking variability: Arctic amplification versus Arctic Oscillation. *Geophys. Res. Lett.* **42,** 8586–8589 (2015).
28. Hurrell, J. *The Climate Data Guide: Hurrell North Atlantic Oscillation (NAO) Index (station-based).* Available at https://climatedataguide.ucar.edu/climate-data/hurrell-north-atlantic-oscillation-nao-index-station-based.
29. Kalnay, E. *et al.* The NCEP/NCAR 40-year reanalysis project. *Bull. Amer. Meteor. Soc.* **77,** 437–471 (1996).
30. Brun, E., David, P., Sudul, M. & Brunot, G. A numerical model to simulate snow-cover stratigraphy for operational avalanche forecasting. *J. Glaciol.* **38,** 13–22 (1992).

Acknowledgements

M.T. and J.J. were supported by the National Science Foundation grants PLR1304807 and ANS 0909388, and NASA grant NNX1498G. T.M. was supported by the NASA grant NNX1498G. M.T. thanks Adriana Basagni for reviewing the draft of the manuscript and for the discussions concerning the Arctic atmospheric variability. We also wish to thank the three anonymous reviewers for their suggestions that led to an improved version of this manuscript.

Author contributions

M.T. conceived the study, analysed the data and prepared the main body of the manuscript. M.T., T.M., E.H., R.D., X.F. and J.F.B. contributed to the analysis of the atmospheric data. X.F. and M.T. generated MAR outputs. J.J. contributed to the extraction and processing of MAR outputs. K.B. and M.T. contributed to the analysis of MAR data. All authors contributed to the editing of the final version of the manuscript.

Additional information

Competing financial interests: The authors declare no competing financial interests.

Rapid intensification and the bimodal distribution of tropical cyclone intensity

Chia-Ying Lee[1], Michael K. Tippett[2,3], Adam H. Sobel[2,4] & Suzana J. Camargo[4]

The severity of a tropical cyclone (TC) is often summarized by its lifetime maximum intensity (LMI), and the climatological LMI distribution is a fundamental feature of the climate system. The distinctive bimodality of the LMI distribution means that major storms (LMI $>$96 kt) are not very rare compared with less intense storms. Rapid intensification (RI) is the dramatic strengthening of a TC in a short time, and is notoriously difficult to forecast or simulate. Here we show that the bimodality of the LMI distribution reflects two types of storms: those that undergo RI during their lifetime (RI storms) and those that do not (non-RI storms). The vast majority (79%) of major storms are RI storms. Few non-RI storms (6%) become major storms. While the importance of RI has been recognized in weather forecasting, our results demonstrate that RI also plays a crucial role in the TC climatology.

[1] International Research Institute of Climate and Society, Columbia University, Palisades, New York 10964, USA. [2] Department of Applied Physics and Applied Mathematics, Columbia University, New York 10027, USA. [3] Center of Excellence for Climate Change Research, Department of Meteorology, King Abdulaziz University, Jeddah 21589, Saudi Arabia. [4] Division of Ocean and Climate Physics, Lamont-Doherty Earth Observatory, Columbia University, Palisades, New York 10964, USA. Correspondence and requests for materials should be addressed to C.-Y.L. (email: clee@iri.columbia.edu).

The question of how climate change will affect tropical cyclone (TC) activity has drawn considerable attention in the past two decades[1–4]. The current expectation is that we can expect a small increase in the global frequency of intense storms along with a small reduction in the total number of storms[4]. However, details of how the TC intensity distribution may change remain uncertain. This uncertainty reflects both the difficulty of simulating the most intense storms in climate change projections[4], as well as an incomplete understanding of what determines the climatological intensity distribution in the current climate[5]. Improved understanding of the TC intensity distribution in the current climate seems necessary if we are to understand TC intensity changes in a warming climate. Among various metrics of TC intensity, lifetime maximum intensity (LMI) is an integrated statistic of TC intensification, and its distribution represents a fundamental property of the TC climatology[6–8].

Several authors have noted that the probability density function (PDF) of global LMI is bimodal. The LMI PDF from International Best-Track Archive for Climate Stewardship for the period 1975–2007 has two local maxima around 40 and 100 kt, and a local minimum at 65 kt (ref. 9). The LMI PDF from a more temporally consistent global data for the period 1982–2009 shows the first maximum at 50 kt and the second one around 120 kt (ref. 10). Considering individual basins, the LMI PDF has its first peak at 50 kt for the North Atlantic, eastern and western North Pacific storms and its second maximum at 110 and 90 kt for eastern and western North Pacific storms[11]. The LMI distribution of Atlantic storms shows no clear secondary maximum, but substantial right skewness[11]. While the precise locations of the LMI PDF peaks vary with the data set and basin, the bimodal feature is quite robust and impossible to overlook. The existence of a second mode means that major storms (sustained winds >96 kt, that is, categories 3–5 TCs in Saffir-Simpson Hurricane Wind Scale) are much less rare than would be expected from the behaviour of the LMI distribution at lower values. The nonmonotonic behaviour of the LMI distribution is in contrast to that of the intensification rate distribution, which is exponential[12].

Although no complete explanation has been offered in the literature for the bimodality of the LMI distribution, there have been suggestions and indications of what processes might be involved. Uncertainties in the best-track winds is one of them. For example, the widely used Dvorak technique for estimating TC intensity from satellite imagery has low resolution at higher intensities, with only one bin in the category 3 range of the Saffir-Simpson scale. It has been argued that this may result in an artificially low number of category 3 hurricanes (LMI between 96–112 kt) in the Atlantic[13]. A recent study, on the other hand, proposed a parameterization of the ratio of surface exchange coefficients C_k/C_d, which appears in potential intensity theory[14,15], as a function of wind speed with a local maximum around 115 kt, and speculated that such a maximum would be favourable for rapid intensification (RI) and might explain the bimodal distribution of LMI[5]. In addition to the slope change in C_k/C_d, observations have suggested an association between RI and eye formation[16]. Organized convective heating results in increased vortex efficiency[17–20], which can potentially increase RI probability and cause intense systems. Perhaps most directly relevant, storms reaching the highest intensities in the western North Pacific[21] and North Atlantic[22] typically do so after undergoing RI at least once.

Here we extend these results to all basins, and show explicitly that RI explains the bimodality of the LMI distribution. We separate storms into two groups: those that undergo RI during their lifetime (RI storms) and those that do not (non-RI storms). We find that the bimodality of the LMI distribution reflects the mixture of these two unimodal distributions, with the higher intensity peak consisting mostly of storms which have undergone RI at some point. In other words, the LMI distribution is unimodal when RI storms are excluded—RI storms are responsible for the bimodality of the LMI distribution. Various thresholds have been used to define RI[23,24]. Here we define RI as an increase of at least 35 kt in the maximum sustained surface wind over a period of 24 h or less. We find this definition of RI to be the most effective in separating the LMI distribution into two unimodal distributions. Sensitivity of the results to other RI thresholds, such as 30 kt, is discussed later.

Results

The bimodal LMI distribution. The global distribution of LMI for the period 1981–2012 has a peak at 45 kt and an indication of a secondary maximum around 120 kt (Fig. 1). Basin distributions of LMI are bimodal with local maxima around 45 and 120–135 kt except for the Atlantic where there is only a hint of a second maximum (Fig. 2), similar to what is shown in the literature[11]. The observed LMI distribution shows that category 3 and 4 storms are more common than category 1 and 2 storms. Most types of natural hazards become more rare as they become more extreme, for example, earthquakes[25] and tornadoes[26]. TCs, as measured by LMI, are unusual in having a range over which frequency increases with intensity. The second peak in the western North Pacific occurs at a higher intensity than in the other basins, consistent with the observation that the stronger storms globally occur more often in that basin[27].

Relation to RI. The LMI distribution of the 2,303 non-RI storms in the global record is unimodal and forms the first peak of the complete distribution (blue curve in Fig. 1). About 6% (141) of the non-RI storms are major storms (LMI >96 kt, categories 3–5), and the largest LMI value in this group is 145 kt (category 5). The second peak in the LMI distribution is formed by the 766 RI storms (red curve in Fig. 1), and 79% (603) of them are major storms. The same separation by RI in individual basins yields

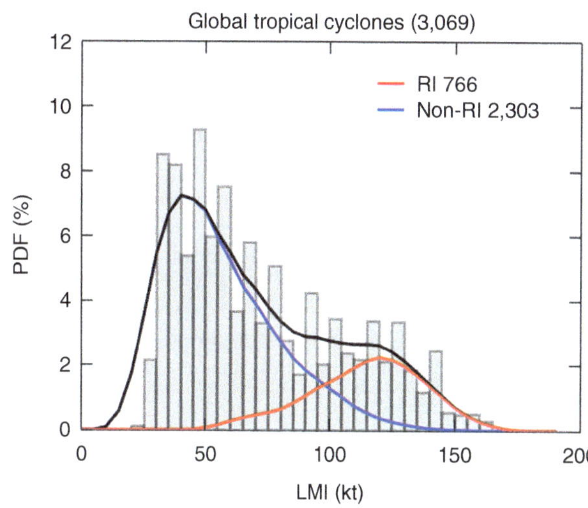

Figure 1 | Distributions of global tropical cyclone LMI. PDFs are calculated using 1981-2012 global tropical cyclone LMI. The grey bars show the raw data binned in 5 kt bins. The black, red and blue lines show the smoothed PDF for all storms, storms those undergo rapid intensification during their lifetime (RI storms), and those do not (non-RI storms), respectively. Smoothing is by moving average with window width of 15 kt. Total number of storms is listed in the title, while the numbers of RI and non-RI storms are given in the legend.

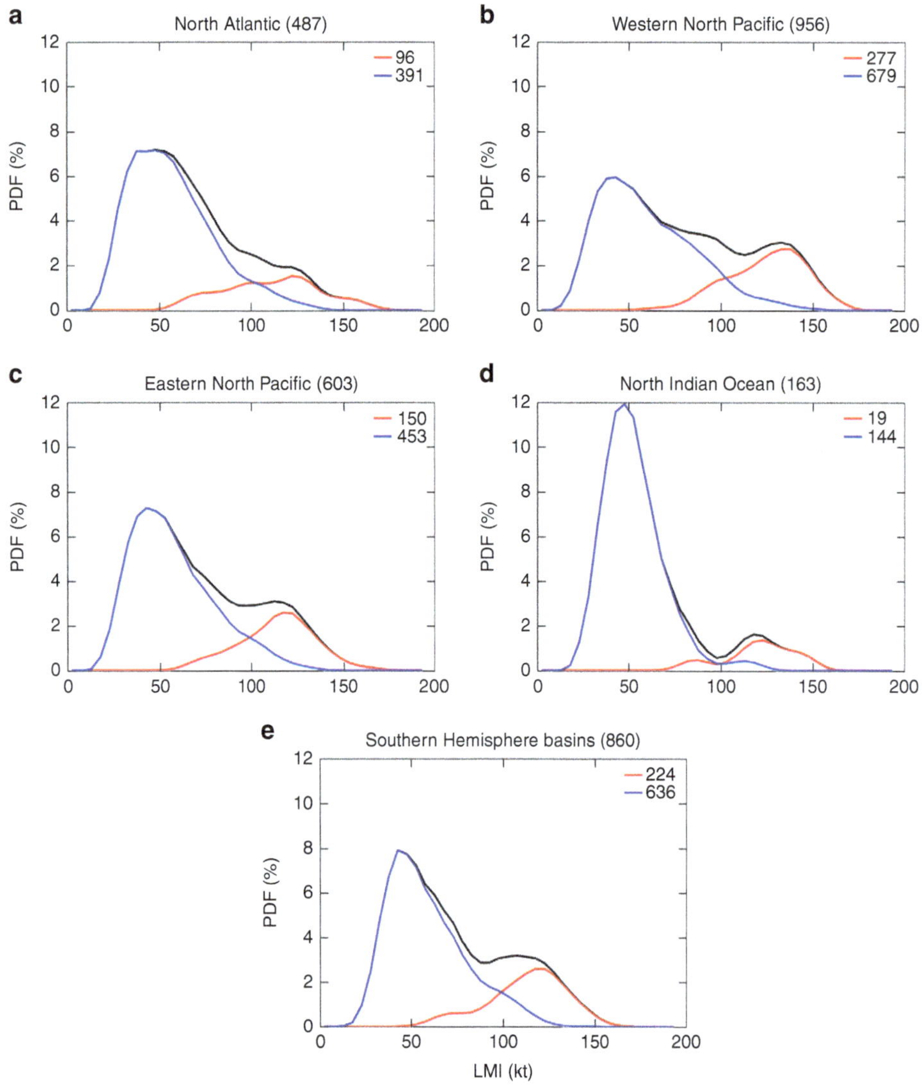

Figure 2 | Distributions of regional tropical cyclone LMI. PDFs are calculated using 1981–2012 tropical cyclone for individual basins: (**a**) North Atlantic, (**b**) Western North Pacific, (**c**) Eastern North Pacific, (**d**) North Indian Ocean, and (**e**) Southern Hemisphere basins. The black, red and blue lines show the smoothed PDFs for all, the subset of storms those undergo rapid intensification during their lifetime (RI storms), and those do not (non-RI storms), respectively. The raw data is binned in 5 kt bins and smoothing is by moving average with window width of 15 kt. The number of storms in each basin is given in the title, and the numbers of RI and non-RI storms are given in the legend.

similar results (Fig. 2). The intensification rate of 35 kt in 24 h (representing the 97th percentile of the intensification rate over that duration) is the optimal RI threshold for explaining the bimodality of the LMI distribution. Using other thresholds, such as 25, 30 or even 40 kt, does not separate the two peaks as clearly, and results a secondary maximum or a hint of it in either the RI or the non-RI LMI distributions (Supplementary Fig. 1). With the 35 kt definition, 80–85% of major storms in the eastern and western North Pacific, North Indian Ocean and southern Hemisphere basins are RI storms, but only 70% of major storms in the Atlantic are RI storms. In the Atlantic, the LMI distribution of RI storms is quite different from that in other basins, being much less peaked (red lines in Fig. 2). The probability that an RI storm in the Atlantic will become a minor hurricane (categories 1–2) is close to the probability that it will become a major hurricane. Still, RI storms are responsible for the rightmost shoulder of the LMI distribution in the Atlantic.

Relation to storm lifetime. Storm lifetime is another factor related to LMI. The distribution of lifetime itself is unimodal (Supplementary Fig. 2). Overall, LMI is positively correlated with storm lifetime with longer lived storms tending to have higher LMI values. The correlation of storm lifetime with LMI is 0.66 globally. Here the idea that the bimodality in LMI can be explained by variations in lifetime is tested by separating storms into two groups, longer and shorter lived, and examining the LMI distributions for the two groups, analogously to what was done for RI and non-RI storms above. Thresholds of lifetime from 7 to 13 days are tested for this purpose (Supplementary Fig. 3). The classification produces less completely unimodal LMI PDFs than were achieved using RI as the criterion, especially for storms in the North Indian Ocean and southern Hemisphere basins. The best threshold of lifetime is also basin dependent, and noticeable departures from uni-modality are found in even when that best value is used. We, therefore, conclude that while lifetime has an impact on LMI, RI is a better criteria for explaining the LMI bimodality.

Discussion

The observed relation of RI with the bimodal LMI distribution is consistent with numerical simulation studies in which higher horizontal resolution results in both higher intensification rates and the appearance of the second LMI peak[9,28]. Such numerical simulations provide evidence that the observed LMI bimodality and its relationship to RI is not an artifact of the uncertainty associated with intensity estimation[13,29,30]. Our findings are also in agreement with the hypothesis of ref. 5 that RI is responsible for the bimodal distribution of LMI, although we cannot comment on the link to the ratio of exchange coefficients. On the other hand, a recent study showed that RI has no apparent signature in the overall intensification rate statistics[12]. The intensification rate distributions of RI and non-RI storms do differ, even when the RI events are excluded, but no distinguishing features result when all storms are considered (Supplementary Fig. 4). A key difference between the distribution of intensification rates and that of LMI is that the LMI distribution accounts for the intensification behaviour over the storm lifetime, and thus is a measure of the joint (across different times) intensification rate statistics.

Another measure of the TC intensity climatology is LMI normalized by local potential intensity. The distribution of normalized LMI for storms whose peak intensities are not limited by declining potential intensity has a linear cumulative distribution with two slopes[6], corresponding approximately to tropical storms (>34 kt) and hurricanes (>64 kt). This separation of the normalized LMI distribution appears to be distinct from the one here based on RI. The subset of storms not limited by declining potential intensity is a relatively small fraction of all storms, and the normalized LMI distribution over all storms (which we consider here) is not uniform (not shown; for example, Fig. 13 in ref. 6).

The basic physical mechanisms of RI are still not completely understood, and whether they are distinct from those of lower intensification rates remain an open question. We do not attempt to answer this question. The message here is that RI is relevant not only to short-term weather forecasting, but also to the relationship between TCs and climate. The most intense storms are those that undergo RI, and the storms that undergo RI are responsible for the observed bimodality of the LMI distribution. This finding suggests that a complete understanding of the most intense storms in either the current climate or future (or past) climates may need to include some understanding of RI. Our results also suggest that numerical models that do not simulate RI are likely to be incomplete in their representation of the LMI distribution and in the frequency of major storms. Therefore, an important research question is to what extent simulations and projections of the frequency of major storm occurrence can be accurate without either resolving RI or accounting for its absence.

Methods

Data. Best-track data from the National Hurricane Center (NHC)[29,31] and the Joint Typhoon Warming Centers (JTWC)[32] from 1981 to 2012 are used in this study. Best-track data include 1-min maximum sustained wind, minimum sea level pressure and location every 6 h.

Calculations. The LMI here is defined using maximum wind speed. Maximum wind speed itself is not an observed quantity, but rather estimated from *in situ* observations, remotely sensed estimates of winds or via satellite-based techniques[29]. The distribution of LMI is calculated globally, as well as for individual basins, following the definitions of the NHC and JTWC, that is, North Atlantic and eastern North Pacific (from NHC), and western North Pacific, North Indian Ocean and southern Hemisphere basins (from JTWC). Correlations between global LMI and storm lifetime are calculated using Spearman's rank correlation coefficient.

Terminology. We refer to storms with LMI >96 kt (categories 3–5 in Saffir-Simpson Hurricane Wind Scale) as major storms.

References

1. Bengtsson, L., Botzet, M. & Esch, M. Will greenhouse-induced warming over the next 50 years lead to higher frequency and greater intensity of hurricanes? *Tellus* **48A,** 57–73 (1996).
2. Emanuel, K. A. Increasing destructiveness of tropical cyclones over the past 30 years. *Nature* **436,** 686–688 (2005).
3. Webster, P. J., Holland, G., Curry, J. A. & Chang, H.-R. Changes in tropical cyclone number, duration, and intensity in a warming environment. *Science* **309,** 1844–1846 (2005).
4. Knutson, T. R. *et al.* Tropical cyclones and climate change. *Nat. Geosci.* **3,** 157–163 (2010).
5. Soloviev, A. V., Lukas, R., Donelan, M. A., Haus, B. K. & Ginis, I. The air-sea interface and surface stress under tropical cyclones. *Sci. Rep.* **4,** 5306 (2014).
6. Emanuel, K. A. A statistical analysis of tropical cyclone intensity. *Mon. Weather Rev.* **128,** 1139–1152 (2000).
7. Kossin, J. P., Emanuel, K. A. & Vecchi, G. A. The poleward migration of the location of tropical cyclone maximum intensity. *Nature* **509,** 349–352 (2014).
8. Park, D.-S. R., Ho, C.-H. & Kim, J.-H. Growing threat of intense tropical cyclones to East Asia over the period 1977-2010. *Environ. Res. Lett.* **9,** 014008 (2014).
9. Manganello, J. V. *et al.* Tropical cyclone climatology in a 10-km global atmospheric GCM: toward weather-resolving climate modeling. *J. Climate* **25,** 3867–3893 (2012).
10. Kossin, J. P., Olander, T. L. & Knapp, K. R. Trend analysis with a new global record of tropical cyclone intensity. *J. Climate* **26,** 9960–9976 (2013).
11. Zhao, M., Held, I. M., Lin, S.-J. & Vecchi, G. A. Simulations of global hurricane climatology, interannual variability, and response to global warming using a 50-km resolution GCM. *J. Climate* **22,** 6653–6678 (2009).
12. Kowch, R. & Emanuel, K. Are special processes at work in the rapid intensification of tropical cyclones? *Mon. Weather Rev.* **143,** 878–882 (2014).
13. Torn, R. D. & Snyder, C. Uncertainty of tropical cyclone best-track information. *Weather Forecast.* **27,** 715–729 (2012).
14. Bister, M. & Emanuel, K. A. Low frequency variability of tropical cyclone potential intensity 1. Interannual to interdecadal variability. *J. Geophys. Res.* **107,** 4801 (2002).
15. Emanuel, K. A. Sensitivity of tropical cyclones to surface exchange coefficients and a revised steady-state model incorporating eye dynamics. *J. Atmos. Sci.* **52,** 3969–3976 (1995).
16. Vigh, J. L., Knaff, J. A. & Schubert, W. H. A climatology of hurricane eye formation. *Mon. Weather Rev.* **140,** 1405–1426 (2012).
17. Shapiro, L. J. & Willoughby, H. E. The response of balanced hurricanes to local-sources of heat and momentum. *J. Atmos. Sci.* **39,** 378–394 (1982).
18. Hack, J. J. & Schubert, W. H. Nonlinear response of atmospheric vortices to heating by organized cumulus convection. *J. Atmos. Sci.* **43,** 1559–1573 (1986).
19. Vigh, J. L. & Schubert, W. H. Rapid development of the tropical cyclone warm core. *J. Atmos. Sci.* **66,** 3335–3350 (2009).
20. Nolan, D. S., Moon, Y. & Stern, D. P. Tropical cyclone intensification from asymmetric convection: energetics and efficiency. *J. Atmos. Sci.* **64,** 3377–3405 (2007).
21. Holliday, C. R. & Thompson, A. H. Climatological characteristics of rapidly intensifying typhoons. *Mon. Weather Rev.* **107,** 1022–1034 (1979).
22. Kaplan, J. & DeMaria, M. Large-scale characteristics of rapidly intensifying tropical cyclones in the North Atlantic basin. *Weather Forecast.* **18,** 1093–1108 (2003).
23. Knaff, J. *et al.* in *Sixth International Workshop on Tropical Cyclones* vol. 1.5 (WMO, Jose, Costa Rica., 2006).
24. Kaplan, J., DeMaria, M. & Knaff, J. A. A revised tropical cyclone rapid intensification index for the Atlantic and Eastern North Pacific basins. *Weather Forecast.* **25,** 220–241 (2010).
25. Hristopulos, D. T. & Mouslopoulou, V. Strength statistics and the distribution of earthquake interevent times. *Physica A* **392,** 485–496 (2013).
26. Feuerstein, B., Dotzek, N. & Grieser, J. Assessing a tornado climatology from global tornado intensity distributions. *J. Climate* **18,** 585–596 (2005).
27. Knapp, K. R., Kruk, M. C., Levinson, D. H., Diamond, H. J. & Neumann, C. J. The international best track archive for climate stewardship (IBTrACS). *Bull. Am. Meteor. Soc.* **91,** 363–376 (2010).
28. Murakami, H. *et al.* Future changes in tropical cyclone activity projected by the new high-resolution MRI-AGCM. *J. Climate* **25,** 3237–3260 (2012).
29. Landsea, C. W. & Franklin, J. L. Atlantic hurricane database uncertainty and presentation of a new database format. *Mon. Weather Rev.* **141,** 3576–3592 (2013).
30. Knaff, J. A., Brown, D. P., Courtney, J., Gallina, G. M. & Beven, J. L. An evaluation of Dvorak technique-based tropical cyclone intensity estimates. *Weather Forecast.* **25,** 1362–1379 (2010).
31. Jarvinen, B. R., Neumann, C. J. & Davis, M. A. S. *A Tropical Cyclone Data Tape for the North Atlantic Basin, 1886-1983: Contents, Limitations, and Uses* pp 21 (NOAA Technical Memorandum NWS NHC 22, 1984).
32. Chu, J.-H., C. R. Sampson, A. S. L. & Fukada, E. *The Joint Typhoon Warning Center tropical cyclone best-tracks, 1945-2000.* Report No. NRL/MR/7540-02-16, 22pp (Washington, D. C., USA, 2002).

Acknowledgements

The research is supported by Office of Naval Research under the research grant of MURI (N00014-12-1-0911). We thank Dr Shuyi S. Chen from University of Miami for her influential suggestion on the possible role of RI on the LMI distribution during 2014 AGU Fall Meeting. Comments and suggestions from Dr John Knaff and the other anonymous reviewer are appreciated.

Author contributions

The study was led by C.-Y.L. and M.K.T., calculations were carried out and manuscript was drafted by C.-Y.L. All authors were involved with designing the research, analysing the results and revising and editing the manuscript. The data were prepared by S.J.C.

Additional information

Competing financial interests: The authors declare no competing financial interests.

PERMISSIONS

All chapters in this book were first published in NC, by Nature Publishing Group; hereby published with permission under the Creative Commons Attribution License or equivalent. Every chapter published in this book has been scrutinized by our experts. Their significance has been extensively debated. The topics covered herein carry significant findings which will fuel the growth of the discipline. They may even be implemented as practical applications or may be referred to as a beginning point for another development.

The contributors of this book come from diverse backgrounds, making this book a truly international effort. This book will bring forth new frontiers with its revolutionizing research information and detailed analysis of the nascent developments around the world.

We would like to thank all the contributing authors for lending their expertise to make the book truly unique. They have played a crucial role in the development of this book. Without their invaluable contributions this book wouldn't have been possible. They have made vital efforts to compile up to date information on the varied aspects of this subject to make this book a valuable addition to the collection of many professionals and students.

This book was conceptualized with the vision of imparting up-to-date information and advanced data in this field. To ensure the same, a matchless editorial board was set up. Every individual on the board went through rigorous rounds of assessment to prove their worth. After which they invested a large part of their time researching and compiling the most relevant data for our readers.

The editorial board has been involved in producing this book since its inception. They have spent rigorous hours researching and exploring the diverse topics which have resulted in the successful publishing of this book. They have passed on their knowledge of decades through this book. To expedite this challenging task, the publisher supported the team at every step. A small team of assistant editors was also appointed to further simplify the editing procedure and attain best results for the readers.

Apart from the editorial board, the designing team has also invested a significant amount of their time in understanding the subject and creating the most relevant covers. They scrutinized every image to scout for the most suitable representation of the subject and create an appropriate cover for the book.

The publishing team has been an ardent support to the editorial, designing and production team. Their endless efforts to recruit the best for this project, has resulted in the accomplishment of this book. They are a veteran in the field of academics and their pool of knowledge is as vast as their experience in printing. Their expertise and guidance has proved useful at every step. Their uncompromising quality standards have made this book an exceptional effort. Their encouragement from time to time has been an inspiration for everyone.

The publisher and the editorial board hope that this book will prove to be a valuable piece of knowledge for researchers, students, practitioners and scholars across the globe.

LIST OF CONTRIBUTORS

Yafang Cheng, Hang Su and Ulrich Pöschl
Multiphase Chemistry Department, Max Planck Institute for Chemistry, 55128 Mainz, Germany

Thomas Koop
Faculty of Chemistry, Bielefeld University, 33615 Bielefeld, Germany

Eugene Mikhailov
Institute of Physics, St. Petersburg State University, 198904 St. Petersburg, Russia

Hyacinth C. Nnamchi
Department of Geography, University of Nigeria, Nsukka 410001, Nigeria

Jianping Li
College of Global Change and Earth System Science, Beijing Normal University, Beijing 100875, China
Joint Center for Global Change Studies, Beijing 100875, China

Fred Kucharski and Riccardo Farneti
Earth System Physics Section, Abdus Salam International Centre for
Theoretical Physics, Strada Costiera, 11, Trieste 34151, Italy

In-Sik Kang
School of Earth and Environmental Sciences, Seoul National University, Seoul 151-742, South Korea

Noel S. Keenlyside
Geophysical Institute, University of Bergen and Bjerknes Centre for Climate Research, Alle´gaten 70, 5007 Bergen, Norway

Ping Chang
Department of Oceanography, Texas A&M University, College Station, Texas 77843-3160, USA
Collaborative Innovation Center of Marine Science and Technology, Ocean University of China, Qingdao 266100, China

María Antiñolo
Department of Chemistry, University of Toronto, 80 St George Street, Toronto, Ontario, M5S 3H6 Canada
Facultad de Ciencias y Tecnologías Químicas, Universidad de Castilla-La Mancha, Avda. Camilo José Cela s/n, Ciudad Real 13071, Spain

Megan D. Willis, Shouming Zhou and Jonathan P.D. Abbatt
Department of Chemistry, University of Toronto, 80 St George Street, Toronto, Ontario, M5S 3H6 Canada

I-I Lin
Department of Atmospheric Sciences, National Taiwan University, No.1, Sec. 4, Roosevelt Road, Taipei 106, Taiwan

Young Soo Joung and Cullen R. Buie
Department of Mechanical Engineering, Massachusetts Institute of Technology, 77 Massachusetts Avenue, Cambridge, Massachusetts 02139, USA

Christopher J. Cox and Matthew D. Shupe
Cooperative Institute for Research in Environmental Sciences, University of Colorado, Boulder, Colorado 80309, USA
NOAA Earth System Research Laboratory, Boulder, Colorado 80305, USA

Von P. Walden
Department of Civil and Environmental Engineering, Washington State University, Pullman 99164,Washington, USA

Penny M. Rowe
Department of Geography, University of Idaho, Moscow 83844, Idaho, USA
Departamento de F´ısica, Universidad de Santiago de Chile, Santiago
9170124, Chile

W. Feng
School of Earth and Environment, University of Leeds, Leeds LS2 9JT, UK
National Centre for Atmospheric Science (NCAS), UK

R.L. McKenzie
National Institute ofWater and Atmospheric Research (NIWA), Lauder Private Bag 50061, New Zealand

G.J.M. Velders
National Institute for Public Health and the Environment, PO Box 1, Bilthoven 3720 BA, The Netherlands

J.A. Pyle
Department of Chemistry, University of Cambridge, Cambridge CB2 1EW, UK

M.P. Chipperfield and S.S. Dhomse
School of Earth and Environment, University of Leeds, Leeds LS2 9JT, UK
National Centre for Earth Observation (NCEO), University of Leeds, Leeds LS2 9JT, UK

Xi Zhang
Department of Earth and Planetary Sciences, University of California Santa Cruz, Santa Cruz, California 95064, USA

Robert A. West
Jet Propulsion Laboratory, California Institute of Technology, 4800 Oak Grove Drive, Pasadena, California 91109, USA

Patrick G.J. Irwin
Atmospheric, Oceanic and Planetary Physics, University of Oxford, Clarendon Laboratory, Parks Road, Oxford OX1 3PU, UK

Conor A. Nixon
NASA Goddard Space Flight Center, Greenbelt, Maryland 20771, USA

Yuk L. Yung
Division of Geological and Planetary Sciences, California Institute of Technology, Pasadena, California 91125, USA

Yoshimitsu Chikamoto
International Pacific Research Center, University of Hawaii at Manoa, 1680 East-West Road, Honolulu, Hawaii 96822, USA

Axel Timmermann
International Pacific Research Center, University of Hawaii at Manoa, 1680 East-West Road, Honolulu, Hawaii 96822, USA
Department of Oceanography, University of Hawaii at Manoa, 1680 East-West Road, Honolulu, Hawaii 96822, USA

Jing-Jia Luo
Centre for Australian Weather and Climate Research, Bureau of Meteorology, GPO Box 1289, Melbourne, Victoria 3001, Australia

Takashi Mochizuki
Japan Agency for Marine-Earth Science and Technology, 3173-25 Showa-machi, Kanazawa-ku, Yokohama, Kanagawa 236-0001, Japan

Masahide Kimoto and Masahiro Watanabe
Atmosphere and Ocean Research Institute, University of Tokyo, 5-1-5 Kashiwanoha, Kashiwa, Chiba 277-8568, Japan

Masayoshi Ishii
Meteorological Research Institute, Japan Meteorological Agency, 1-1 Nagamine, Tsukuba, Ibaraki 305-0052, Japan

Shang-Ping Xie
Scripps Institution of Oceanography, University of California San Diego, 9500 Gilman Drive MC206, La Jolla, California 92093-0206, USA

Fei-Fei Jin
Department of Meteorology, University of Hawaii at Manoa, 2525 Correa Road, Honolulu, Hawaii 96822, USA

Jong-Yeon Park
Max Planck Institute for Meteorology, 20146 Hamburg, Germany
International Max Planck Research School on Earth System Modelling, 20146 Hamburg, Germany

Jürgen Bader
Max Planck Institute for Meteorology, 20146 Hamburg, Germany
Uni Climate, Uni Research & the Bjerknes Centre for Climate Research, 5007 Bergen, Norway

Daniela Matei
Max Planck Institute for Meteorology, 20146 Hamburg, Germany

Johnny C.L. Chan
Guy Carpenter Asia-Pacific Climate Impact Centre, School of Energy and Environment, City University of Hong Kong, Hong Kong, Hong Kong

Sarah Ineson, Jeff R. Knight, James C. Manners, Richard A. Wood Adam A. Scaife and Nick J. Dunstone
Met Office Hadley Centre, Exeter EX1 3PB, UK

Amanda C. Maycock
Centre for Atmospheric Science, Department of Chemistry, University of Cambridge, Cambridge CB2 1EW, UK
National Centre for Atmospheric Science, University of Oxford, Oxford OX1 3PU, UK

Lesley J. Gray
National Centre for Atmospheric Science, University of Oxford, Oxford OX1 3PU, UK
Department of Atmospheric, Oceanic and Planetary Physics,
University of Oxford, Oxford OX1 3PU, UK

Jerald W. Harder
Laboratory for Atmospheric and Space Physics, University of Colorado, Boulder, Colorado 80303 USA

Mike Lockwood
Department of Meteorology, University of Reading, Reading RG6 6BB, UK

K. Van Tricht, N.P.M. van Lipzig, S. Lhermitte and I.V. Gorodetskaya
KU Leuven, Department of Earth and Environmental Sciences, Celestijnenlaan 200E, Leuven 3001, Belgium

J.T.M. Lenaerts, B. Noël and M.R. van den Broeke
Institute for Marine and Atmospheric research Utrecht—Utrecht University, Utrecht 3584CC, The Netherlands

T.S. L'Ecuyer
Atmospheric and Oceanic Sciences, University of Wisconsin-Madison, Madison, Wisconsin 53706, USA

D.D. Turner
National Severe Storms Laboratory, NOAA, Norman, Oklahoma 73072, USA

Angel Fernandez-Cortes
Department of Geology, National Museum of Natural Sciences (MNCN-CSIC), 28006 Madrid, Spain
Department of Earth Sciences, Royal Holloway, University of London, Egham, Surrey TW20 0EX, UK
Geomnia Natural Resources SLNE, 28003 Madrid, Spain

Soledad Cuezva
Department of Geology, National Museum of Natural Sciences (MNCN-CSIC), 28006 Madrid, Spain
Geomnia Natural Resources SLNE, 28003 Madrid, Spain.

Miriam Alvarez-Gallego, Elena Garcia-Anton and Sergio Sanchez-Moral
Department of Geology, National Museum of Natural Sciences (MNCN-CSIC), 28006 Madrid, Spain

Concepcion Pla and David Benavente
Department of Environment and Earth Sciences, University of Alicante, San Vicente del Raspeig, 03690 Alicante, Spain

Valme Jurado and Cesareo Saiz-Jimenez
Department of Agrochemistry, Environmental Microbiology and Soil Conservation, Institute of Natural Resources and Agricultural Biology (IRNAS-CSIC), 41012 Seville, Spain

Didier Swingedouw
Environnements et Paléoenvironnements Océaniques et Continentaux (EPOC), UMR CNRS 5805 EPOC—OASU—Université de Bordeaux, Allée Geoffroy Saint-Hilaire, Pessac 33615, France

Pablo Ortega and Myriam Khodri
LOCEAN/IPSL Sorbonne Universités (UPMC, Univ Paris 06)-CNRS-IRD-MNHN, 4 place Jussieu, Paris F-75005, France

Juliette Mignot
LOCEAN/IPSL Sorbonne Universités (UPMC, Univ Paris 06)-CNRS-IRD-MNHN, 4 place Jussieu, Paris F-75005, France
Climate and Environmental Physics, Physics Institute, University of Bern, Falkenplatz 16, 3012 Bern, Switzerland
Oeschger Centre of. Climate Change Research, University of Bern, Falkenplatz 16, Bern 3012, Switzerland

Eric Guilyardi
LOCEAN/IPSL Sorbonne Universités (UPMC, Univ Paris 06)-CNRS-IRD-MNHN, 4 place Jussieu, Paris F-75005, France
NCAS-Climate, Univeristy of Reading, Reading RG6 6BB, UK

Valérie Masson-Delmotte
Laboratoire des Sciences du Climat et de l'Environnement (Institut Pierre Simon Laplace, CEA-CNRS-UVSQ, UMR8212), 91191 Gif-sur-Yvette, France

Paul G. Butler
School of Ocean Sciences, Bangor University, Menai Bridge, Anglesey LL59 5AB, UK

Roland Séférian
Centre National de Recherches Mètèorologiques-Groupe d'Etude de l'Atmosphère Mètèorologique/ Groupe de Mètèorologie de Grande Echelle et Climat, Toulouse 31100, France

Shang Liu
Earth and Environmental Sciences Division, Los Alamos National Laboratory, Los Alamos, New Mexico 87545, USA
Cooperative Institute for Research in the Environmental Sciences and Department of Chemistry and Biochemistry, University of Colorado, Boulder, Colorado 80309, USA

Allison C. Aiken and Manvendra K. Dubey
Earth and Environmental Sciences Division, Los Alamos National Laboratory, Los Alamos, New Mexico 87545, USA

Kyle Gorkowski
Earth and Environmental Sciences Division, Los Alamos National Laboratory, Los Alamos, New Mexico 87545, USA
Department of Civil and Environmental Engineering, Carnegie Mellon University, Pittsburgh, Pennsylvania 15213, USA

Christopher D. Cappa
Department of Civil and Environmental Engineering, University of California, Davis, California 95616, USA

Leah R. Williams, Scott C. Herndon, Paola Massoli, Edward C. Fortner, William A. Brooks, John T. Jayne and Douglas R. Worsnop
Aerodyne Research, Inc. Billerica, Massachusetts 01821, USA

Puneet S. Chhabra
Aerodyne Research, Inc. Billerica, Massachusetts 01821, USA
Department of Chemical Engineering, University of Texas at Austin, Austin, Texas 78712, USA

Timothy B. Onasch
Aerodyne Research, Inc. Billerica, Massachusetts 01821, USA
Department of Chemistry, Boston College, Boston, Massachusetts 02467, USA

Swarup China, Noopur Sharma and Claudio Mazzoleni
Physics Department and Atmospheric Sciences Program, Michigan Technological University, Houghton, Michigan 49931, USA

Lu Xu
School of Chemical and Biomolecular Engineering, Georgia Institute of Technology, Atlanta, Georgia 30332, USA

Nga L. Ng
School of Chemical and Biomolecular Engineering, Georgia Institute of Technology, Atlanta, Georgia 30332, USA
School of Earth and Atmospheric Sciences, Georgia Institute of Technology, Atlanta, Georgia 30332, USA

Dantong Liu
School of Earth, Atmospheric and Environmental Science, University of Manchester, Manchester M13 9PL, UK

James D. Allan
School of Earth, Atmospheric and Environmental Science, University of Manchester, Manchester M13 9PL, UK
National Centre for Atmospheric Science, University of Manchester, Manchester M13 9PL, UK

James D. Lee
Wolfson Atmospheric Chemistry Laboratory and National
Centre for Atmospheric Science, University of York, York YO10 5DD, UK

Zoë L. Fleming
National Centre for Atmospheric Science, Department of Chemistry, University of Leicester, Leicester LE1 7RH, UK

Claudia Mohr
Department of Atmospheric Sciences, University of Washington, Seattle, Washington 98195, USA
Institute for Meteorology and Climate Research, Karlsruhe Institute of Technology, Eggenstein-Leopoldshafen 76344, Germany

Peter Zotter
Laboratory of Atmospheric Chemistry, Paul Scherrer Institute, Villigen 5232, Switzerland
Lucerne School of Engineering and Architecture, Bioenergy Research, Lucerne University of Applied Sciences and Arts, Horw 6048, Switzerland

Sönke Szidat
Department of Chemistry and Biochemistry and Oeschger Centre for Climate Change Research, University of Bern, Bern 3012, Switzerland

André S.H. Prévôt
Laboratory of Atmospheric Chemistry, Paul Scherrer Institute, Villigen 5232, Switzerland

Minghua Zhang
School of Marine and Atmospheric Sciences, Stony Brook University, Stony Brook, New York 11794-5000, USA

Yanluan Lin
Ministry of Education Key Laboratory for Earth System Modeling, Center for Earth System Science, Tsinghua University, Beijing 100084, China

Ming Zhao
University Corporation for Atmospheric Research/ Geophysical Fluid Dynamics Laboratory, Princeton, New Jersey 08540-6649, USA

Daniel C. Anderson, Timothy P. Canty and Russell R. Dickerson
Department of Atmospheric and Oceanic Science, University of Maryland, College Park, Maryland 20742, USA

Julie M. Nicely
Department of Chemistry and Biochemistry, University of Maryland, College Park, Maryland 20742, USA

Ross J. Salawitch
Department of Atmospheric and Oceanic Science, University of Maryland, College Park, Maryland 20742, USA
Department of Chemistry and Biochemistry, University of Maryland, College Park, Maryland 20742, USA
Earth System Science Interdisciplinary Center, University of Maryland, College Park, Maryland 20742, USA

Thomas F. Hanisco and Anne M. Thompson
NASA Goddard Space Flight Center, Greenbelt, Maryland 20771, USA

Glenn M. Wolfe
NASA Goddard Space Flight Center, Greenbelt, Maryland 20771, USA
Joint Center for Earth Systems Technology, University of Maryland Baltimore County, Baltimore, Maryland 21250, USA

Eric C. Apel, Douglas E. Kinnison, Samuel R. Hall, Teresa L. Campos, Rebecca S. Hornbrook, Kirk Ullmann and Andrew J. Weinheimer
Atmospheric Chemistry Observation and Modeling Laboratory, National Center for Atmospheric Research, Boulder, Colorado 80305, USA

Elliot Atlas and Daniel D. Riemer
Department of Atmospheric Sciences, Rosenstiel School of Marine and Atmospheric Science, University of Miami, Miami, Florida 33149, USA

Thomas Bannan, Carl Percival and Michael Le Breton
Centre for Atmospheric Science, School of Earth, Atmospheric, and Environmental Science, The University of Manchester, Manchester M13 9PL, UK

Stephane Bauguitte
Facility for Airborne Atmospheric Measurements, Cranfield MK43 0JR, UK

Nicola J. Blake
Deparment of Chemistry, University of California, Irvine, California 92697, USA

James F. Bresch
Mesoscale and Microscale Meteorology Laboratory, National Center for Atmospheric Research, Boulder, Colorado 80305, USA

Lucy J. Carpenter
Wolfson Atmospheric Chemistry Laboratories, Department of Chemistry, University of York, York YO10 5DD, UK

Mark D. Cohen and Barbara J.B. Stunder
NOAA Air Resources Laboratory, College Park, Maryland 20740, USA

Mathew Evans
Wolfson Atmospheric Chemistry Laboratories, Department of Chemistry, University of York, York YO10 5DD, UK
National Centre for Atmospheric Science, Department of Chemistry, University of York, York YO10 5DD, UK

Rafael P. Fernandez
Department of Atmospheric Chemistry and Climate, Institute of Physical Chemistry Rocasolano, CSIC, Madrid 28006, Spain
Department of Natural Science, National Research Council (CONICET), FCEN-UNCuyo, Mendoza 5501, Argentina

Brian H. Kahn
Jet Propulsion Laboratory, California Institute of Technology, Pasadena, California 91109, USA

Neil R.P. Harris
Department of Chemistry, Cambridge University, Cambridge CB2 1EW, UK

Jean-Francois Lamarque
Atmospheric Chemistry Observation and Modeling Laboratory, National Center for Atmospheric Research, Boulder, Colorado 80305, USA
Climate and Global Dynamics Laboratory, National Center for Atmospheric Research, Boulder, Colorado 80305, USA

James D. Lee and Adam Vaughan
National Centre for Atmospheric Science, Department of Chemistry, University of York, York YO10 5DD, Uk

Leonhard Pfister
Earth Sciences Division, NASA Ames Research Center, Moffett Field, California 94035, USA

R. Bradley Pierce
NOAA/NESDIS Center for Satellite Applications and Research, Madison, Wisconsin 53706, USA

Alfonso Saiz-Lopez
Department of Atmospheric Chemistry and Climate, Institute of Physical Chemistry Rocasolano, CSIC, Madrid 28006, Spain

Ayako Yamamoto
Department of Atmospheric and Oceanic Sciences, McGill University, 805 Sherbrooke StreetWest, Montreal, Quebec, Canada H3A 2K6

Jaime B. Palter
Department of Atmospheric and Oceanic Sciences, McGill University, 805 Sherbrooke StreetWest, Montreal, Quebec, Canada H3A 2K6
Graduate School of Oceanography, University of Rhode Island, Narragansett Bay Campus, Narragansett, Rhode Island 02882, USA

Hiroki Ando and Takeshi Imamura
Institute of Space and Astronautical Science, Japan Aerospace Exploration Agency, Sagamihara, Kanagawa 252-0222, Japan

Norihiko Sugimoto
Research and Education Center for Natural Sciences, Department of Physics, Keio University, Yokohama, Kanagawa 223-8521, Japan

Masahiro Takagi
Faculty of Science, Kyoto Sangyo University, Kita-ku, Kyoto 603-8555, Japan

Hiroki Kashimura
Japan Agency for Marine-Earth Science and Technology, Yokohama, Kanagawa 236-0001, Japan

Yoshihisa Matsuda
Department of Astronomy and Earth Sciences, Tokyo Gakugei University, Koganei, Tokyo 184-8501, Japan

Max Popp
Max Planck Institute for Meteorology, Bundesstrasse 53, Hamburg 20146, Germany
Program in Atmospheric and Oceanic Sciences, Princeton University, 300 Forrestal Road, Sayre Hall, Princeton, New Jersey 08544, USA

Hauke Schmidt and Jochem Marotzke
Max Planck Institute for Meteorology, Bundesstrasse 53, Hamburg 20146, Germany

R.L. Thompson
Norsk Institutt for Luftforskning (NILU), Kjeller, Norway

P.K. Patra and T. Saeki
Department of Environmental Geochemical Cycle Research, Japan Agency for Marine Earth Science and Technology (JAMSTEC), Yokohama 236-0001, Japan

F. Chevallier and P. Ciais
Laboratoire des Sciences du Climat et l'Environnement (LSCE, CEA-CNRS-UVSQ), Gif-sur-Yvette, France

S. Maksyutov, T. Machida, T. Shirai and M. Ishizawa
National Institute for Environmental Studies (NIES), Tsukuba 305-8506, Japan

R.M. Law and T. Ziehn
Oceans and Atmosphere, Commonwealth Scientific and Industrial Research Organisation (CSIRO), 3195 Aspendale, Australia

I.T. van der Laan-Luijkx
Department of Meteorology and Air Quality, Environmental Sciences Group, Wageningen University (WU), 6708 PB Wageningen, The Netherlands

W. Peters
Department of Meteorology and Air Quality, Environmental Sciences Group, Wageningen University (WU), 6708 PB Wageningen, The Netherlands
University of Groningen, Centre for Isotope Research, 9747 AG Groningen, The Netherlands

A. Ganshin
Department of Upper Atmospheric Layers Physics, Central Aerological Observatory (CAO), Moscow 141700, Russia
National Research Tomsk State University (TSU), 634050 Tomsk, Russia

R. Zhuravlev
Department of Upper Atmospheric Layers Physics, Central Aerological Observatory (CAO), Moscow 141700, Russia
National Research Tomsk State University (TSU), 634050 Tomsk, Russia
Department of Atmospheric Physics and Microwave Diagnostics, Institute of Applied Physics of the Russian Academy of Sciences, Nizhny Novgorod 603950, Russia

T. Maki
Atmospheric Environment and Applied Meteorology Research Department Meteorological Research Institute (MRI), Tsukuba 305-0052, Japan

T. Nakamura
12 Japan Meteorological Agency (JMA), Global Environment and Marine Department, Tokyo 100-8122, Japan

B. Poulter
Institute on Ecosystems and Department of Ecology, Montana State University (MSU), 59717 Bozeman, Montana, USA

J.G. Canadell
Global Carbon Project, Commonwealth Scientific and Industrial Research Organisation (CSIRO), 2601 Canberra, Australia

Ivan A. Janssens
Department of Biology, University of Antwerp, Universiteitsplein 1, 2610 Wilrijk, Belgium

Ana Bastos
Laboratoire des Sciences du Climat et de l'Environnement, LSCE/IPSL, CEA-CNRS-UVSQ, Université Paris-Saclay, F-91191 Gif-sur-Yvette, France
Instituto Dom Luiz, IDL, Faculdade de Ciências, Universidade de Lisboa, Lisboa 1749-016, Portugal

Célia M. Gouveia and Ricardo M. Trigo
Instituto Dom Luiz, IDL, Faculdade de Ciências, Universidade de Lisboa, Lisboa 1749-016, Portugal

Philippe Ciais and Frédéric Chevallier
Laboratoire des Sciences du Climat et de l'Environnement, LSCE/IPSL, CEA-CNRS-UVSQ, Université Paris-Saclay, F-91191 Gif-sur-Yvette, France

Josep Peñuelas
CREAF, Cerdanyola del Vallès, Catalonia, 08193 Barcelona, Spain
CSIC, Global Ecology Unit CREAF-CSIC-UAB, Cerdanyola del Vallès, Catalonia, 08193 Barcelona, Spain

Christian Rödenbeck
Max Planck Institute for Biogeochemistry, Jena 07701, Germany

Shilong Piao
Department of Ecology, College of Urban and Environmental Sciences, Peking University 5 Yiheyuan Road, Haidian District, Beijing 100871, China

Pierre Friedlingstein
College of Engineering, Mathematics and Physical Sciences, University of Exeter, Exeter EX4 4QF, UK

Steven W. Running
Numerical Terradynamic Simulation Group, University of Montana, Missoula, Montana 59812, USA

Øivind Hodnebrog, Gunnar Myhre, Jana Sillmann and Bjørn H. Samset
Center for International Climate and Environmental Research-Oslo (CICERO), P.O. Box 1129 Blindern, N-0318 Oslo, Norway

Piers M. Forster
School of Earth and Environment, University of Leeds, Leeds LS2 9JT, UK

Michael K. Tippett
Department of Applied Physics and Applied Mathematics, Columbia University, New York, New York 10027, USA
Center of Excellence for Climate Change Research, Department of Meteorology, King Abdulaziz University, Jeddah 21589, Saudi Arabia

Joel E. Cohen
Laboratory of Populations, Rockefeller University, New York,
New York 10065, USA
The Earth Institute, Columbia University, New York, New York 10027, USA

M. Tedesco
Lamont-Doherty Earth Observatory, Columbia University, New York, New York 10964, USA
NASA Goddard Institute of Space Studies, New York, New York 10025, USA

T. Mote
University of Georgia, Athens, Georgia 30602-2502, USA

X. Fettweis
University of Liege, Liege 4000, Belgium

E. Hanna
University of Sheffield, Sheffield S10 2TN, UK

J. Jeyaratnam and J.F. Booth
The City College of New York, New York, New York 10031, USA

R. Datta
Lamont-Doherty Earth Observatory, Columbia University, New York, New York 10964, USA
The City College of New York, New York, New York 10031, USA
The Graduate Center of the City University of New York, New York, New York 10016, USA

K. Briggs
University of Leeds, Leeds LS2 9JT, UK

Chia-Ying Lee
International Research Institute of Climate and Society, Columbia University, Palisades, New York 10964, USA

Michael K. Tippett
Department of Applied Physics and Applied
Mathematics, Columbia University, New York 10027, USA
Center of Excellence for Climate Change Research, Department of Meteorology, King Abdulaziz University, Jeddah 21589, Saudi Arabia

Adam H. Sobel
Department of Applied Physics and Applied Mathematics, Columbia University, New York 10027, USA
Division of Ocean and Climate Physics, Lamont-Doherty Earth Observatory, Columbia University, Palisades, New York 10964, USA

Suzana J. Camargo
Division of Ocean and Climate Physics, Lamont-Doherty Earth Observatory, Columbia University, Palisades, New York 10964, USA

Index